Microparticulate Systems for the Delivery of Proteins and Vaccines

DRUGS AND THE PHARMACEUTICAL SCIENCES

James Swarbrick, Executive Editor
AAI, Inc.
Wilmington, North Carolina

Advisory Board

DRUGS AND THE PHARMACEUTICAL SCIENCES

A Series of Textbooks and Monographs

edited by

James Swarbrick
AAI, Inc.
Wilmington, North Carolina

ADDITIONAL VOLUMES IN PREPARATION

Microparticulate Systems for the Delivery of Proteins and Vaccines

edited by

Smadar Cohen
Ben-Gurion University of the Negev
Beer-Sheva, Israel

Howard Bernstein
Acusphere, Inc.
Cambridge, Massachusetts

CRC Press
Taylor & Francis Group
Boca Raton   London   New York

CRC Press is an imprint of the
Taylor & Francis Group, an informa business

FIRST INDIAN REPRINT, 2011

Library of Congress Cataloging-in-Publication Data

Microparticulate systems for the delivery of proteins and vaccines / edited by
  Smadar Cohen, Howard Bernstein.
     p. cm. — (Drugs and the pharmaceutical sciences, v. 77)
  Includes index.
  ISBN 0-8247-9753-1
  1. Microencapsulation. 2. Protein drugs—Controlled release. 3. Vaccines
—Controlled release. I. Cohen, Smadar. II. Bernstein, Howard. III. Series.
RS201.C3M544 1996
615'.6—dc20   96-25983
                                                                         CIP

The publisher offers discounts on this book when ordered in bulk quantities.
For more information, write to Special Sales/Professional Marketing at the address below.

Printed and bound in India by Bhavish Graphics.

Marcel Dekker, Inc.
270 Madison Avenue, New York, New York 10016

FOR SALE IN SOUTH ASIA ONLY

Preface

There are multiple FDA-approved protein therapeutic agents and many more in various stages of preclinical and clinical development. For these molecules to realize their full therapeutic potential, improved ways of delivering these drugs must be developed. Almost 20 years have passed since the initial demonstration that macromolecules, such as proteins, can be delivered at controlled rates from polymeric matrices. Since then, we have witnessed the development of numerous encapsulation techniques and microparticulate-delivery systems using both natural and synthetic polymers. The systems, to date, have been developed almost exclusively for conventional, low molecular weight drugs and peptides. As we look ahead toward advances in protein delivery, the areas of greatest consequence will be understanding and developing ways to stabilize these molecules, both during the microencapsulation processes and while encapsulated within the delivery system. This book offers a review of the recent advances made in developing microparticulate systems for the delivery of therapeutic proteins and vaccines.

The book begins with an account on protein stability issues as related to encapsulation and controlled delivery from biodegradable polymer microspheres. Chapters 2–5 present various methods of protein microencapsulation, including new approaches that do not involve organic solvents, such as supercritical fluids (Chapter 3), ionic cross-linking of the hydrogel-like polyphosphazenes (Chapter 4), and liposphere technology (Chapter 5). In Chapter 6, various spectral and microscopic methods are applied to microsphere characterization.

The next four chapters are devoted to polymeric nanoparticles and liposomes. Because of their smaller size, these carriers are being extensively studied for drug targeting. Recent developments, such as efficient protein encapsulation in these particles (Chapter 7) and increased stability in vivo by using the "stealth" approach (Chapters 8 and 9) have revived interest in these systems. In Chapter 10, two additional approaches to liposome stabilization are presented: polymerization of the liposomal phospholipids and microencapsulation of liposomes within hydrogel microspheres.

The final chapters deal with different pharmaceutical applications of microencapsulation technology. Chapter 11 provides the mathematical basis for the estimation of the in vivo drug-release rate, from a given dosage form, that can be used to analyze microparticulate systems. Microspheres fabricated from a novel synthetic polymer with built-in adjuvanticity are described in Chapter 12. The potential of polymeric microspheres for oral vaccination is discussed in Chapter 13. The use of gelatin microspheres in immunotherapy for tumor treatment is presented in Chapter 14. Chapter 15 discusses the use of bioadhesive liposomes for the topical delivery of growth factors in wound healing. The last two chapters describe two approaches to achieve modulated protein delivery from microparticulates: Chapter 16 discusses using ultrasound as a way to modulate peptide and protein release, and Chapter 17 describes the use of microencapsulated cells as a self-modulating system. Good manufacturing practice (GMP) in the manufacturing of microparticulate systems for protein delivery is not covered in this book. As we are at only the very early stages of commercial manufacturing of such systems, the guidelines and regulations for the manufacturing of microencapsulated proteins have not been well established.

It is our intention in this book to address the various issues of protein encapsulation, stabilization, and delivery from controlled-release systems. We hope that it will be useful to a broad audience of undergraduates, postgraduates, and clinicians in the pharmaceutical and clinical sciences.

Smadar Cohen
Howard Bernstein

Contents

Contributors

Maria J. Alonso, Ph.D. Faculty of Pharmacy, University of Santiago de Compostela, Santiago de Compostela, Spain

Carl R. Alving, Ph.D. Department of Membrane Biochemistry, Walter Reed Army Institute of Research, Washington, D.C.

Shimon Amselem, Ph.D. Department of Pharmaceutics, Pharmos Ltd., Rehovot, Israel

Alexander K. Andrianov, Ph.D. Virus Research Institute, Cambridge, Massachusetts

Julia E. Babensee, Ph.D. Department of Chemical Engineering and Applied Chemistry, University of Toronto, Toronto, Ontario, Canada

Hitesh R. Bhagat, Ph.D. OraVax, Inc., Cambridge, Massachusetts

Malcolm R. Brandon, Ph.D. Centre for Animal Biotechnology, School of Veterinary Science, The University of Melbourne, Parkville, Australia

Anette Brunner, Ph.D. Department of Pharmaceutical Technology, University of Erlangen-Nürnberg, Erlangen, Germany

Michael Cardamone, Ph.D. Department of Chemical Engineering, Massachusetts Institute of Technology, Cambridge, Massachusetts

Pravin R. Chaturvedi, Ph.D.   Vertex Pharmaceuticals Incorporated, Cambridge, Massachusetts

Paresh S. Dalal, Ph.D.   OraVax, Inc., Cambridge, Massachusetts

Pablo G. Debenedetti, Ph.D.   Department of Chemical Engineering, Princeton University, Princeton, New Jersey

Abraham J. Domb, Ph.D.   Department of Pharmaceutical Chemistry, School of Pharmacy—Faculty of Medicine, The Hebrew University of Jerusalem, Jerusalem, Israel

Achim Göpferich, Ph.D.   Department of Pharmaceutical Technology, University of Erlangen-Nürnberg, Erlangen, Germany

Ruxandra Gref, Ph.D.   Laboratoire de Chimie-Physique Macromoleculaire (CNRS-URA 494), Ecole Nationale Supérieure des Industries Chimiques, Nancy, France

Justin Hanes, Ph.D.   Department of Chemical Engineering, Massachusetts Institute of Technology, Cambridge, Massachusetts

Yoshito Ikada, Ph.D.   Research Center for Biomedical Engineering, Kyoto University, Kyoto, Japan

Thomas Kissel, Ph.D.   Department of Pharmaceutics and Biopharmacy, Philipps University, Marburg, Germany

Alexander Klibanov, Ph.D.   Department of Chemistry, Massachusetts Institute of Technology, Cambridge, Massachusetts

Douglas Kline, Ph.D.   Schering-Plough Research Institute, Kenilworth, New Jersey

Barbara L. Knutson, Ph.D.   Department of Chemical Engineering, Princeton University, Princeton, New Jersey

Regina Koneberg, Ph.D.   Department of Pharmaceutics and Biopharmacy, Philipps University, Marburg, Germany

Joseph Kost, Ph.D.   Department of Chemical Engineering, Ben-Gurion University of the Negev, Beer-Sheva, Israel

Robert Langer, Sc.D.   Department of Chemical Engineering, Massachusetts Institute of Technology, Cambridge, Massachusetts

Rimona Margalit, Ph.D.   Department of Biochemistry, Tel Aviv University, Tel Aviv, Israel

Contributors ix

Yoshiharu Minamitake, Ph.D. Department of Pharmaceutical Development, Suntory Limited, Ohra-Gun, Gunma-Ken, Japan

Ranjani Nellore, Ph.D. Department of Pharmaceutics, University of Maryland, Baltimore, Maryland

Jun'ichi Okada, Ph.D. Sankyo Co., Ltd., Shinagawa, Tokyo, Japan

Lendon G. Payne, Ph.D. Virus Research Institute, Cambridge, Massachusetts

Maria Teresa Peracchia, Ph.D. Università degli Studi di Parma, Parma, Italy

Steven P. Schwendeman, Ph.D.* Department of Chemical Engineering, Massachusetts Institute of Technology, Cambridge, Massachusetts

Michael V. Sefton, Ph.D. Department of Chemical Engineering and Applied Chemistry, University of Toronto, Toronto, Ontario, Canada

Yasuhiko Tabata, Ph.D. Research Center for Biomedical Engineering, Kyoto University, Kyoto, Japan

Jean W. Tom, Ph.D. Merck Research Laboratories, Merck & Co., Inc., Rahway, New Jersey

Vladimir P. Torchilin, Ph.D. Center for Imaging and Pharmaceutical Research, Massachusetts General Hospital and Harvard Medical School, Charlestown, Massachusetts

Vladimir S. Trubetskoy, Ph.D. Center for Imaging and Pharmaceutical Research, Massachusetts General Hospital and Harvard Medical School, Charlestown, Massachusetts

---

* *Current affiliation:* Division of Pharmaceutics and Pharmaceutical Chemistry, College of Pharmacy, The Ohio State University, Columbus, Ohio

Microparticulate Systems for the Delivery of Proteins and Vaccines

1

Stability of Proteins and Their Delivery from Biodegradable Polymer Microspheres

Steven P. Schwendeman,* Michael Cardamone, Alexander Klibanov, and Robert Langer

Massachusetts Institute of Technology
Cambridge, Massachusetts

Malcolm R. Brandon

The University of Melbourne
Parkville, Australia

I. INTRODUCTION

Not long after it was understood that molecules of any size, including large globular proteins (e.g., bovine serum albumin; BSA) could be delivered slowly and continuously from biocompatible polymers [1], the field of controlled release of proteins and peptides has grown immensely. If this concept could be combined successfully with polymers that are both biodegradable and have been processed into microspheres having distinct advantages for oral and parenteral administration, the therapeutic value of such dosage forms could be enormous. Development of this field is a very likely requisite for creation of a supervaccine capable of immunizing individuals against several diseases in a single dose [2]. In addition, patient compliance and comfort, as well as control over blood levels, may be improved with the development of controlled-release protein injectables because regular invasive doses can be avoided. This latter concept is the primary objective in developing insulin-delivery systems [3–5]

---

* *Current affiliation*: The Ohio State University, Columbus, Ohio and has been accomplished for delivery of luteinizing hormone-releasing hormone (LH-RH) analogues such as leuprolide, using controlled-release polymer microspheres, (Lupron Depot), which deliver the peptide for 1 month [6].

Although it has become a routine objective to analyze release kinetics of proteins, the number of publications that consider the stability of protein being delivered from polymers is very low [7–10]. Mechanistic analysis of the stability of proteins during processing, storage, and delivery from polymers is also virtually nonexistent. Recombinant DNA technology has been responsible for the current increase in commercial production of proteins for pharmaceutical use. Several hundred investigational new protein drugs are currently undergoing clinical trials [11,12]. Unfortunately, proteins possess intrinsic properties that, to a large extent, are responsible for the low numbers of therapeutic proteins that have received US Food and Drug Administration (FDA) approval [12]. Thus, it is likely that protein stability is one of the most important obstacles for successful formulation of biodegradable polymer microspheres that control the release of proteins. The aim of this chapter is to combine the concepts of biochemistry, polymer science, and microencapsulation to examine the complexity of formulating these dosage forms; particular emphasis is placed on the strategies and experimental methods used to handle the problems associated with protein stability.

Unlike low molecular weight drugs, proteins as biopolymers have very large globular structures (typically 2–8 nm a side or even larger [13]), possess complex internal architecture that defines their unique biological functions, and contain numerous chemically reactive moieties on their side chains, as well as chemically labile bonds. In contrast, small molecules, and even most peptides, do not have a higher-order structure that may be lost. The fundamental concept behind most rational stabilization approaches is that protein stability must be investigated at the molecular level. Thus, knowledge of the amino acid sequence is essential to developing this approach. Proteins may become inactive by chemical alteration, denaturation, and aggregation [14,15]; also, they may undergo complex adsorption processes that can denature them, particularly when the surface is hydrophobic [16]. Finally, the size of the protein will present special mass transport issues. For example, small molecular weight stabilizers in the formulation that can regulate the environment of the protein (i.e., pH, ionic strength, surface tension, and viscosity) will be transported more freely through the degrading polymer than will the protein. This fact, no doubt, will have stability repercussions as well.

The term *stability,* as it relates to proteins, has several definitions. It is important to distinguish between pharmaceutical and conformational stability and to have specific objectives relative to these before carrying out in-depth stability studies of proteins. For example, Creighton states that a protein is usually most stable at its isoelectric point (referring to the following biochemical definition) [17]. Yet, to avoid aggregation or irreversible conformational changes at surfaces, it is suggested that one move the pH away from the iso-electric point [14,16]. Thus, the term stability is often used loosely in a variety of ways. The pharmaceutical definition according to the US FDA considers a *stable pharmaceutical product* as one that deteriorates no more than 10% in 2 years [18]. We define the *conformational* or *physical stability* of a protein as the ability of the protein to retain its tertiary structure (see Section II.A.3). The three-dimensional structure dictates the properties that allow enzymes, for example, to recognize and bind their natural substrates as well as to have the reactive functional groups properly aligned for catalysis. There is some debate among immunologists of whether the conformational stability of proteinaceous vaccine antigens is essential for a long-lasting and neutralizing immune response, although it is becoming clearer that preservation of the native antigenic determinants on the antigen is necessary to attain protective immune responses [19]. *Chemical stability* involves the reactivity of the side chains and lability of the peptide bonds. The alteration of even one crucial amino acid can seriously impair or abolish the protein's function. The single amino acid substitution in the hemoglobins of sickle cell anemics is evidence of this fact [20]. Biochemists often use the term *protein stability* to denote the magnitude of the change in Gibbs free energy between the folded and unfolded state of the protein [21]. This biochemical definition can be distinguished from *physical stability*, in that the latter may depend on both thermodynamic and kinetic factors, whereas the former is a function of thermodynamic factors only.

The stability of proteins encapsulated in biodegradable polymer microspheres may be separated into at least three complex themes and their interrelations: processing polymers into microspheres, hydration and erosion of the polymer during release incubation, and the intrinsic stability of the protein. During preparation of microspheres, the use of organic solvents [7], formation of the microsphere [9], and lyophilization, all are processes able to inactivate proteins [22]. During release incubation, the protein becomes slowly hydrated (i.e., slower than direct reconstitution), a process known to cause inactivation of proteins by so-called moisture-induced aggregation [23,24]. Other processes during release, such as a reduction in pH, resulting from the formation of new carboxylic acid end groups of the polymer; the presence of the hydrophobic polymer surface; and the creation of water-soluble polymer fragments and mo-nomers, all are potential sources of protein inactivation. In addition, during all of these processes there is a potential of chemical reactions between the protein and the polymer [25,26]. The intrinsic stability of the protein will determine whether it is susceptible to processes, such as deamidation, oxidation of methi-onine and cysteine residues, or the many other deleterious reactions encountered during storage and release, the latter of which is carried out at physiological temperature for extended time periods. Thus, there is a time-line of physical–chemical events during processing, lyophilization and storage, and release from the polymer at which protein inactivation can occur (see later discussion).

We choose to examine predominantly the poly(lactide-*co*-glycolide) (PLGA) family of copolymers, because these are the only synthetic biodegradable polymers that are currently US FDA approved for biomedical applications and are the subject of the vast majority of research in this area. We will also discuss other polymers that have pharmaceutical potential, such as the poly(*ortho* esters) and polyanhydrides, when examining the preparation of micropheres and polymer erosion.

The subject of protein stability, as it relates to biodegradable polymer microspheres, is in its infancy. Therefore, for the time being, we must rely on what is known about both protein stability developed for other applications and the general processes during the processing and release from polymers; these principles will be brought into the context of protein stability during preparation and release from biodegradable polymer microspheres. Thus, the objectives of this review are to examine the following as they relate to controlled release of proteins from biodegradable polymer microspheres: (a) physicochemical nature of proteins; (b) adoption of a strategy of protein stabilization; (c) potential causes of protein inactivation during various stages of microsphere preparation, lyophilization and storage, and release incubation; (d) mechanisms of protein inactivation; (e) minimization of inactivating mechanisms; and (f) experimental methods for evaluating protein stability.

II. THE PHYSICAL AND CHEMICAL NATURE OF PROTEINS

A. The Hierarchy of Protein Structure

Globular proteins have internal architecture, comprised of a combination of irregular structures, such as random coils, and of regular structures, such as α-helices, β-sheets, and β-turns. From the taxanomic rules of protein structure, it is possible to describe the structure of proteins hierarchically at four separate, but interrelated or interdependent levels (Fig. 1).

1. Primary Structure

Early work on the reduction and oxidation of ribonuclease laid the foundation for the theory that all the necessary information required to obtain the native state is contained in the amino acid sequence or primary structure of a protein [27–29]. Primary structure refers to the amino acid sequence or, more precisely, all the covalently bonded amino acids that constitute the polypeptide backbone of the protein. The polypeptide backbone consists of a series of planar *trans*-linkages. The dihedral angles (ψ and ϕ), which represent free rota-

Quaternary Structure $n\,(M) \longrightarrow M_n$

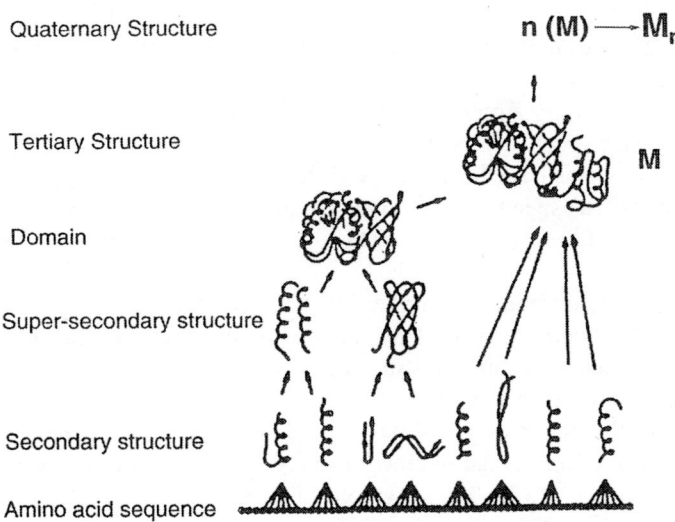

Tertiary Structure

Domain

Super-secondary structure

Secondary structure

Amino acid sequence

Fig. 1 Hierarchy of protein structure and protein folding: The one-dimensional primary structure determines the folding of globular proteins. Next-neighbor (short-range) interactions at the secondary structural level are supplemented by long-range interactions, causing the merging of structural motifs and domains (supersecondary structure and tertiary structure). The quaternary structure defines the association of polypeptide subunits, which constitute the native protein. (Adapted from Ref. 43.)

tions about single bonds between the peptide linkage and the α-carbon of each amino acid, are constrained by steric hindrances. Proline is more restricted in its rotational freedom than other amino acids because the N-Cα bond is connected to a five-membered ring, leaving ϕ as a variable. The peptide bond may, therefore, exist in the *cis*- and *trans*-conformation preceding the pyrrolidine ring [30]. The *cis–trans* isomerization has a characteristic activation energy of approximately 80 kJ/mol, but there is no enthalpy change [30]. The *trans*- form of the peptide bond is favored 1000-fold over the *cis*-form, except when the next amino acid is proline, for which the *trans*- form is favored only fourfold. This leads to a reverse turn and explains why prolines are α-helix and β-sheet breakers [30].

2. Secondary Structure

Secondary structure generally refers to the recurring two-dimensional patterns and regular structures, such as α-helices, β-sheet structures, and β-turns, which are characterized by specific torsional angles [31,32]. The latter serve as nucleation sites for folding because they can "flicker in and out of the conformation that they occupy in the final protein" [28]. They are almost entirely due to the interactions of short-range hydrogen bonds between the amide hydrogen and nitrogen atoms and the carbonyl oxygen atoms of the peptide linkage of one or more polypeptide chains [33].

The formation of secondary structure is thought to occur spontaneously, and many intermediates have been described in the literature. For instance, flickering clusters of amino acids have been observed, using high-resolution nuclear magnetic resonance (NMR) spectroscopy, to have transient secondary structure. In model systems, α-helices are formed extremely rapidly, on the order of 10^{-7} s [34].

Rao and Rossman [35] introduced a further level of structural hierarchy. The term *supersecondary structure* was used by them to describe a recurring fold consisting of several discrete secondary structural elements that do not provide the complete tertiary or domain structure of the protein (see following). Such structures result from the energetic stability of the packed secondary structural elements and can involve long-range interactions between the folding elements on the polypeptide chain [36,37]. The assembly of α-helices, the stacking of β-sheet structures, and the taxonomy of protein structure have been reviewed in detail [38]. These studies summarize extensive investigations of various protein structures that reveal the observed pathways of folding to be subject to definite limitations and preferences. The rules that define the topology of proteins are covered in detail by Chothia [39].

3.   Tertiary Structure

Tertiary structure generally refers to the packing or assembly of recurring, stably folded units, called domains, into an ordered three-dimensional structure. A domain may represent the repetition of sequences, either within the same polypeptide, or in a different, but homologous, protein. They can be spatially separated from each other, or can form a cluster of residues [40–43]. Such domains may have a specific function. For example, they may be a binding site for another molecule. The three-dimensional structure of the "active site" of a protein often consists of the interfaces between domains, which permits the individual function of each to be brought together to form a more sophisticated structure [40]. In the formation of domain structures, portions of the polypeptide chain, which may be linearly or spatially distant apart along this chain, are brought together by a combination of electrostatic forces, dispersion or van der Waals forces, hydrophobic interactions, and hydrogen bonds [42]. However, each domain folds independently, and domain structures have been proposed to be the major intermediates in the folding pathways of proteins [43]. It has also been theorized that each exon, at the level of the gene sequence, encodes for a separate domain [44,45].

4.   Quaternary Structure

Quaternary structure generally refers to the oligomeric assembly formed when folded monomeric polypeptides combine into multisubunit proteins. The same covalent and noncovalent chemical forces that maintain the tertiary structure of each subunit are responsible for maintaining the subunits' geometry in a more stable and ordered oligomeric structure [44]. During the assembly process, association of the individual subunit must not occur until each subunit has attained its correct tertiary structure. Protein aggregation is often kinetically faster than the natural-folding process, especially at high protein concentrations; and it is believed that the formation of insoluble inclusion bodies (protein aggregates formed during overexpression in recombinant organisms) are *in vivo* examples in which nonspecific aggregation pathways have overridden the natural-folding pathways [46].

For a complete protein stability assessment, the amino acid sequence and secondary, tertiary, and quaternary structure of the protein should be known. Such information is available only if a complete X-ray crystallographic analysis has been done. Some information on substrate- or ligand-binding characteristics (activity) of the protein should also be available. For example, the physical properties of albumin are different, depending on its fatty acid and bilirubin composition [16].

B. The Chemical Determinants and Thermodynamics of Protein Conformation

The forces in protein folding arise from covalent bonds (peptide bonds and disulfide bonds) and noncovalent interactions (such as classic charge repulsions and ion-pairing, hydrogen bonding, van der Waals interactions, and the hydrophobic effect). The chemical forces that determine the structure of a protein were early observed to be dependent on distance [47], and their contribution to the free energy of folding has been investigated in detail [48,49].

The structure of the native protein in its functional form is determined by kinetic and thermodynamic constraints: *kinetic* because of the vectorial character of protein biosynthesis from the amino or NH_2-terminus to the carboxy or COOH-terminus of the polypeptide chain, and *thermodynamic* because the driving force of formation depends on a subtle balance of weak noncovalent interactions [50]. In addition, post-translational modifications, such as specific proteolysis, glycosylation, and the binding of cofactors and ligands, are also involved in determining the final native protein structure.

Studies of the thermodynamics of protein folding in vitro have shown that, for many single-domain proteins, folding may be viewed as a two-state process in which intermediate states are negligibly populated, at least thermodynamically [51]. The equilibrium folding transition is given by:

$$\text{Unfolded state} \overset{K_{eq}}{\leftrightarrow} \text{Folded state} \tag{1}$$

where K_{eq} is the equilibrium constant for the folding transition, which is related to the Gibbs free energy difference by the normal relationship

$(K_{eq} = e^{-\Delta G_{stab}/RT})$. A number of folding studies use guanidine hydrochloride (GuanHC1) and urea equilibrium denaturation to estimate the ΔG_{stab} of proteins. In these studies protein is exposed to increasing (and decreasing) amounts of denaturant, and the average free energy of denaturation (ΔG_{H_2O}) is extracted from the data [21]. The ΔG_{H_2O} can give an apparent value of the ΔG_{stab} of a protein.

A wealth of information on the importance of covalent and noncovalent interactions has been obtained in the last decade or so, with the advent of recombinant DNA technology. Experimental findings from site-directed mutagenesis studies and X-ray crystallographic analyses have provided valuable information about the forces that determine the conformational stability of globular proteins [21,52]. The ΔG_{stab} of almost all naturally occurring globular proteins is represented by free energy of denaturation (ΔG_{H_2O}) of some 50 ± 15 kJ/mol [21]. The results of these studies indicate that the contribution of each amino acid residue depends on its structural context within the entire protein, and that many different types of interactions, including disulfide bonds, hydrophobic interactions, hydrogen bonds, electrostatic interactions, and dispersion forces, make quantitatively comparable contributions to protein stability. Moreover, certain amino acids can participate in a wide range of interactions. The contributions of each of these chemical interactions to the ΔG_{stab} of a protein are relatively high individually, sometimes greater than 400 kJ/mol, but the net ΔG_{stab} is far lower. Thus, thermodynamically, the native state is only marginally stable [21]. In addition, many amino acid substitutions do not have large effects on the conformational stability of a protein. This indicates that some substitutions preserve critical interactions, whereas others do not make significant contributions toward reducing the free energy of the native form. It is also possible that the protein is able to compensate for the structural alterations incurred by such amino acid substitutions [52], because proteins are not static, rigid structures, but are highly dynamic.

C. The Denatured State

The denatured state of a protein is considered to be a combination of many different molecular conformations and reflects a major noncovalent and reversible change from the original native structure. Dill and Shortle [53] define two categories of denatured states: the unfolded state and the compact state. The former state refers to the specific subset of conformers that are highly open and solvent-exposed, with little or no residual secondary and domain structure. These conformations are found under strongly denaturing conditions. The compact state is usually obtained under weaker-denaturing conditions, and this gives rise to *molten globule* folding intermediates [53]. The classic experiments of Nozaki and Tanford [54] showed that a number of proteins in 6 *M* GuanHC1 exhibited hydrodynamic properties of random coils. Strong denaturants, such as urea and GuanHC1, lower the conformational stability of a protein in proportion to their concentration in solution [55,56]. Dill and Shortle [53] suggest that Tanford's work with GuanHC1 has been misconstrued to imply that the denatured state is a random coil under all conditions, and there is new evidence that shows that proteins even in 8 M GuanHC1 exhibit significant amounts of internal structure [53]. The intermediates of protein folding or denaturation may lead to the inactivation of proteins by aggregation. Partially unfolded intermediates may retain secondary structure, although the native tertiary structure is mostly lost. Intermediates of folding are believed to be in dynamic equilibrium with the native state, provided they have not undergone covalent modification. In this case, they associate and cannot be refolded to the native form, but inevitably proceed toward protein aggregation.

Denaturants may include chaotropes (substances that reorder the hydrogen bonds between water molecules in aqueous solutions) and detergents and other materials that are capable of changing the spatial configuration or conformation of proteins particularly at the surface of the protein; this is achieved either by altering the state of hydration, the solvent environment, or solvent–surface interactions [53,57]. Urea, GuanHC1, sodium thiocyanate, and detergents, such as sodium dodecyl sulfate (SDS) and Triton X-100, all are denaturants. Nozaki and Tanford [54] demonstrated that the denaturing action of urea on globular proteins is due to stabilization of the unfolded form of the protein by urea, diminishing the hydrophobic interactions between nonpolar side chains, and by the increased affinity of the solvent (water) for peptide groups in the presence of urea [54,57].

The interactions of proteins at interfacial surfaces, such as air–liquid and solid–liquid, have been observed to result in significant conformational changes in secondary, tertiary, and quaternary structures [16]. Transmission infrared spectroscopy and, more specifically, the shifts in the amide I band region absorption, can be used to measure the secondary structure perturbations during the protein adsorption onto a silica surface. Adsorbed protein in a deuterated buffer was observed to have sufficient time and space to accommodate to a new microenvironment by conformational change, resulting in significant hydrogen bonding to a silica surface [16].

III. ADOPTING A STRATEGY FOR PROTEIN STABILIZATION

After one has determined that the encapsulated protein has been inactivated (e.g., protein in the release medium is devoid of biological activity, or some fraction of encapsulated protein is not releasable), it becomes necessary to adopt a strategy suitable to solve the stability problem in a rational fashion. Volkin and Klibanov suggest an approach [14], which has been used successfully by many [14,23–62]. We subscribe to this methodology, which involves (a) *determining the causes of inactivation* (e.g., pH and hydrophobic surfaces) (b) *elucidating the molecular mechanisms of inactivation* (e.g., deamidation of asparagine residues and thiol–disulfide interchange) *for each specific cause*, and (c) *devising methods of inhibiting or avoiding the specific mechanism altogether*.

Thus, the first step in this approach is to identify the critical physical or chemical event(s) that may inactivate the protein during microsphere processing, storage, and release incubation. This is a difficult experimental task, because it involves either simulation of the potential inactivating cause, or monitoring stability directly during the processing and release events. In some circumstances, the inactivating step may require examination to elucidate the particular destabilizing mechanism, or it may be a step that may be bypassed by changing a variable in the formulation or the microencapsulation method. Finally, if the mechanism of inactivation is determined (undoubtedly the most difficult task of the three), then means of stabilization usually become readily apparent. We discuss each of these tasks in the sections to follow.

IV. POTENTIAL CAUSES OF PROTEIN INACTIVATION DURING PREPARATION, STORAGE, AND RELEASE INCUBATION

The fate of the protein from the point of microsphere formation to when the protein is released may be separated into three individual processes: preparation of microspheres, storage, and release. The potential causes of inactivation during these processes are given in Fig. 2–5 using PLGA and the well-known double-emulsion–solvent evaporation technique as an example, since the technique has become popular following the success of the Lupron Depot [6].

A. Microsphere Preparation

One of the most often used techniques of encapsulating proteins is the so-called double-emulsion–solvent evaporation technique [8,63–65]. As outlined in Fig. 2, this method typically involves several steps, including the addition of an aqueous protein solution to an organic polymer solution, followed by the creation of the first emulsion (e.g., by sonication or homogenization); the addition of the first emulsion to a second aqueous phase containing an emulsifier [e.g., poly(vinyl alcohol)]; the formation of a double emulsion by some form of mixing (e.g., by vortex); the addition of the multiple emulsion into a second aqueous solvent (large volume); the removal of the volatile organic solvent with stirring and simultaneous hardening of the particles; and finally, the collection of the particles, washing, and lyophilization. One variation of this method of preparation involves adding the solid protein directly to the polymer solution without water (no first emulsion) and completing the foregoing steps following addition of the polymer solution–protein suspension into the aqueous phase containing an emulsifier. Another alternative is to remove the organic solvent by extraction with a water-miscible solvent (e.g., isopropanol), which decreases the particle hardening time [7].

The general physicochemical events that take place during the foregoing manufacturing process may be characterized as follows: As the first emulsion is formed, the organic solvent and water begin to partition into one another to the extent allowed by the solubilities [64] (e.g., 2% methylene chloride in water at 20°C [66]); thus, the organic solvent becomes an unwanted denaturant in the protein-containing innerwater phase. A large hydrophobic surface is formed around the innerwater phase, and any surface-active components (e.g., proteins) of the two phases will diffuse toward and occupy this water–oil surface; excipients and protein within the innerwater phase begin diffusing into the organic polymer solution if any solubility exists. During homogenization or sonication, large pressure gradients are created, and the solution can become quite warm, particularly for viscous polymer solutions for which the higher flow resistance must favor the thermal route of energy dissipation. During microsphere formation by the second emulsion, some of the innerwater phase droplets containing protein are lost to the second aqueous phase [65]. If the solid protein was suspended directly into the organic phase, the protein may begin hydrating at this time, because very small amounts of water can diffuse through the polymer phase; the volatile organic solvent diffuses into the second aqueous (continuous) phase (if not presaturated with the organic solvent) and begins to evaporate into the surroundings. A phase separation occurs at the surface of the polymer and second aqueous solution, leading to a polymer-rich phase at the microsphere surface and a more dilute polymer solution in the interior. Thus, discrete particles are formed with entrapped protein inside. The hardening must proceed either by particle shrinkage or co-transport of organic phase and second aqueous phase components (mass must be conserved across the interphase if particle volume is fixed); thus, there is the potential of introduction of components (e.g., isopropanol) into the protein-containing innerwater phase. At the time of collection and lyophilization, residual solvent remains and is removed under vacuum, leaving pores behind, which can be visualized by microscopy [67].

Hence, a physicochemical analysis of the events during the solvent evaporation method indicate that the important aspects relating to protein stability during microsphere preparation include the organic solvent used [7], the presence of water (protein mobility) [14,68], and the method of emulsification [9].

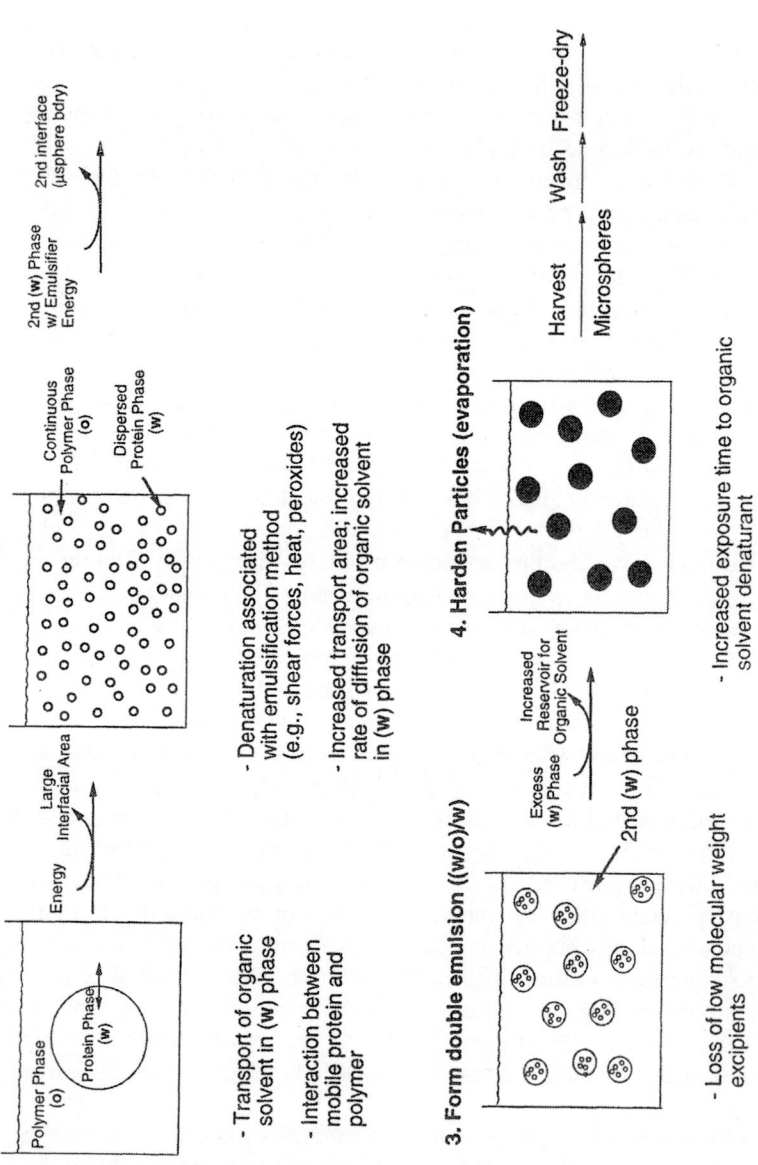

Fig. 2 Analysis of protein stability during preparation of biodegradable polymer microspheres by the double-emulsion–solvent evaporation technique. The diagram depicts the events that occur during preparation, with potential sources of inactivation described underneath.

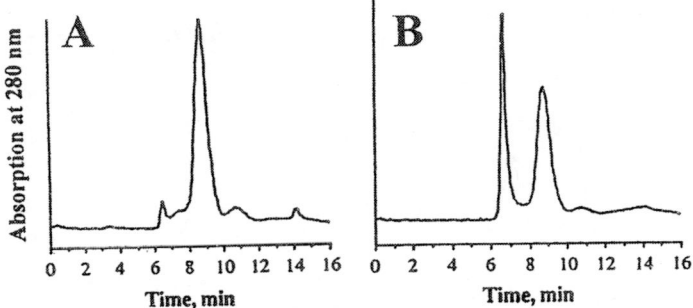

Fig. 3 Formation of soluble aggregate of (A) tetanus toxoid solution alone or (B) in the presence of methylene chloride, as determined by size-exclusion chromatography with UV detection (280 nm). A single interface was formed between 100 μL of vaccine solution and 200μL of the organic solvent. The two-phase system was incubated at 37°C for 3 h and analyzed. The negative control was incubated and analyzed without addition of methylene chloride. Insoluble aggregates were observed after 1 day at the same temperature. Several days were required to form the soluble aggregate at room temperature. (S. P. Schwendeman, unpublished observations.)

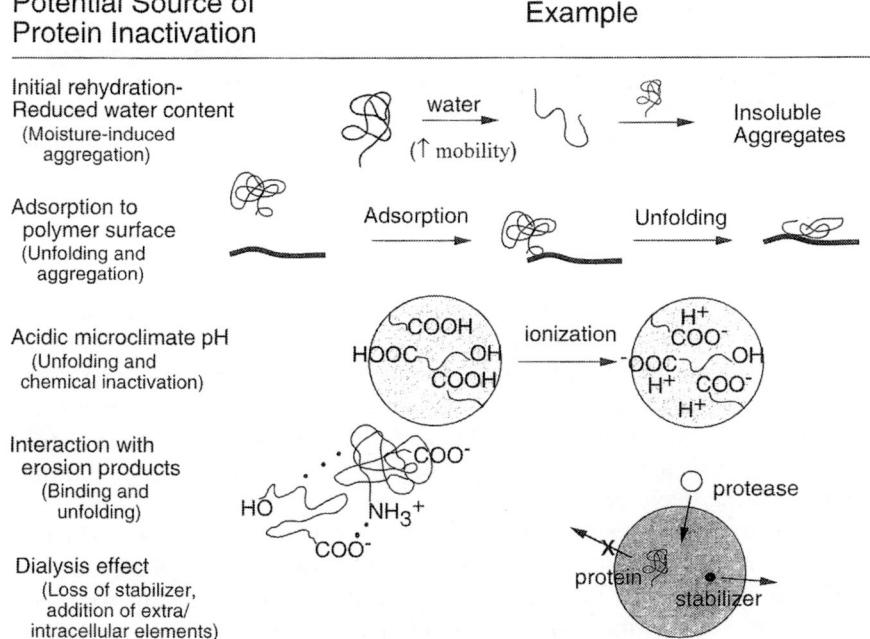

Fig. 4 Potential sources of inactivation during microsphere release incubation: These include initial rehydration and exposure to reduced water content, adsorption to polymer surfaces, an acidic microclimate pH, interaction with erosion products, and a "dialysis effect."

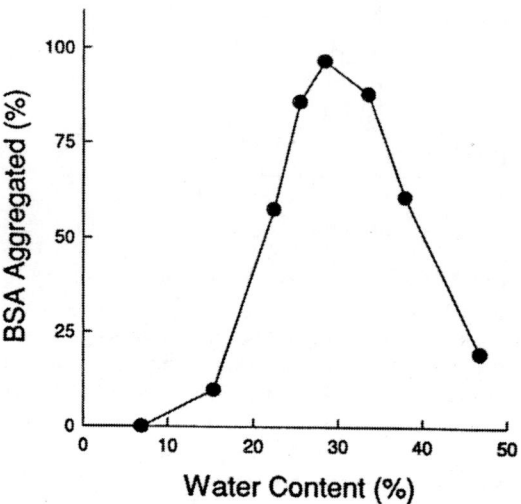

Fig. 5 The dependence of aggregation of bovine serum albumin on incubation of the lyophilized protein in the presence of moisture on the water content: The extent of insoluble aggregates formed is plotted *versus* water content in the protein powder after 24 h at 37°C. The lyophilized protein (6.7% water) was mixed with various amounts of aqueous buffer and sealed in vials. The amount of soluble protein was determined following reconstitution. The mechanism of inactivation was determined to be thiol–disulfide exchange. (Adapted from Ref. 23.)

1. Effects of Organic Solvents

The events immediately following the addition of a lyophilized protein to an organic solvent (relevant to the discussion when the lyophilized protein is encapsulated without the first emulsion) are by no means simple, beginning with the fact that the actual conformation of lyophilized proteins is uncertain (i.e., whether native or reversibly unfolded to some degree). A method recently proposed to evaluate this is infrared spectroscopy [69]. Nonetheless, some important concepts have been established in the enzyme literature. One of the central concepts involves the conformational mobility of the protein, which is particularly restricted in the aprotic and hydrophobic solvents typically used to dissolve polymers (e.g., methylene chloride). As small amounts of water are added to such an enzyme suspended in an organic solvent, thereby increasing the protein's flexibility, the activity increases by orders of magnitude [68]. This initial lack of mobility (particularly in the side chains, as opposed to the backbone) has been predicted to be due to the increased intramolecular interactions of the lyophilized protein (particularly hydrogen bonding) [70]. Thus, conditions during lyophilization, such as pH, the addition of stabilizers, and final water content, have a direct influence on the mobility of the protein when added to the organic solvent [68,71]. The next point for consideration is the stability of the bound water retained following lyophilization. Organic solvents with lower hydrophobicity, as indicated by a decreasing octanol/water partition coefficient, have a greater capability of "stripping" away water molecules and thus disrupting the catalytic centers [72]. Finally, if too large an amount of water is added to the organic solvent, proteins have sufficient mobility to denature, and then aggregate [68].

In contrast with the foregoing case in which the protein has severely hindered conformational mobility, when an aqueous protein solution is added directly to an organic polymer solution, the protein's mobility is maximal. As the organic solvent diffuses into the inner water phase, the former solvent can alter the aqueous ionic strength, or bind directly to the protein, favoring the exposure of the protein's hydrophobic regions [14]. We have observed that organic solvents, such as methylene chloride, cause the formation of first soluble and then insoluble aggregates of tetanus toxoid when a single interface is formed between the solvent and the aqueous protein solution (see Fig. 3).

2. Solubility of Proteins in Organic Solvents

Some organic solvents are capable of solubilizing proteins and, sometimes, without irreversibly denaturing them. These solvents are generally protic and hydrophilic, such as glycerol and formamide [73]. Remarkably enough, the pH of the aqueous solution before lyophilization has a dramatic effect on the protein solubility in organic solvents (i.e., the further away from the isoelectric point, the greater the subsequent solubility [73]).

3. Emulsification Processes and Formation of Large Surfaces

To form discrete microparticles, at some stage of the preparation, a large surface must be created. Emulsion processes are widely used for encapsulating proteins [8,63,65]. Typical methods of preparing emulsions include simple mixing (e.g., vortex), homogenization, and ultrasonication. All of these methods increase the rate of mass transfer between the bulk liquid and the air–liquid interface. This exposure is believed to be responsible for protein inactivation, owing to the hydrophobicity of air, which favors unfolding [14]. Ultrasound has caused highly reactive free radicals following thermal reactions in the microbubbles [74] and was determined to result in 40% loss in activity of lysozyme during the double-emulsion technique [9]. Homogenization involves pumping the liquid past a narrow orifice; both pressure and shear forces at a boundary increase unfolding [14]. Other methods for forming large surfaces, such as air-atomization, have been reported to cause protein denaturation [75].

4. Other Microsphere Preparation Techniques

In addition to solvent evaporation, other common methods of preparing protein-loaded microspheres include phase-separation processes (such as coacervation) and spray-drying [76], as well as less frequently used air-atomization and hot-melt encapsulation [77,78]. During the phase-separation method, the protein is either suspended or emulsified (as an aqueous solution) in a dilute polymer solution (e.g., with methylene chloride), which may have a phase inducer (a chemical that aids in the phase separation). A nonsolvent (e.g., alcohols) is added, causing the phase separation, and the polymer hardens, thus trapping the protein phase. This usually results in monolithic-type microspheres. In the cases of phase separation and spray-drying, once again an organic solvent must be used. With hot-melt encapsulation, no organic solvent is required, as the polymer is simply heated in oil [78]. This method may have potential, because proteins are unusually resistant to heat denaturation when water is not present [62]. However, several deleterious reactions occur during exposure to high temperatures [62]. For hydrophilic polymers, such as alginate and poly[bis(carboxylatophenoxy)phosphazene], air-atomizing apparatus can be used without the necessity of an organic solvent or high temperature [77,79].

B. Lyophilization and Storage

There are many techniques that can be employed to maintain the stability of proteins to afford an acceptable shelf life, the most common of which is freeze-drying or lyophilization [22]. Lyophilization is also used to remove residual organic solvent not removed during hardening to attain acceptable levels for therapeutic use. The procedure consists of two processes; freezing and dehydration. Typically, a dilute aqueous solution of the protein is placed in vials that are then loaded onto temperature-controlled shelves in a freeze-drier chamber. The temperature is lowered to below 0°C to allow the protein solution to be frozen so that most water can be subsequently removed by sublimation in the dehydration steps. Depending on the starting concentration, two possible scenarios can arise on cooling. In 40% (w/v) sucrose, ice will form as the temperature is reduced, resulting in a two-phase system consisting of pure ice and a freeze concentrate. As the temperature is further lowered, more ice will continue to crystallize, and the freeze concentrate will be concentrated until it reaches a sharp increase in viscosity and forms a glass (vitrification). The temperature at which this occurs is called the glass transition temperature of the maximally concentrated solute, T_g' (the glass transition temperature of the protein is denoted T_g), at which the viscosity is such that no more ice can form [22]. In solutions that are initially highly concentrated, such as sucrose at concentrations greater than 80% (w/v), freezing does not occur, and the formulation first forms a rubber, and then a glass [22]. Any material that is in the glass state will be stabilized by immobilization. The pressure chamber is then reduced, and the shelf temperature is increased to supply heat for the sublimation of ice during the primary-drying stage. The temperature used must be less than the T_g' to maintain the protein in the glass state. Once all the ice has been removed, large amounts of unfrozen water that are trapped in the amorphous glass state are removed by desorption during the secondary drying of the product. The drying rate during the secondary drying is much slower compared with primary drying. This effect is not due to the tight binding of water; rather, it is a kinetic problem associated with the very slow rate of movement of water and evaporation from a solid state. The rate of secondary drying is independent of chamber pressure and is ordinarily conducted at moderate pressure (Pikal suggests a normal operating chamber pressure of 0.2 torr [22]). The rate of temperature increase and the duration of processing can be calculated from the rate of diffusion of the residual moisture from the partially dried matrix [80]. By using differential scanning calorimetry, it is possible to measure the T_g' of different formulation additives [80]. The T_g' of proteins is often undetectable using conventional differential scanning calorimetry (DSC) because the ice-melt endotherm often masks the T_g, and more sensitive instruments, such as scanning microcalorimetry, are required [81]. This sensitive analytical technique can be used to define optimum-processing parameters, such as time, temperature, and pressure, which may be more efficient than deriving these parameters by trial and error [81].

Inactivation processes can occur during freezing, drying, and storage. During freezing, the freeze concentrate can reach very high solute concentrations (e.g., normal saline, 0.14 M NaC1, can reach 5 M [22]). High salt concentrations can denature proteins [14]. Also, preferential crystallization of one of the multiple buffering species can sharply alter pH. For example, crystallization of disodium phosphate monohydrate during freezing can shift the pH from 7.4 to as low as 3.5 [22]. During drying, if the temperature is allowed to rise above T_g', some of the ice will melt back into the nonice phase, forming a highly concentrated protein—excipient solution. Under such conditions, the protein will be subject to deleterious processes, owing to an increase in the mobility of the functional side chains in the protein [22]. For the same reason, loss of a glassy protein–excipient phase is an effect also important during storage. Excipients and polymers may also have contaminants, such as metal catalysts, that can inactivate proteins during storage [14]. A final effect is the loss of a native-like structure in the solid state owing to excipients that do not take the place of the normal hydrogen bonds provided by water [69]. Inactivation generally occurs much more rapidly following loss of physical stability (see Section V) [14,15].

C. Release Incubation (Polymer Erosion)

Polymer erosion, which we define as the loss of polymer mass from the initial level, is accompanied or preceded by losses in average molecular weight, by increases in polymer porosity, and by accumulation of monomer in the release medium. These complex processes are often oversimplified in terms of the release behavior (i.e., release kinetics characteristic of bulk [82] and surface erosion [83]). However, recent detailed analysis of erosion processes has provided much more insight about what is happening within these polymers [84–88]. These details are critical to our objectives in understanding the potential causes of inactivation within biodegradable microspheres. These recent findings underscore the importance of monomer and oligomer solubility, microclimate pH, pH-dependent hydrolysis, crystallization, polymer microstructure, and heterogeneous erosion (PLGA). However, the foregoing studies were carried out using macroscopic geometries, so one must be cautious in extrapolating some of these results to the smaller-length scale (microns). Recently, there is evidence that some of these processes (e.g., heterogeneous erosion) also can occur for microspheres [89].

Erosion begins with water uptake. As water penetrates the polymer, changes in *microstructure, plasticization,* and *hydrolytic bond cleavage,* which we define as polymer degradation, begin. The glass transition temperature of PLGA may fall below physiological temperature in the hydrated sample [90]. As degradation proceeds, the polymer may become more hydrophilic. In addition, new functional groups, such as carboxylic acids, will be formed, depending on the monomer, which may alter pH. If the polymer chain becomes small enough, it will become soluble in the release medium [89], or form an insoluble precipitate [88]. The polymer matrix can form cracks, particularly if it is brittle [87]. For PLGA, significant degradation typically occurs before appreciable mass loss from the polymer, indicating the build-up of low molecular weight fragments within the polymer. As the events of degradation, precipitation of insoluble products, and diffusion of smaller soluble material occur, the permeability of the polymer steadily rises, allowing the encapsulated protein to slowly diffuse out.

From the foregoing simple physicochemical analysis, we may hypothesize several important factors that can cause inactivation of proteins during release incubation. Many of these are depicted in Fig. 4 and include protein rehydration (moisture-induced aggregation), interaction with hydrophobic surfaces, exposure to physiological temperature, contact with products of erosion (acid equivalents and low molecular weight fragments), reactivity of the protein with the polymer (not shown), and transport of low molecular weight species in and out of the polymer matrix before the protein is released.

1. Protein Rehydration

As microspheres become hydrated, the protein will be exposed to increasing amounts of water, a process known to cause aggregation of proteins [23,24,60], including vaccine antigens (e.g., tetanus toxoid [91]). The rate and extent of water uptake will depend on the polymer type, porosity, protein loading, and other factors. The time scale of this event can be substantially longer, as compared with direct reconstitution. For example, we observe that PLGA microspheres can float at early stages of release, presumably owing to air-filled pores present initially. After a day or so, they will sink as the density increases above that of water. Many proteins (e.g., bovine serum albumin, insulin, and tetanus toxoid) form insoluble aggregates when the lyophilized powder is exposed either directly to small amounts of water or to elevated levels of humidity. In fact, a *critical* water content within the protein powder was observed to cause most rapid aggregation (see Fig. 5) [23]. The ascending portion of the curve was hypothesized to be due to steadily increasing conformational mobility of the protein, and the decending portion of the curve due to a dilution effect. The effect of stabilizers on the moisture-induced aggregation of recombinant human albumin (rHA) has been evaluated recently [59]. In this study, the ability of the excipients to inhibit the disulfide interchange mechanism at the *critical* relative humidity for rHA (96%) was directly related to the water-sorbing capability of the excipient (i.e., the greater the water sorbed, the greater the stability of rHA). An additional hypothesis was put forward to explain the descending portion of the bell-shaped of curve for rHA (similar to that described in Fig. 5). Because rHA is likely to exist in a structurally altered (partially denatured) state in the lyophilized powder, increasing the water content allows the protein to refold, thereby rendering the heretofore reactive thiols inert in the interior of the protein. Evaluating protein aggregation in this way is reasonable, because the solvent characteristics of water in polymers, such as PLGA, is likely to be vastly reduced, and mechanistic analysis can be readily performed (see Section V.B). If aggregation occurs within the microsphere, the consequences can be sharply reduced biological activity, increased immunogenicity, and insufficient release [92]. Hence, the potential for this deleterious process will always exist for polymers that must go through a rehydration step.

2. Hydrophobic Surfaces

A hydrophobic surface (e.g., the air–water interface) can lead to aggregation of stored protein solutions [14]. Proteins can adsorb strongly to both hydrophilic and hydrophobic materials. The former adsorption is typically reversible, and the latter often results in irreversible conformational changes. Sluzky et al. observed that insulin aggregation rose with increasing surface hydrophobicity and agitation rate [58]. Adsorption of proteins onto polymers is a subject most documented in the area of blood-compatible biomaterials [16]. Large proteins have many binding sites. The kinetics of protein adsorption on short-time scales is always mass transfer-limited. Thus, proteins present in excess, albumin for example, will occupy surfaces first, because the rate of movement by diffusion is greatest (owing to the steepest concentration gradient). At later times, proteins with particularly high-binding constants will begin to occupy a steadily higher fraction of the surface. Generally, the greater the hydrophobicity of the surface and the protein, the tighter is the binding. In addition, the binding is strongest at the protein's isoelectric point, the pH at which electrostatic potential barriers at the surface are minimal [16].

3. Physiological Temperature

If long-term release incubation is required, stability at physiological temperature and a reasonable pH range is one of the most basic prerequisites to consider before encapsulation. The use of elevated temperatures (or accelerated studies) is a convenient method of identifying possible mechanisms of inactivation [93] at physiological temperature. For example, at mildly acidic pH (pH 4) and 90°C, ribonuclease was determined to inactivate by peptide bond hydrolysis at aspartic acid residues and deamidation of asparagine or glutamine residues [93]. At neutral pH (pH 6 and 8) and the same temperature, disulfide interchange, β-elimination of cystine residues, and deamidation of asparagine or glutamine residues were determined to be responsible for loss of enzymatic activity [93].

4. Degradation Products

As the polymer degrades into smaller pieces, the nature of these products may strongly affect the internal environment of the microspheres and, hence, that of the protein.

  a.   Monomer and Oligomeric Products.   For soluble monomers and oligomers, such as those produced by PLGA, the co-solvent effects on the protein (i.e., whether their presence increases or decreases the conformational stability) are unknown [94]. This may become important if these products are sequestered in microspheres [84,85,89]. Oligomers of poly(D,L-lactic acid) in the neighborhood of 15 monomer units have appreciable water solubility [89].

  b.   Internal or Microclimate pH.   If the products of degradation are either a base or an acid, the pH can dramatically change within the polymer. This has been established for large geometries in the cases of PLGA [84,85,95], polyanhydrides [88], and poly(*ortho* esters) [96], and for smaller geometries with PLGA [89]. Irrespective of the crystallinity changes that occur if the L-isomer of lactic acid is used for the PLGA, it appears that all copolymers of this polymer type possess the ability of sequestration of acid [84,85]. Therefore, one must consider a pH range that may be present in the microspheres ($\sim$ 2–7.4) because the actual pH remains unknown for PLGA microspheres. Certainly, a pH in the neighborhood of 2, found for large geometries of PLGA by direct measurement of the internal gel-like phase with a pH electrode [95], would be a substantial hurdle to overcome for delivery of proteins, because acid denaturation of proteins may be rapid at this pH [14].

5. Loss of Low Molecular Weight Excipients

As a general rule, polymers discriminate on the basis of size of the diffusing species to a much greater extent than in free solution [97]. Thus, low molecular weight stabilizers, such as monosaccharides, will diffuse through the polymer pores faster than will a protein. This could pose problems during later stages of release, if these excipients are required to stabilize the native conformation of the protein.

6. Covalent Reactions Between Proteins and Polymers

Covalent chemical reactions between proteins and biodegradable polymers has thus far not been reported. However, under certain conditions, amide formation occurs following reactions between organic amines and polyanhydrides [25], poly (ϵ-caprolactones), and polyesters [26]. Note that the relative reactivity of the carbonyl carbon of various functional groups with nucleophiles follows the trend [98].

$$\text{Acid chlorides} > \text{anhydrides} > \text{esters} \geq \text{carboxylic acids} > \text{amides} > \text{carboxylates} \tag{2}$$

The α-amino group from the NH_2-terminal amino acid and lysine side chains in proteins are potential nucleophiles in the foregoing reaction. The pH is important, since the amine must be nonionized for the reaction to occur (i.e., reaction is favored at neutral or basic pH). Importantly, the transient pH may change during lyophilization (see Section IV. B). During rehydration, the pH is not likely to remain constant over short-time scales, because differing buffering salts may have different dissolution rates.

V. MECHANISMS OF PROTEIN INACTIVATION

A substantial list of known mechanisms of loss of protein activity has been discussed in detail elsewhere [14,15]. These include reversible unfolding, aggregation, and chemical inactivation pathways. Aggregation pathways are discussed in Fig. 6 and chemical inactivation pathways in Fig. 7. We will mention these briefly in the present context of biodegradable microspheres. It is of general use, however, to emphasize the most general of kinetic pathways leading to inactivation [14,15]:

$$\begin{array}{c} N \rightleftarrows pU \rightarrow D \\ \downarrow \\ (pU)_n \end{array} \tag{3}$$

where N is a native protein in equilibrium with a partially unfolded state, pU. The pU has less tertiary structure (e.g., the molten globule state [99]) and is much more reactive to most side reactions, leading to an irreversibly denatured state D or to irreversible aggregates $(pU)_n$.

A. Physical (Conformational) Stability

The native structure is maintained by a delicate balance of noncovalent interactions (e.g., hydrogen bonds, van der Waals interactions, salt bridges, and hydrophobic interactions). The physical stability of the protein is critical, as reflected by Eq. (3). On loss of much of the native tertiary structure to form pU, the protein becomes far more reactive as its hydrophobic regions become exposed and as previously buried reactive side chains interact with the solvent environment. The classic conditions relating to loss of conformational stability are elevated temperature, extremes of pH, denaturants, and adsorption to hydrophobic surfaces, all of which are important in the present context. At any temperature, the native state is in equilibrium with pU and the concentration of the latter will depend on the equilibrium constant between the two. The folded form is typically favored by some 50 kcal/mol at physiological temperature.

As the pH is adjusted away from the protein's isoelectric point, changes in ionization of amino acid residues within the protein may cause unfolding owing to the electrostatic repulsion that results [14]. Denaturants include salts, chelating agents, and organic solvents. Salts that decrease the solubility of the hydrophobic groups in the protein (salting-out salts) favor the native state, whereas salts raising the solubility in water of hydrophobic moieties (salting-in) are destabilizing and favor protein unfolding [14]. These effects follow the Hofmeister series [17]. If the protein is stabilized by a heavy-metal ion, chelating agents, such as ethylenediamine-tetraacetic acid (EDTA), can cause unfolding by removing the metal ion. In organic solvents that do not dissolve proteins (e.g., a protein suspended in ethyl acetate), the solvent can bind directly to hydrophobic portions of the protein, displacing bound water molecules [14]. If the organic solvent is saturated with water, the protein can become more flexible as the water molecules diffuse into the protein molecules [68]. For water-miscible solvents, addition of the solvent to the aqueous protein solution can cause either stabilizing (e.g., glycerol) or destabilizing, depending on whether the solvent increases the amount of water bound to the protein (preferential hydration) or decreases it [100].

Detergents also can cause unfolding by binding directly with the protein. A classic example is SDS (see Section VII.A.6). Nonionic surfactants generally cause less unfolding and are used extensively in some protein assays, such as enzyme-linked immunosorbent assay (ELISA) and in release media because they inhibit adsorption to surfaces.

Following adsorption of a protein to a surface, the protein may either subsequently desorb, without a conformational change (typically true for hydrophilic surfaces), or change its conformation to promote stronger binding (i.e., expose its hydrophobic regions to make additional contacts with the hydrophobic surface).

B. Aggregation Processes

There are several known mechanisms of nonspecific aggregation of proteins by noncovalent interactions between unfolded protein molecules or by covalent cross-links. These mechanisms are discussed for the relevant case of solid-phase aggregation by Costantino et al. [24] and are listed in Fig. 6. Hydrophobic interactions are responsible for noncovalent-bonded aggregation. These aggregates may be dissolved in a denaturing solvent, such as GuanHCl or urea [23]. A common covalent mechanism is thiol–disulfide interchange, which involves the ionized form of the free thiol of cysteine. This thiolate nucleophile of one protein molecule attacks the sulfur atom of a disulfide of a second protein molecule, causing a new disulfide bridge between the two proteins, while forming a new reactive thiolate (which may react with yet another protein molecule) [23]. Another process involves β-elimination (see following) of a disulfide of cystine to form thiocysteine and dehydroalanine. Once formed, the thiocysteine residue can decompose to hydrogen sulfide ion (HS^-) and a cysteine containing the free thiolate. The HS^- can catalyze intermolecular disulfide exchange between two intact disulfides, and the free thiolate is available for thiol–disulfide interchange, as before. The β-elimination–disulfide exchange has been observed for insulin when the solid protein is incubated in the presence of moisture [60]. Additional mechanisms have been observed to involve lysine residues. Dehydroalanine formed from β-elimination can undergo a nucleophilic attack by lysine's ϵ-NH_2 group of another protein molecule to form a lysinoalanine [101]. Here, the two proteins are held together by a carbon–nitrogen bond, which is stable to acid hydrolysis and may be detected by amino acid analysis [93]. A final mechanism is transamidation between glutamine or asparagine residues of one protein and a lysine of another. Indirect evidence for this mechanism has been provided by a simultaneous liberation of ammonia and loss of ϵ-amino group of lysine, both of which can be easily monitored [101].

C. Chemical Inactivation

Proteins, similar to most biochemicals, are susceptible to numerous chemical changes, outlined in the following sections (see Fig. 7).

1. Deamidation

The deamidation of asparagine and glutamine residues in proteins is a common covalent pathway, which can lead to a loss in protein activity. The reaction is catalyzed nonenzymatically by water and possibly by functional groups within the protein. The deamidation is facile at both acidic and alkaline pH and at high tem-

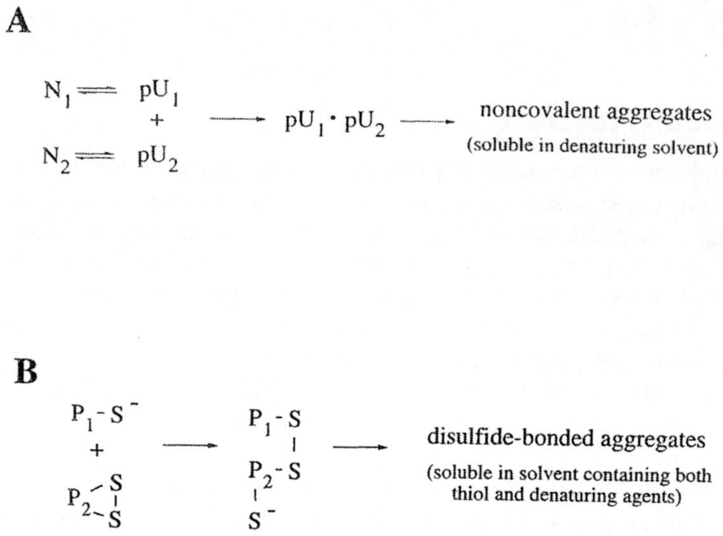

A

$$N_1 \rightleftharpoons pU_1$$
$$+$$
$$N_2 \rightleftharpoons pU_2$$
$$\longrightarrow \quad pU_1 \cdot pU_2 \longrightarrow$$
noncovalent aggregates
(soluble in denaturing solvent)

B

$$P_1\text{-}S^-$$
$$+$$
$$P_2 \overset{\diagup S}{\underset{\diagdown S}{|}}$$
$$\longrightarrow \quad \begin{array}{c} P_1\text{-}S \\ | \\ P_2\text{-}S \\ | \\ S^- \end{array} \longrightarrow$$
disulfide-bonded aggregates
(soluble in solvent containing both
thiol and denaturing agents)

Fig. 6 Mechanisms of protein aggregation: (A) Noncovalent aggregation caused by hydrophobic interactions following partial unfolding of two or more native protein molecules, denoted N_1 and N_2, to pU_1 and pU_2, to form soluble aggregates (e.g., $pU_1 \cdot pU_2$) and then insoluble aggregates, soluble in a denaturing solvent (e.g., 6 M guanidine hydrochloride). (B) Thiol–disulfide interchange is triggered by the formation of a reactive thiolate, which attacks the disulfide bridge to form a new disulfide bond and thiolate ion. The resulting insoluble aggregates are soluble in solvents containing both denaturing and thiol agents. (C) The β-elimination-related aggregation begins with the β-elimination of a cystine to form thiocysteine and dehydroalanine. Thiocysteine then decomposes to thiol-containing products, such as hydrogen sulfide ion, while dehydroalanine can be a target of a nucleophilic attach by a lysine residue from a second protein (to form a lysinoalanine cross-link). Hydrogen sulfide ion can catalyze disulfide exchange between protein molecules to form disulfide-bonded aggregates as in panel B. The lysinoalanine cross-link may be determined by amino acid analysis of the acid hydrolysate. (D) Isopeptide aggregates are formed following the nucleophilic addition of lysine to an asparagine or glutamine residue.

peratures [102]. A rule-of-thumb pH of 6 ensures maximal stability against deamidation; the rate of deamidation accelerates when an aspargine or glutamine residue is adjacent to certain amino acids in the primary structure and is also dependent on the higher order of the protein (e.g., tertiary structure) [102].

2. *Oxidation*

Oxidation of methionine to its sulfone or sulfoxide, can happen simply in the presence of atmospheric oxygen [15]. Other oxidizing agents are hydrogen per-

C

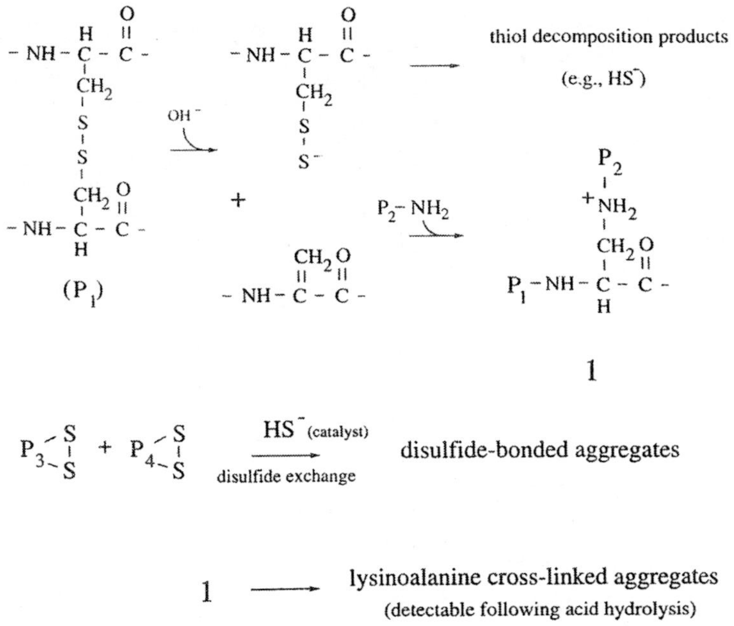

D

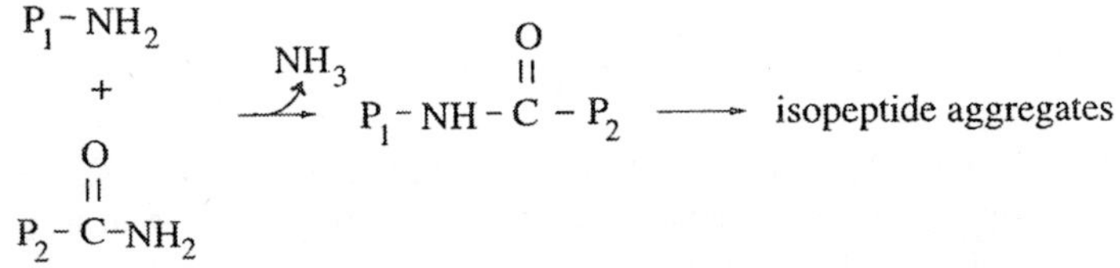

oxide, hydroxyl and superoxide radicals, hypochlorite ion, and visible light [14,15]. Oxidation of methionine is not easy to detect and is typically revealed by the reactivity of the nonoxidized residue with cyanogen bromide, which does not react with the oxidized form [15]. Another relatively unstable amino acid residue, tryptophan, is oxidized at low pH by visible light. At neutral and alkaline pH, the thiol of cysteine can be oxidized by the oxygen in air, especially in the presence of even catalytic amounts of transition metal ions such as Cu^{2+} [15,60]. The products of oxidation of cysteine can be sulfenic acid, cystine, sulfinic acid, or sulfonic acid, or a combination thereof.

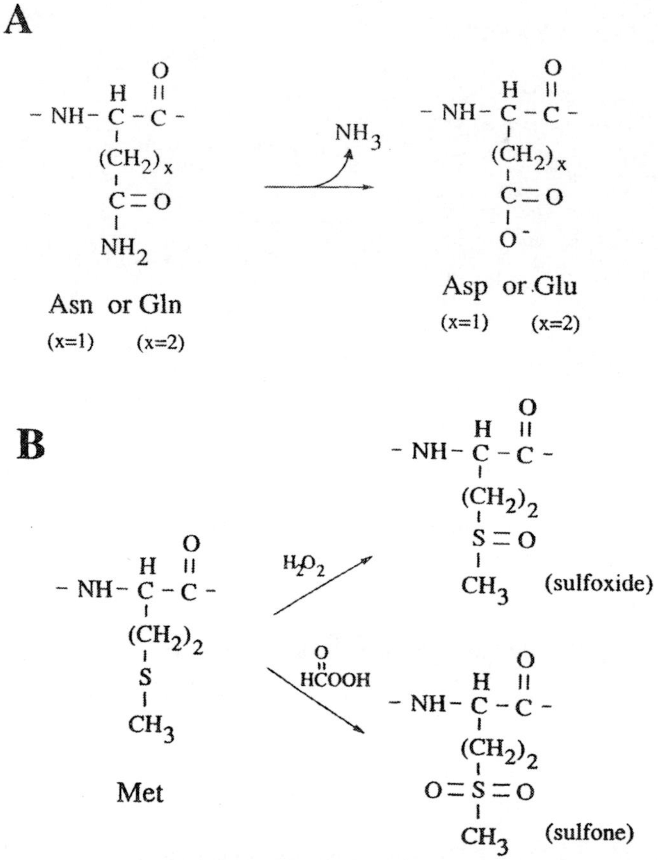

A

B

Fig. 7 Examples of chemical inactivation of proteins: (A) Deamidation of asparagine or glutamine residues results in the evolution of ammonia to form aspartic acid or glutamic acid. (B) Oxidation of methionine residues typically causes formation of its sulfone or sulfoxide. (C) Hydrolysis of peptide bonds, such as apartyl–glycine bonds, forms two new peptides entities. (D) The β-elimination of cystine, serine, and threonine residues is caused by abstraction of a proton of the α-carbon and subsequent loss of the leaving group (X) to form the α-ethenyl amino acid residue. (E) Racemization of an L-amino acid residue (e.g., aspartic acid) to its D-enantiomer is also depicted. For more details concerning these reactions, see Manning *et al.* [15]

3. *Hydrolysis, β-Elimination, and Racemization*

Nonenzymatic hydrolysis of proteins occurs primarily at peptide bonds adjacent to aspartic acid (Asp) residues. The reaction is favored in acidic media [93]. Cyclic intermediates are formed to either the NH_2- or COOH-terminal side of the aspartate side chain, resulting in cleavage of the peptide bond on either side

C

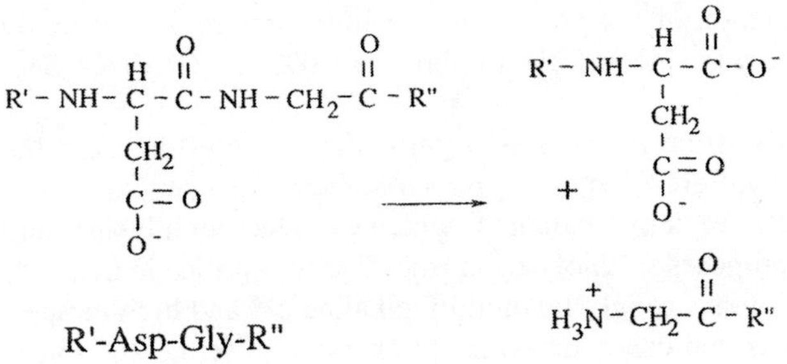

$$R' - NH - \underset{\underset{\underset{\underset{O^-}{|}}{\underset{C=O}{|}}}{\underset{CH_2}{|}}}{\overset{\overset{H}{|}}{C}} - \overset{\overset{O}{\parallel}}{C} - NH - CH_2 - \overset{\overset{O}{\parallel}}{C} - R''$$

R'-Asp-Gly-R"

$\longrightarrow$

$$R' - NH - \underset{\underset{\underset{O^-}{|}}{\underset{C=O}{|}}}{\underset{CH_2}{|}}}{\overset{\overset{H}{|} \;\; \overset{O}{\parallel}}{C - C - O^-}}$$

+

$$\overset{+}{H_3}N - CH_2 - \overset{\overset{O}{\parallel}}{C} - R''$$

D

$$- NH - \underset{\underset{\underset{X}{|}}{\underset{H - C - R}{|}}}{\overset{\overset{H}{|} \;\; \overset{O}{\parallel}}{C - C}} -$$

$$\xrightarrow{\;\; X^-\;\;}$$

$$- NH - \underset{\underset{CH - R}{\parallel}}{\overset{\overset{O}{\parallel}}{C - C}} -$$

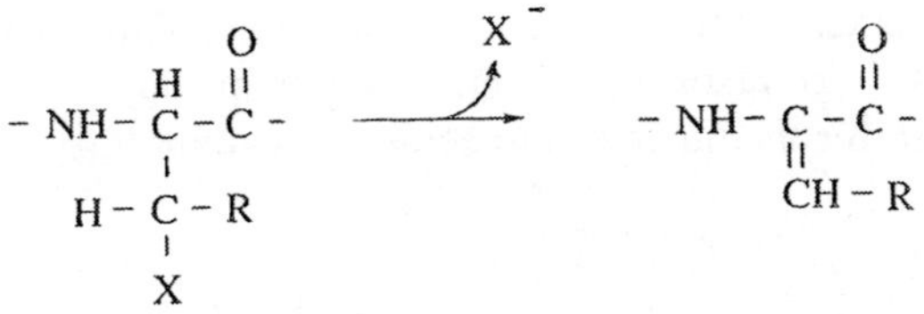

$X_1 = $ S-S-Cys $X_2 = $ OH

$R_1 = $ H $R_2 = $ H, CH_3

E

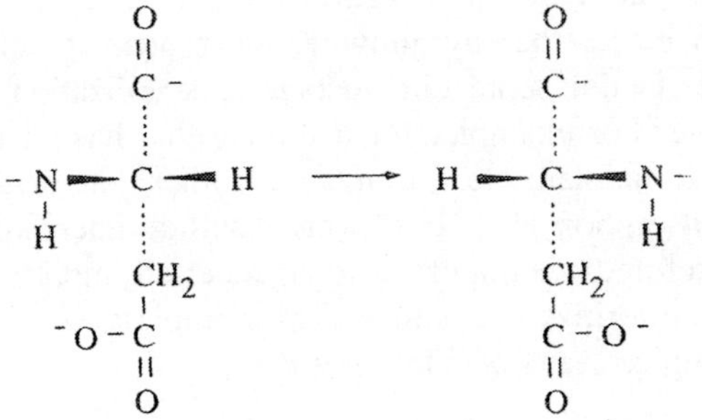

L-Asp

D-Asp

of the residue, with cleavage rates greater than 100-fold of those for typical peptide bonds [15]. Aspartic acid–glycine and aspartate–proline linkages are particularly labile. Evaluation of the progression of the reaction is most easily examined by SDS-polyacrylamide gel electrophoresis (SDS–PAGE) (see Section VII.A.2).

The β-elimination of cystine residues is a particularly interesting case because it can lead to two different types of intra- or intermolecular cross-links (see Fig. 6). The products are dehydroalanine, which can react with lysine, and thiocysteine, which decomposes to thiolate ion (the reactive species in disulfide interchange). This reaction is strongly favored by alkaline pH and high temperatures [93]. Other residues that can undergo β-elimination are serine and threonine [15]. Dehydroalanine, lysinoalanine, and newly formed thiols can be readily detected [60,93].

Natural, L-amino acids are subject to racemization to the corresponding D-enantiomers [15]. This is believed to proceed through a carbanion intermediate, predominantly for aspartic acid, which forms a cyclic imide. This intermediate stabilizes the carbanion through resonance, leading to a frequency of isomerization 10^5 times greater than for the other amino acids that cannot form the cyclic imide [15].

VI. INHIBITING OR AVOIDING MECHANISMS OF PROTEIN INACTIVATION

There are generally two approaches toward stabilization: (a) to minimize reversible unfolding of the protein (loss of physical stability) and the consequent greater exposure of the reactive groups or hydrophobic regions, leading to aggregation; and (b) to prevent irreversible aggregation or chemical reactions once unfolding has taken place, thereby allowing the protein to refold to its native conformation [14]. Under some circumstances, stabilization strategies may be mutually exclusive. For example, for a protein that has an isoelectric point of 4.8, it is better to maintain close to neutral, rather than acidic pH, to prevent aggregation on adsorption [16]. However, disulfide interchange is inhibited at acidic pH. Therefore, it is important to consider all important mechanisms of inactivation, to prioritize them in order of importance for a given case, and to develop multiple means of inhibiting them.

A. General Approaches

Once inactivation mechanisms have been elucidated, or at least partially characterized, there are some general approaches that have been established to protect against these processes. These include increasing the intrinsic stability of the protein, the use of additives, immobilization, and chemical modification

[14]. The intrinsic stability may be increased by expressing the proteins in thermophilic bacteria [103], or by site-directed mutagenesis used to replace labile residues such as asparagine and aspartic acid (see Section V.C.3) [102]. The use of additives, such as salts, nonreducing sugars, and polyols, minimizes the free energy of the native state by increasing the number of bound water molecules around the surface of the protein [100]. Specific additives, such as a substrate for an enzyme, can shift the equilibrium in Eq. (3) toward the native state, and scavengers (e.g., antioxidants) can inhibit reactions such as oxidation [14]. Immobilization methods can be used to change the local environment of the protein by partitioning out toxic species [104]. An example is ion-exchangers, which either reduce (cation-exchangers) or increase (anion-exchangers) the local pH according to Donnan equilibrium [105]. Immobilization can also be used to separate proteins from reacting with one another. Chemical modification such as S-alkylating free thiols by iodoacetamide [23] is a useful alternative to site-directed mutagenesis. To stabilize the native state, cross-linking with bifunctional agents, such as glutaraldehyde [106], can also be used.

B. Stabilization During Encapsulation

Proteins display markedly increased thermostability in anhydrous organic solvents, owing to the reduced conformational mobility in the absence of water [14]. Therefore, the methods of encapsulation on which water is absent or controlled at an optimal level may be helpful to avoid aggregation processes incurred with methods, such as the double-emulsion technique [7], for which the protein is conformationally mobile. There is some evidence that certain solvents are worse than others. Ethyl acetate decreases the soluble aggregates formed and increases antigenicity recovered relative to methylene chloride, following emulsion of tetanus toxoid in the organic solvent [7]. It is not known which among methods of forming the microsphere (e.g., homogenization, sonication, or other) is the least denaturing. Protein microparticulates can be created without the use of organic solvents. For example, a method whereby an aqueous protein solution is air-atomized in liquid nitrogen has been described [75]. After lyophilization, if such microparticulates were encapsulated with a biodegradable polymer, it is reasonable to assume that the protein will be most stable, because exposure to the organic solvent occurs when the protein is immobile in the solid state. Thus, new methods of encapsulation are needed if protein stability is to be optimized.

C. Stabilization During Lyophilization and Storage

Before freeze-drying, stabilizers, such as amino acids, sugars, surfactants, and different polymers can be added. However, the preservation of protein stability during freezing and during drying is mechanistically distinct for each process.

Protection by cryoprotectants appears to be due to preferential exclusion of the solute that stabilizes the native state of the protein in the freeze-concentrate during freezing [107]. However, some cryoprotectants provide little protection during subsequent drying [108]. Lyoprotectants are formulation additives that maintain the stability of the product during freezing, dehydration, and possibly storage. Pikal [22] describes two hypotheses for stabilization during drying and storage: (a) the kinetic *vitrification hypothesis,* which states that the protein is stabilized during storage when lyophilized in the presence of good glass formers (thus, the protein is not mobile to undergo deleterious reactions), and (b) the thermodynamic *water substitute hypothesis,* which states that hydrogen bonding between the protein and the sugar promotes the native conformation once water has been removed. There is evidence that both of these hypotheses can be important [69,109].

D. Stabilization During Release

Several methods have been developed to inhibit moisture-induced aggregation. In evaluating the influence of excipient on the moisture-induced aggregation of recombinant human albumin, lyoprotectants that increased the water content in the lyophilized powder were the most stabilizing [59]. To inhibit disulfide interchange (see Fig. 6B), it is possible to *S*-alkylate the thiol groups (at least on the surface of the protein), thereby rendering thiols unreactive [23]. Reducing the pH before lyophilization can have the same effect [23]. To inhibit another moisture-induced aggregation mechanism, disulfide exchange following β-elimination of a disulfide bridge (see Fig. 6C), trace amounts of divalent metals such as Cu^{2+} were added to insulin [60]. The metal ions act by oxidizing the newly formed thiol decomposition products, such as HS^-, thereby preventing thiolate-catalyzed disulfide exchange.

Adsorption is typically strongest at the pI of the protein and when the polymer is hydrophobic [16]. Therefore, maintenance of the pH of the protein phase away from the isoelectric point and the use of more hydrophilic polymers are called for from this perspective. Minimization of the polymer–protein surface area by the use of microcapsules (i.e., polymer coating of a single protein phase) instead of microspheres (i.e., protein phase dispersed throughout the polymer) is a concept that we have been developing in our research laboratory. Finally, the use of certain nonionic surfactants reduces this denaturing effect for insulin [58].

As in immobilization methods (see Section VI.A.), the polymer that encapsulates the protein will affect the microclimate of the protein and the diffusion of small solutes between the release medium and the protein. The very acidic pH ($\sim$2) inside macroscopic PLGA specimens [95], described earlier, is an illustration of an undesirable immobilization effect. The ability to counter potential factors, such as acidic pH, erosion products, and protein–polymer reactions, awaits better information on the processes of erosion in microspheres and the solution characteristics within them as they erode.

E. Other Biodegradable Polymers

The most "protein-friendly" synthetic biodegradable polymers are likely those that either provide a suitable aqueous environment, or are capable of maintaining the rigidity of the protein by preventing its contact with water. The first class of these is represented by certain polyphosphazenes (e.g., poly[bis(carboxylatophenoxy)phosphazene]), which require no organic solvents or rehydration steps. The encapsulation conditions are so mild that these may even be used to encapsulate liposomes [77]. An example of the second class of polymers may be the polyanhydrides, because extremely hydrophobic polyanhydrides are capable of limiting water penetration owing to their hydrophobicity [10].

VII. EXPERIMENTAL METHODS

A. Methods of Characterizing Protein Conformation and Conformational Changes

In vitro studies depend on biophysical methods for characterizing the structure of proteins and extent of denaturation. Information at low levels of resolution may be obtained from intrinsic viscosity, gel filtration, and quasi-elastic light scattering, all of which assess the size of the protein molecule [56]. Various types of spectroscopy, such as circular dichroism, can monitor the amounts of secondary structure, and fluorescence spectroscopy can monitor the environments around the tyrosine and tryptophan residues. Such methods have relatively low resolution, and they provide little detailed molecular information. High-resolution NMR spectroscopy is a rapidly growing technique that has enormous potential for providing the spatial coordinates of individual protein atoms in aqueous solution. The following section will describe the background for some of the simpler techniques used to study protein structure. In general, these techniques may be divided into three categories: spectroscopy, electrophoresis, and chromatography.

Table 1 depicts different features, sensitivity, and buffer constraints of the methods available to analyze proteins [110]. Numerous test procedures are available. However, a sensible range of these must be chosen to ensure continued equivalence of different batches of a pharmaceutical protein. This is essential from a safety and efficacy viewpoint.

1. Spectroscopy

Several established spectroscopic methods are available for determination of the physical structure of native and unfolded proteins [110]. These methods

Table 1 A Comparison of the Experimental Methods Used in the Analysis of Polypeptide and Proteins

Method	Protein Concentrations Required	Possible Structural Information Obtained	Limitations and Buffer Constraints
UV absorption	nM–μM	Aromatic amino acid residues' microenvironment	Only for protein in solution
Intrinsic fluorescence	nM–μM	Aromatic amino acid residues' microenvironment	Generally protein is in solution state
Circular dichroism	$>1\ \mu M$	Secondary structure	Only for protein in solution
Electrophoresis	$>0.1\ nM$	Molecular weight, isoelectric point, mass/charge ratio, and antigenicity	Can be used with protein of low solubility, sensitive to high salt concentrations
Chromatography	$>10\ nM$	Molecular weight, mass/charge ratio, and hydrophobicity	Only for protein in solution, sensitive to buffer conditions
Fourier transform infrared spectroscopy	$>0.1\ mM$	Secondary structure	Can be used with solid states, some solutes can interfere with measurements
Laser light scattering	$>1\ \mu M$	Molecular weight and shape	Generally not affected by buffers
Mass spectrometry	Varies	Very accurate molecular weight, covalent modifications, and amino acid sequence	Solutes can interfere, protein is normally bound to a solid phase
ELISA	pM–nM	Antigenicity, integrity of active sites, and immunogenicity (in case of vaccines)	Sensitive to buffers
Bioassay	pM–nM	Functional activity and bioactivity	Sensitive to buffers

Source: Adapted from Ref. 110.

include mass spectroscopy, UV absorbance spectroscopy, fluorescence spectroscopy, and circular dichroism. The spectral properties of a protein are dependent on the environment of the intrinsic chromophores, particularly peptide bonds and aromatic amino acid residues [110]. Spectral methods have the advantage of being performed in solution for which, typically, only small amounts of material are required for analysis [110]. These techniques can be used to investigate changes in the structure and behavior of a protein under different solvent conditions, such as in the presence of urea, GuanHCl, or detergents. In addition, they can be used to compare the properties of related proteins, as well as mutated forms and intermediate-folding states of proteins [111]. Information about both the secondary and tertiary structure can be obtained from the spectral properties of a protein. Many comprehensive reviews exist on the theory and practice of spectroscopy for the analysis of protein solutions [111–113].

a. *Identifying Changes in the Primary Structure by Mass Spectroscopy.* The formulation scientist is often faced with the problem of identifying covalent modifications in the protein. The NH_2^- and $COOH^-$ terminal sequencing of a protein product are required to establish the primary structure of the physiologically active protein. However, this is only a starting point, and internal sequencing must be verified to ensure the fidelity of the sequence. Difficulties often arise in Edman degradation-based sequencing because the NH_2-termini of recombinant proteins are frequently blocked. In recent years, mass spectroscopy (MS) has become a powerful tool in the determination of the primary structure, with high sensitivity and accuracy [114]. Only picomoles of a protein sample are required. The procedure is not limited by the presence of blocking groups and, in fact, may characterize the nature of the blocking group on the NH_2-terminus. In electrospray MS, a solution of protein is sprayed into the mass spectrometer. The solvent in the droplet then evaporates, leaving a suspended aerosol of protein molecules with varying mass/charge ratios, as a result of the introduction of positive charges on their surfaces. The resultant spectrum of the charged species consists of adjacent major peaks that differ by 1 unit of charge or by a proton by mass. The mass of a protein as large as 100,000 Da can be measured in this way to an accuracy of 0.01%. Mass spectroscopy can also be used to characterize the nature of the glycosylated protein [115]. One-dimensional MS is incapable of distinguishing between residues that have the same mass, but different structures, such as lysine and glutamic acid or leucine and isoleucine. In tandem MS, a second spectrometer bombards a selected ion with neutral atoms that cause further fragmentation. The resultant spectrum of fragments can provide sequence information, and this procedure is believed to distinguish between leucine and isoleucine residues. Mass spectroscopy and tandem MS, when used in conjunction with conventional chemical and biochemical techniques, can lead to a complete structural covalent structural changes that can occur in a protein [116]. The limitation of this technique is that not all peptides and proteins are analyzed with equal efficiency, and some cannot be analyzed at all. However, advances in MS are rapid, and it soon may become an indispensible complement to high-performance liquid chromatography (HPLC) and electrophoretic methods of analysis [117].

   b. Ultraviolet Absorbance and Difference Spectroscopy. The theory and application of protein absorption are reviewed by Wetlaufer [112]. The absorbance of a substance is directly proportional to its concentration in solution, according to the Beer–Lambert law:

$$A = \epsilon \cdot c \cdot l \tag{4}$$

where A is the absorbance, c is the molar concentration, l is the path length in centimeters, and ϵ is the molar absorption coefficient [112]. The absorption of proteins is characterized by the contributions of peptide bonds that absorb strongly below 230 nm and of the aromatic side chains of tyrosine (Tyr), tryptophan (Trp), and phenylalanine (Phe) residues, which absorb in the 230- to 300-nm range. Disulfides have a weak absorbance band near 250 nm. The molar absorbance coefficients of the three aromatic amino acids differ from one another, such that the ϵ (Trp) $> \epsilon$ (Tyr) $> \epsilon$ (Phe) [112]. The absorption spectra of the aromatic amino acids depend on the nature of the environment of each amino acid within the protein [113]. The sensitivity of the spectra to the solvent is reflected by the shifts in the wavelength, corresponding to maximal absorbance (λ_{max}), and overall changes in intensity [113]. Yanari and Bovey [118] observed red shifts in the absorption spectra of various proteins (that is, an increase in λ_{max}) when the polarity of the solvent decreased, whereas a blue shift (a decrease in λ_{max}) was observed when the polarity of the solvent was increased.

   Measurement of the spectral changes that accompany unfolding provides a powerful tool to study the stability of proteins and to follow the kinetics of conformational changes [112,113]. For instance, UV difference spectroscopy has been employed to study solvent perturbation of proteins (i.e., the effect of denaturants and detergents on the structure of a protein) [112]. The difference spectrum depends on the kind and number of aromatic amino acid residues, as well as on the degree to which their side chains are buried in the interior of the protein [113]. The spectral shifts are attributed to the difference in energy levels between the ground state and the first excited state of the π to π^* electronic transitions in the peptide bonds and the side chains of the aromatic amino acid residues.

   c. Circular Dichroism. The theory and application of circular dichroism (CD) in the study of protein conformation has been reviewed extensively [113,119–121]. When chromophores are located in an asymmetric environment, left-handed and right-handed circularly polarized lights are absorbed to different extents; this behavior is called CD [121]. The CD signals are observed in the same spectral regions at which the absorption of the relevant chromophores of a protein takes place [121]. The far UV or amide region (170–250 nm) is dominated by the contributions of the peptide bonds. Thus, it is possible to estimate the content of secondary structure from the CD spectrum in the amide region, where it can be represented as a linear combination of the different elements of secondary structure:

$$\theta\ (\lambda) = f_\alpha\ [\theta_\alpha(\lambda)] + f_\beta\ [\theta_\beta(\lambda)] + f_t\ [\theta_t(\lambda)] + f_r\ [\theta_r(\lambda)] \tag{5}$$

where $\theta(\lambda)$ is the measured ellipticity of the observed spectra $[\theta_\alpha(\lambda)], [\theta_\beta(\lambda)]$, $[\theta_t(\lambda)]$, and $[\theta_r(\lambda)]$ correspond to reference spectra for α-helical, β-sheet, β-turn, and random or undefined secondary structure, and f_α, f_β, f_t, and f_r represent the respective fractions of the structural elements [113]. The fractions are calculated by solving Eq. (5) simultaneously at a selected set of wavelengths [113]. There are different reference spectra used in calculating secondary structure fractions, ranging from the CD spectra of synthetic poly(amino acids) [121] to those of native proteins for which the three-dimensional structures are known [122]. There are also different mathematical approaches used in calculating secondary structure fractions [113].

A critical analysis of the CD spectra of polypeptides has been described [120]. An α-helix shows a strong and characteristic CD spectrum, characterized by the large value for the mean residue molar ellipticity in the 205- to 230-nm range; with typical values of the order of $(-3$ to $-4) \times 10^4$ deg • cm²/decimole [120]. Therefore, the fraction of α-helix can be determined with a relatively high degree of accuracy [113]. On the other hand, the CD spectral intensities for the β-sheet and β-turn structures are weaker and depend on the length and the spatial configuration or twist of the β-structures. Differentiating between parallel and antiparallel β-sheet configurations is also difficult [120]. Consequently, in practice, an estimation of the β-structure content using CD is far less reliable than for α-helical content [113].

The CD spectra of the aromatic amino acids can also be used as a probe for the native structure, for the CD in the spectral region of 230–300 nm is sensitive to the precise orientation of the aromatic side chains within the folded protein [113]. Also, aromatic residues and disulfides display characteristic CD bands in the far-UV region, but these signals are not as strong as helical CD bands [113]. Aromatic residues, therefore, account for much of the measured CD spectra at about 225 nm in proteins with low helical content and many aromatic amino acid residues [113].

d. Fluorescence Spectroscopy. The theory and practical aspects of fluorescence spectroscopy in the study of proteins have been extensively reviewed [111,113,123–125]. Fluorescence is the light emitted by a molecule when an excited electron radiatively decays to the ground state. As some energy is always lost by nonradiative decay, such as vibrational transitions, the energy of the emitted light is always less than the energy of the absorbed light. Thus, the fluorescence emission spectra generally occur at longer wavelengths than those of excitation. The excitation spectra are coincident with the respective absorption spectra [125]. Other experimental parameters measured include the quantum yield, the polarization of fluorescence, and the lifetime of fluorescence. The *quantum yield* (ϕ) is defined as the ratio of the quanta of energy emitted to the quanta of energy absorbed. Therefore, the maximum possible value approaches unity and, indeed, ϕ of strongly fluorescent dyes is very close to, but never reaches this value [125]. Quantum yield is most often obtained relative to a reference fluorescent compound, such as β-carboline [126] from the following expression:

$$\phi_{protein} = \frac{\phi_{reference}(1 - 10^{-A_{reference}})\,area_{protein}}{(1 - 10^{-A_{protein}})\,area_{reference}} \tag{6}$$

where

$\phi_{reference}$ is the quantum yield of the reference compound; $A_{reference}$ and $A_{protein}$ are the absorbances of the reference and the test protein at the wavelength of excitation; $area_{reference}$ and $area_{protein}$ denote the areas under the emission spectra of the reference and protein, respectively [123]. The lifetime of fluorescence (τ_0) refers to the time required for the fluorescence intensity to fall to $1/e$ of the initial value. This is sometimes referred to as a half-life [125].

The intrinsic fluorescence of proteins originates from the aromatic amino acid residues [124]. Once again, the fluorescence is dominated by the contribution of the tryptophan residues, because both their absorbance at the wavelength of excitation and their quantum yield of emission ($\phi_{Trp} = 0.20$) are considerably greater than the respective values for tyrosine and phenylalanine residues ($\phi_{Tyr} = 0.14$; $\phi_{Phe} = 0.04$) [113]. Another important factor in the interpretation of fluorescence data is the transfer of energy between residues; for instance, fluorescence due to phenylalanine is rarely observed because its emission is efficiently quenched by the energy transfer to other aromatic residues. This is because tryptophan and tyrosine absorb strongly at 280 nm at which phenylalanine emits its fluorescence. Fluorescence due to tyrosine is also poorly represented in folded proteins because the energy of tyrosine fluorescence is usually transferred to tryptophan, and also, because tryptophan fluorescence emission is sometimes shifted to shorter wavelengths.

Fluorescence emission is much more sensitive to changes in the environment of the fluorophore than absorption. The fluorescence emission of the exposed aromatic amino acid residues of a protein depends once again on the solvent;

solvent-quenching studies can be utilized to probe the local environment of aromatic residues within the native protein [127]. Wavelength shifts and changes in fluorescence intensity are generally observed on unfolding of proteins that contain tryptophan and tyrosine, and the direction of the shift depends on the new environment. Thus, the fluorescence intensity of tryptophan in an aqueous solvent is very low compared with that when it is located in an apolar organic environment [113]. Both tyrosine and tryptophan are excited at 280 nm, but it is possible to preferentially excite tryptophan fluorescence at wavelengths above 295 nm. Tyrosine fluorescence may be obtained by subtracting the selected tryptophan fluorescence excited at 295 nm from the protein fluorescence excited at 280 nm [124]. The most important advantage of the fluorescence technique is its sensitivity and, thus, the ability to handle extremely dilute protein solutions.

2. Electrophoresis

Electrophoretic methods, including gel sieving under denaturing and nondenaturing conditions, as well as isoelectric focusing and related variations, have been used for the analysis of protein conformation. The basis for these techniques is that the mobility of a macromolecule in a gel (such as polyacrylamide or agarose, of defined and controlled pore size) in an electric field is dependent on the net charge, the size, and the shape of the molecule [128]. Macromolecules that differ by a single unit charge have distinguishable electrophoretic mobilities, which have been used to determine the number of reactive amino acid residues (e.g., cysteine, in a polypeptide) [129]. Because proteins contain both acidic and basic residues, their net charge is pH-dependent. Under acidic conditions, side chains of basic amino acid residues (e.g., amino groups) are ionized, whereas carboxyl groups are not; therefore, the protein carries a net positive charge. Conversely, the protein carries a net negative charge under alkaline conditions, as carboxyl groups are ionized and the amino groups are not.

Depending on the ratio of acidic and basic amino acid residues in a protein, at some intermediate pH value, the net charge will be zero, and the protein is said to be at its isoelectric point (pI). The pI value is determined by both the type of amino acids and the structure of the protein. Isoelectric focusing (IEF) in narrow pH range gradients is a practical method of detecting charge heterogeneity, theoretically up to a level of a single amino acid change. For example, deamidation of glutamine or asparagine residues, which introduces a mass change of 1 Da is easily detected by IEF, whereas it is poorly detectable by MS [130]. In IEF, the sample is more highly resolved than in conventional electrophoresis because diffusion of the individual bands is minimized. In IEF, as soon as the protein moves away from its isoelectric zone, it becomes charged and, hence, migrates back [131].

Polyacrylamide gel electrophoresis (PAGE) is the most often used method of resolving and quantifying proteins larger than 40 kDa from crude extracts, such as culture supernatants and cell lysates [130]. New electrophoretic methods, such as high-resolution capillary zone electrophoresis (CZE) involving the electrophoretic separation of analytes in a narrow capillary, are proving to be valuable techniques for peptide and protein analysis. High-resolution CZE is currently being interfaced with sophisticated detection methods, such as fluorescence and MS, and should prove to be powerful for protein analysis. Gel electrophoresis methods have also been developed to study unfolding induced by urea or other nonionic denaturants [132]. The gel electrophoresis results are consistent with those of other biophysical methods; thus, this method can provide useful thermodynamic and kinetic information on the unfolding process [128].

3. Chromatography

a. Size-Exclusion. Significant conformational changes associated with the folding–unfolding equilibria of solvent- or temperature-dependent denaturation of proteins may be studied using analytical size-exclusion chromatography [133]. It has been possible to quantitatively assess the concentration and characteristics of partially unfolded, thermodynamically stable species, provided the intermediates are stable enough kinetically to survive within the same time-scale as the chromatographic run [134].

b. Ion-Exchange. Advances in the preparation of pressure-stable, microparticulate ion-exchange material with large pores have led to the high-performance, ion-exchange and reversed-phase chromatography of proteins. The separation may be accelerated 10–100 times if high-performance liquid chromatography (HPLC) is employed. The ion-exchange process is largely governed by electrostatic interactions. In addition to the net charge of the protein, the retention of proteins on HPLC is also dependent on the intramolecular charge asymmetry, which promotes differences in electrical potential on the surface of the protein—separation, therefore, also depends on the conformation of the protein [135]. In addition, intrinsic characteristics of the protein, such as the way it responds to the displacing salt and the way it interacts with the column support, are also important [135].

c. Reversed-Phase. The separation by reversed-phase chromatography is largely determined by the number of hydrophobic groups in a protein. Thus, the more hydrophobic the protein, the more it is retained on the reversed-phase column; hence, the later its peak appears in the elution profile [136]. Hydrophobic interaction chromatography, which is claimed to be more gentle on proteins than reversed-phase chromatography, is another possible method for separating intermediates during unfolding, as the ability of a protein to engage in hydrophobic interactions is dependent on the surface hydrophobicity of the protein which, in turn, is dependent on the tertiary and quaternary structure [137].

4. Calorimetry

Microcalorimetry has been applied to study thermal unfolding of proteins; it also can be used to study protein–protein interactions and the dynamics of the unfolding of complex multisubunit proteins. This technique can be used to evaluate the ability of certain excipients to affect the temperature at which the protein undergoes unfolding. For example, differential scanning microcalorimetry has been used to investigate the stabilizing effect of certain sugars or other additives on protein structure [94]. For most small proteins, the calorimetrically derived enthalpy of unfolding generally agrees with that derived from other methods using GuanHC1 and urea equilibrium denaturation studies [138]. The limitations are that this method cannot distinguish between aggregation versus unfolding of the protein, and that it involves the assumption that unfolding follows a two-state model (thermodynamically stable folding intermediates complicate the analysis) [139].

5. Light Scattering

Protein solutions scatter light, and this property can be used to measure the molecular weight and obtain shape information about the protein [140]. The classic treatment is to measure the angular dependence of the intensity of the scattered light at a series of protein concentrations, and then to extrapolate the data to zero protein concentration and zero angle. Low-angle laser light scattering (or photon correlation spectroscopy) has been coupled to various chromatography systems as detectors, which has yielded accurate molecular weights, with no size constraints of the protein. The chromotagraphy boosts the signal generated and eliminates interferences from solvents and co-solvents. This method has also been used to determine the diffusion coefficient of a protein, giving valuable information about its unfolding transition by autocorrelation methods [141]. The limitations of this method include the difficulty in examining heterogeneous mixtures, primarily the result of the presence of exponentials in the autocorrelation calculations. However, monomers and oligomers can be quantitated, and this method has been used to study the cosolvent-assisted refolding of proteins [141].

B. Biological Assays

The final assessment of the efficacy of the drug product is biological testing. This is especially important when *in vitro* testing does not correlate with in vivo assessment of the protein pharmaceutical. This has often been true for growth hormones (GH) [142]. Bioassay tests are often arduous and expensive exercises. For instance, there is a need for handling and housing animals, large variability in the data generated, and often a lack of quantitative interpretation of the biological response [142]. The original method for measuring the bioactivity of GH was based on measuring the weight gain of an experimental animal (e.g., of a young 70- to 100-g hypophysectomized rat during a course of nine or ten daily subcutaneous injections of GH extracts [142]). Another method relies on measuring the increase in the width of the proximal tibial epiphysis of these rats using the same course of GH treatment [143]. In both cases, the biological response increased proportionally to the log of the dosage. However, as is common with bioassays that use experimental animals, the method was subjected to certain statistical limitations, Even with more than ten animals in each dosing group, it is often difficult to reduce the variability to less than $\pm 30\%$. Bioassays for insulin involve measuring the decrease in blood glucose in rabbits. This bioassay may have been replaced by a chromatographic-based test that allows precise quantitation and reduces the number of animals required for testing, although it does not eliminate them completely [141]. The advent of ELISA [144] and the radioimmunoassay (RIA) [142] has simplified the measurement of many protein drugs, including insulin and GH [141].

Immunochemical approaches may be useful for assessing the biological activity of different proteins, because the specificity and high affinity of antibodies render them useful as analytical tools [144]. However, they have long been underused owing to a lack of a truly quantitative interpretation of antigen–antibody interaction with polyspecific immunosera. Hybridoma technology makes it possible to produce highly pure, monospecific antibodies that circumvent the problem of antibody heterogeneity [145]. Both polyclonal and monoclonal antibodies have been used for measuring the concentration of a specific protein in complex mixtures, such as cell supernatants, blood serum, plasma, urine, and others. They have been used for estimating expression levels and for following the recovery and pharmacokinetics of different proteins [141].

It is advantageous to know the epitope (i.e., the surface configuration or the region on the protein molecule at which the antibody binds). For instance, this information can be useful in carrying out sandwich assays, in which a signal is generated only if the protein antigen possesses two distinctive epitopes [141]. Here, small proteolytic degradation products will not be detected unless they have the correct conformation and combination of the peptides that make up the epitope. However, immunoreactivity may be unrelated to biological activity (i.e., a protein may lose all of its biological activity, without losing any reactivity of its epitopes) [141]. Therefore, the ultimate estimate of biological activity based on in vivo bioassays is necessary.

With recent advances in recombinant DNA technology and cell biology, in vitro assays have been developed for protein pharmaceuticals. Manipulated cell lines respond to drugs by proliferation or cell death, or by secretion of some cellular product into the culture supernatant. In vitro assays reduce the need for animal subjects because they depend on cell culture techniques, excised tissues, and biological reagents. Examples of cell-based bioassays for interferons include antiviral [146], antiproliferative assays [147], and ELISA-based tests that measure the induction of certain cell surface anigens in response to interferon [148]. Radioreceptor assays have also been developed for several hormones. These assays depend on the availability of a receptor source. For example, plasma membrane or microsomal membrane fractions obtained by differential centrifugation of tissue homogenates have been used as a source of receptors [149]. Cultured human hepatocytes [150] and lymphocytes [151] have also been used for measuring the receptor binding of GH.

VIII. CONCLUSIONS

The study of protein stability, microencapsulation, and biodegradable polymers, each pose complex challenges in its own right. From the analyses presented herein it becomes evident that advances in these research areas are necessary for successful formulation of biodegradable polymer microspheres capable of deliverying stable proteins over extended time periods. The evaluation of proteins as complex macromolecules requires careful consideration of their structure and their biochemical–biophysical analyses. Protein mobility during microencapsulation, storage, and release is an extremely important concept for maintaining the native state of the protein, as a rigidified protein molecule is rendered mostly unreactive. Once the protein is hydrated within the polymer, its fate will be dictated by the environment created by the excipient–polymer system and the mass-transfer events as polymer erosion proceeds. A rational approach of analyzing protein stability problems, such as discussed in this fundamental review, is required to evaluate the many complex issues in an efficient manner. In the future, this approach may provide insight into the development of novel encapsulation techniques, the improved use of excipients, and the creation of new biodegradable polymers, all of which are designed to optimize protein stability from biodegradable polymer microspheres.

ACKNOWLEDGMENTS

This work was supported by grants from the National Institutes of Health (GM 26698 and AI 33575), the Biotechnology Process Engineering Center at MIT, and the World Health Organization. Individual support was provided to S. P. S. by a National Institutes of Health postdoctoral fellowship (AI 08965) and to M. C. by the Australian Government Department of Industry, Science, and Technology (G.I.R.D. grant no. 15069).

REFERENCES

1. R. Langer and J. Folkman, Polymers for the sustained release of proteins and other macromolecules, *Nature. 263:*797–800 (1976).
2. A. Gibbons, A booster shot for children's vaccines, *Science 255:*1351 (1992).
3. J. Brange and S. Havelund, Insulin pumps and insulin quality—requirements and problems, *Acta Med. Scand. Suppl. 671:*135–138 (1993).
4. L. Brown, C. Munoz, L. Siemer, E. Edelman, and R. Langer, Controlled release of insulin from polymer matrices. Control of diabetes in rats, *Diabetes 35:*692–697 (1986).
5. H. M. Creque, R. Langer, and J. Folkman, One month of sustained release of insulin-containing surfactants, in contact with different materials, *Diabetes 34:*37–40 (1980).
6. Y. Ogawa, H. Okada, M. Yamamoto, and T. Shimamoto, In vivo release profiles of leuprolide acetate from microcapsules prepared with polylactic acids or co-poly(lactic/glycolic acids) and in vivo degradation of these polymers, *Chem. Pharm. Bull. 36:*2576–2581 (1988).
7. M. J. Alonso, R. K. Gupta, C. Min, G. R. Siber, and R. Langer, Microspheres as controlled-release tetanus toxoid delivery systems, *Vaccine 12:*299–306 (1994).
8. S. Cohen, T. Yoshioka, M. Lucarelli, L. H. Hwang, and R. Langer, Controlled delivery systems for proteins based on poly(lactic/glycolic acid) microspheres, *Pharm. Res. 8:*1991 (1991).
9. Y. Tabata, S. Gutta, and R. Langer, Controlled delivery systems for proteins using polyanhydride microspheres, *Pharm. Res. 10:*487–496 (1993).
10. E. Ron, T. Turek, E. Mathiowitz, M. Chasin, M. Hageman, and R. Langer, Controlled release of polypeptides from polyanhydrides, *Proc. Natl. Acad. Sci. USA 90:*4176–4180 (1993).
11. M. M. Struck, Biopharmaceutical R&D success rates and development times: A new analysis provides benchmarks for the future, *Bio/Technol. 12*: 674–677 (1994).
12. J. E. Talmadse, The pharmaceutics and delivery of therapeutic polypeptides and proteins, *Adv. Drug. Del. Rev. 10:*247–299 (1993).
13. P. G. Squire and M. E. Himmel, Hydrodynamics and protein hydration, *Arch. Biochem. Biophys. 196:*165–177 (1979).
14. D. B. Volkin and A. M. Klibanov, Minimizing protein inactivation, *Protein Function: A Practical Approach* (T. E. Creighton, ed.), Oxford University Press, Oxford, 1985, pp. 1–24.
15. M. C. Manning, L. Patella, and R. T. Borchardt, Stability of protein pharmaceuticals, *Pharm. Res. 6:*903–917 (1989).
16. J. D. Andrade and V. Hlady, Protein adsorption and materials biocompatibility: A tutorial review and suggested hypotheses, *Adv. Polym. Sci. 79:*1–63 (1985).
17. T. E. Creighton, *Proteins: Structures and Molecular Properties*, W. H. Freeman & Co., New York, 1993.
18. J. L. Cleland and R. Langer, Formulation and delivery of proteins and peptides design and development strategies, *Formulation and Delivery of Proteins and Peptides* (J. L. Cleland and R. Langer, eds.), American Chemical Society, Washington DC, 1994, pp. 1–19.

19. T. P. Levine and B. M. Chain, The cell biology of antigen processing, *Crit. Rev. Biochem. Mol. Biol. 26:*439–473 (1991).

20. W. A. Eaton and J. Hofrichter, Sickle-cell haemoglobin polymerization, *Adv. Protein Chem. 40:*63–279 (1990).

21. C. N. Pace, Conformational stability of globular proteins, *Trends Biochem. Sci. 10:*14–17 (1990).

22. M. J. Pikal, Freeze-drying of proteins, process, formulation, and stability, *Formulation and Delivery of Proteins and Peptides* (J. L. Cleland and R. Langer, eds.), American Chemical Society, Washington DC, 1994, pp. 120–133.

23. W. R. Liu, R. Langer, and A. M. Klibanov, Moisture-induced aggregation of lyophilized proteins in the solid state, *Biotechnol. Bioeng. 37:*177–184 (1991).

24. H. R. Costantino, R. Langer, and A. M. Klibanov, Solid-phase aggregation of proteins under pharmaceutically relevant conditions, *J. Pharm. Sci. 83:*1662–1669 (1994).

25. A. J. Domb, L. Turovsky, and R. Nudelman, Chemical interactions between drugs containing reactive amines with hydrolyzable insoluble biopolymers in aqueous solutions, *Pharm. Res. 11:*865–868 (1994).

26. W.-J. Lin, D. R. Flanagan, and R. J. Linhardt, Accelerated degradation of poly(ϵ-caprolactone) by organic amines, *Pharm. Res. 11:*1030–1034 (1994).

27. C. B. Anfinsen, The limited digestion of ribonuclease with pepsin, *J. Biol. Chem. 221:*405–412 (1956).

28. C. B. Anfinsen, Principles that govern the folding of protein chains, *Science. 181:*223–230 (1973).

29. C. B. Anfinsen and H. A. Scheraga, Experimental and theoretical aspects of protein folding, *Adv. Protein Chem. 29:*205–300 (1975).

30. J. F. Brandts, H. R. Halvorson, and M. Brennan, Consideration of the possibility that the slow step in protein denaturation reaction is due to *cis–trans* isomerism of proline residues, *Biochemistry 14:*4953–4963 (1975).

31. O. B. Ptitsyn and A. V. Finkelstein, Similarities of protein topologies: evolutionary divergence, functional convergence or principles of folding, *Q. Rev. Biophys. 13:*339–386 (1980).

32. O. B. Ptitsyn and A. V. Finkelstein, Self-organisation of proteins and the problem of their three-dimensional structure prediction, *Protein Folding* (R. Jaenicke, ed.), Elsevier/North Holland, Amsterdam, 1980, pp. 101–115.

33. M. G. Rossman and P. Argos, Protein folding, *Annu. Rev. Biochem. 50:*497–532 (1981).

34. B. Gruenewald, C. U. Nicola, A. Lustig, G. Schwarz, and H. Klump, Kinetics of the helix–coil transition of a polypeptide with non-ionic side groups, derived from ultrasonic relaxation measurements, *Biophys. Chem. 9:*137–147 (1979).

35. S. T. Rao, and M. G. Rossman, Comparison of super-secondary structures in proteins, *J. Mol . Biol. 76:*241–256 (1973).

36. M. Levitt and C. Chothia, Structural patterns in globular proteins, *Nature 261:*552–558 (1976).

37. J. L. Crawford, W. H. Lipscomb, and C. G. Schellman, The reverse turn as a polypeptide conformation in globular proteins, *Proc. Natl. Acad. Sci. USA 70:*538–542 (1973).

38. J. S. Richardson, Describing patterns of protein tertiary structure, *Methods Enzymol. 115*:341–380 (1985).

39. C. Chothia, The classification and origins of protein folding patterns, *Annu. Rev. Biochem. 59*:1007–1039 (1990).

40. M. G. Rossman and A. Liljas, Recognition of structural domains in globular proteins, *J. Mol. Biol. 85*:177–181 (1974).

41. D. B. Wetlaufer, Nucleation, rapid folding, and globular intrachain regions in proteins, *Proc. Natl. Acad. Sci. USA 70*:697–701 (1973).

42. J. Janin and S. J. Wodak, Structural domains in proteins and their role in the dynamics of protein function, *Prog. Biophys. Mol. Biol. 42*:21–78 (1983).

43. R. Jaenicke, Folding and association of proteins, *Prog. Biophys. Mol. Biol. 49*:117–237 (1987).

44. C.-I., Branden, H. Eklund, C. Cambillan, and A. J. Pryor, Correlation of exons with structural domains in alcohol dehydrogenase, *EMBO J. 3*:1307–1310 (1984).

45. W. Gilbert, Genes-in-pieces revisited, *Science 233*:823–824 (1983).

46. T. Kiefhaber, R. Rudolph, H.-H. Kohler, and J. Buchner, Protein aggregation in vitro and in vivo: A quantitative model of the kinetic competition between folding and aggregation, *Bio/Technol. 9*:825–872 (1991).

47. L. Pauling, *The Nature of the Chemical Bond.* Cornell University Press, Ithica, NY, 1960.

48. W. Kauzmann, Some factors in the interpretation of protein denaturation, *Adv. Protein Chem. 14*:1–64 (1959).

49. K. A. Dill, Dominant forces in protein folding, *Biochemistry 29*:7133–7155 (1990).

50. R. Jaenicke, Protein folding: Local structures, domains, subunits, and assemblies, *Biochemistry 30*:3147–3161 (1991).

51. C. N. Pace, Determination and analysis of urea and guanidine hydrochloride denaturation curves, *Methods Enzymol. 131*:266–280 (1986).

52. T. Alber, Mutational effects on protein stability, *Annu. Rev. Biochem. 58*:765–798 (1989).

53. K. A. Dill and D. Shortle, Denatured states of proteins, *Annu. Rev. Biochem. 60*:795–825 (1991).

54. Y. Nozaki and C. Tanford, The solubility of amino acids and related compounds in aqueous urea solutions, *J. Biol. Chem. 238*:4074–4081 (1963).

55. M. M. Santoro and D. W. Bolen, Unfolding free energy changes determined by the linear extrapolation method. 1. Unfolding of phenylmethanesulfonyl-α-chymotrypsin using different denaturants, *Biochemistry 27*:8063–8068 (1988).

56. D. Shortle, A. K. Meeker, and S. L. Gerring, Effects of denaturants at low concentrations on the reversible denaturation of staphylococcal nuclease, *Arch. Biochem. Biophys. 272*:103–113 (1989).

57. P. P. Kamoun, Denaturation of globular proteins by urea: Breakdown of hydrogen or hydrophobic bonds, *Trends Biochem. 13*:424–425 (1988).

58. V. Sluzky, J. A. Tamada, A. M. Klibanov, and R. Langer, Kinetics of insulin aggregation in aqueous solutions upon agitation in the presence of hydrophobic surfaces, *Proc. Natl. Acad. Sci. USA 88*:9377–9381 (1991).

59. H. R. Costantino, R. Langer, and A. Klibanov, Aggregation of a lyophilized

pharmaceutical protein, recombinant human albumin: Effect of moisture and stabilization by excipients, *Bio/Technol. 13*:493–496 (1995).

60. H. R. Costantino, R. Langer, and A. M. Klibanov, Moisture-induced aggregation of lyophilized insulin, *Pharm. Res. 11*:21–29 (1994).

61. T. J. Ahern, J. I. Casal, G. A. Petsko, and A. M. Klibanov, Control of oligomeric enzyme thermostability by protein engineering, *Proc. Natl. Acad. Sci. USA 84*:675–679 (1987).

62. D. B. Volkin, A. Staubli, R. Langer, and A. M. Klibanov, Enzyme thermoinactivation in anhydrous organic solvents, *Biotechnol. Bioeng. 37*:843–853 (1991).

63. M. J. Alonso, S. Cohen, T. G. Park, R. K. Gupta, G. R. Siber, and R. Langer, Determinants of release rate of tetanus vaccine from polyester microspheres, *Pharm. Res. 51*:945–953 (1993).

64. C. Thies, Formation of degradable drug-loaded microparticles by in-liquid drying processes, *Microcapsules and Nanoparticles in Medicine and Pharmacy* (M. Donbrow, ed.), CRC Press, London, 1992, pp. 47–71.

65. Y. Ogawa, M. Yamamoto, H. Okada, T. Yashiki, and T. Shimamoto, A new technique to efficiently entrap leuprolide acetate into microcapsules of polylactic acid or copoly (lactic/glycolic) acid, *Chem. Pharm. Bull. 36*:1095–1103 (1988).

66. R. H. Perry, D. W. Green, and J. O. Maloney, eds., *Perry's Chemical Engineers' Handbook*, 6th ed., McGraw-Hill Book Co., New York, 1984.

67. E. Mathiowitz and R. Langer, Polyanhydride microspheres as drug delivery systems, *Microcapsules and Nanoparticles in Medicine and Pharmacy*, (M. Donbrow, ed.), CRC Press, Ann Arbor, 1992, pp. 99–124.

68. A. Zaks and A. M. Klibanov, The effect of water on enzyme action in organic media, *J. Biol. Chem. 263*:8017–8021 (1988).

69. S. J. Prestrelski, Dehydration-induced conformational transitions in proteins and their inhibition by stabilizers, *Biophys. J. 65*:661–671 (1993).

70. D. S. Hartsough and K. M. Merz, Jr., Protein dynamics and solvation in aqueous and nonaqueous environments, *J. Am. Chem. Soc. 115*:6529–6537 (1993).

71. K. Dabulis and A. M. Klibanov, Dramatic enhancement of enzymatic activity in organic solvents by lyoprotectants, *Biotechnol. Bioeng. 41*:556–571 (1993).

72. P. A. Burke, R. G. Griffin, and A. M. Klibanov, Solid-state NMR assessment of enzyme active center structure under nonaqueous conditions, *J. Biol. Chem. 267*:20057–20064 (1992).

73. J. T. Chin, S. L. Wheeler, and A. M. Klibanov, On protein solubility in organic solvents, *Biotechnol. Bioeng. 44*:140–145 (1994).

74. W. T. Coakley, R. C. Brown, C. J. James, and R. K. Gould, The inactivation of enzymes by ultrasonic cavitation at 20 kHz, *Arch. Biochem. Biophys. 159*:722–729 (1973).

75. W. R. Gombotz, Process for producing small particles of biologically active molecules, International Publication Number WO 90/13285 (1990).

76. H. V. Maulding, Prolonged delivery of peptides by microcapsules, *J. Controlled Release 6*:167–176 (1987).

77. S. Cohen, M. C. Bano, K. B. Visscher, M. Chow, H. R. Allcock, and R. Langer, Ionically cross-linkable polyphosphazene: A novel polymer for microencapsulation, *J. Am. Chem. Soc. 112*:7832–7833 (1990).

78. E. Mathiowitz and R. Langer, Polyanhydride microspheres as drug carriers I. Hot-melt microencapsulation, *J. Controlled Release* 5:13–22 (1987).

79. K. K. Kwok and A. R. Groves, Production of 5–15 microns diameter alginate-polylysine microcapsules by an air-atomization technique, *Pharm. Res.* 8:341–344 (1991).

80. R. H. M. Hatley and F. Franks, Applications of DSC in the development of improved freeze-drying processes for labile biologicals, *J. Thermal Anal.* 37:1905–1914 (1991).

81. P. L. Privalov and S. A. Potekhin, Scanning microcalorimetry in studying temperature-induced changes in proteins, *Methods Enzymol.* 131:4–51 (1986).

82. H. Okada, Y. Inoue, T. Heya, H. Ueno, Y. Ogawa, and H. Toguchi, Pharmacokinetics of once-a-month injectable microspheres of leuprolide acetate, *Pharm. Res.* 8:787–791 (1991).

83. K. Leong, A. Domb, E. Ron, and R. Langer, eds., Polyanhydrides, *Encyclopedia of Polymer Science and Engineering*, John Wiley & Sons, New York, 1989, pp. 648–665.

84. S. Li, H. Garreau, and M. Vert, Structure–property relationships in the case of the degradation of massive aliphatic poly-(α-hydroxy acids) in aqueous media, Part 1: Poly(DL-lactic acid), *J. Mater. Sci. Mater. Med.* 1:123–130 (1990).

85. S. M. Li, H. Garreau, and M. Vert, Structure–property relationships in the case of the degradation of massive aliphatic poly-(α-hydroxy acids) in aqueous media, Part 2: Degradation of lactide–glycolide copolymers: PLA37.5GA25 and PLA75GA25, *J. Mater. Sci. Mater. Med.* 1:131–139 (1990).

86. A. Göpferich, Mechanisms of polymer degradation and erosion, *Biomaterials* 17:103–114 (1996).

87. A. Göpferich and A. Langer, The influence of microstructure and monomer properties on the erosion mechanism of a class of polyanhydrides, *J. Polym. Sci. A* 31:2445–2458 (1993).

88. A. Göpferich and R. Langer, Modeling of polymer erosion, *Macromolecules* 1993:4105–4112 (1993).

89. T. G. Park, Degradation of poly(D,L-lactic acid) microspheres: Effect of molecular weight, *J. Controlled Release* 30:161–173 (1994).

90. S. S. Shah, Y. Cha, and C. G. Pitt, Poly(glycolic acid-*co*-DL-lactic acid): Diffusion or degradation controlled drug delivery? *J. Controlled Release* 18:261–270 (1992).

91. S. P. Schwendeman, J. H. Lee, R. K. Gupta, H. R. Costantino, G. R. Siber, and R. Langer, Inhibition of moisture-induced aggregation of tetanus toxoid by protecting thiol groups, *Proc. Int. Symp. Controlled Release Bioact. Mater.* 21:54–55 (1994).

92. J. L. Cleland, M. F. Powell, and S. J. Shire, A close look at protein aggregation, deamidation, and oxidation, *Crit. Rev. Ther. Drug Carrier Syst.* 10:307–377 (1993).

93. S. E. Zale and A. M. Klibanov, Why does ribonuclease irreversibly inactivate at high temperatures? *Biochemistry* 25:5432–5444 (1986).

94. T. Arakawa and S. N. Timasheff, Stabilization of protein structure by sugars, *Biochemistry* 21:6536–6544 (1982).

95. M. Herrlinger, In Vitro Polymerabbau un Wirksttofffreigabe von Poly-DL-Lakid-Formlingen. PhD Thesis. University of Heidelberg, 1994.
96. J. Heller, D. W. Penhale, B. K. Fritzzinger, and S. Y. Ng. The effect of copolymerized 9,10-dihydroxystearic acid on erosion rates of poly(*ortho* esters) and its use in the delivery of levonorgestrel, *J. Controlled Release* 5:173–177 (1987).
97. J. Crank, *Diffusion in Polymers*, Academic Press, New York, 1968.
98. S. N. Ege, *Organic Chemistry*. D. C. Heath & Co., Toronto, 1984.
99. O. B. Pititsyn, R. H. Pain, G. V. Semisotnov, S. E. Zerovnik, and O. I. Razgulyaev, Evidence for a molten globule state as a general intermediate in protein folding, *FEBS Lett.* 262:20–24 (1990).
100. S. N. Timasheff and T. Arakawa, Stabilization of protein structure by solvents, *Protein Structure: A Practical Approach* (T. E. Creighton, ed.), IRL Press, New York, 1990, pp. 331–345.
101. J. Bjarnason and K. J. Carpenter, Mechanisms of heat damage in proteins 2. Chemical changes in pure proteins, *Br. J. Nutr.* 24:313–329 (1970).
102. H. T. Wright, Nonenzymatic deamidation of asparaginyl and glutaminyl residues in proteins, *Crit. Rev. Biochem. Mol. Biol.* 26:1–52 (1991).
103. T. D. Brock, *Thermophiles: General Molecular and Applied Microbiology*, J. Wiley & Sons, New York, 1986.
104. O. R. Zarbosky, *Immobilized Enzymes*, CRC Press, Cleveland, 1973.
105. F. Helfferich, *Ion Exchange*, McGraw-Hill, New York, 1962.
106. G. E. Means, and R. E. Feeney, *Chemical Modification of Proteins*, Holden-Day, San Francisco, 1971.
107. J. F. Carpenter, S. J. Prestrelski, T. J. Anchordoguy, and T. Arakawa, Interaction of stabilizers with proteins during freezing and drying, (J. L. Cleland and R. Langer, eds.), *Formulation and Delivery of Proteins and Peptides*, American Chemical Society, Washington DC, 1994, pp. 134–147.
108. J. Carpenter, J. Crowe, and T. Arakawa, *J. Dairy Sci.* 73:3627–3636 (1990).
109. L. Slade and H. Levine, *Crit. Rev. Food Sci. Nutr.* 30:115–359 (1991).
110. C. H. Schein, Physical methods and models for the study of protein aggregation, *Protein Refolding* (G. Georgiou and E. De Bernardez-Clark, eds.), American Chemical Society, Washington DC, 1991, pp. 21–34.
111. C. R. Cantor and S. N. Timasheff, *Optical spectroscopy of proteins*, The Proteins, Vol. V, Academic Press, New York, 1982, pp. 145–306.
112. D. B. Wetlaufer, Ultraviolet spectra of proteins and amino acids, *Adv. Protein Chem.* 17:303–390 (1962).
113. F. X. Schmid, Spectral methods of characterising protein conformation and conformational changes, *Protein Structure: A Practical Approach* (T. E. Creighton, ed.), IRL Press, Oxford, 1989, pp. 251–285.
114. S. A. Carr, Recent advances in the analysis of peptides and proteins by mass spectroscopy, *Adv. Drug Deliv. Rev.* 4:113–147 (1990).
115. H. R. Morris and F. M. Greer, Mass spectrometry of natural and recombinant proteins and glycoproteins, *Trends Biotechnol.* 6:140–147 (1988).
116. J. B. Smith, Elucidation of the primary structures of proteins by mass spectrometry, *Anal. Biochem.* 193:118–124 (1991).

117. H. A. Scoble and S. A. Martin, Mass spectroscopy of proteins, *Methods Enzymol. 193*:519–56 (1990).

118. S. Yanari and F. A. Bovey, Interpretation of the ultraviolet spectral changes of proteins, *J. Biol. Chem. 235*:2818–2826 (118).

119. A. J. Adler, N. J. Greenfield, and G. D. Fasman, Circular dichroism and optical rotatory dispersion of proteins and polypeptides, *Methods Enzymol. 27*:675–735 (1973).

120. D. W. Urry, Absorption, circular dichroism and optical rotatory dispersion of polypeptides, proteins, prosthetic groups and biomembranes, *Modern Physical Methods in Biochemistry; part A* (A. Neuberger and L. L. M. Van Deenen, eds.), Elsevier, Amsterdam, 1985, pp. 275–346.

121. N. Neumann and G. Snatzke, Circular dichroism of proteins, *Proteins: Form and Function* (R. A. Bradshaw and M. Purtan, eds.), Elsevier Trends Journals, Cambridge, 1990, pp. 107–116.

122. J. T. Yang, C.-S. C. Wu, and H. M. Matinez, Calculation of protein conformation from circular dichroism, *Methods Enzymol. 140*:228–233 (1986).

123. C. A. Parker and W. T. Rees, Fluorescence spectrometry: A review, *Analyst 87*:83–111 (1962).

124. M. J. Kronman and L. G. Holmes, The fluorescence of native, denatured and reduced-denatured proteins, *Photochem. Photobiol. 14*:113–134 (1971).

125. L. Brand and B. Witholt, Fluorescence measurements, *Methods Enzymol. 27*:776–856 (1973).

126. K. P. Ghiggino, P. F. Skilton, and P. J. Thistlethwaite, β-Carboline as a fluorescence standard, *J. Photochem. 31*:113–121 (1985).

127. M. R. Eftink and C. A. Ghiron, Fluorescence quenching studies with proteins, *Anal. Biochem. 114*:199–227 (1981).

128. D. P. Goldenberg, Analysis of protein conformation by gel electrophoresis, *Protein Structure: A Practical Approach* (T. E. Creighton, ed.), IRL Press, Oxford, 1989, pp. 225–250.

129. T. E. Creighton, Counting integral numbers of amino acid residues per polypeptide chain, *Nature 284*:487–489 (1980).

130. M. J. Geisow, Characterising recombinant proteins, *Bio/Technol. 9*:921–924 (1991).

131. H. Gordon, Electrophoresis of proteins in polyacrylamide and starch gels, *Laboratory techniques in Biochemistry and Molecular Biology* (T. S. Work and R. H. Burden, eds.), Elsevier, Amsterdam, 1979.

132. T. E. Creighton, Electrophoretic analysis of the unfolding of proteins by urea, *J. Mol. Biol. 129*:235–264 (1979).

133. W. W. Fish, J. A. Reynolds, and C. Tanford, Gel chromatography of proteins in denaturing solvents: Comparison between sodium dodecyl sulfate and guanidine hydrochloride as denaturants, *J. Biol. Chem. 245*:5166–5168 (1970).

134. A. M. Al-Obeidi and A. Light, Size exclusion high performance liquid chromatography of native trypsinogen, the denatured protein, and partially refolded molecules—further evidence that non-native disulphide bonds are dominant in refolding the completely reduced protein, *J. Biol. Chem. 263*:8642–8645 (1988).

135. W. Kopaciewicz, M. A. Rounds, J. Fausnaugh, and F. E. Regnier, Retention

model for high performance ion-exchange chromatography, *J. Chromatogr.* *266*:3–21 (1983).

136. P. H. Corran, Reversed-phase chromatography of proteins, *HPLC of Macromolecules—A Practical Approach* (R. W. A. Oliver, ed.), IRL Press, Oxford, 1989, pp. 127–156.

137. R. M. Kennedy, Hydrophobic chromatography, *Methods Enzymol.* *182*:339–343 (1990).

138. P. L. Privalov, Thermal investigations of biopolymer solutions and scanning microcalorimetry, *FEBS Lett.* *40*:5140–5153 (1974).

139. A.J.S. Jones, Analytical methods for the assessment of protein formulations and delivery systems, (J. L. Cleland and R. Langer, eds.), *Formulation and Delivery of Proteins and Peptides*, American Chemical Society, Washington DC, 1994, pp. 22–45.

140. C. Tanford, *Physical Chemistry of Macromolecules*, J. Wiley & Sons New York, 1961.

141. A.J.S. Jones, Analysis of polypeptides and proteins, *Adv. Drug Deliv. Rev.* *10*:29–90 (1993).

142. D. Rudman and R. K. Clawla, Somatotrophic assays in the rat and in man, *Hormone Drugs* (J. L. Gueriguian and E. D. Bransome, Jr., eds.), USP, Rockville, MD, 1982, pp. 287–295.

143. A. E. Wilhelmi, Chemistry of growth hormone. *Handbook of Physiology*; Sec. 7: *Endocrinology* (E. Knobil and W. H. Sawyer, eds.), American Physiological Society, Washington DC, 1974, pp. 59–78.

144. W. Harlow and D. Lane, *Antibodies—A Laboratory Manual.* Cold Spring Harbor Press, Cold Spring Harbor, New York, 1988.

145. B. Friguet, L. Djavadi-Ohaniance, and M. E. Goldberg, *Protein Structure—A Practical Approach* (T. E. Creighton, ed.), IRL Press, Oxford, 1989, pp. 287–310.

146. J. A. Armstrong, Cytopathic effect inhibition assay for interferon: Microculture plate assay, *Methods Enzymol.* *78*:381–387 (1981).

147. R. Eife, T. Hahn, M. DeTavera, F. Schertel, H. Oltman, G. Eife, and S. Levin. A comparison of the antiproliferative and antiviral activities of alpha, beta and gamma interferons: Desription of a unified assay for comparing both effects simultaneously, *J. Immunol. Methods 47*:339–347 (1981).

148. U.E.M. Gibson and S. M. Dramer, Enzyme-linked bio-immunoassay for IFN-gamma by HLA-DR induction, *J. Immunol. Methods 125*:105–113 (1989).

149. T. Tsushima and H. G. Friesen, Radioreceptor assay for growth hormone, *J. Clin. Endocrinol. Metab.* *37*:334–337 (1973).

150. P. E. Mullis, T. Lund, M. S. Patel, and C.G.D. Brook, Regulation of human growth hormone receptor gene expression by human growth hormone in a human hepatoma cell line, *Mol. Cell. Endocrinol.* *76*:125–133 (1991).

151. M. A. Laeniak, P. Gorden, and J. Roth, Reactivity of non-primate grwoth hormones and prolactins with human growth hormone receptors on cultured human lymphocytes, *J. Clin. Endocrinol. Metab.* *44*:835–839 (1977).

2

Injectable Biodegradable Microspheres for Vaccine Delivery

Thomas Kissel and Regina Koneberg

Philipps University
Marburg, Germany

I. INTRODUCTION

A. The Need for New Vaccine Delivery Systems

Immunization against infectious diseases has saved innumerable lives and contributed significantly to today's increased life expectancy. In spite of these impressive results, there is still considerable potential for improved vaccines. Recent estimates by the World Health Organization (WHO) illustrate that in 1991 over 8 million children younger than 5 years of age died of infectious diseases, mostly in underdeveloped countries. Bacterial respiratory infections, measles, poliomyelitis, tetanus, pertussis, malaria, tuberculosis, and bacterial diarrhea claim countless lives [1]. Many of these diseases could be prevented by mass vaccination, but licensed vaccines do not fulfill all necessary requirements:

> Efficacy
> Safety
> Convenient application
> Cost

The most important prerequisite of a vaccine is the protection against the infecting agent or its toxic products. For several infectious diseases, development of an effective vaccine is faced by considerable difficulties (e.g., vaccines against AIDS, cholera, leprosy, malaria, and typhus). One obstacle is that the cellular and molecular mechanisms responsible for establishing immunological memory are still poorly understood [2]. New strategies to achieve more effective immunization are under investigation, among others mucosal vaccines [3], new adjuvants [4,5], and genetically and biochemically engineered antigens [6,7].

The aspect of safety and minimization of adverse reactions is a complex issue, which will not be discussed here in detail [8]. Some aspects of safety are closely related to the mode of application, such as granuloma formation or allergic reactions at the injection site. An additional technical problem is the necessity of maintaining proper storage conditions for vaccines to retain their safety and effectiveness. Serious failures of smallpox and measles immunization have resulted from inadequate refrigeration before use. All live and attenuated agents are particularly sensitive to this requirement. Thermostable vaccines would be not only of benefit in less-developed countries, but could also increase acceptability and reduce cost in industrialized nations. Delivery systems for vaccines are also crucial for patient compliance. Many vaccination guidelines call for multiple-dosing schedules. Reduction of injection frequency, preferably to one, or home administration of vaccines not requiring parenteral administration, such as peroral or nasal vaccine-delivery systems, could lead to a better immunological protection of the population and might facilitate eradication of some infectious pathogens.

Biodegradable delivery systems for vaccines that are administered by a parenteral application route (i.e., subcutaneously or intramuscularly) may have the potential of overcoming some shortcomings of conventional vaccines [2,9,10]. Parenteral microsphere carrier systems for vaccines have met with considerable interest, because they offer the following features:

Improvement of adjuvanticity
Modulation of antigen release
In vitro and in vivo stabilization of antigens

B. Traditional Adjuvants for Vaccines

The adjuvant effect (i.e., potentiation of the immune response) was discovered by Le Moignac and Pinay in 1916, who noted that a suspension of *Salmonella typhimurium* in mineral oil yielded an increased antibody formation [5]. Since then, a host of substances and formulations have been studied for adjuvanticity [see reviews Refs. 4,11–14]. Inactivated bacteria or fragments thereof, such as *Bordella pertussis* [12] or lipid A and derivatives [15], are known to potentiate the immune response of many antigens after parenteral administration. Muramyldipeptide (MDP; *N*-acetylmuramyl-L-alanine-D-isoglutamine) and its derivatives [16], as well as avridine, a lipoidal amine [3,17] are currently under intensive investigation. Lipophilic immunostimulating complexes (ISCOMS) are formed by mixing saponins (Quil A) and cholesterol with protein antigens [18]. Surfactants, on the basis of nonionic block copolymers [19] and cytokines, such as interleukin (IL-2) [20] and interferon gamma (IFN-α) [21] have shown some potential [12].

Lipophilic injection vehicles, such as nonbiodegradable mineral oils or biodegradable oils (e.g., triglycerides or squalane) are formulated into emulsions using surfactants [13]. Freund's adjuvant is an emulsion consisting of the surfactant Arlacel A, the mineral oil Drakeol 6VR (IFA; incomplete Freund's adjuvant) and inactivated mycobacteria (CFA; complete Freund's adjuvant) [22]. The CFA elicits high levels of antibodies and cell-mediated immunity and, therefore, is considered in laboratory animals as the gold standard to define maximal adjuvant activity. Because of severe local adverse reactions at the injection site, IFA and CFA are currently not approved for human use.

A second important step in the search for an acceptable adjuvant was the observation of Glenny et al. [23] in 1926 that alum-precipitated diphtheria toxoid was a more powerful antigen than toxoid in solution. This led to the development of aluminum preparations as adjuvants. In humans, aluminum salts are most widely used either as $Al(OH)_3^-$ or $AlPO_4^-$ precipitates to which a solution of the antigen is added to form *aluminum-adsorbed* vaccines [14]. Although aluminum-adsorbed vaccines have been used for some decades, they are not undisputed [24]. A major problem has been the variability of aluminum-containing vaccines, owing to differences in composition and mode of preparation. The pH and choice of buffer ions influence the resulting product [25,26]. Although aluminum-adsorbed vaccines are well established, discussion of adverse reactions persists, and modified gel structures based on calcium phosphate [27] are currently under investigation. The mechanism of action of these established adjuvants is still incompletely understood. One important factor seems to be the formation of an antigen depot at the injection site, from which the antigen is slowly released into the surrounding tissue [2,4]. Second, these adjuvants induce a local inflammatory reaction that attracts macrophages and other immunocompetent cells. The uptake of the antigen into the antigen-presenting cells (APC) is facilitated, and the presentation to the immune system leads to an activation of both the humoral and cell-mediated branches of the immune system. A more detailed discussion of interactions between adjuvants and immune system is given by Waksman [28].

Injectable biodegradable microspheres for vaccine delivery are thought to combine both effects; namely, the formation of an antigen reservoir and the induction of a localized inflammatory reaction, in a more sophisticated and reproducible manner than traditional adjuvant would allow.

C. Rationale for "Single-Shot" Vaccines

Many vaccines for active immunization require at least two-repeated administrations to induce effective protection against infectious agents (e.g., tetanus, diphtheria, hepatitis B, and polio). Single-shot vaccine delivery systems approximate a conventional IM dosage scheme by one single injection. The parenteral delivery system for the vaccine releases its content at predetermined time intervals in a pulsatile fashion, mimicking two or three injections over a period of up to 12 months. This aspect is of particular importance for vaccination programs in less-developed countries, where dropout rates can approach 70% [29]. An additional feature of single-shot vaccine delivery systems is the possibility to modulate the antigen release. Theoretically, continuous-release profiles over weeks to months can be attained [29]. It is not yet clear whether continuous antigen presentation by microspheres has any effect on the immune response. Combinations of different antigens or adjuvants, such as MDP or IL-2 [20], are potential applications of injectable microspheres for vaccine delivery, so far unexploited. Shelf life and storage condition, on one hand, and stability of the parenteral vaccine delivery under in vivo conditions, on the other hand, are aspects requiring careful investigation. It is hoped that single-shot vaccines would not require refrigeration [30]. These claims, however, are at present purely speculative, and clearly, more information is needed for their substantiation.

II. ANTIGENS FOR VACCINE DELIVERY SYSTEMS

Single-shot vaccines have not reached the stage of commercialization. In fact, most of the systems under investigation are still in an experimental phase. Table 1 summarizes some of the relevant properties of these delivery systems, which will be explained in later sections. Interestingly, most of the studies have been conducted with bacterial toxins, such as staphylococcal enterotoxin B (SEB), tetanus toxoid (TT), or diphtheria toxoid (DT).

In 11 of the 23 listed publications, tetanus toxoid was used as (model) antigen for microencapsulation. Tetanus toxoid is a 150-kDa protein that has been safely used in humans for several decades as aluminum-adsorbed vaccine, requiring three injections, at 0, 3, and 6–12 months, to induce protective immunity. The investigation of a vaccine delivery system for tetanus toxoid was sponsored by WHO, because in an estimated annual number of 500,000 children died of neonatal tetanus owing to insufficient immunization of their mothers.

Diphtheria toxoid is a 62 kDa protein and requires—similar to tetanus toxoid—three injections in a vaccination protocol. Therefore, DT is a likely candidate for a single-shot vaccine, either alone, or preferably, in combination with TT.

Staphylococcal enterotoxin B is member of the group of enterotoxins produced by *Staphylococcus aureus*, causing food poisoning in humans. Eldridge et al. [31] used SEB in microspheres because its effects on T-cell induction are quite well understood, and it is a likely candidate for mucosal immunization.

One problem with all of the foregoing bacterial toxoids is their complex protein composition, resulting from their production methods that use detoxification with formaldehyde. A protein mixture consisting of partially cleaved and cross-linked fractions of the original antigen is obtained, as shown for TT [32]. Assessment of purity and stability of these vaccines during or after microencapsulation is complicated by this feature. Therefore, ovalbumin (OVA; 43 kDa) and bovine serum albumin (BSA; 67 kDa) have frequently been used to study microencapsulation parameters. Ovalbumin is sensitive to aggregation [33]; moreover, it is a poor antigen, allowing adjuvant effects of microencapsulated OVA to be tested [34].

Until now only a few viral antigens, such as hepatitis B surface antigen (HBsAg) [35] or influenza virus [36] have been investigated. Viral vaccines could benefit from a parenteral delivery system because they are regarded as weakly immunogenic [2].

Somewhat outside the scope of this review is the concept of an antifertility vaccine, using microencapsulating human chorionic gonadotropin (hCG) fragments to raise antibodies against hCG [37].

Table 1 gives an overview of the literature on microencapsulated antigens, the focus being injectable microspheres from biodegradable polyesters.

III. BIODEGRADABLE POLYMERS FOR VACCINE DELIVERY SYSTEMS

A key factor in the design of injectable vaccine delivery systems is the choice of an appropriate polymer. Biodegradable polymers (i.e., polymers that are degraded under physiological conditions) are preferred because surgical removal of the spent device is unnecessary. Nondegradable polymers, such as poly(ethylene-*co*-vinyl alcohol) [38] and poly(methyl methacrylate) [39] were used in experimental studies. The role of the polymeric material in the design of parenteral delivery systems exceeds that of an inert excipient. This integral component influences not only the biodegradation kinetics, as one would expect, but also the mode and rate of antigen presentation, toxicity, and tissue compatibility, as well as antigen stability under in vitro and in vivo conditions. Mechanical and physiochemical properties also impinge on the selection of an appropriate microencapsulation technology. A variety of synthetic and naturally occurring polymers have been intensively studied over the past 30 years [40], but polyesters have found more widespread use. Thermoplastic polyesters of poly(lactic acid) (PLA), poly(glycolic acid) (PGA) and their copolymers poly(-lactic-*co*-glycolic acid) (PLGA) have many advantages, including an excellent track record for biocompatibility and lack of toxicity. These materials have been used as synthetic resorbable sutures, and their approval by regulatory authorities as polymeric excipient for microparticles is regarded as less costly and

Table 1 Overview of Injectable Microencapsulated Vaccine Systems with Biodegradable Polymers of the Poly(lactide-*co*-glycolide) Type

Antigen	Polymer Type	Polymer Mw (Da)	Technique	Particle Size (μm)	Yr	Ref.
Staphylococcal enterotoxin B (SEB)	DL-PLGA 50:50	n. m.	Solvent evaporation	1–10	1991	113
SEB	DL-PLGA 50:50	n. m.	Emulsion process	20–50 1–10	1991	31
Ovalbumin (OVA)	PLGA 50:50	9,000	O/W emulsion (antigen suspended)	5,34	1991	34
OVA	PLGA 50:50	40,000	W/O/W emulsion	5,34	1991	111
Diphtheria toxoid (DT)	DL-PLA	49,000	W/O/W emulsion	30–100	1991	97
SEB	DL-PLGA 53:47 DL-PLGA 85:15 L-PLA	n. m.	n. m.	1–10 >10	1992	115
DT	PLA	49,000	W/O/W emulsion	40–50	1992	99
Hepatitis B surface antigen (HBsAg)	PGA	128,000	Phase separation (suspension)	1–10	1992	35
Tetanus toxoid (TT)	DL-PLGA 50:50 65:35 85:15 100	n. m.	n. m.	20–60 15 85	1992	76
TT	PLA PLGA 50:50	PLA 2,000 50,000 PLGA 100,000	Emulsion	30 or 60	1992	106
TT	PLGA–glucose 55:45	40,000	W/O emulsion	n. m.	1992	75
Synthetic peptide of the subunit of β-human chorionic gonadotropin	PLA PLGA 50:50	n. m.	n. m.	n. m.	1992	37
DT	PLA PLGA 65:35 + 50:50 75:25 + 85:15	n. m.	Solvent evaporation	<50 <100 <200	1993	121

Antigen	Polymer	MW / viscosity	Preparation method	Size (μm)	Year	Ref.
TT	DL-PLGA 65:35	75,000	W/O/W emulsion	5–70	1993	105
TT	Various PLGA	n. m.	n. m.	n. m.	1993	112
TT	PLGA 50:50 PLA	3,000 + 100,000 3,000 + 100,000	W/O/W emulsion	6–9 or 60	1993	94
TT	DL-PLGA 50:50	Intr. visc. 0.8	W/O/W emulsion	From 9 to 55 (given for each preparation)	1993	98
TT	PLGA 50:50 PLGA 75:25 DL-PLA	n. m.	Coacervation method	1–15 or 10–60 or 20–90	1993	30
TT, BSA	PLA	2,000	Adsorption on microparticles	0.8	1993	122
Influenza virus type A	DL-PLGA	n. m.	Emulsion-based solvent evaporation	2.2–10.8	1993	36
TT	PLA PLGA 50:50	3,000 + 100,000 3,000 + 100,000	W/O/W emulsion	3 or 9 or 60	1994	77
P30B2	PLGA 50:50 PLGA 75:25 DL-PLA	n. m.	Spray-drying coacervation	1–15 or 20–90	1994	123
Branched V3 peptide from HIV-1	PLGA PLA	n. m.	Solvent evaporation	1	1994	124
DT	PLA PLGA	n. m.	Solvent evaporation	<45 45–90 90–150	1994	125
TT	PLGA 50:50	75,000	W/O/W emulsion	n. m.	1994	126

PLA, poly(lactide); PLGA, poly(lactide-*co*-glycolide); n. m., not mentioned.

more straightforward than with new polymers. Homo- and copolymers of PLA contain an asymmetric carbon atom. Usually, the amorphous, racemic poly(D,L-lactic acid) and poly(D,L-lactic-*co*-glycolic acid) are preferred, because L-PLA is a stereoregular semicrystalline biomaterial that degrades at a slower rate than DL-PLA. Homo- and copolymers of lactic and glycolic acids are synthesized by a ring-opening polymerization of the cyclic dimers, lactide and glycolide [41]. Direct condensation of lactic acid and glycolic acid yields homo- or copolymers with comparatively low weight average molecular weight (Mw) in the range of 10–15 kDa [42–45]. Because of its additional methyl group, PLA is more hydrophobic than PGA. Both PLA and PLGA are soluble in organic solvents, such as chloroform, dichloromethane, acetone, and ethylacetate, to a variable extend, depending on copolymer composition and molecular weight. Copolymerization of lactic and glycolic acid is a very powerful method to manipulate biodegradation and antigen release of microspheres by controlling monomer stereochemistry, copolymer composition, and polymer molecular weight [41].

Degradation of aliphatic polyesters occurs by a random, nonenzymatic hydrolytic cleavage of ester linkages, usually referred to as bulk erosion mechanism. The degradation products of PLGA, lactic and glycolic acids, are physiologically occurring metabolites. As outlined in Fig. 1 the mass loss of the polymer at the injection site is preceded by a decrease of the average molecular weight of the polyester. When the Mw of PLGA reaches the threshold level of

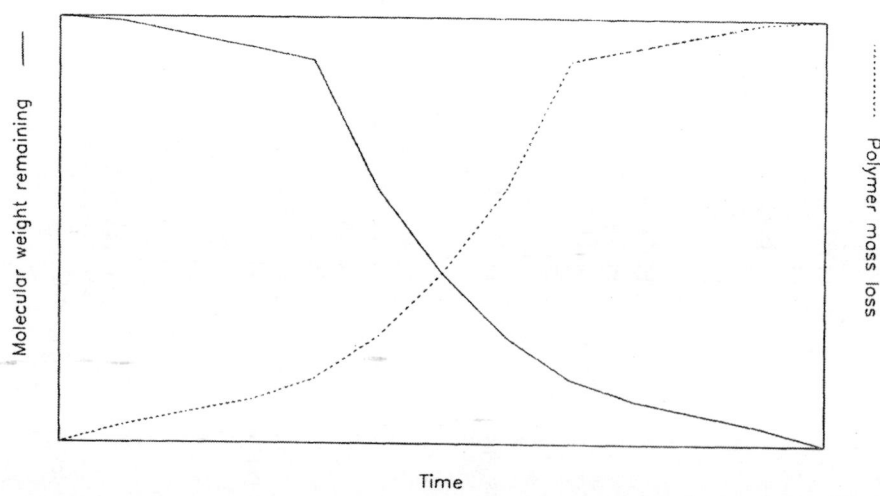

Fig. 1   Decrease of average molecular weight and increase of mass loss of poly(lactide-*co*-glycolides) in relation to time (schematic).

Table 2 In Vivo Biodegradation Times of Poly(lactide-*co*-glycolide) Polymers

Polymer	Approximate Time for Biodegradation (mo)
Poly(L-lactide)	18–24
Poly(glycolide)	12–16
Poly(lactide-*co*-glycolide) 85:15	5
Poly(lactide-*co*-glycolide) 50:50	2

Source: Adapted from Ref. 41.

water solubility of the oligomeric breakdown products, a rapid mass loss is observed. Rate of hydrolysis depends mainly on comonomer ratio and molecular weight of PLGA. As shown in Table 2, many investigators have studied the degradation kinetics of polyesters both under in vitro and in vivo conditions [46–48]. The half-life of polymer mass loss for DL-PLGA with a comonomer composition of 50% mol LA and 50% GA, frequently abbreviated in the literature as PLGA (50:50), is in the order of 14 days, leading to a complete resorption of the microspheres within 50–60 days [41]. A shift of comonomer ratio toward either component leads to a significant decrease in biodegradation. The well-known biodegradation and biocompatibility profiles of PLGA have favored this type of biomaterials considerably. On the other hand, PLGA also has some inherent shortcomings. It is quite hydrophobic compared with most of the antigens to be microencapsulated. A lack of antigen–polymer compatibility may lead to stability problems of the antigen during storage or under in vivo release conditions. Because both hydration and degradation of PLGA are prerequisites for a release of the antigen during bioerosion phase, the proteins need to be sufficiently stable at low pH over an extended period (1–12 months) at body temperature. Both hydrolytic degradation of the antigen and formation of aggregates are to be expected. Very little information on antigen–polymer compatibility and stability is now available from the literature.

One approach to improve antigen–polymer compatibility consists in coencapsulating buffer salts and stabilizers for proteins, which are thought to modify the internal pH of the microspheres and to accelerate swelling. It remains to be seen if this formulation approach increases antigen stability and reduces compatibility problems. Another way to influence swelling and antigen–polymer compatibility can be realized by modification of the PLA or PLGA structure itself. The ABA triblock copolymers containing hydrophobic A-blocks of PLA or PLGA and hydrophilic B-blocks of poly(oxyethylene) (PEO; Fig. 2) form physically entangled biodegradable hydrogels after swelling in a physiological milieu and may provide a more gentle environment for proteins [49,50].

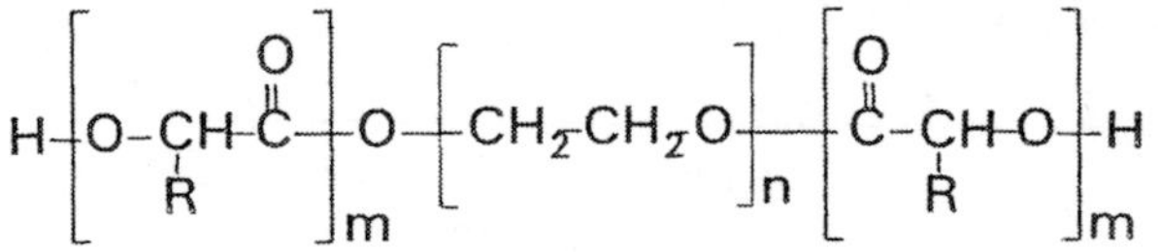

$$H \left[O-\underset{\underset{R}{|}}{CH}-\overset{\overset{O}{\|}}{C} \right]_m \left[O-CH_2-CH_2O \right]_n \left[\overset{\overset{O}{\|}}{C}-\underset{\underset{R}{|}}{CH}-O \right]_m H$$

A - Block :   PLA or PLGA
B - Block :   PEO (polyoxyethylene)

Fig. 2   Structure of ABA-triblock copolymers.

Figure 3 shows protein-loaded microspheres of PLGA- and ABA-triblock copolymers before and after release studies. The high porous structure of the
ABA microparticles is created by the fast-swelling PEO units that facilitate
intrusion of water into the microspheres and enable fast hydration of the encapsulated antigens, whereas PLGA is slowly hydrated through small pores. Further studies are necessary to demonstrate that parenteral delivery systems will
provide a sufficient stabilization of antigens over the intended duration.

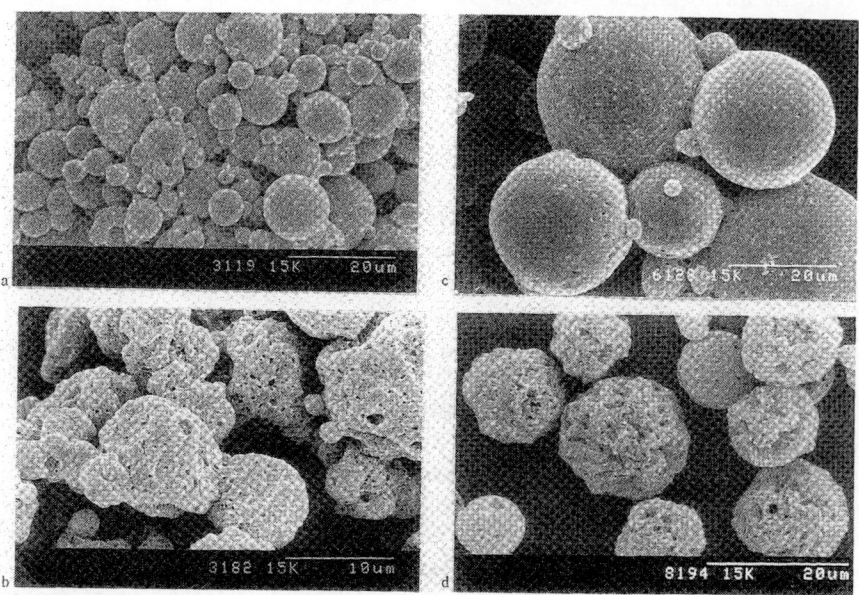

Fig. 3   Protein-loaded microspheres made of (a, b) poly(lactide-*co*-glycolide) and (c,
d) ABA-triblock copolymer before (a, c) and after release (b, d).

The adjuvanticity of the vaccine-delivery system is another aspect influenced by the polymer structure and composition. Recently, L-tyrosine was incorporated into a biodegradable poly(anhydride-*co*-imide) polymer to increase the adjuvant effect of parenteral antigen application [51,52].

IV. BIODEGRADABLE MICROSPHERES

Parenteral depot formulations for antigens have met with increasing interest because several aspects, such as protection of sensitive proteins from degradation, prolonged or modified release of the antigen, pulsatile release patterns, and intrinsic adjuvant effects of the carrier system itself, might provide a promising platform for the development of single-shot vaccines. From a formulation point of view the following systems can be distinguished:

Implants
Nanospheres
Microparticles
Liposomes

The term *microspheres*, or *microparticles*, is sometime used rather liberally in the literature to describe particulate-delivery systems, irrespective of size, chemical composition, and manufacturing technique. In the following, we will restrict our discussion to microspheres of biodegradable polyesters, mainly because this mode of antigen delivery has been intensively investigated in recent years (see Table 1). Liposomes [53,54] and nanoparticles [39] have shown potential as adjuvants for vaccines after parenteral application, but will be discussed elsewhere.

Implants are rod- or disk-shaped devices for subcutaneous application. These devices have been manufactured by compression molding, injection molding, and screw extrusion. Sizes of 1–1.5 mm in diameter and 1–2 cm in length can be applied subcutaneously using a trocar. Disk- or tablet-shaped implants require a small surgical incision for application. Experimental studies in mice have been carried out using a range of model antigens in nondegradable ethylene–vinyl acetate copolymers, demonstrating that long-lasting antibody formation is elicited [55]. Although the fabrication of implants is rather straightforward and may be applied to a wide range of biodegradable polymers, the necessity to resort to melt-processing techniques for large-scale production became a limiting factor. For example, lactide–glycolide copolymers require temperatures of close to 80–100°C for screw extrusion and 140–180°C for injection molding. Many antigens are unstable at these manufacturing conditions; therefore, the focus of interest has moved to alternative-delivery systems.

A. Microparticles

A host of microencapsulation techniques has emerged in the past 63 years, following the discovery of the complex coacervation of gelatine by Bungenberg de Jong and Kaas [56]. In contrast with film-coating techniques, the spherical particles are formed in one step. The size range covered by microparticles is, according to definition, between 1 and 1000 μm. In contrast, smaller particles, designated as *nanoparticles*, with sizes ranging from 1 to 1000 nm. The payload in gaseous, liquid, or solid form is encapsulated in a polymeric material; leading to typical structures, shown in Fig. 4. Depending on process parameters and physicochemical properties of payload and polymer used, either "true microcapsules" are obtained, containing a core of the payload surrounded by a continuous coat of polymer (see Fig. 4A), or *microspheres* that contain the payload dispersed (see Fig. 4B), or dissolved (see Fig. 4C) in a polymeric matrix. The distribution of drug, or payload, in the carrier system has profound consequences on the release and degradation properties of microparticles, as will be shown later. The term *microcapsule* should be reserved for reservoir type devices, whereas *microspheres* are monolithic or matrix-type microparticles.

Pharmaceutically acceptable microencapsulation techniques for hydrophilic macromolecules, such as peptides and proteins, using biodegradable polyesters as matrix materials can be classified into three categories:

1. Spray-drying methods
2. Water–oil–water (W/O/W) triple-emulsion methods
3. Phase-separation methods

The *spray-drying method* is used to manufacture bromocriptine mesylate (Parlodel LAR), a parenteral depot formulation recently commercialized [57,58]. In principle the biodegradable polyester is dissolved in a volatile or-

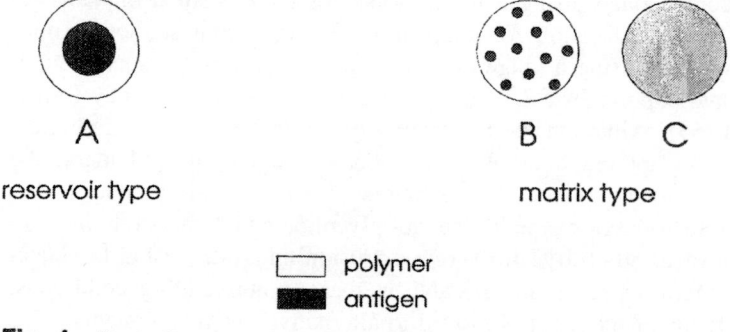

A
reservoir type

B   C
matrix type

☐ polymer
■ antigen

Fig. 4 Typical structures of microparticles.

ganic solvent, such as dichloromethane or acetone, the drug in solid form is dispersed in the polymer solution by high-speed homogenization, and this dispersion is atomized in a stream of heated air. From the droplets formed, the solvent evaporates instantaneously, yielding microspheres in typical size ranges from 1 to 100 μm, depending on the atomizing conditions. The microspheres are collected from the airstream by a cyclone separator. Residual solvents are removed by vacuum drying. The process can be operated under aseptic conditions, and in closed loop configurations, spray-drying in a nitrogen atmosphere is technically feasible [59]. The process scheme is outlined in Fig. 5. The important advantages of the spray-drying technique over other encapsulation techniques are: its reliability under production conditions, the proven reproducibility, and the well-defined control of particle size, as well as drug release properties of the resulting microspheres. The process is quite tolerant to small changes of the polymer specifications. Disadvantages are high capital investment and residual organic solvents, a feature shared by all microencapsulation processes that use biodegradable polyesters. Thermal stress to the drug substance is usually not a limiting factor, as the product is typically subjected to 40–60°C for only a short time. Encapsulation of proteins using this technique requires lyophilization of the antigen before the dispersion and homogenization in the organic polymer solution. These processing conditions are likely to induce aggregation and denaturation to sensitive antigens. The factors influencing protein stability are many and diverse (see Section V). Therefore, stability of the microencapsulated antigen during processing, release, and storage becomes a major concern. Some of these problems could be avoided if liquid antigen preparations were emulsified into the organic polymer solution. So far, data on antigen encapsulation using the spray-drying method are still very scarce.

Frequently, the *W/O/W double-emulsion technique* has been used to encapsulate a variety of peptides and proteins, including antigens. This process has also been termed as *in water drying method*, by Ogawa et al. [43,45,60], who applied this method to microencapsulate a luteinizing hormone-releasing hormone (LH-RH) agonist into a PLGA matrix. The product has recently become commercially available [61]. As shown schematically in Figure 6, the antigen in an aqueous solution is emulsified with the nonmiscible organic solution of the polymer to form a water-in-oil (W1/O) emulsion. The organic solvent dichloromethane is mainly used, and the homogenization step is carried out using either high-speed homogenizers, ultrasound, or vortex mixing. This primary (W1/O) emulsion is then rapidly transferred to a vast excess of an aqueous medium, containing a stabilizer, usually poly(vinyl alcohol). Again homogenization or intensive stirring is necessary to initially form a triple emulsion of W1/O/W2. The protein containing W1 phase is separated from the continuous W2 phase by the organic polymer solution O. The solvent dichloromethane is only slightly water-soluble (ca. 1%), but through the large excess of water, the

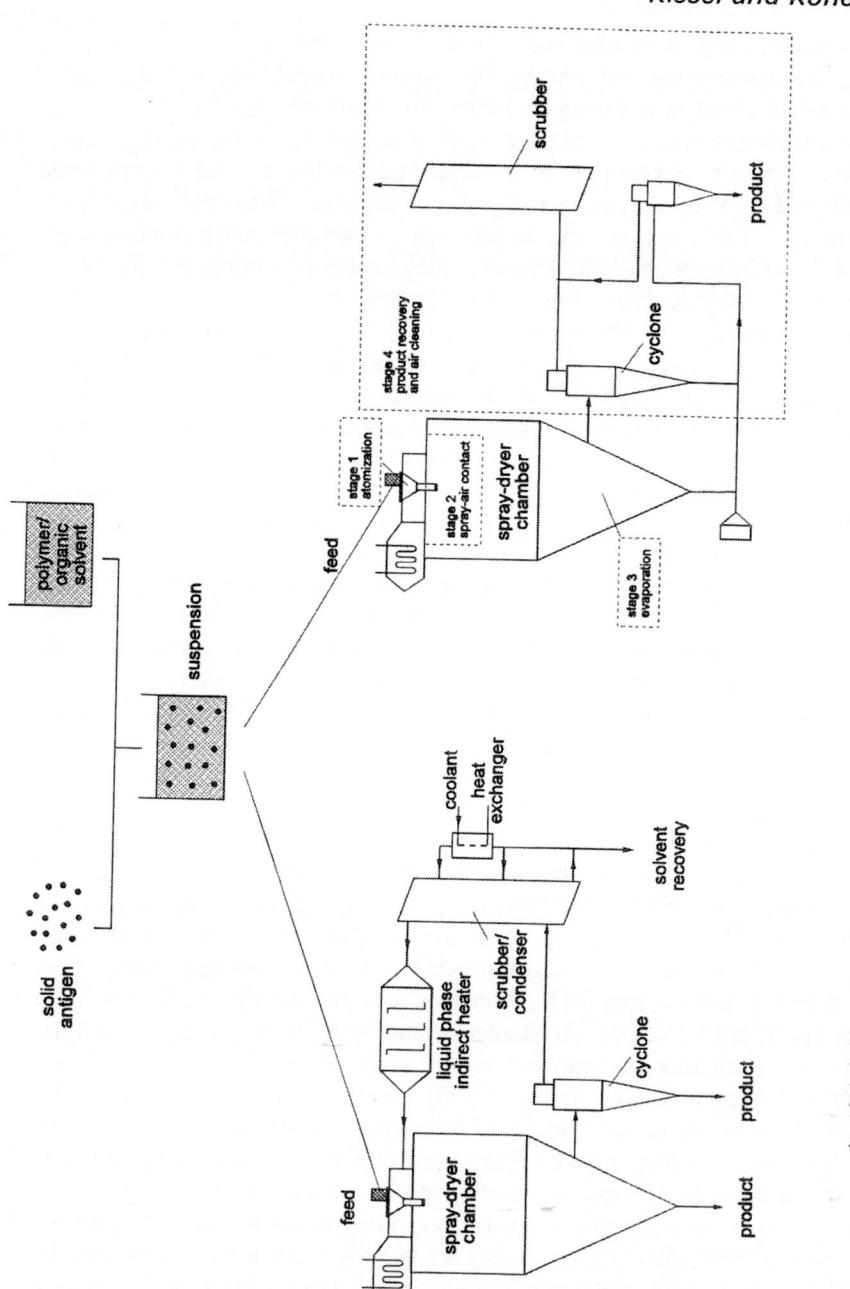

Fig. 5 Manufacturing of microspheres: spray drying.

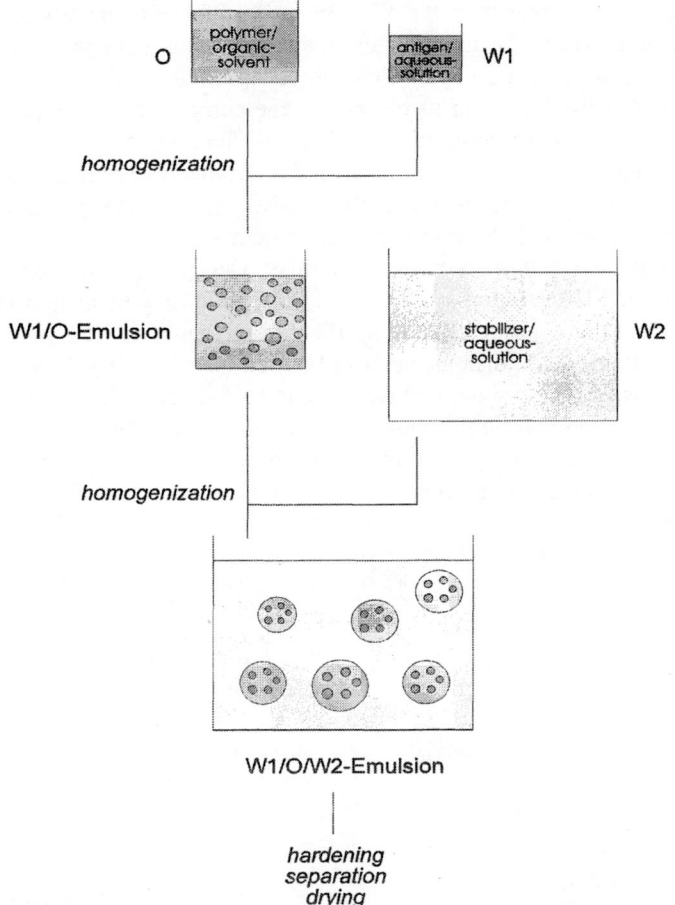

Fig. 6 Manufacturing of microspheres: W/O/W triple-emulsion method.

organic solvent is rapidly extracted from the O-phase, yielding solid microparticles that contain antigen in a polymeric matrix. In the hardening step, residual
amounts of solvent are extracted and evaporated (solvent extraction or solvent
evaporation). Modifications of the W/O/W method for peptides and proteins
have been studied by several groups [62–69]. Model proteins, such as BSA
and OVA seem to retain their integrity. The internal structure of the microspheres depends on the method of homogenization and the volume ratio W1/
O. Drug loading for peptides and proteins in the range of 1–10% is frequently
reported with good encapsulation efficiencies (80–100%) and yields (>80%).
The control of the drug-release properties, however, is an issue requiring fur-

ther investigations. The advantages of the W/O/W process for antigen delivery are the following: proteins can be used for encapsulation as an aqueous solution, scaling-down is possible, and high yields and encapsulation efficiency are obtained. The main disadvantages are to be seen in the complexity of the process, the sensitivity to polymer properties, and the difficulties in modifying release profiles of drugs from these microspheres. The arguments pertaining to stability and shelf life of antigens in microspheres also apply to this process, and more information is needed to come to a final conclusion.

The *phase-separation technique* is used to fabricate Decapeptyl, the parenteral depot form of an LH-RH agonist [70,71]. The peptide or protein is dispersed in solid form into a solution containing dichloromethane, and the polymer (Fig. 7A). Alternatively, emulsions can also be used (see Fig. 7B). Silicon oil is added to this dispersion at a defined rate, reducing the solubility of PLG in its solvent. The polymer-rich liquid phase (coacervate) encapsulates the dispersed drug particles and the "embryonic" microspheres are subjected to a hardening and washing step using organic solvents, such as heptane (see Fig. 7) [72–74]. The process is quite sensitive to polymer properties. Residual solvents are also an important issue. The phase-separation technique was used to prepare

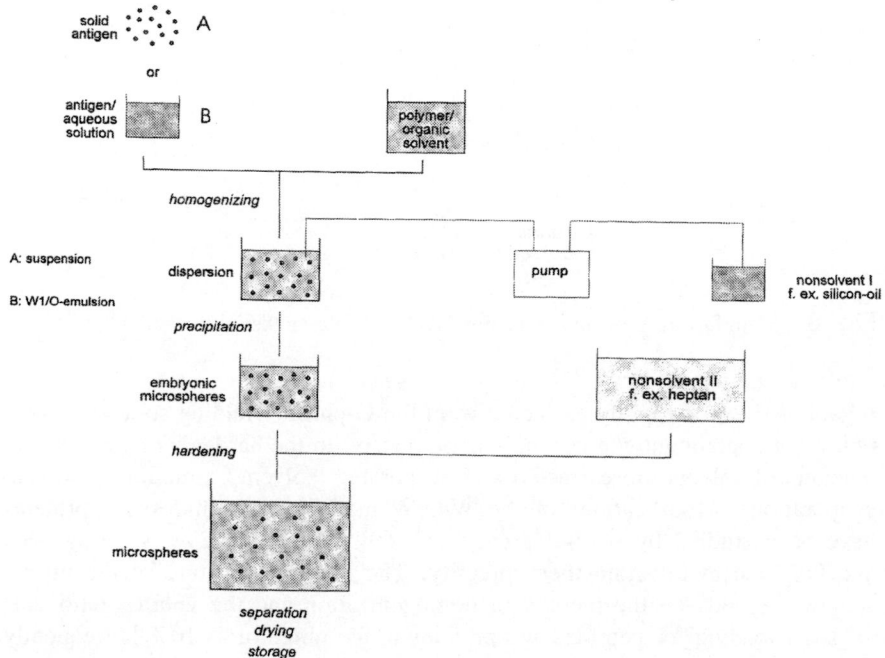

Fig. 7 Manufacturing of microspheres: phase-separation technique.

tetanus toxoid- [75] and HBsAG- containing microspheres [35], but little information is available from the literature on stability and control of protein release in antigen-delivery systems. The influence of the additional process steps necessary to obtain a commercially acceptable dosage form, such as sizing of the microspheres, removal of residual solvents, sterilization of microspheres, and packaging, are issues requiring further investigations.

Especially gamma-sterilization techniques were studied to determine their applicability. Hazrati and co-workers [76] treated TT microspheres and TT alum microspheres with gamma-irradiation using a dose of 10 kGy. The TT microspheres yielded antibody titers in mice similar to nonirradiated controls, whereas the TT alum microspheres showed reduced antibody titers. Esparza et al. [75] demonstrated a significant decrease in immunoreactivity of encapsulated tetanus toxoid by an enzyme-linked immunosorbent assay (ELISA) after exposition to gamma-irradiation, suggesting that TT may be susceptible to degradation by gamma-sterilization.

Solvent residues were frequently removed by lyophilization of the semifinished product, and this process step was critical for antigen stability [77].

V. ANTIGEN STABILITY TO MICROENCAPSULATION CONDITIONS

Antigens are mostly proteins that must adopt specific, folded, three-dimensional structures (conformations) to be biological active. The inactivation of proteins is often connected with a destruction of the conformation. The literature on peptide and protein stability indicates that denaturation can be caused by a variety of factors, such as temperature [78–81], pH [79,82,83], organic solvents [78], freeze-drying [82–87], moisture [33,88,89], salt concentration [82,90], buffer ions [79], and spray-drying [91] (Fig. 8).

Hora et al. [92], who were the first to investigate the structural integrity of human serum albumin (HSA) released from microspheres by sodium dodecyl sulfate–polyacrylamide gel electrophoresis (SDS–PAGE) and isoelectric focusing, showing that neither degradation nor aggregation of HSA was induced by the encapsulation technique. Cohen and co-workers [66] confirmed these results using fluorescein isothiocyanate (FITC)-labeled bovine serum albumin (FITC-BSA) and FITC-horseradish peroxidase (FITC-HRP). The FITC-HRP was used to assess the enzyme activity of HRP after microencapsulation. About 18% of HRP activity was lost in the microencapsulation process. Moreover, FITC-HRP within the PLGA microspheres retained more than 55% of its activity after 21 days in solution, whereas FITC-HRP in solution degraded rapidly under the same conditions. Although these results point to some protein-stabilizing effects by microencapsulation, the immunological consequences of partially degraded antigens need to be addressed. There is only scant information on anti-

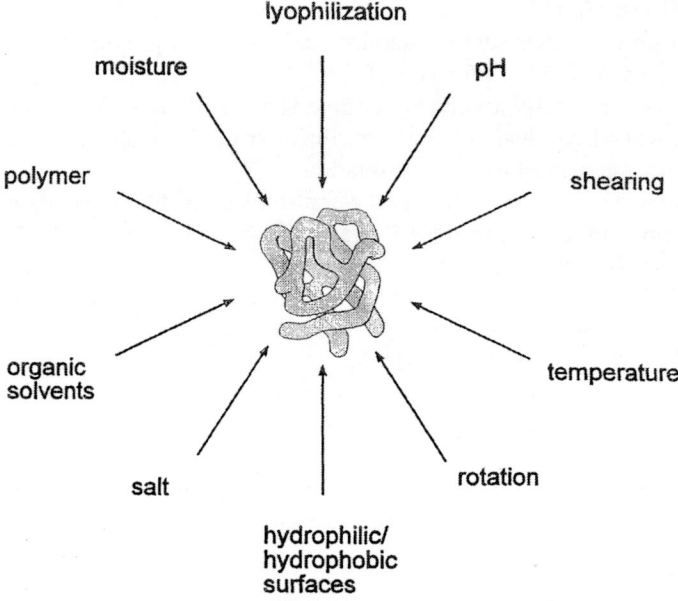

Fig. 8   Factors that influence the nativity of proteins and antigens.

gen stability within microparticles. Alonso et al. [93] investigated TT
microspheres, detecting only a slight formation of aggregates (<10%) by im-
munodiffusion and SDS–PAGE. Gupta et al. [94] conducted in vitro-release
studies using TT microspheres, by measuring protein release using a micro-
BCA assay. The in vitro release medium was assayed for immunoreactive TT
by ELISA, yielding only very small amounts of intact TT. In our own investigations [32,95], we employed SDS–PAGE, native PAGE, size-exclusion chromatography (SEC–HPLC), ELISA, and BCA assay to characterize the influence of several factors relevant for TT stability during manufacture of
microspheres, such as organic solvents, homogenization conditions, pH, and
temperature effects. Critical factors turned out to be exposure to dichloromethane, the most widely used solvent for PLGA, pH conditions lower than 5
[30,32], and mechanical stress at aqueous–air interfaces. The factors pH and
agitation are also important considerations for in vitro release and degradation
studies of TT microspheres. The selection of suitable conditions is crucial for
in vitro release studies (e.g., Fig. 9 shows the influence of different agitation
modes on the stability of TT). Overhead rotation of TT solutions caused significant aggregation (SEC–HPLC) and loss of ELISA activity. Surprisingly,
high-speed homogenizers, creating large air–solvent interfaces, did not affect
molecular weight distribution nor antigen activity, whereas ultrasound homo-

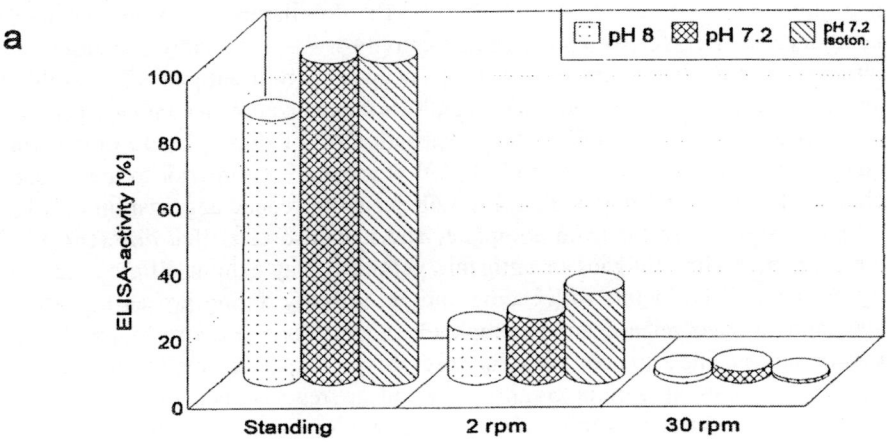

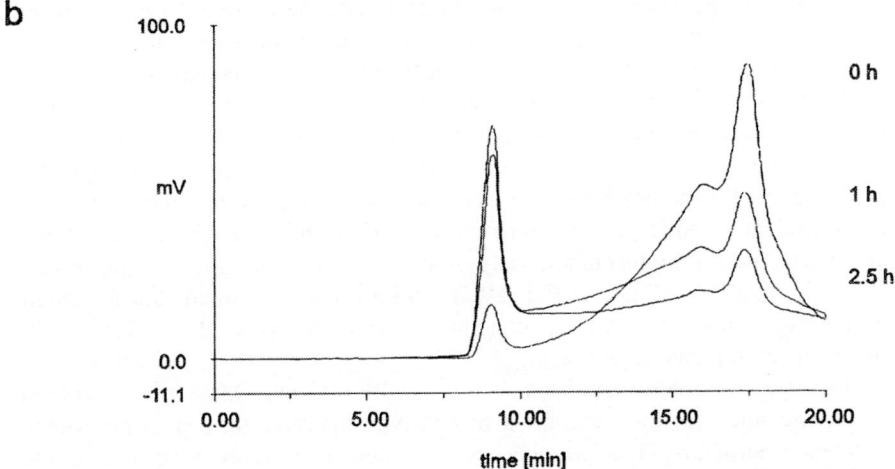

Fig. 9 Stability of tetanus toxoid in aqueous solution influenced by different release conditions: (a) ELISA activity after different agitation modes; (b) SEC–HPLC chromatograms after rotation (37°C/30 rpm) after 0, 1, and 2.5 h. [for details see Refs. 32, 95].

genization induced severe aggregation of TT. Similar results were obtained using trypsin [96], for which enzymatic activity was substantially decreased by ultrasonication. Sonication was used for encapsulation of antigens [35,97–99], and it remains to be seen if more gentle homogenization techniques can be developed. Alonso et al. [77] noted aggregation of TT during freeze-drying, in analogy with insulin and trypsin [100]. Ethyl acetate was superior to methylene chloride for TT aggregation. Pluronic F68 seems to reduce aggregation of TT.

These studies are far from complete, and it is also clear that different proteins, such as viral antigens or antigenic peptides, may exhibit different stability problems. The factors influencing antigen stability during processing, storage, and in vivo release are incompletely understood. Because many of the traditional vaccines are complex mixtures of proteins, the analytical tools to describe changes of protein composition and aggregation become a matter of concern. Very little is known about the effect of these changes on vaccine efficiency. The aspect of antigen stability will determine if single-shot vaccines are technically feasible; therefore, they should not be underestimated.

VI. ANTIGEN RELEASE FROM MICROSPHERES

The release of macromolecules from biodegradable microspheres is influenced both by the structure, or micromorphology, of the microparticles and properties of the biodegradable polymer itself. True microcapsules, consisting of a solid or liquid core, in which the antigen is incorporated, and a biodegradable coating could hypothetically release proteins or antigen by a osmotically driven burst mechanism. The biodegradable coating of the microcapsules would be impermeable for the protein, but water could slowly diffuse into the core, creating sufficient osmotic pressure for rupturing the membrane. The release of the antigen would follow instantaneously, leading to a pulsatile in vitro and in vivo release profile, as outlined in Fig. 10. Therefore, reservoir microcapsules might be able to mimic conventional injection schemes for vaccines, which require two to three separate injections.

This type of microparticles has not yet be realized. Most of the antigen delivering microparticles exhibit a matrix-type internal, solid dispersion morphological structure. The proteins are insoluble in the polymeric matrix, and the macromolecules are released by a mechanism that combines pore diffusion and polymer erosion. Initially, water diffuses into the matrix, dissolving drug particles adjacent to the surface of the device. The resulting osmotic pressure is relieved by forming a tortuous channel to the surface, releasing a defined amount of antigen in the initial drug burst. This burst effect is controlled mainly by three factors: the protein/polymer ratio, the particle size of the dispersed protein, and the particles size of the microspheres. As the penetrating water front continues to diffuse in the direction of the microparticulate core,

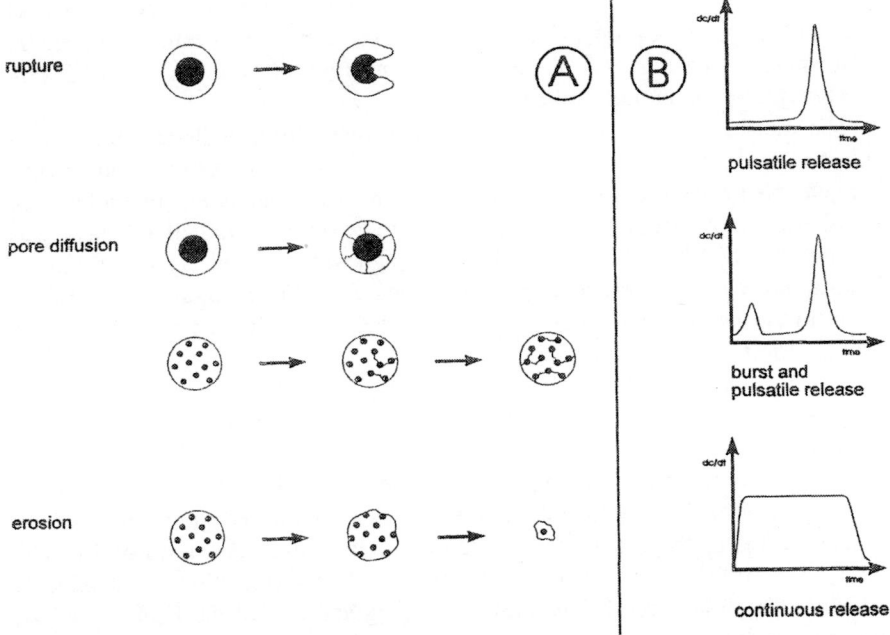

Fig. 10   Release of antigens from microspheres: (A) mechanisms, (B) release profiles.

the dispersed protein particles are dissolved, creating a network of water-filled pores through which the antigen diffuses in a controlled manner; hence, the name *pore-diffusion mechanism* [63,101–103]. The PLGA microspheres release macromolecules by this mechanism in the initial phase. Additional factors, such as swelling phenomena and osmotic effects have to be taken into account to describe the release behavior of PLGA microspheres within the first 3–10 days. The in vitro and in vivo release decreases considerably, reaching very low levels (see Fig. 10). Depending on the composition and molecular weight of the PLGA, protein release recommences when the polymer degradation has reached to stage of rapid mass loss. This phase is controlled by the degradation of the polymer, leading to a degradation or erosion of the matrix. The erosion may occur at the surface of the device as a consequence of enzymatic processes, or by hydrolysis of the polymer in the bulk of the microspheres, as observed with biodegradable polyesters of the PLGA type. Physicochemical properties of protein and polymer, as well as the drug loading, will define the length and depth of the "valley" between the initial drug burst and the antigen release as a consequence of matrix erosion. A relatively low level of drug loading (ca. 0.1–1%) will favor this pulsatile release profile, as has been shown for BSA [104]. Continuous release of macromolecules from PLGA microspheres (see Fig. 10) requires synchronization of both the pore-diffusion and polymer-erosion mechanisms.

The mechanism by which antigens are released from biodegradable microspheres are not known with certainty. So far, only very few studies on in vitro antigen release have appeared in the literature. Continuous in vitro release of diphtheria toxoid from DL-PLA microspheres over a period of 60 days was reported by Singh et al. [97], whereas Raghuvanshi et al. [105], using an ELISA assay, demonstrated continuous release of TT for more than 60 days under in vitro conditions. Gander and co-workers [30] obtained pulsatile release profiles for TT microspheres.

Alonso et al. studied the release characteristics of tetanus toxoid from microspheres as a function of encapsulation technology and polymer properties, using PLA and PLGA (3 kDa and 100 kDa) [77,98,106]. The W/O/W emulsion technique was superior to dispersion of lyophilized TT in the O-phase, giving better control of the in vitro release rate [106]. As expected, polymer viscosity influenced the microparticle size, affecting the in vitro release rates of TT. The increase in surface area of small particles caused an acceleration of TT release, which is also influenced by the antigen loading and Mw of the PLA or PGLA. By manipulating the fabrication parameters, the initial antigen burst could be reduced to 5–20% [77]. Moreover, the micromorphology of TT microspheres and antigen purity seem to influence the release of TT from the microspheres under in vitro conditions [98]. For a high-purity vaccine, a good correlation between polymer degradation and release properties was observed, whereas, for unknown reasons, impure TT led to an unexpectedly high in vitro release rate.

Many issues are still open for debate, ranging from the most suitable in vitro release setup, to methods for quantitation. Mass balance studies, both under in vitro and in vivo conditions, would be desirable to define predictive in vitro release conditions. Given the susceptibility of TT to aggregation and degradation, one is surprised to see that continuous release of close to 80% over 60 days seems to be feasible [105]. More work is clearly necessary to establish the appropriated in vitro-release models and to demonstrate their predictive power for in vivo antigen release.

VII. INTERACTION OF MICROSPHERES WITH THE IMMUNE SYSTEM

Microspheres are capable of forming antigen depots from which the antigen is slowly released at the injection site, comparable with the situation in CFA emulsions or alum-adsorbed systems. The interaction of particulate carrier systems for antigens is presented schematically in Fig. 11. Interestingly, micro-

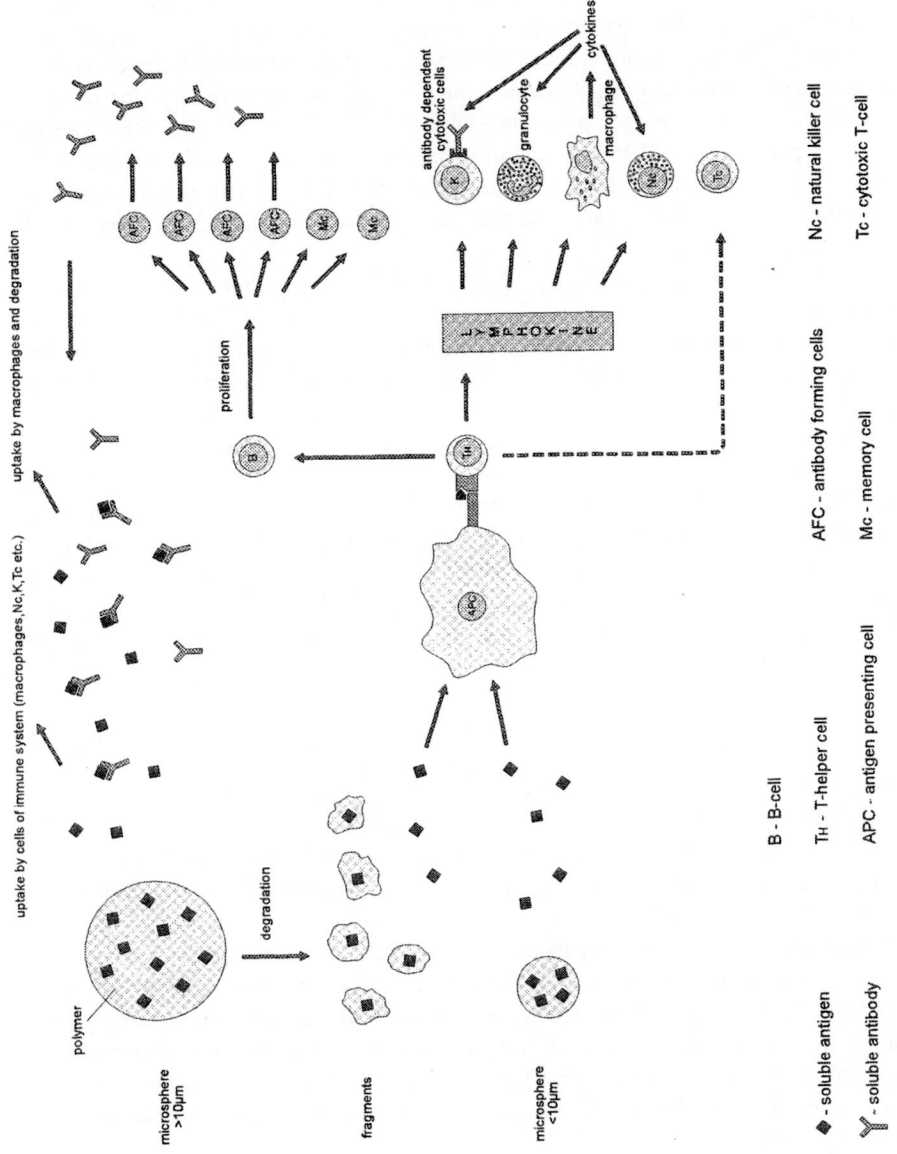

Fig. 11 Interactions of particulate carrier systems with the immune system.

B - B-cell

♦ - soluble antigen	TH - T-helper cell	AFC - antibody forming cells	Nc - natural killer cell
Y - soluble antibody	APC - antigen presenting cell	Mc - memory cell	Tc - cytotoxic T-cell

sphere size is an important design parameter. Small particles, with sizes smaller than 10 μm can be directly taken up into macrophages by phagocytosis, whereas larger microspheres (>10 μm) need to undergo biodegradation before phagocytosis can occur. In this case, microspheres are covered by one or several layers of macrophages as a consequence of the wound-healing response to injected particles [107]. Consequently, degradation, antigen release, location, and antigen presentation of microparticles larger than 10 μm are expected to be different from smaller ones. Antigen presentation is mediated by specialized macrophages (APC).

Both the activation of T cells and of B cells is important for an effective immunogenic reaction, because, in a complex way, they interact with one another, either directly or through interleukins. The T-helper (T_H) cells are the "principal orchestrators" [108] of the immune response because they are needed for the activation of the major effector cells in this response [i.e., cytotoxic T (T_C) cells and antibody-producing B cells]. Antigen presentation is mediated by specialized macrophages (APC). After internalization of the antigen-loaded particle or free antigen, immunorelevant epitopes are presented on the surface of the APC, in combination with a major histocompatibility complex (MHC). The T_H cells are attracted and activated by two signals: binding of the T-cell antigen receptor to the MHC complex and production of interleukin-1 by the APC. The activated T_H cells trigger a complex cascade. They release lymphokines (a) stimulating B lymphocytes to proliferate and produce specific antibodies after differentiation as well as memory cells, and (b) activating granulocytes, macrophages, and natural killer cells. Cytotoxic T cells are activated directly. Moreover, some long-lived T_H cells seem to provide a memory function in the T-cell compartment of the immune system [108,109].

Certain molecules appear to trigger B lymphocytes directly (thymus-independent antigens). This mechanism is still incompletely understood because the antibodies produced are mainly of the IgM class, and little or no immunological memory is effected.

The release of the antigen from microparticles is controlled by a variety of factors, such as microparticle morphology, microparticle size, and the polymer used for microencapsulation. In this section we will attempt to summarize some of the in vivo results obtained with experimental vaccine-delivery systems after parenteral application, although one has to bear in mind, that such a retrospective comparison is obscured by the unknown effect of experimental differences on the biological response.

For one, different species were used (e.g., mice [34,76,77,97–99,110], rats [105,111], and guinea pigs [35,94]). Injection sites and vehicles for microparticle administration were different, such as IP, IM, or SC. Therefore, it is not surprising that antibody titers, duration of immune response, and type of immune reaction varied considerably from study to study. Second, there is no consensus on a generally accepted standard dose for the respective animal model used. Only two groups reported efforts in establishing an adequate dose for tetanus toxoid in mice [75,76,112]. A dose–response relation for microencapsulated antigens has not been established to our knowledge; therefore, interspecies comparisons and in vivo–in vitro correlations of antigen release from animal studies are subject of speculation. Third, and most importantly, there seems to be little agreement in the literature on a standard immunization protocol for in vivo studies. Positive and negative controls, time schedule for boostering, and time intervals for blood sampling differ considerably from group to group.

Several groups reported that for a variety of antigens, such as SEB [113], OVA [34], HBsAg [35], DT [97], and TT [106] in microspheres, caused the formation of higher antibody titers in animal models than injected reference solutions. Eldridge and co-workers, for instance, demonstrated a 30 to 60-fold higher IgG antibody response after SC immunization of mice using microencapsulated SEB than with an SEB solution [113]. The toxin-neutralizing antibody titers were comparable with those obtained by FCA in the same animal model [31]. Antibody titers after microparticle administration remained constant for up to 90 days. These results are encouraging and indicate that vaccine delivery with microparticles seems to be a useful approach.

A necessary requirement for enhancing the immune response is the fixation of the antigen onto or within the microparticles. Adsorpion of the antigen (OVA [110]) to the microparticle surface showed antibody titers similar to encapsulation within microspheres, whereas free solutions gave a lower response. Similar effects have also been observed with TT adsorbed to particles [114].

The immune response seems to be related to the particle size of the vaccine-delivery system [31,35,110,113,115–118]. Particles with sizes smaller than 10 μm are thought to be taken up by macrophages more efficiently, leading to a higher antibody response in mice [113,115]. These microparticles, or fragments of larger ones, can be engulfed by macrophages, which transport the antigen directly to the lymph nodes. Larger particles (>10 μm) remain at the injection site and release the antigen by pore diffusion or polymer erosion. The molecular mechanism by which microparticles enhance immune response is not well understood, but it seems to be related to an intracellular delivery of antigen to APCs. However, particle size and other factors (e.g., antigen release and degradation of the carrier system) are not independent variables. Smaller particles show faster release rates and degrade more rapidly, especially when clearance of the microspheres by phagocytosis becomes prominent. More data are clearly necessary to elucidate the influence of particle size on these parameters.

In this context, the surface hydrophobicity of the vaccine-delivery system has to be taken into account, which is known to influence phagocytotic uptake by macrophages. The adjuvantivity [119] and phagocytotic index [117] of microspheres increased with increasing surface hydrophobicity. This effect has also consequences for the selection of polymeric excipients.

Antigen-release profiles from parenteral vaccine-delivery systems are a matter of dispute. Although some immunologists insist on a truly pulsatile-release profile, characterized by defined, short-lasting intervals of antigen release, followed by absolutely no release at all, to avoid induction of immunological tolerance to the antigen, others claim that a continuous exposure to a small amount of antigen may be beneficial in terms of maximizing the immune response. Experimental evidence is still too scant to settle this fundamental question.

Singh and co-workers [97] demonstrated that continuous release of DT from PLA microspheres led to antibody titers in mice comparable with those achieved by the conventional injection therapy, using three SC injections, on days 0, 30, and 60, of a DT adsorbate vaccine. The antibody titers increased after one injection of DT microspheres up to 70 days and, then, steadily declined, up to day 240 [99]. These in vivo results were compared with in vitro release data, showing a more or less continuous release up to day 60, when close to 88% of the antigen was released. The SEM micrographs demonstrated a massive erosion of the polymer matrix already after 21 days.

Raghuvanshi et al. [105] encapsulated TT in microspheres using PLGA (65:35; Mw 75,000) and investigated both the in vitro release and in vivo antibody response in rats (Fig. 12). Under in vitro conditions, an initial burst of TT of about 12% was followed by a continuous release of ELISA-reactive antigen for 60 days. At that point almost 80% of the initial TT loading had been removed from the microspheres. The in vitro release profile was biphasic, showing a slower release rate until day 30 and, afterward, an accelerated release up to 60, probably caused by the degradation of the PLGA. This biphasic in vitro release was not observed in rats, in which the anti-TT antibody titers immediately reached a plateau level, lasting up to days 60–90. This level was comparable with the positive control (TT adsorbate vaccine).

In both studies, the antigen was released in a more or less continuous fashion from PLA and PLGA microspheres under in vitro conditions, leading to the induction of IgG antibodies against TT and DT, which increased slowly, but steadily. Neither a booster effect at the time point of massive polymer degradation, nor immune tolerance was observed in the animal models used (mice or rats).

Alonso et al. [77,98] and Gupta et al. [94] studied TT microspheres in mice, confirming the results of Raghuvanshi et al. [105] that the immune response, measured as IgG anti-TT titers, was significantly higher than that obtained for an antigen reference solution. In contrast with Hazrati et al. [112], however, none of the preparations induced higher antibody titers nor a more prolonged response compared with the conventional injection scheme using the aluminum-

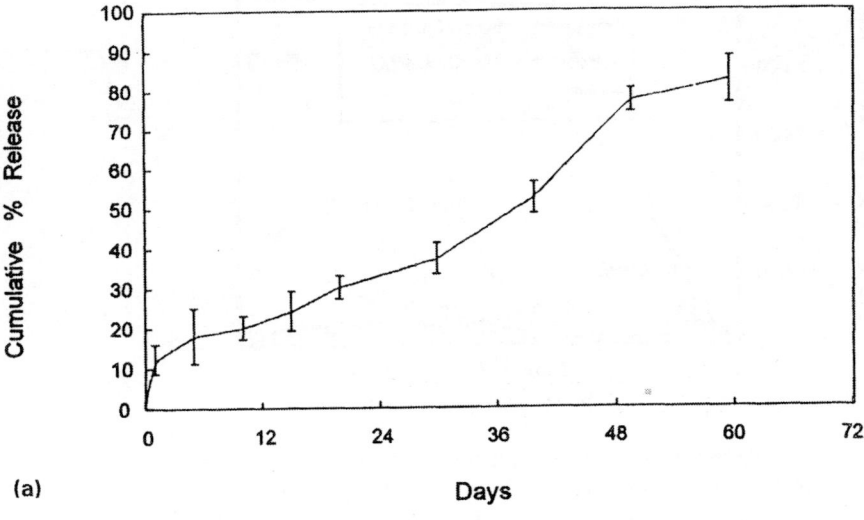

(a)

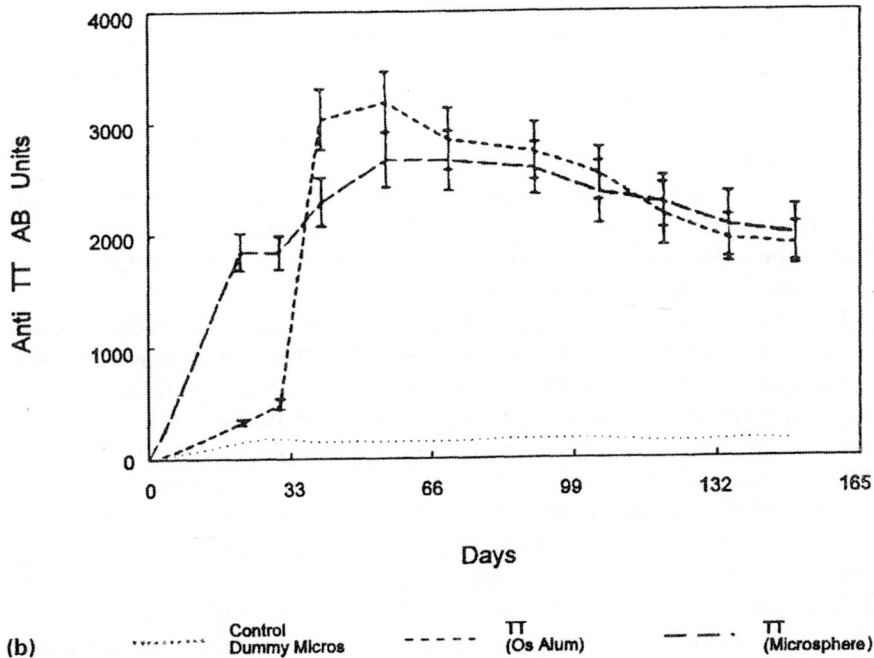

(b)

Fig. 12 Release of TT from PLGA 65:35 microspheres (a) in vitro and (b) in vivo [for details see Ref. 105].

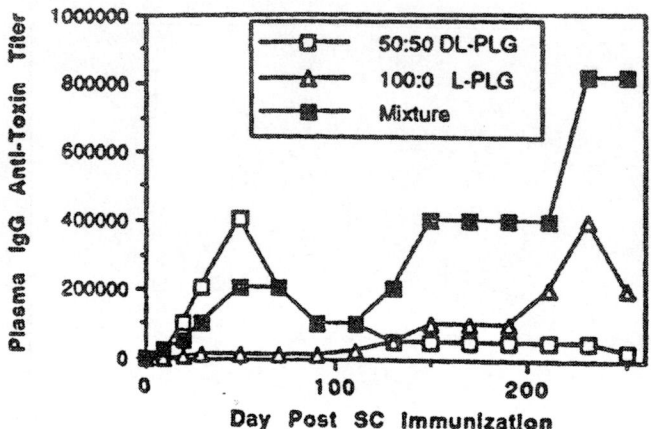

Fig. 13 Plasma IgG antitoxin antibody response resulting from immunization with SEB toxoid microspheres made of PLGA 50:50 (fast degradation) or PLA (slow degradation), or a mixture of both preparations (pulsatile release) [see Ref. 120].

adsorbed TT as control [77]. The TT microspheres induced high-affinity neutralizing antibodies, and substantially higher levels were attained compared with the antigen solution. This effect was attributed to progressive affinity maturation of tetanus antibodies induced by prolonged exposure to low concentrations of TT. The authors suggest that induction of a long-lived antibody response, with high-affinity neutralizing activity of modest magnitude, is a desirable feature for parenteral vaccine delivery systems, because protection would be maintained and immunologically mediated reactions may be minimized. Long-term studies are clearly indicated to substantiate these claims. In these studies [94,98], no correlation between in vitro release profiles and in vivo antibody titers could be established.

The issue of a pulsatile-release pattern of antigens from biodegradable parenteral microspheres was investigated by Eldrigde et al [115,120] using SEB microencapsulated into polyesters of different compositions (DL-PLGA 50:50, 85:15, and 100:0), degrading at a fast, medium, and slow rate. Plasma IgG antitoxin titers shown for the respective SEB microspheres peaked at day 50, 150, and 230, using a single SC injection in mice. A mixture of the fast-releasing SEB microspheres (DL-PLGA 50:50) and the slower version (PLGA 100:0) induced an IgG antitoxin antibody response as shown in Fig. 13. This pattern tends to suggest, that a booster effect is observed, leading to an additional augmentation of the IgG response. Gander et al. [30] obtained different results with TT microspheres using PLGA 50:50 and 75:25. In this study both particle size effects and polymer properties were investigated. Although the in vitro-release studies showed pulsatile-release profiles of the respective formulations, the in vivo studies in mice failed to reach antibody titers comparable with the positive controls, such as IFA or alum-adsorbed TT.

Given the information presently available, a firm conclusion on the "ideal"-release properties of a parenteral vaccine delivery system cannot be drawn, and more investigations are needed addressing this issue, both from an immunological and from a drug-delivery point of view.

VIII. CONCLUSIONS AND PROSPECTS

Despite considerable research efforts and impressive progress made in recent years, the question of feasibility for injectable biodegradable microspheres as vaccine-delivery system remains open to debate. Microencapsulation techniques have been developed that allow incorporation of sensitive proteins into biodegradable polymers under mild conditions. There seems to be a general agreement that the W/O/W triple-emulsion technique is most suitable for this purpose. So far, biodegradable polyesters of PLGA were mainly used as matrix material, making use of their well-known degradation properties. The biphasic or pulsatile release of bacterial endotoxins was observed under both in vitro and in vivo conditions. Biodegradable microspheres that release antigens in a continuous pattern have elicited a sustained immune response for DT and TT. The efficiency of both release patterns in generating a protective effects is a subject of further investigations. An area requiring additional efforts is the analytical characterization of vaccine microspheres. The compatibility of proteins with biodegradable polymers and the stability of antigens during fabrication, storage, and under in vivo release are only incompletely known for most of the vaccines yet studied. Toxoid vaccines are complex mixtures of proteins, and assessment of their purity and stability is notoriously difficult. Therefore, advanced methods for protein characterization and "pure" antigens will be needed to approach the problem of antigen stability in vaccine microspheres. Another issue closely related to the latter is the development of in vitro–in vivo correlations for antigen release from vaccine microspheres. Mass balance and antigen integrity need to be assessed in a more profound way. Finally, the question of antigen release and presentation to the immune system requires more intensive interactions between immunologists and drug-delivery specialists.

Parenteral vaccine-delivery systems based on biodegradable microspheres have been subject of research for only little more than 5 years. The next 5 years will demonstrate, if the hope for a single-shot vaccine-delivery system remains a dream, or will become therapeutically relevant approach.

REFERENCES

1. P. H. Lambert, New vaccines for the world: Needs and prospects, *Rev. Med. Suisse Rom.* *113*:193 (1993).
2. A. C. Capron, C. Locht, and G. N. Fracchia, Safety and efficacy of new generation vaccines, *Vaccine 12*:667 (1994).
3. R. I. Walker, New strategies for using mucosal vaccination to achieve more effective immunization, *Vaccine 12*:387 (1994).
4. R. Edelman, Vaccine adjuvants, *Rev. Infect. Dis. 2*:370 (1980).
5. D. A. Eppstein, N. E. Byars, and A. C. Allison, New adjuvants for vaccines containing purified protein antigens, *Adv. Drug Deliv. Rev. 4*:233 (1990).
6. C. A. Gilligan and A. Li Wan Po, Oral vaccines: Design and delivery, *Int. J. Pharm. 75*:1 (1991).
7. R. Bawa, R. A. Siegel, B. Marasca, M. Karel, and R. Langer, An explanation for the controlled release of macromolecules from polymers, *J. Controlled Release 1*:259 (1985).
8. H. H. Fudenberg, D. P. Stites, J. L. Caldwell, and J. V. Wells, *Basic and Clinical Immunology*, 2nd ed., Lange Medical, Los Altos CA, 1978.
9. R. Edelman, Vaccines: New technologies and applications, *Vaccine 11*:1361 (1993).
10. R. E. Spier, Meeting on vaccine strategies for tomorrow, *Vaccine 11*:1450 (1993).
11. J. Drews, Imunostimulantien, *Klin. Wochenschr. 62*:254 (1984).
12. H. S. Warren, F. R. Vogel, and L. A. Chedid, Current status of immunological adjuvants, *Annu. Rev. Immunol. 4*:369 (1986).
13. A. Altman and F. Dixon, Immunomodifiers in vaccines, *Adv. Vet. Sci Comp. Med. 33*:301 (1989).
14. R. K. Gupta, E. H. Relyveld, E. B. Lindblad, B. Bizzini, S. Ben-Efraim, and C. K. Gupta, Adjuvants—a balance between toxicity and adjuvanticity, *Vaccine 11*:293 (1993).
15. R. L. Richards, M. D. Hayre, W. T. Hockmeyer, and C. R. Alving, Liposomes, lipid A, and aluminum hydroxide enhance the immune response to a synthetic malaria sporozoite antigen, *Infect. Immun. 56*:682 (1988).
16. L. A. Chedid, M. A. Parant, F. M. Audibert, G. J. Riveau, F. J. Parant, E. Lederer, J. P. Choay, and P. L. Lefrancier, Biological activity of a new synthetic muramyl peptide adjuvant devoid of pyrogenicity, *Infect. Immun. 35*:417 (1982).
17. K. Dalsgaard, Adjuvants, *Vet. Immunol. Immunpathol. 17*:145 (1987).
18. K. Lovgren and B. Morein, The ISCOM: An antigen delivery system with built-in adjuvant, *Mol. Immunol. 28*:285 (1991).
19. R. L. Hunter and B. Bennett, The adjuvant activity of nonionic block polymer surfactants, *J. Immunol. 133*:3167 (1984).
20. J. H. Nunberg, M. V. Doyle, S. M. York, and C. J. York, Interleukin 2 acts as an adjuvant to increase the potency of inactivated rabies virus vaccine, *Proc. Natl. Acad. Sci. USA*(June) *86*:4240 (1989).
21. J.H.L. Playfair, and J. B. De Sanza, Recombinant gamma-interferon is a potent adjuvant for a malaria vaccine in mice, *Clin. Exp. Immunol. 67*:5 (1987).

22. J. Freund, The mode of action of immunologic adjuvants, *Adv. Tuberc. Res.* 7:130 (1956).
23. A. T. Glenny, G.A.H. Buttle, and M. F. Stevens, Rate of disappearance of diphtheria toxoid injected into rabbits and guinea-pigs: toxoid precipitated with alum, *J. Pathol. Bacteriol. 34*:267 (1931).
24. M. A. Aprile and A. C. Wardlaw, Aluminum compounds as adjuvants for vaccines and toxoids in man: A review, *Can J. Pub. Health 57*:343 (1966).
25. P. M. Callahan, A. L. Shorter, and S. L. Hem, The importance of surface charge in the optimization of antigen–adjuvant interactions, *Pharm. Res. 8*:851 (1991).
26. R. Al-Shakhshir, F. Regnier, J. L. White, and S. L. Hem, Effect of protein adsorption on the surface charge characteristics of aluminium-containing adjuvants, *Vaccine 12*:472 (1994).
27. N. Goto, H. Kato, J. Maeyama, K. Eto, and S. Yoshihara, Studies on the toxicities of aluminium hydroxide and calcium phosphate as immunological adjuvants for vaccines, *Vaccine 11*:914 (1993).
28. B. H. Waksman, Adjuvants and immune regulation by lymphoid cells, *Springer Semin. Immunopathol*: 5 (1979).
29. M. T. Aguado and P.-H. Lambert, Controlled-release vaccines—biodegradable polylactide/polyglycolide (PL/PG) microspheres as antigen vehicles, *Immunbiology 184*:113 (1992).
30. B. Gander, C. Thomasin, H. P. Merkle, Y. Men, and G. Corradin. Pulsed tetanus toxoid release from PLGA-microspheres and its relevance for immunogenicity in mice, *Proc. Int. Symp. Controlled Release Bioact. Mater. 20*:65 (1993).
31. J. H. Eldridge, J. K. Staas, J. A. Meulbroek, T. R. Tice, and R. M. Gilley, Biodegradable and biocompatible poly(D,L-lactide-*co*-glycolide) microspheres as an adjuvant for staphylococcal enterotoxin B toxoid which enhances the level of toxin-neutralizing antibodies. *Infect. Immun. 59*:2978 (1991).
32. R. Koneberg, A. Hilbert, and T. Kissel, Parameters affecting the stability of tetanus toxoid, *Eur. J. Pharm. Biopharm. 40*:545 (1994).
33. W. R. Liu, R. Langer, and A. M. Klibanov, Moisture induced aggregation of lyophilized proteins in the solid state, Biotechnol. Bioeng. *37*:177 (1991).
34. D. T. O'Hagan, D. Rahman, J. P. McGee, J. Jeffery, M. C. Davies, and S. S. Davis, Biodegradable microparticles as controlled release antigen delivery systems, *Immunology 73*:239 (1991).
35. R. V. Nellore, P. G. Pande, D. Young, and H. R. Bhagat, Evaluation of biodegradable microspheres as vaccine adjuvant for hepatitis B surface antigen, *J. Parent. Sci. Technol. 46*:176 (1992).
36. Z. Moldoveanu, M. Novak, W. Huang, R. M. Gilley, J. K. Staas, D. Schafer, R. W. Compans, and J. Mestecky, Oral immunization with influenza virus in biodegradable microspheres, *J. Infect. Dis. 167*:84 (1993).
37. G. P. Talwar, O. Singh, R. Pal, N. Chatterjee, S. N. Upadhyay, C. Kaushic, S. Garg, R. Kaur, M. Singh, S. Chandrasekhar, and A. Gupta, A birth control vaccine is on the horizon for family planning, *Ann. Med. 25*:207 (1993).
38. R. Langer, Polymers for the sustained release of macromolecules: Their use in an single-step method of immunisation, Methods Enzymol.: *73*:57 (1981).

39. J. Kreuter and P. Speiser, New adjuvants on a polymethylmethacrylate base, *Infect. Immun. 13*:204 (1976).
40. M. Chasin and R. Langer, *Biodegradable Polymers as Drug Delivery Systems*, Marcel Dekker, New York, 1990.
41. D. H. Lewis, Controlled release of bioactive agents from lactide/glycolide polymers, *Biodegradable Polymers as Drug Delivery Systems* (M. Chasin and R. Langer, eds.), Marcel Dekker, New York, 1990, pp. 1–42.
42. M. Yamamoto, H. Okada, Y. Ogawa, and T. Miyagawa, EP 202 065, 06. 05, 1986.
43. H. Okada, Y. Ogawa, and T. Yashiki, EP 145240, 02. 11, 1984.
44. H. Okada, Y. Inoue, and Y. Ogawa, EP 0 442 671, 11.02, 1991.
45. Y. Ogawa, M. Yamamoto, H. Okada, T. Yashiki, and T. Shimamoto, A new technique to efficiently entrap leuprorelin acetate into microcapsules of polylactide acid or copoly(lactide/glycolide)acid. *Chem. Pharm. Bull. 36*:1095 (1988).
46. R. A. Miller, J. M. Brady, and D. E. Cutright, Degradation rates of oral resorbable implants (polylactates and polyglycolates): Rate modification with changes in PLA/PGA copolymer ratios, *J. Biomed. Mater. Res. 11*:711 (1977).
47. G. E. Visscher, R. L. Robinson, H. W. Maulding, J. W. Fong, J. E. Pearson, and G. J. Argentieri, Biodegradation of and tissue reaction to 50:50 poly(DL-lactide-*co*-glycolide) microspheres, *J. Biomed. Mater. Res. 19*:349 (1985).
48. T. R. Tice and R. M. Gilley, Preparation of injectable controlled-release microcapsules by a solvent-evaporation process. *J. Controlled Release 2*:343 (1985).
49. Li. Youxin and T. Kissel, Synthesis and properties of biodegradable ABA triblock copolymers consisting of poly(L-lactic acid) or poly (L-lactic–glycolic acid) A-blocks attached to central poly(oxyethylene) B-blocks, *J. Controlled Release 27*:247 (1993).
50. T. Kissel, Y. X. Li, C. Volland, and S. Görich, Release properties of macromolecules from microspheres of an ABA triblock copolymer consisting of poly(lactic-*co*-glycolic acid) and polyoxyethylene, *Proc. Int. Symp. Controlled Release Bioact. Mater. 21*:286 (1994).
51. J. Hanes, M. Chiba, and R. Langer, degradation and erosion of tyrosin-containing polyanhydrides: implications for vaccine delivery. Proc. Int. Symp. Controlled Release Bioact. Mater. *21*:44 (1994).
52. M. Chiba, J. Hanes, and R. Langer, Polyanhydride microspheres as potential vaccine delivery systems, Proc. Int. Symp. Controlled Release Bioact. Mater. *21*:877 (1994).
53. A. C. Allison and G. Gregoriadis, Liposomes as immunological adjuvants, *Nature 252*:252 (1974).
54. G. Gregoriadis, D. Davis, N. Garcon, L. Tan, V. Weissig, and Q. Xiao, The immunoadjuvant action of liposomes, *Liposomes in the Therapy of Infectious Diseases and Cancer*, Marcel Dekker, New York, 1989, p. 35.
55. I. Preis, and R. S. Langer, A single-step immunization by sustained antigen release, *J. Immunol. Methods: 28*:193 (1979).
56. Bungenberg, de Jong, and Kaas, *Biochem. Z. 232*:338 (1931).
57. T. Kissel, A. Rummelt, and W. Schmutz, Technical aspects of microspheres for

depot injection, Palodel LAR—a once-a-month delivery system for bromocriptine, Triangle *29*:205 (1990).

58. T. Kissel, Z. Brich, S. Bantle, I. Lancranjan, F. Nimmerfall, and P. Vit, Parenteral depot-systems an the basis of biodegradable polyesters, *J. Controlled Release 16*:27 (1991).
59. K. Masters, *Handbook of Spray-Drying*, 3rd ed., John Wiley & Sons, New York, 1979.
60. H. Okada, EP 145249, 1984.
61. I. Deris, Fortschritte in der Therapie der Endometriose und des Uterus myomatosus, *Gynäkologe 24*:6 (1991).
62. D. Bodmer, J. W. Fong, T. Kissel, H. V. Maulding, O. Nagele, and J. E. Pearson, DE 4021517 A1, 06.07.90.
63. D. Bodmer, T. Kissel, and E. Traechslin, Factors influencing the release of peptides and proteins from biodegradable parenteral depot systems, *J. Controlled Release 21*:129 (1992).
64. H. Jeffery, S. S. Davis, and D. T. O'Hagan, The preparation and characterization of poly(lactide-*co*-glycolide) microparticles: I. Oil-in-water emulsion solvent evaporation, *Int. J. Pharm. 77*:169 (1991).
65. H. Jeffery, S. S. Davis, and D. T. O'Hagan, The preparation and characterization of poly(lactide-*co*-glycolide) microparticles. II. The entrapment of a model protein using a (water-in-oil)-in-water emulsion solvent evaporation technique, *Pharm. Res. 10*:362 (1993).
66. S. Cohen, T. Yoshioka, M. Lucarelli, L. H. Hwang, and R. Langer, Controlled delivery systems for proteins based on poly(lactic/glycolic acid) microspheres, *Pharm. Res. 8*:713 (1991).
67. H. T. Wang, E. Schmitt, D. R. Flanagan, and R. J. Linhardt, Influence of formulation methods on the in vitro controlled release of protein from poly(ester) microspheres, *J. Controlled Release 17*:23 (1991).
68. H. K. Sah, R. Toddywala, and Y. W. Chien, The influence of microcapsule formulations on the controlled release of a protein, *J. Controlled Release 30*:201 (1994).
69. S. Görich and T. Kissel, Influence of different stabilizers on the properties of microspheres prepared using a modified W/O/W-emulsion method, *Eur. J. Pharm. Biopharm. 40*:105 (1994).
70. V. J. Csernus, B. Szende, and A. V. Schally, Release of peptides from sustained delivery systems (microcapsules and microparticles) in vivo, *Int. J. Peptide Protein Rev. 35*:557 (1990).
71. T. W. Redding, A. V. Schally, T. R. Tice, and W. E. Meyers, Long-acting delivery systems for peptides: Inhibition of rat prostate tumors by controlled release of D-Trp6 luteinizing hormone-releasing hormone injectable microcapsules, *Proc. Natl. Acad. Sci. USA 81*:5845 (1984).
72. L. M. Sanders, J. S. Kent, G. I. McRae, B. H. Vickery, T. R. Tice, and D. H. Lewis, Controlled release of a luteinizing hormone-releasing hormone analogue from poly(D,L-lactide-*co*-glyclide) microspheres, *J. Pharm. Sci. 73*:1294 (1984).
73. J. M. Ruiz, B. Tissier, and J. P. Benoit, Microencapsulation of peptide: A study

of the phase separation of poly(D,L-lactic acid-*co*-glycolide acid) copolymers 50/ 50 by silicone oil, *Int. J. Pharm. 49*:69 (1989).

74. J. S. Kent, L. M. Sanders, D. H. Lewis and T. R. Tice, EP 052 510, 17.11, 1981.
75. I. Esparza and T. Kissel, Parameters affecting the immunogenicity of micrencapsulated tetanus toxoid, *Vaccine 10*:714 (1992).
76. A. M. Hazrati, D. H. Lewis, T. J. Atkins, R. C. Stohrer, and L. Meyer, In vivo studies of controlled-release tetanus vaccine. *Proc. Int. Symp. Controlled Release Bioact. Mater. 19*:114 (1992).
77. M. J. Alonso, R. K. Gupta, C. Min, G. R. Siber, and R. Langer, Biodegradable microspheres as controlled-release tetanus toxoid delivery systems, *Vaccine 12*:299 (1994).
78. S. A. Charman, K. L. Mason, and W. N. Charman, Techniques for assessing the effects of pharmaceutical aggregation excipients on the aggregation of porcine growth hormone, *Pharm. Res. 10*:954 (1993).
79. L. C. Gu, E. A. Erdös, H.-S. Chiang, T. Calderwood, K. Tsai, G. C. Visor, J. Duffy, W.-C. Hsu, and L. C. Foster, Stability of interleukin 1 (beta) (IL-1beta) in aqueous solution: Analytical methods, kinetics, products and solution formulation implications, *Pharm. Res. 8*:485 (1991).
80. J. Brange, S. Havelund, and P. Hougaard, Chemical stability of insulin. 2. Formation of higher molecular weight transformation products during storage of pharmaceutical preparations, *Pharm. Res. 9*:727 (1992).
81. W. R. Porter, H. Staack, K. Brandt, and M. C. Manning, Thermal stability of low molecular weight urokinase during heat treatment. I. Effects of protein concentration, pH and ionic strength, *Thromb. Res. 71*:265 (1993).
82. M. J. Pikal, K. M. Dellerman, M. L. Roy, and R. M. Riggin, The effects of formulation variables on the stability of freeze-dried human growth hormone, *Pharm. Res. 8*:427 (1991).
83. M. S. Hora, R. K. Rana, and F. W. Smith, Lyophilized formulations of recombinant tumor necrosis factor, *Pharm. Res. 9*:33 (1992).
84. S. Yoshioka, Y. Aso, K. Izutsu, and T. Terao, Aggregates formed during storage of beta-galactosidase in solution and in the freeze-dried state, *Pharm. Res. 10*:687 (1993).
85. A. W. Ford, and Z. Allahiary, The adverse effect of glycation of human serum albumin on its preserving activity in the freeze-drying and accelerated degradation of alkaline phosphatase, *J. Pharm. Pharmacol. 45*:900 (1993).
86. G. D. J. Adams and L. I. Irons, Some implications of structural collapse during freeze-drying using *Erwinia cratovora* L-asparaginase as a model, *J. Chem. Technol. Biotechnol. 58*:71 (1993).
87. G. M. Jordan, S. Yoshioka, and T. Terao, The aggregation of bovine serum albumin in solution and in the solid state, *J. Pharm. Pharmacol. 46*:182 (1994).
88. B. S. Chang, C. S. Randall, and Y. S. Lee, Stabilization of lyophilized porcine pancreatic elastase, *Pharm. Res. 10*:1478 (1993).
89. V. Sluzky, J. A. Tamada, A. M. Klibanov, and R. Langer, Kinetics of insulin aggregation in aqueous solutions upon agitation in the presence of hydrophobic surfaces, *Proc. Natl. Acad. Sci. USA 88*:9377 (1991).

90. S. Yoshioka, Y. Aso, K.-I. Izutsu, and T. Terao, The effect of salts on the stability of beta-galctosidase in aqueous solution as related to the water mobility, *Pharm. Res. 10*:1484 (1993).

91. M. Mumenthaler, C. C. Hsu, and R. Pearlman, Feasibility study on spray-drying protein pharmaceuticals: Recombinant human growth hormone and tissue-type plasminogen activator, *Pharm. Res. 11*:12 (1994).

92. M. S. Hora, R. K. Rana, J. H. Nunberg, T. R. Tice, R. M. Gilley, and M. E. Hudson, Release of human serum albumin from poly(lactide-*co*-glycolide) microspheres, *Pharm. Res. 7*:1190 (1990).

93. M. M. Bradford, A rapid and sensitive method for the quantitation of microgram quantities of protein utilizing the principle of protein–dye binding, Anal. Biochem. *72*:248 (1976).

94. R. K. Gupta, M. J. Alonso, R. Langer, and G. R. Siber, Development of a single dose tetanus toxoid based on controlled release from biodegradable and biocompatible polyester microspheres. *Vaccines 93*, Cold Spring Harbor Laboratory Press, Cold Spring Harbor, NY, 1993, p. 391.

95. A. K. Hilbert, Analytische Charakterisierung und Stabilitätsprüfung von Tetanus Toxoiden, *Hinblick auf die Herstellung Parenteraler Depotformen*, Diplomarbeit, Marburg, 1993.

96. Y. Tabata, S. Gupta, and R. Langer, Controlled delivery systems for proteins using polyanhydride microspheres, *Pharm. Res. 10*:391 (1993).

97. M. Singh, A. Singh, and G. P. Talwar, Controlled delivery of diphtheria toxoid using biodegradable poly(DL-lactide) microcapsules, *Pharm. Res. 8*:958 (1991).

98. M. J. Alonso, S. Cohen, T. G. Park, R. K. Gupta, G. R. Siber, and R. Langer, Determinants of release rate of tetanus vaccine from polyester microspheres, *Pharm. Res. 10*:945 (1993).

99. M. Singh, O. Singh, A. Singh, and G. P. Talwar, Immunogenicity studies on diphtheria toxoid loaded biodegradable microspheres, *Int. J. Pharm. 85*:R5 (1992).

100. Y. Tabata, Y. Takebayashi, T. Ueda, and Y. Ikada, A formulation method using D,L-lactic acid oligomer for protein release with reduced initial burst, *J. Controlled Release 23*:55 (1993).

101. R. Brown, C. L. Wei, and R. Langer, In vivo and In vitro release of macromolecules from polymeric drug delivery systems, *J. Pharm. Sci. 72*:1181 (1983).

102. R. Langer, Polymeric delivery systems for the controlled drug release, *Chem. Eng. Commun. 6*:1 (1980).

103. R. A. Siegel and R. S. Langer, Controlled release of polypeptides and other macromolecules, *Pharm. Res. 1*:1 (1984).

104. S. S. Shah, Y. Cha, and C. G. Pitt, Poly(glycolid acid-*co*-DL-lactic acid): Diffusion or degradation controlled drug delivery? J. Controlled Release *18*:261 (1992).

105. R. S. Raghuvanshi, M. Singh, and G. P. Talwar, Biodegradable delivery system for single step immunization with tetanus toxoid, *Int. J. Pharm. 93*:R1 (1993).

106. M. J. Alonso, S. Cohen, T. W. Park, R. K. Gupta, G. R. Siber, and R. Langer, Controlled release of tetanus toxoid from poly(lactic/glycolic acid) microspheres, *Proc. Int. Symp. Controlled Release Bioact. Mater. 19*:122 (1992).

107. J. M. Anderson, In vivo biocompatibility of implantable delivery systems and biomaterials, *Eur. J. Pharmacol. Biopharm. 10*:1 (1994).

108. D. P. Stites and A. I. Terr, eds., *Basic and Clinical Immunology*, 2nd ed. Appleton & Lange, Norwalk CT, 1991.

109. I. M. Roitt, J. Brostoff, and D. K. Male, *Kurzes Lehrbuch der Immunologie*, Georg Thieme Verlag, Stuttgart, 1991.

110. D. T. O'Hagan, H. Jeffery, and S. S. Davis, Long-term antibody responses in mice following subcutaneous immunization with ovalbumin entrapped in biodegradable microparticles, *Vaccine 11*:965 (1993).

111. D. T. O'Hagan, J. Jeffery, M. J. J. Roberts, J. P. McGee, and S. S. Davis, Controlled release microparticles for vaccine development, *Vaccine 9*:768 (1991).

112. A. M. Hazrati, D. H. Lewis, T. J. Atkins, R. C. Stohrer, J. E. Little, and L. Meyer, Studies of controlled delivery tetanus vaccine in mice, *Proc. Int. Symp. Controlled Release Bioact. Mater. 20*:67 (1993).

113. J. H. Eldridge, J. K. Staas, J. A. Meulbroek, J. R. McGhee, T. R. Tice, and R. M. Gilley, Biodegradable microspheres as a vaccine delivery system, *Mol. Immunol. 28*:287 (1991).

114. A. J. Almeida, H. O. Alpar, and M. R. W. Brown, Immune response to nasal delivery of antigenically intact tetanus toxoid associated with poly(L-lactic acid) microspheres in rats, rabbits and guinea-pigs, *J. Pharm. Pharmacol. 45*:198 (1993).

115. R. M. Gilley, J. K. Staas, T. R. Tice, J. D. Morgan, and J. H. Eldridge, Microencapsulation and its application to vaccine development, *Proc. Int. Symp. Controlled Release Bioact. Mater. 19*:110 (1992).

116. D. T. O'Hagan, H. Jeffery, and S. S. Davis, Poly(lactide-*co*-glycolide) microparticles as controlled release vaccines, *Proc. Int. Symp. Controlled Release Bioact. Mater. 20*:59 (1993).

117. Y. Tabata and Y. Ikada, Effect of the size and surface charge of polymer microspheres on their phagocytosis by macrophage, *Biomaterials 9*:356 (1988).

118. J. Kreuter, U. Berg, E. Liehl, M. Soliva, and P. P. Speiser, Influence of the particle size on the adjuvant effect of particulate polymeric adjuvants, Vaccine *4*:125 (1986).

119. J. Kreuter, E. Liehl, U. Berg, M. Soliva, and P. P. Speiser, Influence of hydrophobicity on the adjuvant effect of particulate polymeric adjuvants, *Vaccine 6*:253 (1988).

120. J. H. Eldridge, Pulsatile delivery of vaccines, *Pulsatile Drug Delivery*—Current Applications and Future Trends, Third APV-CRS Joint Workshop, Königswinter/Bonn, Germany, May 20–22, 1992.

121. J. Singh and A. Bhowmick, In vitro release of diphtheria toxoid vaccine from biodegradable microcapsules, *Proc. Int. Symp. Controlled Release Bioact. Mater. 20*:396 (1993).

122. A. J. Almeida, H. O. Alpar, and M. R. W. Brown, Formulation studies on particulate carriers as immunological adjuvants, *Proc. Int. Symp. Controlled Release Bioact. Mater. 20*:390 (1993).

123. Y. Men, G. Corradin, C. Thomasin, H. P. Merkle, and B. Gander, Immunopotentation of a synthetic antigen by incorporation into biodegradable microspheres, *Proc. Int. Symp. Controlled Release Bioact. Mater.* *21*:50 (1994).

124. J. P. McGee, W. C. Koff, C. Y. Wang, P. Potts, R. C. Kennedy, and D. T. O'Hagan, Controlled release microparticles with an entrapped branched V3 peptide from HIV-1: Assessment of immunogenicity in baboons, *Proc. Int. Symp. Controlled Release Bioact. Mater.* *21*:871 (1994).

125. M. Singh, D. Thassu, and G. P. Talwar, Vaccine delivery systems—effect of microsphere size on the immune response, *Proc. Int. Symp. Controlled Release Bioact. Mater.* *21*:881 (1994).

126. R. S. Raghuvanshi, A. Misra, and G. P. Talwar, Strategies for improving immune response, *Proc. Int. Symp. Controlled Release Bioact. Mater.* *21*:883 (1994).

3

Preparation of Microparticulates Using Supercritical Fluids

Barbara L. Knutson and Pablo G. Debenedetti

Princeton University
Princeton, New Jersey

Jean W. Tom

Merck Research Laboratories
Merck & Co., Inc.
Rahway, New Jersey

I. INTRODUCTION

The use of supercritical fluids as media for the formation of microparticles for therapeutic applications is a very recent development. Interest in this new field is driven by the important advantages that supercritical fluids offer over conventional microparticulate formation routes: namely, the mildness of the operating temperatures, the purity of the products, and the avoidance of organic solvents. In this chapter we summarize work to date on the application of supercritical fluids to the production of microparticles of medical interest. Although the potential benefits of using supercritical fluids are numerous, fundamental understanding of supercritical particle formation is lacking. Hence, research in this area is still exploratory.

A *supercritical fluid* is any substance the temperature and pressure of which are simultaneously higher than the critical point values. In practice, the term is reserved for the description of fluids in the relatively narrow range of conditions $1 \leq T/T_c \leq 1.1$, $1 \leq P/P_c \leq 2$, where subscript c denotes the critical point. It is within this region that most of the changes in thermophysical properties associated with the transformation of a dilute gas into a dense fluid occur. Hence, in the supercritical region, thermophysical properties exhibit very high

Table 1 Critical Points of Some Common Supercritical Solvents

Substance	Critical Temperature (°C)	Critical Pressure (bar)
Carbon dioxide	31.1	73.8
Ethane	32.3	48.8
Ethylene	9.3	50.4
Water	374.2	220.5
Sulfur hexafluoride	45.5	37.6
Chlorotrifluoromethane	28.8	39.2
Fluoroform	26	48.8

rates of change with respect to pressure and temperature. Table 1 lists the critical temperatures and pressures of some common supercritical solvents.

Not all thermophysical properties change at the same rate; consequently, supercritical fluids have unique combinations of properties [1]. The most important property of a supercritical fluid is its large compressibility. This is because all fluids are infinitely compressible at the critical point. Figure 1 [2] shows a schematic phase diagram of a fluid, including vapor–liquid coexistence, the critical point (CP), and the supercritical region. The critical isotherm $(T_r = T/T_c = 1)$ has vertical slope at the critical point, which means that the rate of change of density with respect to pressure is infinite. In contrast, the portions of isotherms corresponding to liquid states have almost horizontal slopes, be-

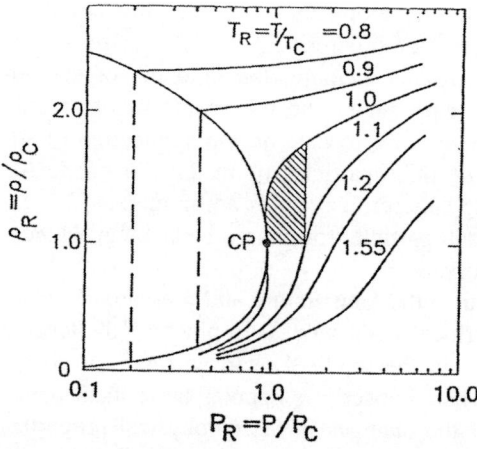

Fig. 1 Schematic density–pressure phase diagram of a pure fluid in the relative vicinity of the critical point (CP). (From Ref. 2.)

Table 2 Physical Properties of Gases, Liquids, and Supercritical Fluids

	Density (kg/m^3)	Viscosity (Ns/m^2)	Diffusivity (cm^2/s)
Gases	1	10^{-5}	10^{-1}
Liquids	10^3	10^{-3}	10^{-5}
Supercritical fluids	700	10^{-4}	10^{-4}

cause liquids have very small compressibilities. In the vicinity of the critical point, then, fluids are arbitrarily compressible: small changes in pressure cause large changes in density. Therefore, supercritical fluids are typically hundreds of times denser than gases at ambient conditions, but they are arbitrarily more compressible.

Table 2 compares typical values for the density, viscosity, and diffusivity of gases, liquids, and supercritical fluids (for diffusivities, the relevant quantity is the diffusivity of small molecules in each fluid). Supercritical fluids can be almost liquid-like in density, but they have viscosities that are intermediate between gas- and liquid-like. Different applications of supercritical fluids exploit different thermophysical properties. For example, in mass transfer applications, it is the combination of high density, low viscosity, and high diffusivity that is of interest. For particle formation, however, it is the combination of liquid-like density and large compressibility that is of interest. The large density means that supercritical fluids have solvent powers that are comparable with those of the same substances in the liquid state [3]. The high compressibility means that this solvent power is continuously adjustable between gas- and liquid-like extremes with small changes of pressure. Thus, with supercritical fluids, pressure is a fundamental degree of freedom that is absent with liquids.

There are two routes to particle formation with supercritical fluids: the rapid expansion of supercritical solutions (RESS), and the supercritical antisolvent (SAS) route [4]. In RESS [5], the solute of interest is dissolved in a supercritical fluid at high pressure and precipitated by rapid decompression. The dissolution exploits the supercritical fluid's liquid-like solvent power; the decompression, its compressibility. Expansion can be total (i.e., down to atmospheric pressure) or partial, and it is done by flowing the supercritical solution through a capillary, nozzle, or calibrated restriction. Because of the large compressibilities involved, small pressure drops cause large density changes. These are accompanied by correspondingly large drops in solvent power; hence, the dissolved solute precipitates. This is illustrated in Fig. 2 [6], which shows the solubility of acridine in carbon dioxide at three supercritical temperatures. Note the sensitivity of the solubility to small changes in pressure in the relative vicinity of the solvent's critical point ($T_c = 304$ K; $P_c = 73.8$ bar). Expansions

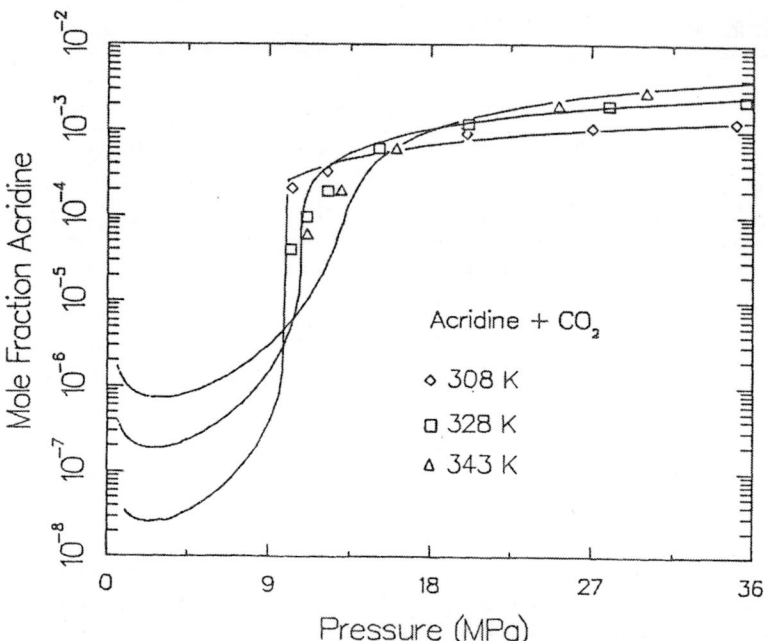

Fig. 2 Equilibrium mole fraction of acridine (y_1) in supercritical carbon dioxide ($T_c = 31.1°C$; $P_c = 73.8$ bar) as a function of pressure, at three supercritical temperatures. Symbols are experimental measurements, lines are theoretical calculations. (From Ref. 6.)

during nozzle or capillary flow in RESS are very fast (typically $< 10^{-5}$ s); hence, large supersaturations can be attained [7]. Furthermore, pressure drop is a mechanical perturbation that travels at the speed of sound, giving rise to uniform conditions within the nucleating fluid. Thus, RESS is capable of producing very small and uniform particles. Because the supercritical fluid is a dilute gas after expansion, the solid product is obtained dry, solvent-free, and requires no downstream purification.

The use of RESS as an alternative to comminution for the production of ultrafine pharmaceutical powders is very promising, especially for specialized applications, such as pulmonary delivery of aerosolized drugs. Also of interest is the coprecipitation of drugs and low molecular weight bioerodible polymers by RESS to form drug-loaded microspheres and microparticles for controlled release. Work on this application of RESS is still at a preliminary stage.

The chief limitation of RESS is the very low solvent power of common supercritical solvents toward potentially useful solutes, such as proteins and high polymers. The SAS method overcomes this limitation by using the supercritical fluid as an antisolvent. The solute of interest is dissolved in a liquid, generally an organic solvent, and a supercritical fluid that has a low solvent power with respect to the solute, but miscible with the liquid, is added to precipitate the solid. When this process is operated in batch mode, the supercritical fluid is dissolved in an excess of the liquid phase. This can cause appreciable volumetric expansion (Fig. 3), and the resulting decrease of the liquid's cohesive energy density causes the solute to precipitate [8,9]. When the process is operated in continuous mode, the liquid and supercritical phases are fed continuously into a precipitator. Small, submillimeter droplets are formed by flowing the liquid solution through a calibrated restriction. The pressure drop across this restriction is not large, as in RESS: here, the pressure drop is used merely to form the small droplets. The droplets are contacted with an excess of supercritical fluid. Two processes then occur: dissolution of the supercritical fluid in the liquid droplet and, under suitable-operating conditions, evaporation of the liquid into the supercritical fluid. This causes the precipitation of the solute [9–11]. The most important biomedical application of SAS to date involves the formation of biologically active protein powders. Other potential applications include the formation of bioerodible microspheres and the encapsulation of drugs in polymer matrices by SAS coprecipitation.

Of the two routes to particle formation with supercritical fluids, RESS is used to form fine particles of substances that are appreciably soluble in a supercritical solvent; SAS is used for sparingly soluble materials. Together, these techniques allow the processing of a wide range of materials into solid phases, with useful, often unique properties and morphologies.

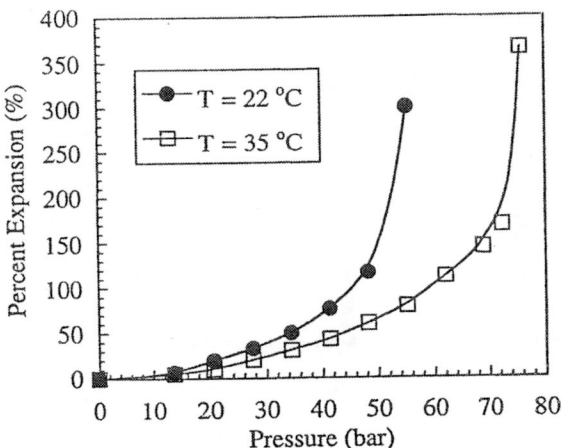

Fig. 3 Volumetric expansion (percentage relative change in volume) of dimethyl sulfoxide (DMSO) in the presence of compressed carbon dioxide. (From Ref. 9.)

II. RAPID EXPANSION OF SUPERCRITICAL SOLUTIONS

A. Experimental Techniques

Experimentally, RESS consists of extracting (or dissolving) solutes in a supercritical solvent and then expanding (or depressurizing) the solution by flowing through an expansion device. The extraction, precipitation, and particle collection are usually continuous in RESS, but can also be done in a batch mode. Figure 4 shows a typical schematic flow sheet of a laboratory-scale RESS apparatus. The equipment consists of two main units, extraction (or dissolution) and precipitation. The pure solvent is pumped to the desired pressure and preheated to extraction temperature by circulating through a preheater (consisting either of a coil or a vessel) in a controlled temperature environment. The supercritical fluid is then passed through an extraction unit. This unit is typically a stainless steel column (e.g., 1″ o.d. × 12″ L) [12,13], or a high-pressure autoclave [14]. At such laboratory scales, up to 100 g of solute can be packed with glass wool or beads in an extraction column. A premixed solution can be prepared if the solvent is a liquid at ambient conditions and the solute's solubility does not

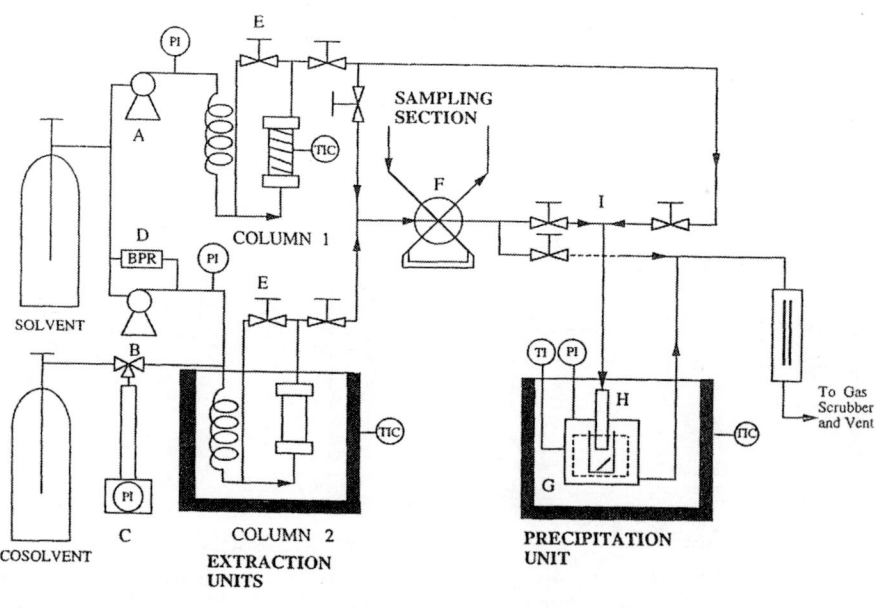

Fig. 4 Schematic of RESS apparatus.

decrease to the point of precipitation upon compression and heating to supercritical conditions [14]. Typical scales of operation are 1–4 mol h^{-1} of solvent and up to 1 g h^{-1} of solid product, the latter figure being a function of solubility.

Several methods for achieving the simultaneous precipitation of multiple solutes have been investigated. A solute can be added to a liquid solvent at ambient conditions and the resulting solution can then be used to extract another solute [15]. Extraction of solutes from separate extraction units and subsequent mixing of the supercritical solutions before expansion allows control over the composition of the resulting powders [15,16]. Packing the extraction unit with multiple solutes is an alternative to multiple extraction columns [12]. However, the precipitate's composition is then set by equilibrium solubilities at extraction.

The distinguishing feature of the apparatus shown in Fig. 4 is the possibility of independently controlling the extraction and precipitation pressures and temperatures, as well as the solute's concentration before precipitation. Temperature control is an important aspect of experimental design. Because precipitation can occur with slight changes in temperature, it is important to maintain the lines connecting the extraction and precipitation units at the same temperature as the extraction unit. Use of low-critical–temperature solvents, such as carbon dioxide (31.1°C), permits the use of constant-temperature liquid baths (water, silicone oil) and temperature-controlled heating tape. For high-critical–temperature solvents, temperature control is achieved by placing the various units and their lines inside a circulating air oven [17]. Although the extraction unit's pressure and temperature determine the solute concentration, this quantity can also be varied by diluting the saturated solution exiting the extraction unit with pure solvent delivered by another pump.

In the precipitation unit, the supercritical solution is expanded across a capillary or an orifice. Some of the expansion configurations tested have been 2- to 15-mm–length, 25- to 60-μm–i.d. tubing [15–17]. Laser-drilled pinpoint orifices of 15–30 μm in diameter and 0.25 mm in length [13,17,18] have also been investigated. The expansion device's temperature is controlled with a variable-resistance cable heater. Heating of the expansion device up to an adequate preexpansion temperature is necessary to prevent phase changes in the solvent (i.e., condensation or freezing) on expansion. Particle collection by impacting onto a plate [15], into a glass beaker [18], or dispersion into an aqueous colloidal gel [13] has been employed.

B. Size Reduction of Pharmaceuticals

The particle size and uniformity of many therapeutic drugs is vital in pharmaceutical formulations, for which dosages can be as low as milligrams. Small-particle sizes are often needed to meet high surface area, bioavailability, and dissolution requirements of pharmaceuticals [19]. Small particles are also necessary in pharmaceutical formulations of multiple drugs and compounds, to maximize the uniformity of the mixture. Drugs are often prone to degradation during processing owing to the presence of heat or oxygen, and the production of high-purity batches requires the use of mild and inert process conditions. Conventional methods of size reduction, such as milling and grinding, adversely affect the crystallinity and chemical stability of pharmaceuticals [19]. In applications for which fine powders are required, RESS with low-critical-temperature solvents allows processing at lower temperatures than conventional particle reduction methods (milling or crystallization) while, at the same time, providing the opportunity to form very fine and monodisperse particles. In drugs delivered as aerosols to the lungs, uniform and small particles ($1-3$ μm) are needed to ensure effective deposition. Larger particles tend to fall out of the gas stream before reaching the lungs, and smaller, submicron particles tend to be difficult to deposit [20]. The potential of RESS for making intimate mixtures can be applied to pharmaceutical formulations incorporating multiple drugs (for treatment of different symptoms). Lastly, supercritical fluids are easily separated from the solute after precipitation, because the supercritical fluid becomes a low-density gas after expansion, leaving a pure solid product. In contrast, conventional crystallization requires liquid solvents, and separation of the precipitated product from the liquid requires centrifugation or filtration, and extensive drying to remove trace quantities of the solvent. Furthermore, a solvent such as carbon dioxide is nontoxic, nonflammable, inexpensive, and has a low critical temperature, which allows processing at mild temperatures.

The first comprehensive study of RESS was reported by Krukonis [21], who obtained microparticles of a wide variety of materials, including pharmaceuticals, biologicals, and polymers. Although Krukonis focused on using RESS as an alternative to comminution, his study initiated an active area of research aimed at exploring further applications, as well as understanding the process quantitatively.

Among the compounds examined by Krukonis [21] was β-estradiol, a steroid. Krukonis extracted β-estradiol, using carbon dioxide at 55°C and 345 bar, and expanding the supercritical solution to atmospheric and ambient conditions across a valve. Remarkably small particles were produced by RESS. The parent particles showed sizes ranging from a few microns to several hundred microns, whereas the precipitated particles constituted a uniform submicron powder.

Larson and King [22] investigated the extraction and precipitation of pharmaceuticals. The solubility of lovastatin (a cholesterol-reducing drug) in supercritical carbon dioxide was enhanced from 0.04 wt% to 0.45 wt% at 379 bar and 40°C when CO_2 was saturated with methanol. The starting crystalline material contained $100-200$ μm needles and $50-100$ μm fragments; the precipitate contained both needles and irregular-shaped particles, ranging in size from 10 to 50 μm. Although X-ray diffraction of the nucleated material showed retention of crystallinity, small variations in the intensity of the diffraction pattern indicated the possibility of some degraded or amorphous material. Later work with lovastatin and CO_2–methanol mixed solvent appeared to show that lovastatin was being recrystallized from methanol, which condensed following expansion [23]. Smaller particles were produced by using pure CO_2 as solvent [23]. Extraction of milled lovastatin (10–50 μm) at 379 bar, preexpansion temperatures ranging from 105° to 135°C and, expansion to 2 bar resulted in aggregates of 0.1–1 μm. Sonication for 3 min in heptane broke up the aggregates, and particle size analysis revealed average particle sizes of 126–255 nm. The crystallinity of these lovastatin powders of loosely aggregated submicron-sized units was retained. Mohamed et al. [23] examined the effects of process conditions on the powder characteristics of lovastatin. They examined the three process variables (solute concentration, pre- and postexpansion temperature) previously found to have significant effects on the morphology and size of a model organic compound, naphthalene [18]. In contrast, the size and morphology of lovastatin particles were found to be insensitive to changes in these process conditions.

Loth and Hemgesberg [24] studied RESS of phenacetin, an analgesic. They compared phenacetin powders produced from RESS with pure carbon dioxide and pure trifluoromethane ($T_c = 26.2$°C; $P_c = 48.6$ bar), with jet-milled material. Melting points and differential-scanning calorimetry results showed no difference among the three powders. Phenacetin powders produced by RESS had greater surface areas than the jet-milled phenacetin, based on their dissolution rate, permeability measurements, and laser light diffraction evaluation. Both the jet-milled and the precipitated powders had particles ranging in size from 3 to 10 μm.

Chang and Randolph [13] examined RESS of β-carotene, a naturally occurring pigment and vitamin A precursor. They precipitated β-carotene by free expansion and by expansion into gelatin solution, after extraction with ethylene ($T_c = 9.3$°C; $P_c = 50.4$ bar) at 306 bar and 70°C. Expansion into a gel solution decreased the amount of agglomeration, but introduced a separation step. The mean particle size of the feed β-carotene, the precipitate from atmospheric expansion, and the precipitate from expansion into gelatin solution were 20.0, 1.0, and 0.3 μm, respectively. Chang and Randolph showed, both theoretically and experimentally, that it was possible to use cosolvents to enhance the solubility of β-carotene in the supercritical phase (ethylene with up to 1.4 mol% toluene) and maintain a single solvent phase after expansion. β-Carotene microparticles precipitated from the ethylene–toluene supercritical solution were submicron (0.5 μm).

In a RESS study with stigmasterol, Ohgaki et al. [25] investigated the growth of stigmasterol microparticles from supercritical CO_2 solutions. They obtained amorphous particles by using preexpansion pressure below 130 bars, and whisker-like crystals above that preexpansion pressure, at 373 K. The whisker-like crystals grew atop primary particles of 10 nm to form crystals of 3–4 μm in length and 0.5 μm in diameter. The size of the microparticles increased as collection time increased, indicating that the crystals were aggregates of primary particles.

The major application of RESS in the processing of pharmaceutical compounds has been as an alternative to comminution of sensitive molecules. The size of particles formed by RESS is smaller or equal to that of particles obtained with conventional size-reduction methods. In addition, RESS offers the ability to produce small particles and mixtures of pharmaceuticals in one processing step. The possibility of milder and more inert operating conditions with RESS makes this technique particularly attractive for pharmaceuticals, which must meet stringent federal regulations on purity. For low molecular weight organic compounds, processing with RESS is dependent only on finding an appropriate supercritical solvent. Because most of the work on solubility in supercritical fluids has been on organic compounds [26–28], there already exists considerable information. However, the typical solubility of organics in carbon dioxide is still much lower than in conventional organic solvents, so there are limitations in the processing of large volumes. In some cases, complex pharmaceutical compounds may show no detectable solubility in the supercritical fluid [22].

C. Polymer–Pharmaceutical Composite Microspheres and Microparticles

The production of polymeric microspheres and microparticles for controlled drug delivery is an area in which RESS offers real promise for improving on existing approaches. A wide variety of techniques exist for fabricating polymer microsphere matrices for controlled drug delivery [29–32]. One method is to dissolve both polymer and drug in a common solvent, which is subsequently evaporated [33]. Organic mixtures of solvent, drug, and polymer can be emulsified in an aqueous medium; removal of the solvent then leaves a drug-containing microsphere [34]. A technique known as phase separation involves addition of a nonsolvent to a single phase, containing polymer, drug, and solvent. The nonsolvent precipitates out microspheres containing polymer and drug, and the microspheres are subsequently hardened by addition of more nonsolvent [34]. Alternatively, the microspheres can be prepared first, and the drug subsequently imbibed into the matrix from a drug-containing solution [35]. All of these methods use organic solvents, which must be removed before in vivo use of the microspheres. Heating, which can affect the drug's stability, is often needed for solvent removal. Surfactants, often used to prevent agglomeration of the solid particles being formed, must also be removed before in vivo use. Emulsification in an aqueous medium can suffer from low drug incorporation when water-soluble drugs are used, because these are continuously leached from the matrix by the aqueous medium. In addition, microsphere batches produced by any of the aforementioned techniques generally possess a wide size distribution [34]. Thus, there are significant disadvantages associated with the methods currently being used to make polymer–drug microspheres. In particular, these methods introduce solvents and surfactants that are not biocompatible. In contrast, the microparticles formed with RESS are completely solvent-free because the solvent expands to a gas and is separated from the solid particle by gravity. Furthermore, with RESS, production of microspheres with a narrow size distribution can be realized [14].

In all techniques currently used to form polymer–drug microspheres, multiple-processing steps are involved. RESS has the ability to produce intimate mixtures of materials in a single processing step. By adjusting the concentration of drug and polymer in two supercritical fluid streams, intimate drug–polymer mixtures of desired composition are produced on expansion of the combined streams. With RESS, the production of polymer–drug microspheres involves dissolution of the polymer and drug, followed by expansion of the solution. However, many of the process variables affect the morphology of RESS powders in ways not yet fully understood. Thus, the morphology of the precipitate can be difficult to control and predict. In addition, the polymers used in controlled-delivery applications are not appreciably soluble in desirable supercritical fluids, such as CO_2. The solubility of low molecular weight poly(hydroxy acids) is typically less than 0.5 wt% [12,16]. Possible solutions to this problem are the use of cosolvents to enhance the solubility, or the processing of low molecular weight polymers, which are suitable for shorter-release periods. Such cosolvents tend to be organic solvents (i.e., acetone) or fluorohalocarbons (i.e., $CHClF_2$), both less desirable than CO_2 from an environmental perspective. Thus, the low throughputs and unsatisfactory control over particle morphology are the main technical challenges to be solved before RESS can be widely applied to the production of composite drug–polymer microspheres. The purity of the product, the mildness of the operating conditions, and the simplicity of the process suggest that further research in this area could lead to an important advances in the production of injectable, controlled-delivery devices. We now summarize work to date on the RESS processing of biodegradable polymers, and on the coprecipitation of such polymers and low molecular weight substances with RESS.

1. Biodegradable Polymers

One of the most widely studied classes of biodegradable polymers has been the poly(hydroxy acids): poly(L-lactic acid) (L-PLA), poly(D,L-lactic acid) (DL-

PLA), poly(glycolic acid) (PGA), and copolymers of these materials. These biocompatible and biodegradable polymers are hydrolyzed by body fluids to metabolic products (lactic acid or glycolic acid), and have been approved by the US Food and Drug Administration (FDA) for in vivo use as sutures and bone repair implants [36]. The use of these polymers for controlled drug-delivery applications has been reviewed in recent journal articles [36–39] and in several books and monographs on controlled drug delivery [30,40,41].

Tom and Debenedetti [12] found that commercial L-PLA (Mw = 5000–6000; Mw/Mn = 0.2, where Mw and Mn are the respective weight-average and number-average polymer molecular weight) was soluble in supercritical CO_2 (0.05 wt% polymer) and in CO_2 with 1 wt% of acetone (0.37 wt% polymer). After sufficient preconditioning in pure CO_2, the molecular weight of the extracted L-PLA was typically 2000–3000, and the solubility was typically less than 0.1 wt%. The range of extraction conditions investigated was 150–300 bar and 45°–60°C.

Tom and Debenedetti [12] obtained 10- to 25-μm microparticles, 10- to 20-μm microspheres, and dendrites several hundred microns in size. The extreme sensitivity of particle morphology to small changes in process conditions is typical of experiments using orifices, from which virtually all of the expansion takes place in an uncontrolled free jet.

In the same study, Tom and Debenedetti [12] found that commercial DL-PLA and PGA (Mw = 5000–6000; Mw/Mn = 2) were also soluble in supercritical CO_2. RESS processing of these polymers, using extraction conditions of 55°C and 180–200 bars, and preexpansion temperatures of 80°–85°C, produced microparticles smaller than 40 μm. The molecular weight of PGA extracted and precipitated was not examined as gel permeation chromatography (GPC) of these polymers requires the use of fluorinated organic solvents.

Although Tom and Debenedetti's study [12] suggests that a range of poly-(hydroxy acids) can be processed by RESS, it also shows that polymers behave quite differently from low molecular weight pharmaceuticals in RESS. Polymer polydispersity had an important effect on the evolution of the experiment over time, because the supercritical fluid fractionates the material in the extraction unit continuously. The extracted and precipitated poly(hydroxy acids) tended to be in the lower range (ca. 6000) of commercially available molecular weights (up to 100,000) for this class of materials.

The problems associated with polydispersity were addressed by Tom et al. [16] in a recent study, using monodisperse L-PLA (Mw = 10,000; Mw/Mn = 1.15), and CO_2–$CHClF_2$ mixtures as the supercritical fluid. After preconditioning, the solubility was less than 0.5 wt%, and it increased with the higher concentration of $CHClF_2$ (up to 55 wt% $CHClF_2$) in the solvent's mixture. Significant processing (150–600 total standard liters of solvent per gram polymer initially in the extraction column) was necessary to extract material with molecular weights similar to that of the starting polymer. Tom et al. [16] examined two types of devices: orifices (L/D < 10) and capillaries (L/D > 100).

In experiments with orifices (L/D = 9.4) at extraction conditions of 55°C and 165–230 bar, and preexpansion temperatures from 80° to 140°C, Tom et al. observed three types of morphology for RESS-processed L-PLA: microparticles, microspheres, and dendrites, as well as agglomerates of microparticles (Fig. 5). Here, we define *microparticles* as nonspherical objects of polyhedral shape, and *agglomerates* as clusters of microparticles without dendritic branching. The formation of dendrites occurred at conditions favoring high polymer solubility: high pressure or high $CHClF_2$ content in the solvent. Microspheres and microparticles were formed over a range of conditions: at all pressures examined, at preexpansion temperatures of 100°–120°C, and at CO_2 contents greater than 75 wt%. Control over particle morphologies was extremely difficult in orifice experiments. Thus, it is important to understand the free-jet expansion after leaving the orifice to relate process conditions to particle morphology. Significant challenges in controlling morphology arise because

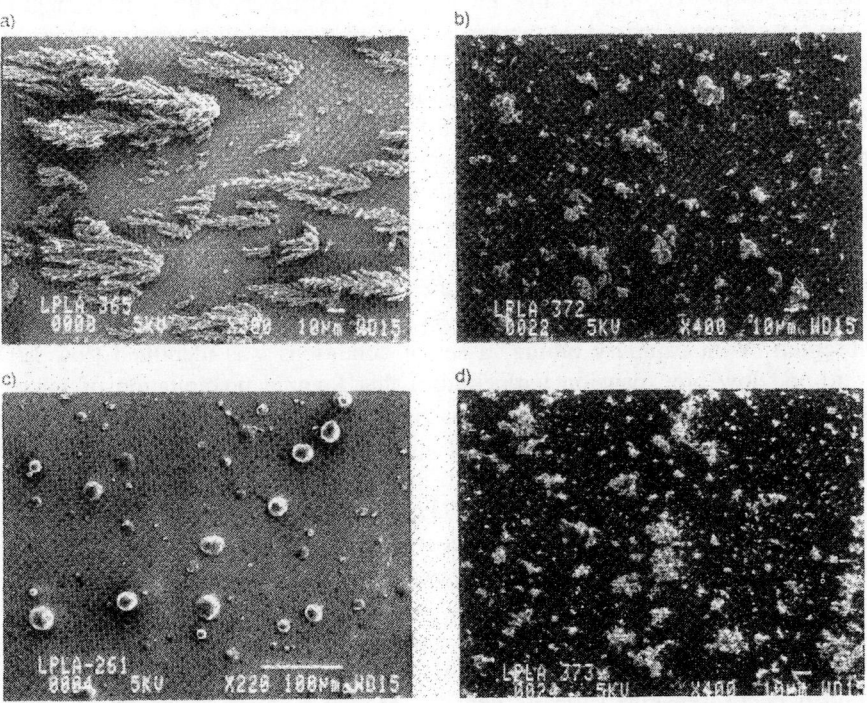

Fig. 5 (a) L-PLA dendrites, (b) microparticles, (c) microspheres, and (d) agglomerates produced by RESS using orifices as the expansion device. (From Ref. 16.)

precipitation occurs after the orifice, where the fluid undergoes a free-jet expansion and then mixes with background gas [42–44]. As the expansion jet decelerates to subsonic velocities, a series of shock waves occur. It is estimated that a significant fraction of the fluid's density drop occurs in that region [14]. This was further verified by Tom et al. [16], using a one-dimensional hydrodynamic model of the expansion within the orifice. Small changes in conditions (initial temperature, pressure, solubility) were found to affect the density and temperature profiles in the expansion device and, thus, in principle, the morphology of the precipitate.

In the same study [16], the relation between capillary geometry and L-PLA precipitate morphology was examined. The higher L/D ratio of capillaries relative to orifices leads to a larger density drop within the expansion device and, hence, to particle formation under more controlled conditions. Capillaries with inner diameters of 30 and 50 μm were used. In these experiments, the extraction pressure was varied from 175 to 220 bar, the extraction temperature was maintained at 55°C, the concentration of CO_2 in CO_2–$CHClF_2$ solvent mixture was varied from 40 to 80 wt%, and the preexpansion temperature was varied from 85° to 145°C. When using the 30-μm–i.d. capillary with lengths ranging from 5 to 15 mm ($167 < L/D < 500$), a range of morphologies was produced by RESS (Fig. 6). At high L/D ratios (500), only microparticles and agglomerates of microparticles were obtained, regardless of the preexpansion temperature (see Fig. 6a). At an L/D ratio of 350, microparticles were formed at high preexpansion temperatures, and microspheres were formed at low preexpansion temperatures (see Fig. 6b and 6c). This transition occurred at 110°–120°C. At low L/D ratios (< 300), microspheres were the dominant morphology. Similar trends were found using 50-μm capillaries with L/D ratios between 120 and 300, with microparticles dominating the morphology at higher L/D ratios and microspheres common at the lower L/D ratios. Dendrite formation was rarely observed. With capillary tubing, a set of conditions was identified (low L/D ratio and low preexpansion temperature) that favored precipitation of microspheres, and another set of conditions (high L/D ratio and high preexpansion temperature) that led to formation of microparticles. The work of Tom and co-workers [16] with monodisperse L-PLA provided the first comprehensive experimental study on the effect of RESS process variables on microparticle and microsphere formation using biodegradable polymers.

These authors [16] interpreted the experimental results using a one-dimensional model to compute the solvent's properties (pressure, temperature, density, and velocity) within the expansion device. This model was originally proposed and used by Lele and Shine [45]. Because the solvent power of a given fluid depends sensitively on its density [3], the point at which precipitation occurs can be correlated with the solvent's density in the capillary. The fluid's density at the exit of the capillary was calculated and correlated with the re-

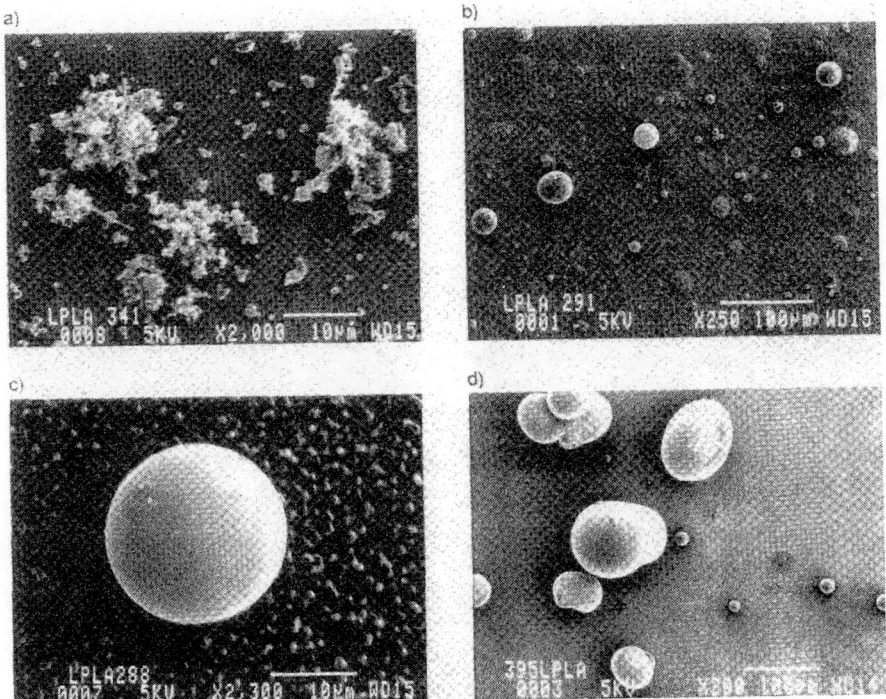

Fig. 6 (a) L-PLA microparticles and (b–d) microspheres produced by RESS using capillaries as the expansion device. (From Ref. 16.)

sulting morphology (Fig. 7). At low exit densities, only microparticles were formed, whereas at high exit densities, only microspheres were formed. The lower exit density corresponded with the solubility detection limit for L-PLA. This means that the quantity of polymer still solubilized at the capillary's exit is minimal and that significant precipitation will have occurred within the capillary. The correspondence of microparticle morphology with low solvent exit density suggests that microparticles are formed within the capillary. On the other hand, the solubility data at solvent densities corresponding to microsphere formation are not inconsistent with the hypothesis that material is still dissolved at these higher densities (and easily measurable), and that some precipitation must thus occur with further expansion downstream. The microspheres are most likely formed outside the capillary. Further evidence for the formation of microspheres outside the capillaries can be found in their sizes, which are often greater than the capillary's diameter (see Fig. 6d). In contrast, microparticles were smaller than the capillary's diameter. The appearance of some micropar-

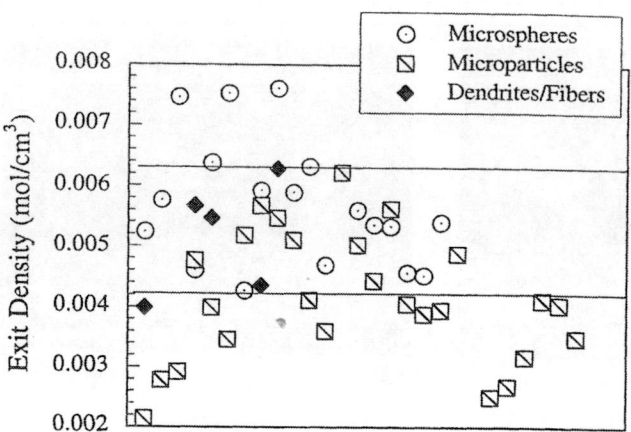

Fig. 7 Relation between particle morphology and calculated solvent density at the exit of the capillary for a series of experiments covering a wide range of conditions. (From Ref. 16.)

ticles with the microspheres suggests that, under certain conditions, precipitation may occur both within and outside the capillary. This is supported by the existence of an intermediate range of exit densities, within which the formation of both microspheres and microparticles occurred.

2. Coprecipitation

As described earlier, an interesting potential application of RESS is in the formation of polymer–pharmaceutical composite particles for controlled drug-delivery applications. With an improved understanding of the effect of RESS process variables on the formation of microparticles and microspheres of L-PLA [16], coupled with previous work on RESS with pharmaceutical compounds (i.e., Mohamed et al. [23]; Larson and King [22]), a considerable body of knowledge has accumulated that allows investigation of the next important problem: the formation of composite polymer–pharmaceutical particles.

Tom et al. [46] reported the first composite polymer–drug microparticles using commercial DL-PLA (Mw = 5000) and lovastatin, an anticholesterol drug. In this work, DL-PLA was mixed with lovastatin in one extraction column, while a second column was packed with lovastatin. By using CO_2 to extract the first column only (200 bar and 55°C and preexpansion temperatures of 75°–80°C), microspheres containing single lovastatin needles (Fig. 8a), egg-shaped polymer particles enveloping lovastatin needles (see Fig. 8b), as well as polymer microspheres without lovastatin, and lovastatin needles, were produced using a 30-μm–calibrated orifice. Nuclear magnetic resonance (NMR) analysis of the precipitate showed the lovastatin concentration to be 20 wt%. An experiment that used the same extraction conditions with lovastatin only (no polymer), produced only needles (see Fig. 8c). A second coprecipitation experiment using both columns with CO_2 at the same extraction conditions, produced microspheres with multiple lovastatin needles (see Fig. 8d). An NMR analysis of this precipitate showed about 27 wt% lovastatin in the coprecipitate. A third experiment, using recrystallized DL-PLA–lovastatin mixture (DL-PLA Mw = 10,000) and CO_2, at the same extraction conditions as the first two experiments, resulted in a solid network, consisting of intertwining fibers of 2- to 5-μm width, with 36 wt% lovastatin.

Tom et al. [16] used their understanding of the behavior of monodisperse L-PLA in supercritical CO_2–$CHClF_2$ solutions (polymer solubility and precipitate morphology) to examine coprecipitation. In this study, a model compound, pyrene, was used instead of lovastatin. The main advantage of pyrene is its fluorescence, allowing the use of fluorescence microscopy to examine the distribution of pyrene in particles of nonfluorescent L-PLA.

By using two extraction columns (one for L-PLA and one for pyrene), Tom et al. [16] were able to control the concentrations of pyrene and L-PLA in the coprecipitate by diluting the CO_2-extracted pyrene solution and using CO_2 (60 wt%)–$CHClF_2$ mixtures to extract L-PLA. The RESS-processed pyrene consisted of microparticles ($< 1–2$ μm). As the desired polymer morphology for controlled-release applications consists of microspheres, a capillary of 50-μm i.d. and an L/D ratio of 200 was used. Extraction conditions of 200–210 bar and 55°C, and preexpansion temperatures of 107°–110°C were used in these experiments. As the pyrene concentration in CO_2 was increased in these coprecipitation experiments, the quantity of fluorescence also increased. At pyrene concentrations of 0.0015–0.002 wt% in CO_2 and polymer concentrations of close to 0.04 wt% in 60% CO_2/40% $CHClF_2$, the resulting coprecipitate showed fluorescent microspheres (Fig. 9). Because L-PLA does not fluoresce and pyrene does not precipitate as microspheres, the particles clearly were L-PLA microspheres containing pyrene. Scanning electron microscopy (SEM) of the coprecipitate shows smooth microsphere surfaces, occasionally in contact with a few submicron particles. Thus, the uniform fluorescence observed cannot be attributed to pyrene particles on the microsphere's surface. At lower pyrene concentrations (0.0013 wt%), fluorescence was not observed in the microspheres, and at higher pyrene concentrations (> 0.002 wt%) in CO_2, fluorescence from pyrene particles outside of microspheres was observed. These results showed the feasibility of uniform incorporation of a solute into a polymer matrix by RESS coprecipitation. This result is quite promising for development of controlled drug-release applications.

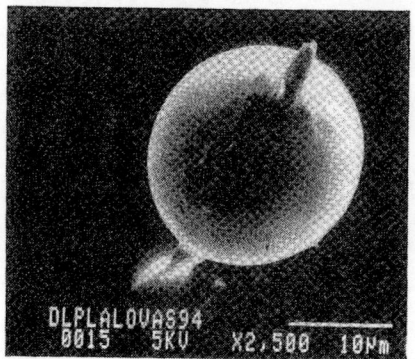

a

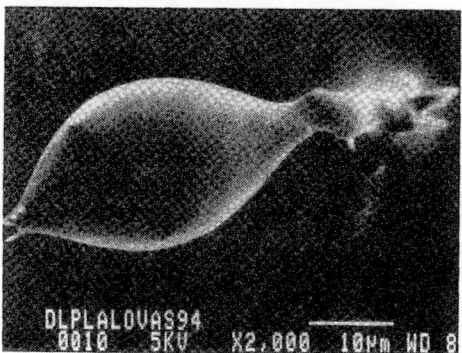

b

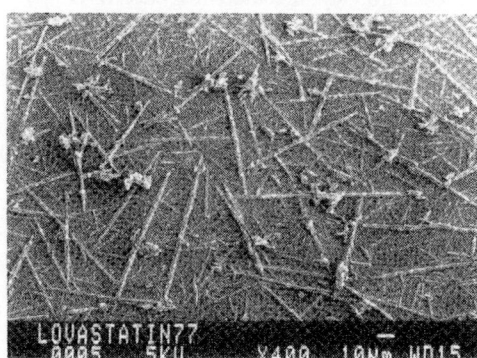

c

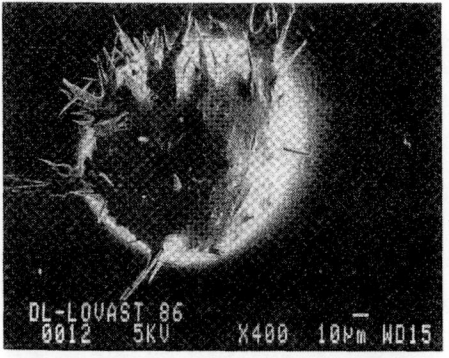

d

Fig. 8   (a) DL-PLA microsphere containing lovastatin needle, (b) DL-PLA polymer-coated lovastatin needle, (c) lovastatin needles, and (d) DL-PLA microspheres containing multiple lovastatin needles. These particles were produced following extraction with CO_2 at 200 bar and 55°C from an extraction column containing a mixture of DL-PLA and lovastatin and at preexpansion temperatures of 75–80°C (a,b); an extraction column containing only lovastatin (c); and two extraction columns in parallel containing a mixture of DL-PLA and lovastatin and only lovastatin, respectively (d). (From Ref. 46.)

III.   SUPERCRITICAL ANTISOLVENT PROCESSING

Supercritical antisolvent processing (SAS) exploits the low solvent power of supercritical fluids with moderate critical temperatures (e.g., CO_2) toward high polymers, proteins, and many biological molecules. In SAS, the solutes of interest are dissolved in a suitable organic phase. This organic phase is then contacted with a supercritical fluid with a low affinity for the solutes and appreciable mutual solubility with the organic phase. The particle nucleation and growth from this solute–organic–antisolvent system is governed by two fundamental processes: organic solvent expansion by the compressed antisolvent and organic solvent evaporation into the antisolvent phase. Only the former mechanism is important when SAS is carried out in semibatch mode (see later discussion); both mechanisms can be important in continuous operation (see later). Solvent expansion decreases the solubility of the precipitate within the organic phase. Simultaneously, solvent evaporation increases the concentration of the solute within the organic phase. Traces of the organic solvent can be removed from the resulting particulate phase by drying with the pure supercritical fluid stream; thus SAS yields solvent-free particles in a single processing step.

In addition to pharmaceutical applications, SAS has been applied to several monomeric and polymeric systems. The resulting particle morphologies highlight the versatility of this technique. For instance, Gallagher and co-workers

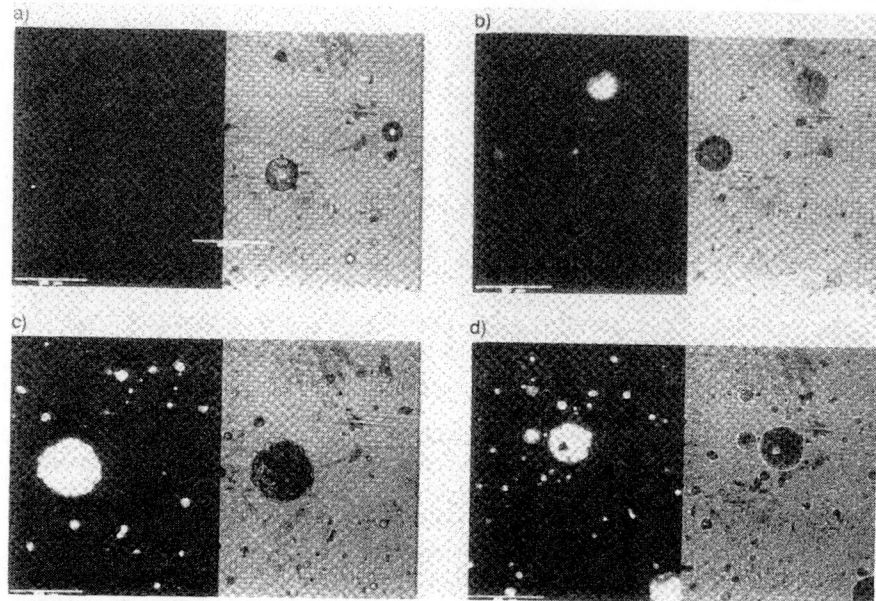

Fig. 9 Fluorescence and transmission images of L-PLA–pyrene microspheres precipitated at pyrene concentrations in CO_2 (a) <0.001 wt%, (b)≈0.0013 wt% and (c,d) >0.002 wt%. The L-PLA was extracted with 60 wt% CO_2–40 wt% $CHClF_2$ at 200–210 bar and 55°C, and pyrene was extracted with CO_2 at 200–210 bar and 65°C and diluted to above concentrations. A preexpansion temperature of 107–110°C and a capillary of $D=50$ μm were used. (From Ref. 16.)

[47,48] recrystallized explosives by SAS to small (100-μm range) microparticulates free of voids, thereby increasing the bulk density of the explosive. To maximize solid loadings of explosives, a high bulk density is desirable. However, high bulk densities cannot be achieved by conventional comminution techniques, which produce voids caused by liquid inclusion (e.g., solvent evaporation and antisolvent recrystallization), or are dangerous when applied to explosives (such as crushing, grinding, and ball-milling). By varying the CO_2 antisolvent's density and pressure, Dixon et al. [10] produced several distinct polystyrene microstructures: uniform microspheres (1–20 μm in diameter), porous and nonporous fibers, and films. Yeo et al. [11] formed molecularly oriented microfibers of a substituted *para*-linked aromatic polyamide by continuous SAS processing. Such polymers are used as high-modulus, heat-resistant fiber precursors. The microfibers produced by SAS consisted of stacks of microfibrils, ranging from 0.1 to 1.0 μm in diameter. In addition, semibatch expansion of polyamide–dimethylsulfoxide (DMSO) or *N,N*-dimethylform-

amide (DMF) organic phases with CO_2 created polycrystalline spherulites ranging from 1 to 10 μm in diameter.

A. Experimental Techniques

Both semibatch and continuous modes of operation have been used to produce microparticles by SAS processing. A typical laboratory-scale apparatus capable of both of these contacting schemes is shown in Fig. 10 [9]. In the semibatch process [9,11], the solute is dissolved in the organic phase and loaded into a 75-ml precipitator–view cell (E). The compressed antisolvent is bubbled through the organic phase from the bottom of the precipitator to promote maximum contacting. Typically, 20 ml of organic solution are processed during each semibatch experiment. Pressurization of the antisolvent is achieved with a high-pressure pump (C; rated to 420 bar), and the entire process is thermostated in an air chamber. Semibatch expansion of the solute–organic solution has been investigated from room temperature to 40°C in our laboratory. The system is typically pressurized to 100 bar or less; both slow (0.5 bar/min) and fast (25 bar/min) pressurizations have been used [46]. Expansion of the organic solvent by the supercritical fluid and the resulting loss of solvent power is the driving force for precipitation. The thermodynamics of this process has been discussed

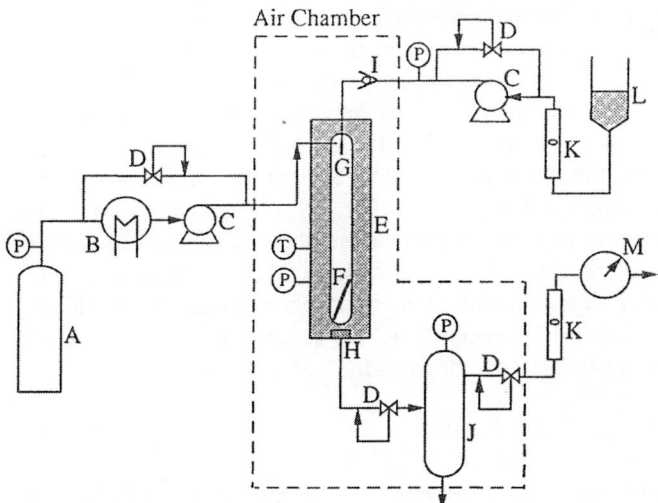

Fig. 10 Schematic diagram of the experimental apparatus for batch and continuous SAS processing. (A) Carbon dioxide; (B) cooler; (C) high-pressure pump; (D) backpressure regulator; (E) precipitator; (F) glass sampler; (G) nozzle; (H) metal filter; (I) check valve; (J) depressurizing tank; (K) rotameter; (L) organic solution; (M) dry test meter. (From Ref. 9.)

by Dixon and Johnston [8]. The resulting particles are collected on a glass slide (F) or a metal filter (H) located in the precipitator. Dry particles are recovered following the flushing of the organic phase from the system by positive supercritical fluid pressure, additional drying by the flow of pure supercritical fluid through the precipitator, and subsequent system depressurization. Drying is achieved from the downward flow of pure supercritical fluid solvent at 5–20 standard liters per minute (SLPM) for several hours while the precipitator is maintained at the final system temperature and pressure.

The apparatus in Fig. 10 can also produce microparticulates by contacting the organic phase and the supercritical fluid in a continuous, downflow cocurrent pattern [9,11,46]. In this mode of operation, the organic solution is pumped from a feeding tank (L) to a nozzle (G; 20- to 30-μm i.d.) located at the top of the precipitator (E). The pressure drop of the organic solution across the nozzle is approximately 15 bar and is required for droplet formation. The nozzle discharges submillimeter liquid droplets into a cocurrently flowing supercritical stream, which also enters from the top of the precipitator. With the correct choice of solvent–antisolvent systems and operating conditions, the organic solvent is rapidly dissolved in the supercritical phase, leaving behind solute particles. In continuous operation, the organic solution and the supercritical fluid streams, the respective flow rates of which are 0.05–0.5 ml/min and 5–20 SLPM, are contacted for several hours in the 75-ml precipitator. By using CO_2 as the antisolvent, the typical operating pressure ranges from 75 to 135 bar. A wide range of solute concentrations in the organic phase has been used to date: 0.1 mg/ml (e.g., precipitation of insulin or catalase from 90% ethanol/ 10% DI water [46]) to 50 mg/ml (e.g., precipitation of D,L-poly(lactic acid) from DMSO). The resulting particles are collected on glass slides (F) or on a filter (H) located before the precipitator outlet. System pressure is maintained through a series of backpressure regulators (D). The outlet stream is partially expanded in a depressurization vessel (J), where the organic solvent can be recovered. The antisolvent phase then passes through a flowmeter (K) and a dry test meter(M) after depressurization to atmospheric pressure. With only supercritical fluid flowing in the precipitator, this same flow pattern is used to evaporate the remaining organic solvent from the particles.

B. Proteins

Controlled-release systems for polypeptides have many advantages over conventional means of peptide distribution. A single administration maintains the drug concentration within a narrow, therapeutically desirable range over an extended time period, thus minimizing extreme variabilities of drug levels. This also results in reduced injection frequency, maximized drug efficiency per dose, and improved patient compliance [49]. Most controlled-release systems currently under investigation for proteins are injectable microparticulates or implantable devices comprised of protein microparticles within a polymeric matrix [32]. The drug-release characteristics of the composite are dependent on the distribution and loading of the drug in the matrix. In general, the smaller the protein particles within the polymeric matrix, the higher the percentage of drug that can be incorporated, and the more uniform the particle distribution within the composite structure. Drug particles in the 1- to 5-μm–size range are ideal for incorporation into injectable microspheres [9].

Many pharmaceuticals are not readily processed into useful morphologies or size ranges by current micronization methods. Proteins, in particular, are susceptible to denaturation caused by the conditions arising during many of the conventional ultrafine-particle formation methods (e.g., temperature, shear, or mechanical stress) [9]. Spray-drying can reduce the particle size to 5 μm or less [50]. However, this technique is characterized by low product yield and often results in denaturation [50]. Milling, which produces particles in the 10- to 50-μm range, can also denature proteins [51,52]. In addition, this process, which uses high-velocity compressed air, is more suitable for brittle materials than for characteristically soft protein powders [53]. Lypholization, a process that must be tailored to each protein, produces particles with the desired size, but with a wide distribution [54]. Controlled precipitation of protein powders from aqueous solutions has been achieved using miscible organic antisolvents or through the addition of salts [53,55], but this requires an additional drying step. In view of the limitations of conventional protein comminution techniques, the recent production of solvent-free, biologically active, insulin powders by SAS is a promising development [9]. By allowing the processing of proteins with very low solubility in supercritical fluids, SAS bypasses the main limitation of RESS.

Carbon dioxide has been the antisolvent of choice for use in protein processing owing to its nontoxicity, mild critical temperature, and environmental acceptability as a solvent. Investigations of protein powder formation by SAS have involved bovine insulin and bovine liver catalase [9,46].

In initial investigations of protein powder formation by SAS, Tom and coworkers [46] precipitated both catalase and insulin, separately, from 90% ethanol/10% DI water solutions, using compressed CO_2 antisolvent. The pH of the organic-phase solution, typically at a protein concentration of 0.1 mg/ml, was adjusted to 3.0–3.5 with 1 N HCl. The antisolvent and organic phase were contacted by spraying the protein solution into a cocurrently flowing CO_2 stream through a 20-μm–i.d. nozzle. All experiments were conducted at 35°C and 90 bar.

In the catalase experiment, the protein solution was introduced at 0.35 ml/min into CO_2 flowing at 16.9 SLPM. The resulting particles, collected on a slide within the precipitator, were approximately 1 μm. Insulin particles were

a

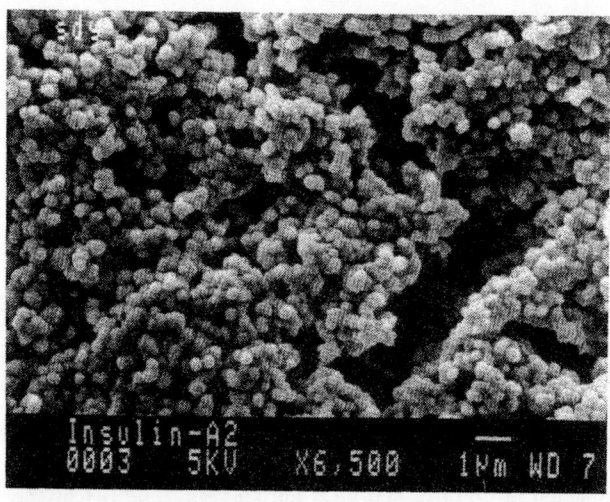

b

Fig. 11   SEM photomicrographs of (a) unprocessed insulin particles; (b) insulin particles formed from 5-mg/ml–insulin–DMSO solution at 25°C; (c) insulin particles formed
from 15-mg/ml–insulin–DMSO solution at 35°C; (d) insulin particles formed from 5-
mg/ml–insulin–DMF solution at 35°C; and (e) of semibatch-formed insulin particles
precipitated from 5-mg/ml–insulin–DMSO solution at 35°C. (From Ref. 9.)

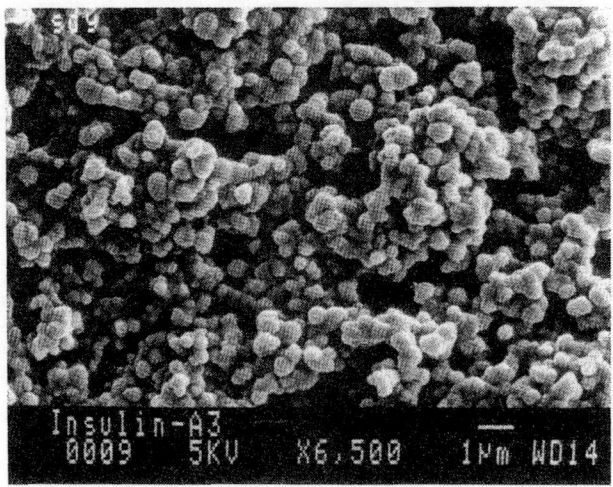

c

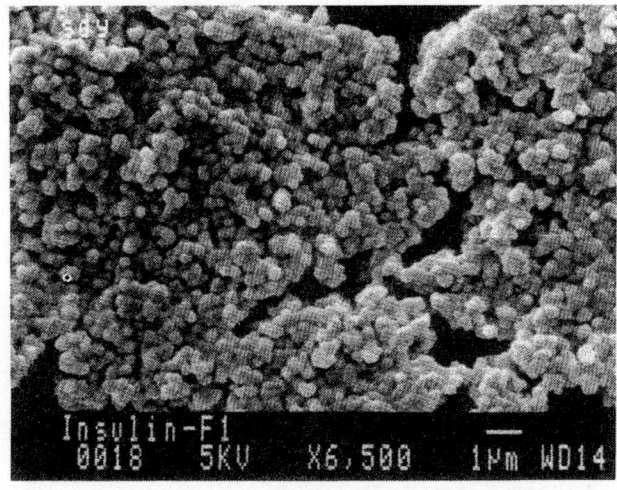

d

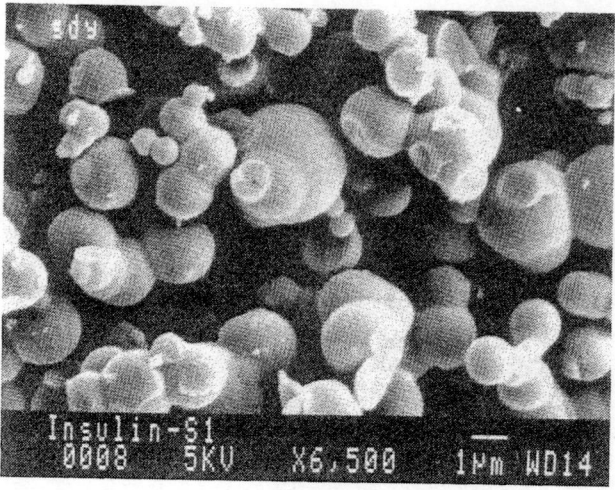

e

Fig. 11 Continued

produced at protein solution and antisolvent flow rates of 0.39 ml/min and 18.2 SLPM, respectively. Two types of microstructures were observed in the insulin particles: spheroids (smaller than 1 μm) and needles (approximately 5 by 1 μm). Attempts to measure the biological activity of these SAS-processed protein particles were hampered by the dissolution of the proteins in residual water in the precipitator. The low solubility of water in CO_2 precluded the recovery of dry particles in a single process step.

In subsequent experiments, Yeo et al. [9] used DMSO or DMF as the organic solvent, and produced biologically active insulin microparticles. Particle formation from 5- and 15-mg/ml–insulin–DMSO solutions and 5-mg/ml–insulin–DMF solutions was investigated. In preparing insulin–DMF solutions, solubilization of the protein was achieved by adjusting the pH to 7.3 with 1 N HCl. The CO_2 antisolvent and protein solutions were contacted at 86.2 bar and 25° or 35°C by the cocurrent technique of Tom et al. [46]. Solvent-free protein particles were recovered from glass slides and filters within the precipitator following drying with a pure CO_2 stream.

Over this range of experimental conditions and solvent systems, spherical and spheroidal insulin particles were produced, as shown in Fig. 11. The morphology of the SAS-produced powders differs significantly from the polyhedral aspect of the crystalline commercial insulin. The size distribution of insulin particles formed from a 15-mg/ml–insulin–DMSO solution at 35°C is shown in Fig. 12. Comparison of Figs. 11 and 12 suggests that there is still room for improvement in the production of small particles, for the primary units of Fig.

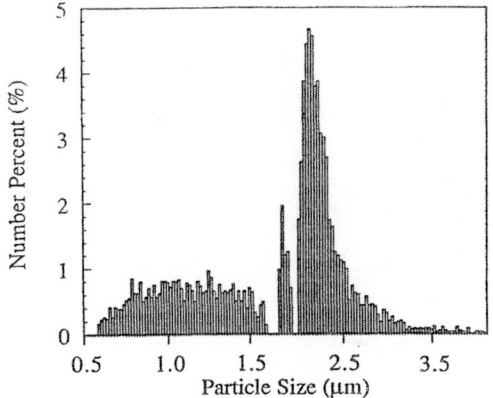

Fig. 12 Particle size distribution of SAS-processed insulin precipitated at 35°C from a 15-mg/ml–insulin–DMSO solution. Coulter counter measurements following sonication for 4 min in CH_2Cl_2. (From Ref. 9.)

11 are in the submicron range, whereas 2- to 4-μm particles are observed in Fig. 12. Under all SAS conditions tested, 90% of the resulting dry protein particles were smaller than 4 μm, and 10% were smaller than 1 μm. These results suggest a robust process for the formation of protein microspheres of a very desirable size and size distribution.

The slow expansion of a 5-mg/ml–insulin–DMSO solution was also performed at 35°C. This semibatch process resulted in the formation of larger spherical particles (see Fig. 11e). The pressure was increased at a rate of 0.57 bar/min by bubbling CO_2 into the precipitator loaded with 20 ml of the protein solution. The system was pressurized to 86.2 bar, with visible particle formation occurring at 66.8 bar. The onset of particle formation occurs at a lower pressure in a more concentrated protein solution. Comparison of the continuous and semibatch results suggest the ability to regulate particle size by controlling the organic solvent evaporation and expansion rates. Rapid organic solvent evaporation, associated with continuous-mode SAS processing, favors the formation of smaller particles than those produced by batch-mode processes.

In vivo tests on rats were used to ascertain the activity of these insulin powders. The precipitated and the commercial insulin were dissolved separately in sterile normal saline at a concentration of 0.5 mg/ml. This solution was injected subcutaneously on the dorsum of Sprague-Dawley rats (100–200 g). The blood glucose levels were monitored by tail vein blood sampling. Control injections were also done. For all processing conditions, the tests revealed no difference in the ability of the original and SAS-precipitated insulin to lower the blood glucose level (Fig. 13).

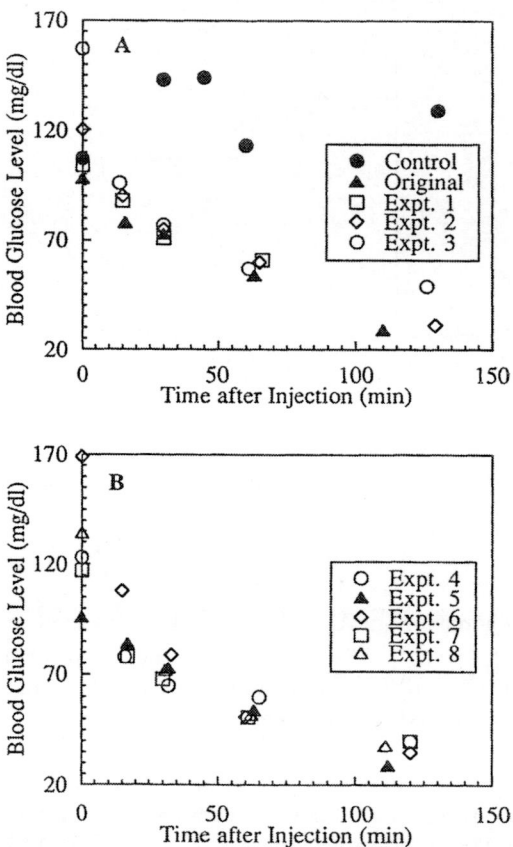

Fig. 13 Blood glucose level in rats as a function of time after injection. Saline solution, unprocessed insulin, and experiments 1 (25°C, 15-mg/ml–insulin–DMSO solution); 2 (25°C, 5-mg/ml–insulin–DMSO solution); 3 (35°C, 15-mg/ml–insulin–DMSO solution); 4 (35°C, 5-mg/ml–insulin–DMSO solution); 5 (25°C, 5-mg/ml–insulin–DMF solution); 6 (35°C, 5-mg/ml–insulin–DMSO solution, semibatch experiment; CO_2 addition rate 0.57 bar/min); 7 (35°C, 15-mg/ml–insulin–DMSO solution); 8 (35°C, 15-mg/ml–insulin–DMSO solution). (From Ref. 9.)

In spite of the full activity of these protein powders after reconstitution, recent Raman spectroscopy results suggest significant disruption of the secondary structure of the SAS-processed insulin [56]. A decrease in α-helix content of the SAS-processed powders was accompanied by an increase in the β-sheet content relative to the original sample. Although solution-phase Raman spectroscopy suggests a perturbation of the insulin native conformation by DMSO, exposure to the organic phase does not entirely explain the changes in secondary structure. The observed increase in β-sheet content is consistent with an increase in intermolecular interactions in the precipitate state, consistent with the formation of irreversible protein aggregate–fibril states. However, unlike irreversible protein aggregates, the insulin powders recover full biological activity on reconstitution. Thus, the rapid solvent evaporation associated with continuous SAS may produce unique, but reversible, protein conformations not usually accessible by conventional processing methods. This intriguing recovery of biological activity, in spite of extensive structural distortion of a magnitude comparable with that which accompanies irreversible denaturation, suggests that SAS may have considerable potential as a "protein-friendly" finishing step in commercial protein production [56]. SAS could also be of use in biopreservation applications, by permitting the storage of labile polypeptides in reversibly denatured (inactive) forms. Further research is necessary to ascertain whether this phenomenon occurs with proteins other than insulin.

C. Bioerodible Polymers

On the basis of the formation of nonbioerodible polymeric films, fibers, and microspheres by Dixon et al. [10] and Yeo et al. [11], SAS technology is very promising for the production of useful bioerodible polymer morphologies. As suggested by Randolph et al. [57], bioerodible polymeric microspheres may be further encapsulated into novel drug-delivery particles, or compacted with drug particles to form a larger drug-delivery matrix. As a first step in the investigation leading to direct microencapsulation of pharmaceuticals in a bioerodible polymeric matrix, studies have been made on the processing of pure bioerodible polymers by SAS. We will describe these investigations, as well as ongoing work in our laboratory.

Microencapsulated drug–bioerodible polymer systems have applications as controlled-release systems for intravenous, intramuscular, and nasal drug delivery [58]. The desired microsphere size, dependent on the application and target organ, ranges from smaller than 1 μm to approximately 40 μm [58]. Narrow-particle–size distributions and consistent drug dispersal within the polymeric matrix are required for reproducible drug-release profiles. Drawbacks of current techniques for the formation of bioerodible polymeric microspheres include large particle sizes and wide size distributions, the presence of residual solvent, and high processing temperatures necessary for solvent evaporation [57]. For pharmaceutical applications, the SAS formation of ultrafine bioerodible polymeric microspheres with a narrow size distribution is of primary interest.

Many of the same advantages are associated with RESS and SAS polymer processing. However, the SAS technique is applicable to a broader spectrum of polymers and molecular weights then RESS, owing to the limited solubility of many polymeric systems in supercritical CO_2. Therefore, polymer microparticles with a wider range of release kinetics can be formed by SAS. Most SAS studies of bioerodible polymer precipitation to date have focused on FDA-approved bioerodible materials, such as the previously mentioned family of poly(hydroxy acids), including PLA, PGA, and PLA–PGA copolymers [57]. As in the production of protein powders, all SAS investigations of bioerodible polymer formation have been conducted using compressed CO_2 antisolvent.

Randolph et al. [57] precipitated L-PLA microparticles from methylene chloride, using near-critical and supercritical CO_2. Semibatch- and continuous-mode experiments were conducted in two different units. In the semibatch experiments, L-PLA (100,000 Mw) was precipitated from 0.3, 0.6, and 1 wt% solutions in methylene chloride at 31°, 36°, and 40°C. The solutions were pressurized and sprayed through a 75-μm–i.d. capillary tube into stagnant CO_2 antisolvent. After approximately 1 min of spraying at 1 ml/min, the particles were allowed to settle onto an aluminum stub, located below the nozzle outlet. Following depressurization, the particles on the aluminum stub were studied by SEM.

In continuous-mode operation, the L-PLA–methylene chloride solution (0.3-wt% polymer) was sprayed into concurrently flowing CO_2 through an ultrasonic atomizer. This atomizer acted to break the polymer solution into droplets by means of a 120-kHz vibration. Polymer solution flow rates were adjusted between 0.02 and 2.0 ml/min, while the CO_2 flow rate was maintained at 151 SLPM. Continuous experiments were conducted at 36° and 40°C. The suspended particle–methylene chloride–CO_2 system formed in the precipitator was expanded across a micrometering valve. The collection of dry samples on a glass slide placed at the micrometering valve outlet was possible because of the low boiling point of methylene chloride (40°C).

In both modes of operation, the effect of pressure on the L-PLA microstructure was explained in terms of methylene chloride evaporation and expansion. At 55.1 bar and 31°C, incomplete evaporation of the methylene chloride from the falling droplets resulted in an irregular film on the aluminum stub removed from the batch reactor. Increasing the pressure to 62.0 bar evaporated the methylene chloride sufficiently to produce cusped particles with rough surfaces. The uneven microparticle surface is consistent with the formation of an L-PLA "skin" on the evaporating droplet caused by a relatively slow particle-drying rate. Nearly spherical particles were produced at 75.8, 82.7, and 96.5 bar. The smallest particles (0.61 μm) were produced at 75.8 bar, slightly above the critical pressure of CO_2. The mean particle diameter increased with pressure to 0.95 and 1.4 μm at 82.7 and 96.5 bar, respectively. Similar pressure trends were observed at 36° and 40°C.

The increase of particle size with pressure was attributed to a decrease in methylene chloride and CO_2 mass-transfer rates in higher-density systems.

High mass-transfer rates allow for more rapid swelling of the polymer droplet and lead to greater supersaturation within the droplet. This results in higher nucleation rates and, therefore, smaller particles. The similarity of particles formed from the capillary and atomizing nozzles also suggests that mass-transfer rates, and not initial droplet size, govern the resulting particle size.

The feasibility of precipitating poly(hydroxy acids) from DMSO solutions using compressed CO_2 antisolvent is currently being studied in our laboratory. Microparticles of DL-PLA (5000 Mw) were formed by spraying a 36-mg/ml–polymer–DMSO solution into cocurrently flowing supercritical CO_2 at 104 bar and 36°C. The flow rates of the polymer solution and CO_2 streams were 0.25 ml/min and 15 SLPM, respectively. The resulting particles were collected on Teflon strips placed in the precipitator. This process produced ellipsoidal particles, with a typical mean diameter of less than 2μm. Ellipsoidal particles with a greater degree of agglomeration were also observed. The free particles are similar to the nearly spherical L-PLA particles produced by Randolph et al. [57]. The effect of varying experimental parameters, such as temperature, pressure, polymer solution concentration, and relative DMSO and CO_2 flow rates, is currently being investigated. The shape of these ellipsoid particles may be influenced by secondary effects following particle formation, such as interactions with the collection surface or polymer swelling and plasticization by CO_2 [59–60].

The SAS technology has also been applied to the formation of tyrosine-derived polycarbonate and polyarylate microspheres. This series of pseudopoly-(amino acids) are currently being studied as biodegradable materials [61]. The pendant chain on the polymer backbone can be systematically varied to affect polymer properties, such as glass transition temperature, mechanical strength, degradation rate, and hydrophobicity. Current investigations of SAS precipitation of these polymers complement investigations of conventional solvent evaporation techniques for the formation of microspheres [62].

Microspheres of a tyrosine-based polyarylate, poly(DTH sebacate), were obtained by the fast-batch expansion (20 bar/min) of a 5.5-mg/ml–DMSO solution with CO_2. At 35°C, visible particle formation occurred at 73.4 bar. Microspheres, with an average particle diameter of approximately 200 μm, were sampled from a glass slide within the precipitator. However, SEM results suggest that these particles may be hollow.

A very different morphology results from the precipitation of a related pseudopoly(amino acid), poly(DTB carbonate) in a continuous, cocurrent SAS process. In a precipitator maintained at 35.1°C and 101 bar, a 6.1-mg/ml–polymer–DMSO solution was sprayed cocurrently into flowing CO_2. The polymer solution and CO_2 flow rates were 0.07 ml/min and 16 SLPM, respectively. The resulting microspheres, collected on a Teflon strip in the precipitator, are shown in Fig. 14. The polymeric microspheres are typically smaller than 4 μm

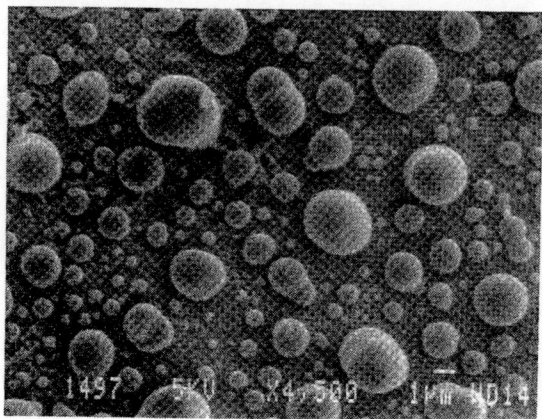

Fig. 14 SEM photomicrographs of the pseudopoly(amino acid), poly(DTB carbonate), precipitated in continuous, countercurrent mode from DMSO using CO_2 as the supercritical antisolvent ($T = 35.1°C$; $P = 101$ bar).

in diameter and show no indication of being hollow. Again, slight deformations in the spheroidal particles suggest interactions or wetting at the polymer–particle collection surface interface or polymer plasticization by CO_2 following particle formation.

D. Microencapsulation

In principle, it should be possible to encapsulate drugs in a polymeric matrix by SAS: this would entail dissolving the appropriate relative proportions of polymer and drug in the organic phase and contacting with a supercritical fluid in the usual way. With many of the same advantages of RESS, SAS is again applicable to a broader range of drug–polymer systems than RESS (which requires that both the drug and polymer phase be appreciably soluble in the supercritical fluid phase). In addition, previous protein and polymer studies of SAS precipitation have shown the possibility of varying particle morphology and size with process conditions. The inherent versatility of the SAS technique suggests a high likelihood for the successful microencapsulation of a variety of drug–bioerodible polymer systems. Work on SAS microencapsulation is in progress in our laboratory.

To date, Muller and Fischer [63] have reported the only coprecipitation of drug and polymer by SAS techniques. The drug clonidine hydrochloride was microencapsulated in an L-PLA matrix from a solution of the drug and polymer in methylene chloride. The organic phase was contacted with CO_2 in continuous, countercurrent flow. Microspheres approximately 100 μm in diameter were formed at 100 bar and 60°C. In vitro-release studies showed that over 40% of the drug was released within the first 1.5 h.

IV. CONCLUSIONS

The application of supercritical fluids to the production of microparticulates of biomedical interest is a recent development, and the field is still in its infancy. The main advantages of using supercritical fluids are the purity of the products, which avoids or minimizes downstream processing; the mildness of the operating temperatures; and the simplicity of the processes. As an alternative to conventional comminution techniques, RESS is attractive, and it could also prove effective as a route to the formation of composite drug-loaded, low molecular weight polymeric microparticulates. SAS shows promise as a technique for making biologically active protein powders, and may be useful as a process for direct encapsulation of drugs in polymer matrices. In all of the foregoing examples, progress in the design and interpretation of experiments would be greatly enhanced by improvements in our very limited understanding of RESS and SAS fundamentals. Not even the effects of individual process variables on particle size and morphology are as yet completely understood. Until this happens, the exploratory nature of much of RESS and SAS research to date is likely to continue.

Given the increasingly stringent regulations limiting the emission of organic solvents and aqueous waste streams, supercritical fluids, and in particular supercritical carbon dioxide, are attractive alternatives to conventional solvents. The environmental and economic incentives for the development and improvement of carbon dioxide-based alternatives to existing technologies are strong, and the production of biomedical microparticulates is no exception. Thus, it is likely that most of the concepts outlined in this chapter will remain active areas of research in the foreseeable future. A particularly promising recent development is a novel dispersion polymerization process in which amphiphilic molecules that are interfacially active in carbon dioxide are used to form a stable polymer colloid, within which the polymerization takes place [64]. This technique has been used to make micron-sized poly(methyl metacrylate) microspheres with narrow size distributions, and it may prove suitable for the direct microencapsulation of therapeutic drugs inside the growing polymer particles. This technique has not yet been applied to bioerodible polymers.

ACKNOWLEDGMENT

PGD gratefully acknowledges the support of the National Science Foundation (Grant SGER 93-9321978), the Air Force Office of Scientific Research (Grant AASERT F49620-93-1-0454), the Associated Institutes for Materials Sciences, and the Princeton Materials Institute.

REFERENCES

1. P. G. Debenedetti and R. C. Reid, Diffusion and mass transfer in supercritical fluids, *AIChE J. 32*:2034 (1986).
2. M. McHugh and V. J. Krukonis, *Supercritical Fluid Extraction, Principles and Practice*, Butterworths, Boston, 1986.
3. S. K. Kumar and K. P. Johnston, Modelling the solubility of solids in supercritical fluids with density as independent variable, *J. Supercrit. Fluids 1*:15 (1988).
4. P. G. Debenedetti, Supercritical fluids as particle formation media, *Supercritical Fluids—Fundamentals for Application* (E. Kiran and J. M. H. Levelt Sengers, eds.), *NATO ASI Ser. E 273*:719 (1994).
5. J. W. Tom and P. G. Debenedetti, Particle formation with supercritical fluids—a review, *J. Aerosol Sci. 22*:555 (1991).
6. P. G. Debenedetti and S. K. Kumar, The molecular basis of temperature effects in supercritical extraction, *AIChE J. 34*:1211 (1988).
7. P. G. Debenedetti, Homogeneous nucleation in supercritical fluids, *AIChE J. 36*:1289 (1990).
8. D. J. Dixon and K. P. Johnston, Molecular thermodynamics of solubilities in gas antisolvent crystallization, *AIChE J. 37*:1441 (1991).
9. S.-D. Yeo, G.-B. Lim, P. G. Debenedetti, and H. Bernstein, Formation of microparticulate protein powders using a supercritical fluid antisolvent, *Biotechnol. Bioeng. 41*:341 (1993).
10. D. J. Dixon, K. P. Johnston, and R. A. Bodmeier, Polymeric materials formed from precipitation with a compressed fluid antisolvent, *AIChE J. 39*:127 (1993).
11. S.-D. Yeo, P. G. Debenedetti, M. Radosz, and H.-W. Schmidt, Supercritical antisolvent (SAS) process for substituted aromatic polyamides: Phase equilibrium and morphology study, *Macromolecules 26*:6207 (1993).
12. J. W. Tom and P. G. Debenedetti, Formation of bioerodible polymeric microspheres and microparticles by rapid expansion of supercritical solutions, *Biotechnol. Prog. 7*:403 (1991).
13. C. J. Chang and A. D. Randolph, Precipitation of microsize organic particles from supercritical fluids, *AIChE J. 35*:1876 (1989).
14. D. W. Matson, J. L. Fulton, R. C. Petersen, and R. D. Smith, Rapid expansion of supercritical fluid solutions: Solute formation of powders, thin films and fibers, *Ind. Eng. Chem. Res. 26*:2298 (1987).
15. D. W. Matson, R. C. Petersen, and R. D. Smith, Formation of silica powders from the rapid expansion of supercritical solutions, *Adv. Ceramic Mater. 1*:242 (1986).
16. J. W. Tom, P. G. Debenedetti, and R. Jerome, Precipitation of poly(*l*-lactic acid) and composite poly(*l*-lactic acid)–pyrene particles by rapid expansion of supercritical solutions, *J. Supercrit. Fluids 4*:429 (1994).
17. D. W. Matson, R. C. Petersen, and R. D. Smith, Formation of silica powders from the rapid expansion of supercritical solutions, *Mater. Lett. 1*:242 (1986).
18. R. S. Mohamed, P. G. Debenedetti, and R. K. Prud'homme, Effects of process conditions on crystals obtained from supercritical mixtures, *AIChE J. 35*:325 (1989).

19.   M. Otsuka and N. Kaneniwa, Effect of grinding on the crystallinity and chemical stability in the solid state of cephalothin sodium, *Int. J. Pharm. 62*:65 (1990).

20.   P. K. Gupta and A. J. Hickey, Contemporary approaches in aerosolized drug delivery to the lung, *J. Controlled Release 17*:129 (1991).

21.   V. Krukonis, Supercritical fluid nucleation of difficult-to-comminute solids, paper 140f, AIChE Meeting, San Francisco, November 1984.

22.   K. A. Larson and M. L. King, Evaluation of supercritical fluid extraction in the pharmaceutical industry, *Biotechnol. Prog. 2*:73 (1986).

23.   R. S. Mohamed, D. S. Halverson, P. G. Debenedetti, and R. K. Prud'homme, Solids formation after the expansion of supercritical mixtures, *Supercritical Fluid Science and Technology. ACS Symp. Ser. 406* (K. P. Johnston and J. M. L. Penninger, eds.), American Chemical Society, Washington DC, 1989.

24.   H. Loth and E. Hemgesberg, Properties and dissolution of drugs micronized by crystallization from supercritical gases, *Int. J. Pharm. 32*:265 (1986).

25.   K. Ohgaki, H. Kobayashi, T. Katayama, and N. Hirokawa, Whisker formation from jet of supercritical fluid solution, *J Supercrit. Fluids 3*:103 (1990).

26.   J. Chrastil, Solubility of solids and liquids in supercritical gases, *J. Phys. Chem. 86*:3016 (1982).

27.   E. Stahl and K. W. Quirin, Dense gas extraction on a laboratory scale: A survey of recent results, *Fluid Phase Equilib. 10*:269 (1986).

28.   W. J. Schmitt, *The Solubility of Monofunctional Organic Compounds in Chemically Diverse Supercritical Fluids*, Ph.D. Thesis, Massachusetts Institute of Technology, Cambridge MA, 1984.

29.   J. Heller, *Medical Applications of Controlled Release*, Vol. 1 (R. S. Langer and D. L. Wise, eds.), CRC Press, Boca Raton FL, 1984.

30.   R. S. Langer and D. I. Wise, eds., *Medical Applications of Controlled Release*, *Vol. 1*, I. Classes of Systems, CRC Press, Boca Raton FL, 1984.

31.   H. B. Rosen, J. Kohn, K. Leong, and R. Langer, *Controlled Release Systems*: *Fabrication Technology*, Vol. 2 (D. Hsieh ed.), CRC Press, Boca Raton FL, 1988.

32.   R. Langer, New methods of drug delivery, *Science 249*:1473 (1990).

33.   R. S. Langer, D. S. T. Hsieh, A. Peil, R. Bawa, and W. Rhine, *Polymers for the Controlled Release of Macromolecules: Kinetics, Applications and External Control, AIChE Symp. Ser. 206*, (K. Chandrasekaran, ed.), American Institute of Chemical Engineers, New York, 1981.

34.   J. W. Fong, *Controlled Release Systems: Fabrication Technology* (D. Hsieh, ed.), CRC Press, Boca Raton FL, 1988.

35.   W. Good, *Polymeric Delivery Systems* (R. Kastelnick, ed.), Midland Macromolecular Monograph Vol. 5., Gordon & Breach, New York, 1978.

36.   K. Juni and M. Nakano, Poly(hydroxy acids) in drug delivery, *CRC Crit. Rev. Ther. Drug Carrier Syst. 3*:209 (1987).

37.   R. Arshady, Preparation of biodegradable microspheres and microcapsules. 2. Polylactides and related polyesters, *J. Controlled Release 17*:1 (1991).

38.   R. Jalil, Biodegradable poly(lactic acid) and poly(lactide-*co*-glycolide) polymers in sustained drug delivery, *Drug . Dev. Ind. Pharm. 16*:2353 (1990).

39.   S. J. Holland, B. J. Tighe, and P. L. Gould, Polymers for biodegradable medical

devices. 1. The potential of polyesters as controlled macromolecular release systems, *J. Controlled Release 4*:155 (1986).

40. D. Hsieh, ed., *Controlled Release Systems: Fabrication Technology, Vol. 2*, CRC Press, Boca Raton FL, 1988.

41. J. R. Robinson and V. H. L. Lee, eds., *Controlled Drug Delivery, Fundamentals and Applications*, Marcel Dekker, New York, 1987.

42. J. I. Brand and D. R. Miller, Ceramic beams and thin film growth, *Thin Solid Films 166*:139 (1988).

43. H. R. Murphy and D. R. Miller, Effects of nozzle geometry on kinetics in free-jet expansions, *J. Phys. Chem. 88*:4474 (1984).

44. L. G. Randall, *Chemical Engineering at Supercritical Fluid Conditions* (M. E. Paulaitis, J. M. L. Penninger, R. D. Gray, and P. Davidson, eds.), Ann Arbor Science: Ann Arbor MI, 1983.

45. A. K. Lele and A. D. Shine, Morphology of polymers precipitated from a supercritical solvent, *AIChE J. 38*:742 (1992).

46. J. W. Tom, G. Lim, P. G. Debenedetti, and R. K. Prud'homme, Applications of supercritical fluids in controlled release of drugs, *Supercritical Fluid Engineering Science ACS Symp. Ser. 514* (E. Kiran and J. F. Brennecke, eds.), American Chemical Society, Washington DC, 1992.

47. P. M. Gallagher, M. P. Coffey, V. J. Krukonis, and N. Klasutis, Gas antisolvent recrystallization: New process to recrystallize compounds insoluble in supercritical fluids, *Supercritical Fluid Science and Technology ACS Symp. Ser. 406* (K. P. Johnston and J. M. L. Penninger, eds.), American Chemical Society, Washington DC, 1989.

48. P. M. Gallagher, M. P. Coffey, V. J. Krukonis, and W. W. Hillstrom, Gas antisolvent recrystallization of RDX: Formation of ultra-fine particles of a difficult-to-comminute explosive, *J. Supercrit. Fluids 5*:130 (1992).

49. Y. W. Chien, Polymer-controlled drug delivery systems: Science and engineering, *Polymeric Materials in Medication* (C. J. Gebelein and C. E. Carraher, eds.), Plenum Press, New York, 1985.

50. S. Riegelman, Application of spray drying techniques to pharmaceutical powders, *J. Am. Chem. Soc. 39*:444, (1950).

51. L. Hixon, M. Prior, H. Prem, and J. Van Cleef, Sizing materials by crushing and grinding, *Chem. Eng. 97*:94 (1990).

52. J. Van Cleef, Powder technology, *Am. Sci. 79*:304 (1991).

53. R. Nash, Pharmaceutical suspensions, *Pharmaceutical Dosage Forms* (H. A. Lieberman, M. Rieger, and G. Banker, eds.), Marcel Dekker, New York, 1988, p. 151.

54. A. Briggs and T. Maxwell, Method of preparation of lyophilized biological products, U.S. Patent 3,928,566, 1975.

55. R. M. Cohn and D. M. Skauen, Controlled crystallization of hydrocortisone by ultrasonic irradiation, *J. Pharm. Sci. 53*:1040 (1964).

56. S.-D. Yeo, P. G. Debenedetti, S. Y. Patro, and T. M. Przybycien, Secondary structure characterization of microparticulate insulin powders, *J. Pharm. Sci. 83*:1651 (1994).

57. T. W. Randolph, A. D. Randolph, M. Mebes, and S. Yeung, Sub-micrometer-

sized biodegradable particles of poly(L-lactic acid) via the gas antisolvent spray precipitation process, *Biotechnol. Prog. 9*:429 (1993).

58. S. S. Davis, L. Illum, D. Burgess, J. Ratcliffe, and S. N. Mills, Microspheres as controlled-release systems for parenteral and nasal administration, *Controlled-Release Technology: Pharmaceutical Applications ACS Symp. Ser. 348* (P. I. Lee and W. R. Good, eds.), American Chemical Society, Washington DC, 1987.

59. G. K. Fleming and W. J. Koros, Dilation of polymers by sorption of carbon dioxide at elevated pressures. 1. Silicone rubber and unconditioned polycarbonate, *Macromolecules 19*:2285 (1986).

60. R. G. Wissinger and M. E. Paulaitis, Swelling and sorption in polymer–CO_2 mixtures at elevated pressures, *J. Polym. Sci. B 25*:2497 (1987).

61. S. I. Ertel and J. Kohn, Evaluation of a series of tyrosine-derived polycarbonates as degradable biomaterials, *J Biomed. Mater. Res. 28*:919 (1994).

62. J. Kohn, C. Yu, S.-D. Yeo, and P. G. Debenedetti, Microspheres from tyrosine-based polymers: Evaluation of processing parameters and initial release data, *Proc. Int. Symp. Controlled Release Bioact. Mater. 21* (1994).

63. B. W. Muller and W. Fisher, Manufacture of sterile sustained-release drug formulations using liquified gas, W. Germany Patent 3,744,329, July 6, 1989.

64. J. M. DeSimone, E. E. Maury, Y. Z. Menceloglu, J. B. McClain, T. J. Romack, and J. R. Combes, Dispersion polymerizations in supercritical carbon dioxide, *Science 265*:356 (1994).

4

Polyphosphazene Hydrogel Microspheres for Protein Delivery

Alexander K. Andrianov and Lendon G. Payne

Virus Research Institute
Cambridge, Massachusetts

I. POLYORGANOPHOSPHAZENE HYDROGELS AS MATRICES FOR DRUG DELIVERY

Polyorganophosphazenes are attracting attention as biomedical polymers, membranes, hydrogels, bioactive and biodegradable polymers [1]. Phosphazene polymers have a long-chain backbone of alternating nitrogen and phosphorus atoms with two side groups attached to each phosphorus. Molecular structure and physicochemical properties, such as water solubility and biodegradability, of these polymers can be tailored widely by small changes in the synthetic design.

Of the many synthetic and natural materials available, polymer hydrogels present opportunities and advantages in the development of controlled-release drug-delivery systems. The accumulated evidence shows that hydrogels possess unique properties, such as a low interfacial tension with surrounding biological fluids, a superhydrophilic diffuse surface, a simulation of the hydrodynamic properties of natural cells and tissues, and other properties that make them highly biocompatible [2].

Recently, a new, mild microencapsulation method was developed based on the ability of polyphosphazene polyelectrolytes to form soft and highly swollen gels on contact with bivalent cations, such as calcium ions, at room temperature or below [3]. This process does not involve harsh conditions, such as elevated temperatures or organic solvents, and is suitable for encapsulation of biologically labile entities, such as proteins and viruses, that usually present unique problems in their therapeutic delivery.

This chapter outlines our present state of knowledge on water-soluble poly-phosphazene polyelectrolytes relative to their synthesis, physicochemical behavior in solution, gelation mechanism, and applications in microencapsulation and protein delivery.

II. SYNTHESIS AND PHYSICOCHEMICAL CHARACTERIZATION OF WATER-SOLUBLE POLYPHOSPHAZENE POLYELECTROLYTES

A. Synthesis and Characterization

The most extensively developed method for polyphosphazene synthesis requires a thermal ring-opening polymerization of hexachlorocyclotriphosphazene to prepare a reactive macromolecular intermediate poly(dichlorophosphazene). This intermediate can be modified with a wide range of nucleophiles to yield the stable derivatized polymers poly(organophosphazenes) [1].

$$\left[\begin{array}{c} R \\ | \\ P = N \\ | \\ R \end{array} \right]_n$$

The synthetic route for polyphosphazenes containing reactive functionalities, including ionic groups, was designed to avoid the reaction of poly(dichloro-phosphazene) with difunctional reagents, which cause an inevitable cross-link-ing of the polymer [4]. Functional residues are introduced onto polyphospha-zene molecules in a two-step process in which the initial attachment of an organic "spacer" side group to the inorganic skeleton is followed by subsequent construction reactions carried out on the organic unit. The feasibility of this approach for polyphosphazenes was first demonstrated by the synthesis of de-rivatized poly(organophosphazenes) by lithiophenoxy intermediates [4]. The re-action sequence included modification of poly(dichlorophosphazene) with so-dium *p*-bromophenoxide and sodium phenoxide. Then, the bromophenoxy-substituted polymer was allowed to undergo a metal–halogen-exchange reac-tion with *n*-butyllithium. Finally, carboxylic acid functions were introduced to the (aryloxy)phosphazene structure by treatment of (lithiophenoxy)phospha-zenes with an aqueous solution of HCl or carbon dioxide. Poly[(carboxylato-phenoxy)(phenoxy)phosphazenes] (**I**) of different composition were obtained, but were not soluble in water.

$$\left[\begin{array}{c} O\!-\!\!\langle\bigcirc\rangle\!-\!COOH \\ | \\ P = N \\ | \\ O\!-\!\langle\bigcirc\rangle \end{array} \right]_n \qquad \textbf{(I)}$$

The same "side group transformation" approach was later simplified by All-cock for the synthesis of water-soluble carboxylic acid-bearing polyphosphazene [5]. Poly(dichlorophosphazene) was allowed to react with a sodium salt of ethyl *p*-hydroxybenzoate to form the ester-type aryloxyphosphazene. Poly-[di(carboxylatophenoxy)phosphazene] (PCPP) **(II)** was obtained by a mild hy-

$$\left[\begin{array}{c} O\!-\!\!\langle\bigcirc\rangle\!-\!COOH \\ | \\ P = N \\ | \\ O\!-\!\langle\bigcirc\rangle\!-\!COOH \end{array} \right]_n \qquad \textbf{(II)}$$

drolysis of the ester function to expose the carboxylic acid without decomposition of the polymer skeleton. High molecular weight polymer was reportedly insoluble in acidic or neutral aqueous media, but was soluble in aqueous base.

A biodegradable water-soluble phosphazene polyelectrolyte was synthesized by incorporation of hydrolytically labile amino linkages in the polymer structure [6]. Poly[(carboxylatophenoxy)(glycinato)phosphazene] (PCGPP) **(III)** was prepared by the two-step modification of poly(dichlorophosphazene) with a mixture of the sodium salt of propyl *p*-hydroxybenzoate and ethyl glycinate hydrochloride, followed by hydrolysis to the corresponding poly(carboxylic acids). Polymer structure was proved by ^{1}H and ^{31}P nuclear magnetic resonance (NMR); however, the molecular weight of the copolymer determined by gel permeation chromatograpy (GPC) was lower than the molecular weight of PCPP itself, which indicates the occurrence of some polymer degradation during the modification reactions.

$$\left(O\!-\!\!\langle\bigcirc\rangle\!-\!COOH \right)_x$$
$$\left. \begin{array}{c} | \\ P = N \\ | \end{array} \right\}_n$$
$$\left(NHCH_2COOH \right)_y \qquad \text{(III)}$$

Water-soluble polyphosphazenes containing strongly acidic groups, such as sulfonic acid, were synthesized by direct sulfonation of high molecular weight poly(aryloxyphosphazenes) [7]. Reaction of poly[bis(phenoxy)phosphazene] with SO_3 yielded a water-soluble sulfonic acid derivative, with the functional group in the *meta*-position (**IV**). The authors reported that polymer solubility

$$\left[P\left(O\text{-}\underset{SO_3H}{\bigcirc} \right)_{1.33} \left(\text{-}O\text{-}\bigcirc \right)_{0.67} = N \right]_n \quad \text{(IV)}$$

was mostly related to the content of acidic functions and that the synthesized sulfonated polyphosphazenes were not hydrolytically stable. A significant molecular weight decrease was observed during sulfonation of poly[bis(phenoxy)-phosphazene], and the attempt to obtain a sulfonic acid derivative of poly-[bis(4-benzoylphenoxy)phosphazene] failed because of the degradation of the polyphosphazene backbone.

Although all of the discussed phosphazene polyelectrolytes had the side groups attached to the phosphorus by oxygen or nitrogen linkages, poly(alkyl/arylphosphazenes) (i.e., phosphazene polymers in which all of the side groups are attached by direct P–C bonds) are of particular interest because of their chemical and thermal stability and unique chemical and solubility properties [8]. Poly(alkyl/arylphosphazenes) can be prepared by a polycondensation reaction that starts from *N*-silylphosphoranimine "monomers" [8]. The advantage of this approach over the ring-opening or substitution route is that it includes the introduction of the desirable alkyl or aryl substituents at the monomer stage before polymerization. This eliminates the formation of the poly(dichlorophosphazene) and, thereby, avoids the difficulties of the organic nucleophile substitution reactions.

The first water-soluble P–C-substituted phosphazene polyelectrolytes were synthesized by the modification of preformed poly(methylphenylphosphazene) [9]. A series of carboxylated derivatives (**V**) was prepared by deprotonation of the appropriate number of methyl groups in the parent polymer with *n*-BuLi, followed by treatment with anhydrous, gaseous carbon dioxide. This reaction could be modulated so that the final product had 10, 25, or 50% of the repeating units in either the lithium salt or free acid form.

$$\left[\left(\underset{CH_3}{P=N} \right)_x \left(\underset{CH_2COOH}{P=N} \right)_y \right]_n \quad \text{(V)}$$

The preparation of an interesting water-soluble cationic polymer that combines the properties of both polyphosphazenes and organic polymers was reported recently [10]. A six-armed star-branched polymer, with a hydrophobic cyclotriphosphazene core and hydrophilic *N*-protonated poly(ethyleneimine) branches, was synthesized. This was achieved by a multistep modification of hexachlorocyclotriphosphazene, followed by a graft copolymerization with 2-methyl-2-oxazoline, with consequent hydrolysis. Star-branched polymers are generally known for their enhanced segment density and unique solution prop-

erties. The authors described polymers with molecular weights up to 23,000, which were soluble in water, but precipitated in solutions at pH 8 and higher.

B. Solution Properties

The physicochemical behavior of phosphazene polyelectrolytes in aqueous solutions has been the focus of very few studies. PCPP was investigated because it formed gels when treated with salts of divalent and trivalent cations, such as calcium, copper, or aluminum [5]. This property was successfully used to prepare biocompatible microspheres [3]. The hydrogels formed by this process were described as soft and highly swollen materials. When PCPP was cross-linked with aluminum ions, the gel contained more than 95% water. Similar cross-linking with dications was reported for polymer (**IV**) [9]. The ionic gelation phenomenon and its application to microsphere preparations will be the subject of more detailed discussion later.

Spectroscopic, viscometric, and potentiometric characterization of PCPP in dilute aqueous solutions has been undertaken to establish a correlation between the unique structure of this polymer and its solution behavior [11]. Fluorescence probe studies showed that PCPP interacts more favorably with counterions of a hydrophobic character. The ability of PCPP to bind hydrophobic counterions, such as a cationic fluorescent probe, was notably larger compared with other polyelectrolyte systems. Size exclusion chromatography of polyphosphazene electrolytes in neutral aqueous solutions in the presence of salt [6] showed no evidence of the abnormal behavior that was reported for solutions of polyphosphazenes in organic solvents [8].

The NMR and infrared (IR) analyses of carboxylic acid derivatives of poly (methylphenylphosphazene) suggested that the acid polyphosphazenes may exist in a self-protonated, zwitterionic form [9].

Most phosphazene polyelectrolytes are stable in aqueous solutions, although sulfonated polyphosphazenes showed considerable degradation of the polymer backbone [7]. An attempt to isolate poly[bis-(4-benzoyl)phosphazenesulfonic acid] using aqueous solutions was unsuccessful because of a breakdown of the polymer backbone. Thermal gravimetric analytic studies showed that the thermal stability of carboxylic acid derivatives appears to become less with incorporation of more COOH moieties [9].

The reactivity of phosphazene polyacids is severely restricted, largely by the limited choice of solvents suitable for the reaction and the salt formation that precedes covalent coupling [5]. Treatment of the carboxylic acid derivative of poly(methylphenylphosphazene) with alcohols in the presence of N,N'-dicyclohexylcarbodiimide showed no evidence that reaction had occurred, even after several days of mild heating [9]. Attempts to cross-link PCPP by chemical condensation of the carboxylic acid groups with diamines or diols using carbodiimide method were also unsuccessful [5]. Degradation of the phosphazene backbone can also be an obstacle to polymer derivatization. It was impossible to obtain poly(acid chlorides) by treatment with thionyl chloride or phosphorous chloride, even in the presence of amine scavengers of HCl by-products, because of polymer breakdown [9].

III. IONIC GELATION OF POLYORGANOPHOSPHAZENES: PREPARATION AND PROPERTIES OF POLYPHOSPHAZENE MICROSPHERES

A. Mechanism of Gelation

Phosphazene polyelectrolytes form highly swollen ionotropic gels in the presence of calcium and other multivalent ions in aqueous media. This property allowed the development of an extremely mild method for encapsulation of biological entities. Nevertheless, little information is available on the mechanism of phosphazene polyelectrolyte gelation.

A general overview of ionic gelation phenomena in macromolecular systems can be helpful to understand the peculiarities of the microencapsulation process for polyphosphazenes. The phase separation of different synthetic and natural polycarboxylates, induced by addition of divalent cations, has been extensively investigated [12–16]. It was reported that different polyanions show the same macroscopic behavior that can be interpreted through the same general scheme of interactions. Affinity constants in this scheme depend on the characteristics of each macromolecular chain and on the electronic configuration of the ions [16]. Phase diagrams for these systems usually consist of two main regions. The homogeneous phase can be either a solution or a gel (low-salt concentration range), whereas the heterogeneous phase contains a transparent polymer-poor phase in equilibrium with a polymer-rich phase composed of precipitates, large aggregates of microgels, or a dense uniform gel (high-salt concentration range).

Three major types of polymer behavior, based on the position and shape of the demixing line in phase diagrams, were identified, regardless of the chemical structure of polycarboxylates [16].

1. Type H

Type H phase separation occurs when specific interactions between ions and polyions can be neglected. That usually takes place with weakly charged polyelectrolytes and divalent cations, or strong polyelectrolytes and monovalent cations. High-salt concentration is required to induce polymer precipitation. Polymer–solvent interactions can be described thermodynamically by the second virial coefficient, which contains nonelectrostatic and electrostatic components. The electrostatic term tends to zero as the ionic strength increases and, if the nonelectrostatic term is negative, then polyelectrolyte precipitation can take place. This effect is related to the electrostatic repulsions between the like charges fixed along the macromolecular chain and their screening by the low molecular weight salt ions in the solution. As the ionic strength increases, the electrostatic repulsions decrease, and allow the hydrophobic interactions to take place with a phase separation as a final result.

2. Type L

Type L phase separation describes the behavior of strongly charged polymers that interact in a specific manner with counterions (polycarboxylates with Ca^{2+}, Ba^{2+}, and so on). The precipitation is observed at low values of cations and can be attributed to the formation of ion–polyion complexes.

3. Type I

Type I behavior lies intermediate between weak ion–polyion interactions, corresponding to H type, and stronger interactions, corresponding to the L type.

The existence of a gel phase in polymer-phase diagrams required the subdivision of this classification into I type, for which homogeneous gels or gels with syneresis were formed, and L type, for which homogeneous gels were obtained.

A gel phase forms from a semidilute solution of polymer into an infinite network of polymer molecules. Gel formation results from intermolecular binding, with at least one cross-link per macromolecular chain. The actual concentration of the cross-links that cause gelation of a poly(acrylamide-*co*-acrylic acid)–Cr^{III} system was established, by the comparison of rheologic and spectroscopic studies [15], to be approximately ten cross-links per chain. This was significantly higher than predicted by the classic theory, which indicated the existence of intramolecular complexes that did not contribute to the gelation.

The presence of the gel phase was observed more frequently in the natural polymers, such as alginate or pectine (PEC). For these polymers the local configuration of the backbone facilitates interchain associations and formation of junction zones, resulting in the interchain chelation of cations on the specific-binding sites ("egg-box" model).

Synthetic copolymers of acrylamide and acrylic acid (PAM), seemed to have mostly type H and type I (gel with syneresis) behavior, with lower affinity toward divalent cations than natural polymers [16]. The Cu^{2+}–PEC and Al^{3+}–PAM systems showed the same types of phase diagrams. The distribution of carboxylic groups along the macromolecular chain in synthetic polymers is critical for prediction of their gelation properties, since polymers with block-type distribution favor generally intramolecular cation binding, which leads preferentially to precipitation without gel formation. Addition of a low molecular weight salt of monovalent cations to these systems, as well as pH changes, could dramatically affect phase diagrams. This may be due to ion-exchange reactions, with gel resolubilization, or to changes in the hydrophobic interactions owing to the changes in the charge-density parameter.

An attempt to quantitatively predict the sol–gel transition in a polycarboxy-late–divalent cation system showed a good correlation for at least some of the experimental systems [16]. The following expression predicts the concentration of salt ions $[Me^{2+}]_{gel}$ at the gelation point,

$$[Me^{2+}]_{gel} = N^{-1} K_{BInter}^{-1} C_p^{-0.5} \ (\alpha\tau)^{-1} [1 + K_B(\alpha\tau C_p)^2]$$

where N is a degree of polymerization; C_p is a polymer concentration; α and τ are ionization degree and molar fraction of carboxylic acid groups, respectively; K_B and K_{BInter} are equilibrium constants of the ion–polyion interactions and intermolecular binding, respectively.

Polyphosphazene polyelectrolytes, such as PCPP, seem to have an advantage to form soft and highly swollen hydrogels on cross-linking with calcium, copper, or aluminum, in a wide range of concentrations. This is important for their application as materials for microencapsulation [5]. Mild conditions of ionic gelation and formation of a hydrogel with desirable physicochemical properties made polyphosphazene polyelectrolytes an attractive class of polymers for use in microencapsulation of sensitive biological molecules, such as proteins, peptides, and even living cells.

B. Microencapsulation

1. Microencapsulation Techniques

Polyphosphazene microspheres were initially prepared by using a droplet-forming apparatus, which produced spherical gel particles, mainly in sizes ranging from 0.5 to 1.5 mm. In this procedure, a 2.5% polymer solution, containing protein, was sprayed as microdroplets into a 7.5% calcium chloride aqueous gelation solution. The microdroplets were produced by simply pumping the solution through a needle located inside a custom-made nozzle, with an inlet for pressurized air [3]. The shear forces generated by the flowing airstream inside the nozzle detaches the droplets forming at the needle tip and sprays them into the calcium chloride solution, where they gel and harden for 30 min. This method was successfully used to encapsulate fluorescein-labeled bovine serum albumin (FITC-BSA) and β-galactosidase, liposomes, and hybridoma cells [3].

Second-generation polyphosphazene microspheres, containing encapsulated antigens, have since been produced in the size range of 1–10 μm to target Peyer's patch M cells and subepithelial macrophages [17,18]. In this process, a pump forces a polymer solution, containing dispersed antigens, through a small orifice in an ultrasonic (Medsonic, Inc., Farmingdale, NY) spray nozzle. Inside the nozzle, the solution is subjected to a sterile airstream at approximately 40 psi (Fig. 1). The configuration of the nozzle results in the generation of a spray cloud containing micron-sized microdroplets that impact a calcium chloride bath, where the microdroplets gel into microspheres. A typical particle size distribution of polyphosphazene microspheres obtained by this method is shown in Fig. 2.

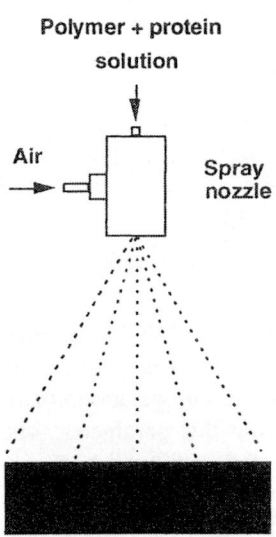

Fig. 1 Microencapsulation process using spray technique.

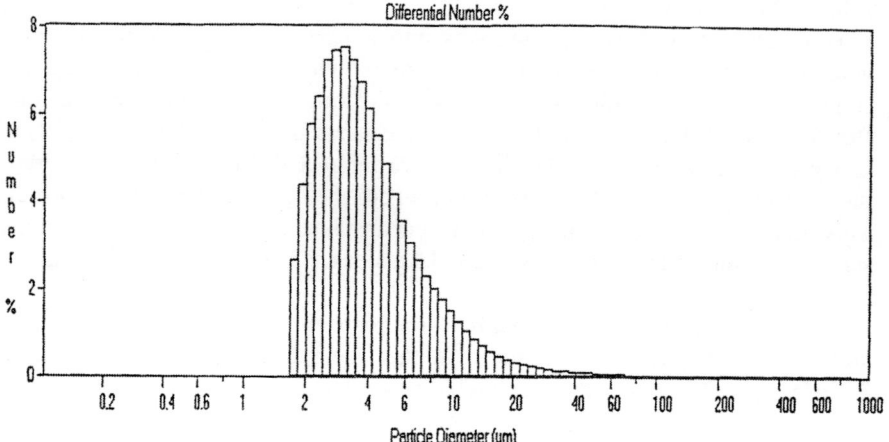

Fig. 2　Particle size distribution of polyphosphazene microspheres. (From Ref. 17.)

2.　Morphology of Microspheres

Visual and microscopic examinations of calcium–PCPP microspheres and their cross sections revealed an overall opaque appearance [19] that was different from the translucent microspheres formed from the calcium salt of alginate obtained under the same conditions. This suggested a heterogeneous structure for the calcium–PCPP gels that would be classified as an I-type gel with syneresis. The morphology of calcium–PCPP gel microspheres was also highly dependent on the concentration of PCPP and calcium ions used in the microencapsulation process. Irregularly shaped microparticles were formed at high and low polymer concentrations [19]. Other parameters, such as composition of polymer solution and the presence of culture media also affected gelation and led to the increase in the formation of broken spheres [20].

3.　Efficiency of Encapsulation and Effect of Polymer Matrix on the Bioavailability and Bioactivity of Substrate

The *efficiency of microencapsulation* into ionically cross-linked PCPP matrices was usually defined as the percentage of the input substrate that is recovered in the polyphosphazene matrices. The nature of the substrate and gelation conditions are usually the most important factors determining this parameter. The recovery of FITC–BSA and β-galactosidase was reported as 60 and 80%, respectively [3]. The activity of β-galactosidase enzyme was comparable with its aqueous solution. Hybridoma cells could be encapsulated with a 25–31% entrapment efficiency and a viability of more than 70% [20]. In these studies, the gelation conditions had significant effect on both the efficiency of encapsulation and cell productivity. When trivalent ions, such as aluminum, were used, higher efficiency and lower productivity were observed. The combination of calcium ions and low pH gave the best results in terms of a high-encapsulation percentage of cells and maintaining the viability and productivity of the hybridoma cells.

4. Microsphere Stabilization and Modification

Ionotropic gels are sensitive to the ionic environment; consequently, most divalent cation–polyanion systems will disintegrate with time in basic and high ionic strength solutions. PCPP gel matrix microspheres can be stabilized by reaction with positively charged polyelectrolytes, such as poly(L-lysine) [3]. Interpolymer complexes, presumably forming on the microsphere surface and possibly in the microsphere interior, contribute to microsphere stability in physiological saline solution [21,22]. This also enables liquefaction and swelling of the internal core of the microcapsules in saline solution by ion-exchange reactions with monovalent salts and chelating agents. The hydrophobicity of the microspheres can also be regulated by subsequent reaction of the PLL surface with a polycation solution [19].

5. Safety

The safety of PCPP was tested both in vitro and in vivo. Hybridoma cells were encapsulated in polyphosphazene microspheres having a diameter between 150 and 200 μm. The encapsulated hybridoma cells were able to undergo cell divisions, and by 10 days after encapsulation the microspheres were essentially filled with living cells (20). This demonstrated the innocuous nature of the polyphosphazenes in cell culture. In vivo acute toxicity of PCPP microspheres was evaluated in 6- to 8-week-old Sprague–Dawley rats [17]. The study demonstrated that an oral dose of 5000 mg/kg of calcium–PCPP microspheres was not acutely toxic. There were no significant differences in body weight gain between the rats that received microspheres and those in the control group. There were no treatment-related abnormalities observed in any organ at necropsy, and the results of hematology and clinical chemistry were normal for all rats.

C. Controlled-Release Using Polyphosphazene Microspheres

The PCPP hydrogel microspheres have potential as materials for controlled-release drug delivery. Diffusion controlled release remains the most important mechanism for hydrogel-delivery systems [2]. To examine the ability of the PCPP hydrogel matrix to effect a controlled release of macromolecular substrates, proteins with varying molecular weights and 24-nm fluorescent polystyrene beads were encapsulated in polyphosphazene microspheres [19]. The release kinetics from these microspheres were characterized as a function of the gelation conditions, such as the ionic cross-linker and polymer concentrations. In addition, the permeability properties of membranes formed by reacting PCPP gel microspheres with the positively charged polyelectrolyte, poly(L-lysine) (PLL), were evaluated as a function of polycation molecular weight (MW), concentration, and reaction time with the microspheres.

1. Effect of PCPP and Calcium Concentrations on Microsphere Permeability

Physical properties, such as the permeability of ionically cross-linked polyions by polyvalent metals, are usually strongly dependent on the polymer concentration and distribution of the gelation-inducing ions [23]. Thus, the effects of polyphosphazene and calcium ion concentrations used in microcapsule preparation on the outward diffusion of several substrates were investigated. Proteins having molecular weights, ranging from 14.3 to 200 kDa, that are commonly used as molecular weight markers in polyacrylamide gel electrophoresis, and polystyrene beads with a diameter of 24 nm, were studied as model substrates. Increasing both the polymer and ionic cross-linker concentrations over a wide range leads to significant decreases in the outward diffusion of the proteins (Fig. 3) and PS beads (Figs. 4 and 5). For example, for a given protein (BSA) of molecular weight 69 kDa, increasing the PCPP concentration from 1.5% to 3.3% (w/w) led to a more than fourfold decrease in protein release after 24 h.

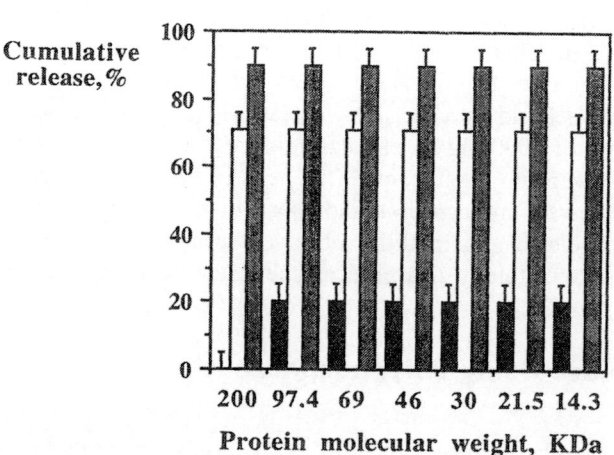

Fig. 3 Cumulative release of Rainbow protein markers of varying molecular weight from PCPP microspheres as a function of polymer concentration: (■) 3.3, (□) 2.5, (▧) 1.5% (w/w). Microspheres were incubated in PBS pH 7.4 at room temperature for 24 h before the amount of protein released in the supernatant was spectrophotometrically measured. (From Ref. 19.)

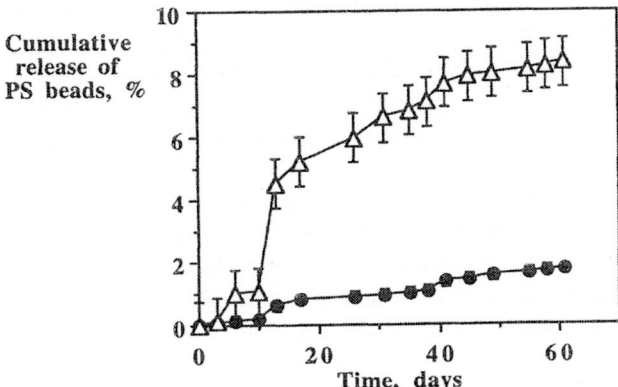

Fig. 4 Cumulative PS bead release from PCPP microspheres as a function of polymer concentration: (●) 2.5 and (△) 2% (w/w). Microspheres were incubated in deionized water at 37° C. (From Ref. 19.)

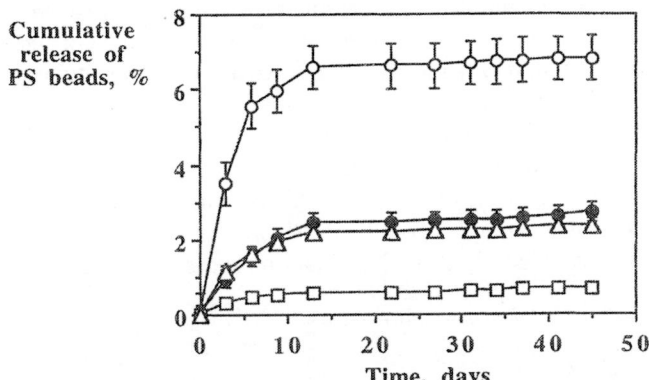

Fig. 5 Effect of calcium chloride concentration on release of PS beads. Concentration of $CaCl_2$: (○) 1.5, (●) 5, (□) 12, (△) 20% (w/w). Microspheres were prepared from 2.5% (w/w) polymer solution and then incubated under the same conditions as in Fig. 4. (From Ref. 19.)

Increasing calcium ion concentrations also tended to decrease substrate release rates.

2. Effect of PLL Coating

An effective way to control the permeability and substrate diffusion rates from polyphosphazene microspheres is the creation of a semipermeable membrane around the capsule by the formation of a stable polyelectrolyte complex between the PCPP microsphere surface and PLL [19]. The release of FITC–BSA (68 kDa), β-galactosidase (540 kDa), and PS beads encapsulated in either PCPP microspheres and PCPP microspheres treated with PLL, were studied in phosphate-buffered saline (PBS; pH 7.4) at 37°C (Fig. 6). The results showed that PLL coating always prolonged the outward substrate diffusion. Uncoated microspheres released more than 40% of the FITC–BSA protein during the first half hour, compared with approximately 5% of the protein from microspheres coated with low molecular weight PLL (21.5 kDa). The permselectivity of PCPP-PLL membranes for macromolecules of different molecular weights was also demonstrated. Although almost 70% of the FITC–BSA diffused out during the first 5 h, only 20% of β-galactosidase was released over the same time period.

The effects of PLLs' molecular weight, concentration, and time of interpolymer complex formation on the permeability of PLL-coated microspheres were investigated using fluorescent polystyrene beads as model substrates (Fig. 7). An increase in the molecular weight of PLL above 60 kDA not only significantly increased the release kinetics, but also changed the release profile to a biphasic pattern, by adding an initial "burst" from microspheres coated with high molecular weight PLL. A similar burst effect was found when the concentration of PLL in the coating step was increased. To explain these results, one must remember that there are two different types of cross-linker in the polymer matrix: (a) "ionic bridges" of divalent metal (type I gel), which are extremely sensitive to the ionic environment, owing to the ion-exchange reactions between divalent and monovalent ions; and (b) cross-links caused by formation of interpolymer polyelectrolyte complexes (PCPP–PLL), which are relatively stable under the conditions studied [21,22]. The gelation process requires diffu-

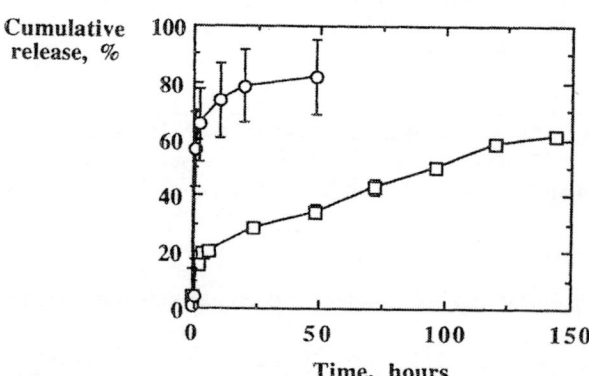

Fig. 6 Cumulative release of FITC–BSA ($\bigcirc$) and β-galactosidase ($\square$) from microspheres coated with 0.025% PLL solution (21.5 kDa) during 30 min. Protein release was measured in PBS (pH 7.4) at 37°C. (From Ref. 19.)

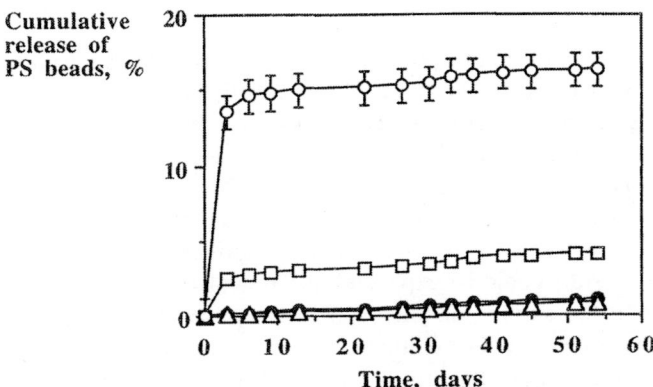

Fig. 7 Release of PS beads from PCPP microspheres coated with PLL of different molecular weight: (●) 12 kDa, (△) 62.5 kDa, (□) 140.8 kDa, (○) 295 kDa. Concentration of PLL solution was 0.04% (w/w), and the reaction time 6 min. The coated microspheres were then incubated in PBS, as in Fig. 6. (From Ref. 19.)

sion of the cross-linking reagents from the surface into the microsphere matrix and, thus, a difference in the molecular size of the cross-linkers leads to a different cross-linker distribution in the matrix. Obviously, as the molecular weight of the PLL increases the greater the diffusion limitations. As the diffusion rate of the cross-linker decreases, the superficial layer of the microsphere that is cross-linked with stable interpolymer complex will be thinner, and the larger part of the microsphere will consist of an ionically sensitive matrix. As a result, microspheres coated with high molecular weight PLL swell dramatically in PBS (diameter in PBS/diameter in water ratio is 1.65 for PLL with an molecular weight 295 kDa), whereas no swelling was observed for microspheres coated with low molecular weight PLL. This swelling, or so-called liquefaction of the internal core of microspheres, [20] is responsible for changes in the release profiles [19]. The greater the degree of microsphere swelling, the higher the initial burst release of substrate. There is evidence that the osmotic pressure in these systems, caused by partial removal of the cross-linker and swelling of the capsules, led to a rupture of the relatively rigid PCPP–PLL membrane "shell" in the same way that is observed for solvent-activated drug-delivery capsules [24]. Following membrane rupture, and after termination of polymer swelling, redistribution of poly(L-lysine) on the surface of microspheres can occur. This leads to the reestablishment of a uniform coating that allows only a slow diffusion of fluorescent beads through the PCPP–PLL membrane, with release rates similar to those of nonswollen microspheres (see Fig. 7).

3. Effect of PCPP–PLL Double Coating

The surface properties of PCPP microspheres, such as hydrophobicity, can be modified by coating with a second layer of PCPP (unpublished data). *Dual-membrane* coating of polyphosphazene microspheres (PLL—first coating; polyphosphazene—second coating) resulted in a further decrease in permeability, as measured by the outward diffusion of encapsulated PS beads. However, variations in PCPP concentrations or increases in the reaction time between PCPP and PLL used in the second coating, did not significantly change the release profiles. These results could be explained in terms of PCPP–PLL interactions being limited to the surface of microspheres because of the diffusion limitations for high molecular weight PCPP.

It was demonstrated in these experiments that hydrogel permeability and release patterns can be engineered by selecting different formulation parameters, such as polyphosphazene and ionic cross-linker concentration, or to a greater extent, by varying the conditions of their coating with the polycation.

D. Degradation of Polyphosphazene Microspheres

The degradation of drug-delivery devices is regarded as highly desirable from a pharmacokinetic standpoint and is often used as a powerful tool to control drug-release profiles [2]. The hydrolytic degradability of ionotropic gel microspheres composed of either PCPP or poly[(carboxylatophenoxy)(glycinato) phosphazene] (PCGPP) was evaluated in aqueous solution (pH 7.4, HEPES-buffered saline, 37°C) by monitoring mass loss, decrease in the molecular weights of polymer matrices, and formation of soluble products [6]. The PCGPP was synthesized and used for microsphere preparation because it was anticipated that introduction of a hydrolysis-sensitive pendant amino acid group would provide sites for rapid hydrolytic chain cleavage and allow the increase of the degradation rate in an aqueous environment [25].

Analysis of erosion profiles for polyphosphazene microspheres showed significant differences in stability, not only for the two types of polymers, but also for the same polymers with different molecular weights. (Fig. 8). No detectable mass loss was observed during a 6-month incubation of microspheres composed of high molecular weight PCPP (3900 kDa). However, introduction of as little as 10% of the hydrolytically unstable glycinato groups in the polymer matrix (PCGPP) resulted in a dramatic increase in microsphere erosion. PCGPP with a weight average molecular weight of 130 kDa degraded into completely water-soluble products within a 3-day period. Microspheres prepared from PCGPP having a higher molecular weight of 170 kDa degraded more slowly, which is similar to the results obtained for PCPP microspheres.

Gel-permeation chromatographic analysis of matrix material and water-soluble products, revealed significant decreases in polymer molecular weight for all samples, including high molecular weight PCPP (Fig. 9). It was initially as-

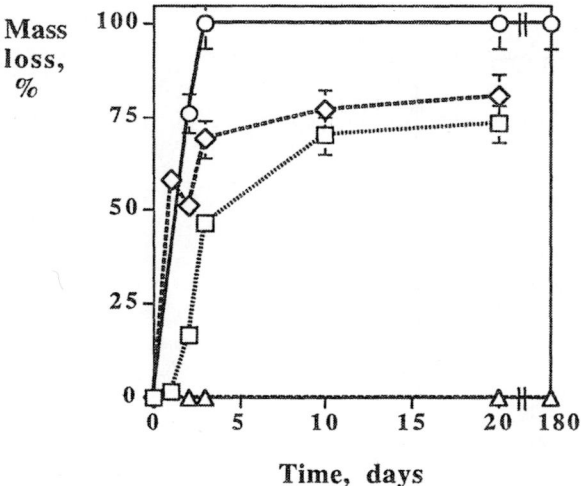

Fig. 8 Erosion profiles of calcium polyphosphazene microspheres. Microspheres were obtained by ionic cross-linking of PCPP with molecular weight 3900 kDa ($\triangle$) and 400 kDa ($\square$); and PCGPP with molecular weight 130 kDa ($\bigcirc$) and 170 kDa ($\diamondsuit$). From Ref. 6.)

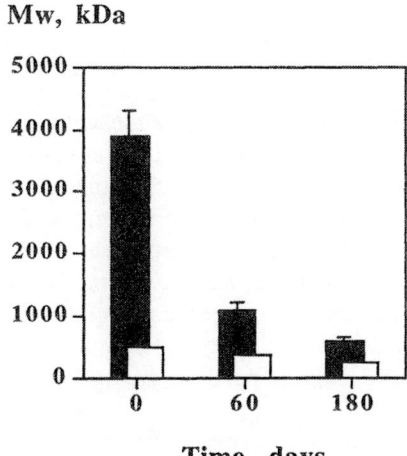

Fig. 9 Molecular weight degradation profiles for polyphosphazene hydrogel microspheres formed by ionic cross-linking of PCPP with molecular weight 3900 kDa. Microspheres were incubated in HEPES-buffered saline (pH 7.4). Matrix samples were isolated and then dissolved to measure weight-average ($\blacksquare$) and number-average ($\square$) molecular weights. (From Ref. 6.)

Mw, kDa

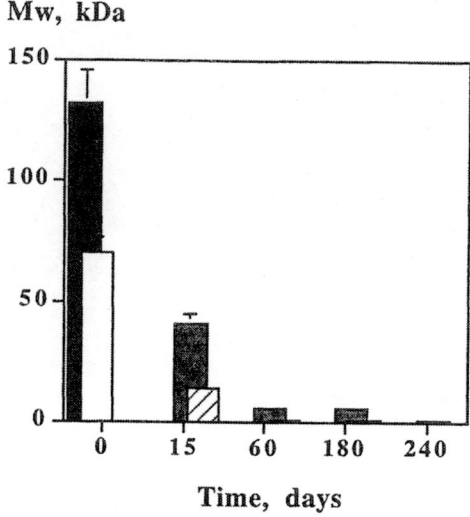

Time, days

Fig. 10 Molecular weight degradation profiles for polyphosphazene hydrogel microspheres formed by ionic cross-linking of PCGPP with molecular weight 130 kDa. Weight-average (▓) and number-average (▨) molecular weights of water-soluble degradation products in supernatant were determined. Matrix samples were isolated and then dissolved to measure weight-average (■) and number-average (□) molecular weights. (From Ref. 6.)

sumed that the mechanism of degradation for PCPP could involve intramolecular carboxylic group catalysis. However, preliminary experiments indicate that this degradation can be caused by the residual hydrolytically labile chlorine atoms attached to the phosphazene backbone. Chloro-containing derivatives of PCPP were synthesized, and the correlation was found between the content of chlorine atoms and the degradation kinetics. Hydrolysis of PCPP "activated" with glycinato groups (130-kDa PCGPP), in an aqueous environment for 8 months, resulted in breakdown of the polymer backbone, leading to fragments with molecular weights lower than 1 kDa (Fig. 10). These results demonstrate that the degradation properties of PCPP can be controlled by incorporation of hydrolysis-sensitive chlorine atoms or amino acids as pendant groups in the polymer structure.

As was discussed earlier, the permeability and release rates from polyphosphazene microspheres can be affected by the creation of semipermeable interpolymer complex membranes around the capsule. Also, coating of the PCGPP hydrogel microspheres with PLL led to a dramatic decrease of the erosion rate (Fig. 11), apparently because of steric hindrance and an increase of matrix hydrophobic properties. This appears to provide an additional tool to control the degradation and stability of polyphosphazene microspheres.

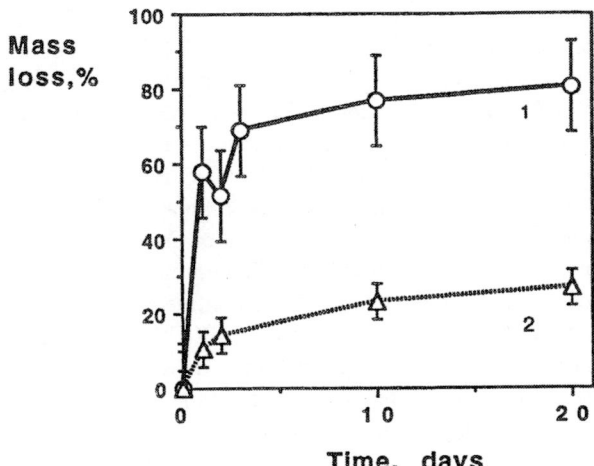

Fig. 11 Erosion profiles for (1) calcium PCGPP hydrogel microspheres and (2) the same microspheres coated with PLL (62 kDa). (From Ref. 6.)

IV. IMMUNE RESPONSE TO ANTIGENS IN MICROSPHERES

Traditionally, most injected nonreplicating vaccines have required multiple doses to achieve sufficient serum antibody titers to be protective. For obvious reasons, it would be much more desirable to achieve protection with a single immunization. The immunogenicity of influenza antigens formulated in polymeric microspheres composed of alginate or polyphosphazene was compared with influenza antigens formulated in either PBS or in the standard adjuvants, alum and complete Freund's adjuvant [17] (Table 1). The PCPP microspheres

Table 1 Effect of Adjuvants on the Antibody Response to Influenza as Measured by ELISA

Influenza Virus in	Specific Titer at Wk				
	3	5	7	9	13
Water	256	1,024	1,024	512	512
Alginate MS	512	1,024	2,048	2,048	2,048
Alum	<256	512	1,024	2,048	2,048
CFA	8,192	16,384	32,768	32,768	16,384
Polyphosphazene MS	8,192	32,768	32,768	8,192	16,384

Mice were immunized SC with 5 μg of whole formalin-inactivated influenza virus particles.
Source: Ref. 17.

were as efficient as complete Freund's adjuvant at inducing a very high titer anti-influenza immune response, measured in enzyme-linked immunosorbent assay (ELISA). In contrast, alum-adjuvanted influenza, influenza in PBS, and alginate microencapsulated influenza elicited rather low titer anti-influenza responses. The mouse sera were also tested for the presence of functional antibodies by hemagglutination inhibition (HAI) and neutralization assays (data not shown). As measured by the HAI assay, the PCPP microspheres containing influenza elicited a high antibody titer. Influenza formulated in water, alum, complete Freund's adjuvant, or alginate microspheres elicited either no detectable or very low HAI titers. Antibodies that neutralize influenza infectivity were assayed in a 50% plaque reduction assay. Influenza in PCPP microspheres induced a detectable neutralization titer, whereas influenza in water, alum, or complete Freund's adjuvant did not elicit detectable neutralizing antibody titers. These results were very encouraging, because the HAI and neutralization assays are sensitive functional antibody assays for influenza. Taken together, these results demonstrate that PCPP microspheres containing an antigen provoke an antibody response that is superior to commonly used adjuvants.

REFERENCES

1. H. R. Allcock, Polyphosphazenes as new biomedical and bioactive materials, *Biodegradable Polymers as Drug Delivery Systems* (M. Chasin and R. Langer, eds.), Marcel Dekker, New York, 1990, p. 163.
2. K. Park, W. S. W. Shalaby, and H. Park, *Biodegradable Hydrogels for Drug Delivery*, Lancaster-Basel: Technomic Publishing Co., 1993, p. 252.
3. S. Cohen, M. C. Baño, K. B. Visscher, M. Chow, H. R. Allcock, and R. Langer, Ionically cross-linkable polyphosphazene: A novel polymer for microencapsulation, *J. Am. Chem. Soc. 112*:7832–7833 (1990).
4. H. R. Allcock, T. J. Fuller, and T. L. Evans, Side-group construction in high polymeric phosphazenes via lithiophenoxy intermediates, *Macromolecules 13*:1325–1332 (1980).
5. H. R. Allcock and S. Kwon, An ionically cross-linkable polyphosphazene: Poly[bis(carboxylatophenoxy)phosphazene] and its hydrogel and membranes, *Macromolecules 22*:75–79 (1989).
6. A. K. Andrianov, L. G. Payne, K. B. Visscher, H. R. Allcock, and R. Langer, Hydrolytic degradation of ionically cross-linked polyphosphazene microspheres, *J. Appl. Polym. Sci. 53*:1573–1578 (1994).
7. E. Montoneri, M. Gleria, G. Ricca, and G. C. Pappalardo, New acid-polyfunctional water-soluble phosphazenes: Synthesis and spectroscopic characterization, *J. Macromol. Sci. Chem. A26*:645–661 (1989).
8. R. H. Neilson and P. Wisian-Neilson, Poly(alkyl/arylphosphazene) and their precursors, *Chem. Rev. 88*:541–561 (1988).
9. P. Wisian-Neilson, M. S. Islam, S. Ganapathiappan, D. L. Scott, K. S. Raghu-

veer, and R. R. Ford, Carboxylic acid, ester, and lithium carboxylate derivatives of poly(methylphenylphosphazene), *Macromolecules* 22:4382–4384 (1989).

10. J. Y. Chang, H. J. Ji, M. J. Han, S. B. Rhee, S. Cheong, and M. Yoon, Preparation of star-branched polymers with cyclotriphosphazene cores, *Macromolecules* 27:1376–1380 (1994).

11. G. Masci, S. Contadini, V. Crescenzi, and M. Dentini, Solution characterization of a polyphosphazene polyelectrolyte, *Polym. Preprints 34*:1067–1068 (1993).

12. J. Ricka and T. Tanaka, Phase transition in ionic gels induced by copper complexation, *Macromolecules 18*:83–85 (1985).

13. A. C. Habert, C. M. Burns, and R. Y. M. Huang, Ionically crosslinked poly-(acrylic acid) membranes. II. Dry technique, *J. Appl. Polym. Sci. 24*:801–809 (1979).

14. A. C. Habert, R. Y. M. Huang, and C. M. Burns, Ionically crosslinked poly-(acrylic acid) membranes. I. Wet technique, *J. Appl. Polym. Sci. 24*:489–501 (1979).

15. C. Allain and L. Salomé, Gelation of semidilute polymer solutions by ion complexation: Critical behavior of the rheological properties versus cross-link concentration, *Macromolecules 23*:981–987 (1990).

16. M. A. V. Axelos, M. M. Mestdagh, and J. Francois, Phase diagrams of aqueous solutions of polycarboxylates in the presence of divalent cations, *Macromolecules* 27:6594–6602 (1994).

17. L. G. Payne, S. A. Jenkins, A. K. Andrianov, and B. E. Roberts, Water soluble phosphazene polymers for parenteral and mucosal vaccine delivery, *Vaccine Design* (M. F. Powell and M. J. Neuman, eds.), Plenum Press, New York, 1995, p. 473.

18. L. G. Payne, S. A. Jenkins, A. K. Andrianov, R. Langer, and B. E. Roberts, Xenobiotic polymers as vaccine vehicles, *Adv. Mucosal Immunology* (J. Mestecky et al., eds.), Plenum Press, New York, 1995, p. 1475.

19. A. K. Andrianov, S. Cohen, K. B. Visscher, L. G. Payne, H. R. Allcock, and R. Langer, Controlled release using ionotropic polyphosphazene hydrogels, *J. Controlled Release 27*:69–77 (1993).

20. M. C. Baño, S. Cohen, K. B. Visscher, H. R. Allcock, and R. Langer, A novel synthetic method for hybridoma cell encapsulation, *Biotechnology 9*:468–471 (1991).

21. V. A. Kabanov and A. B. Zezin, Water-soluble nonstoichoimetric polyelectrolyte complexes: A new class of synthetic polyelectrolytes, *Sov. Sci. Rev., Sec B Chem. Rev. 4*:207–282 (1982).

22. E. Tsuchida and K. Abe, Polyelectrolyte complexes, *Developments in Ionic Polymers—2* (H. J. Prosser and A. D. Wilson, ed.), Elsevier Applied Science New York, 1986, pp. 191–266.

23. A. Martinsen, G. Skjak-Brak, and O. Smidsrod, Alginate as immobilization material: 1. Correlation between chemical and physical properties of alginate gel beads, *Biotechnol. Bioeng. 33*:79–89 (1989).

24. R. Langer, New methods of drug delivery, *Science 249*:1527–1533 (1990).

25. H. R. Allcock, T. J. Fuller, and K. Matsumura, Hydrolysis pathways for aminophosphazenes, *Inorg. Chem. 21*:515–521 (1982).

5

Lipospheres for Vaccine Delivery

Shimon Amselem

Pharmos Ltd.
Rehovot, Israel

Carl R. Alving

Walter Reed Army Institute of Research
Washington, D.C.

Abraham J. Domb

The Hebrew University of Jerusalem
Jerusalem, Israel

I. INTRODUCTION

In the past, the risks of whole-pathogen vaccines and limited supplies of useful antigens posed barriers to development of practical vaccines. Today, the tremendous advances of genetic engineering and the ability to obtain many synthetic recombinant protein antigens, derived from parasites, viruses, and bacteria, has revolutionized the development of new-generation vaccines.

Although the new, small synthetic antigens offer advantages in the selection of antigenic epitopes and safety, a general drawback of small antigens is poor immunogenicity. Unfortunately, the body's immune system does not respond strongly to small peptides. In particular, macrophages do not readily ingest and process the small antigens, resulting in low antibody titers and the need for repeated immunizations. This lack of immunogenicity has created an urgent need to identify pharmaceutically acceptable delivery systems for these new antigens.

Several reports describing the improvement of immune response achieved by the association of antigens with lipid carriers, such as liposomes [1,2], or with microparticles, such as polymeric biodegradable microcapsules [3,4], have been published. The ability of these delivery systems to enhance immunogenicity was related to the physicochemical characteristics of the particles.

Lipospheres, which represent a new type of fat-based encapsulation technology developed for parenteral drug delivery [5,6], have also been used successfully as carriers of vaccines and adjuvants [7,8]. Lipospheres consist of water-dispersible, solid microparticles composed of a solid hydrophobic fat core, stabilized by one layer of phospholipid molecules embedded in their surface. Manufacture of liposphere–vaccine formulations is accomplished by gently melting neutral fat, in the presence of phospholipid, and dispersing the mixture in aqueous solution containing the antigen by vigorous shaking. After cooling of this mixture a phospholipid-stabilized solid hydrophobic fat core containing the antigen forms spontaneously.

This chapter describes the physicochemical properties and immunogenic activity of different liposphere–vaccine formulations containing a recombinant malaria antigen, R32NS1, derived from the circumsporozoite protein of *Plasmodium falciparum* as the model antigen.

II. PREPARATION OF LIPOSPHERES

A. Lipid Composition

In contrast with certain oil emulsions (including incomplete Freund's adjuvant; IFA), the liposphere approach uses only naturally occurring, biodegradable lipid constituents. The internal hydrophobic core of lipospheres is composed of fats, mainly triglycerides, whereas the surface activity of liposphere particles is provided by the surrounding lecithin layer composed of phospholipid molecules.

The neutral fats used in the preparation of the hydrophobic core of the several liposphere–vaccine formulations described here included tricaprin and tristearin (Dynasan 110 and Dynasan 118, microcrystalline triglycerides; Hüls Troisdorf AG, Germany), stearic acid (Aldrich Chemical Co., Milwaukee WI), ethyl stearate (Sigma Chemical Co., St. Louis MO), olive oil (Pompeian extra virgin; Pompeian Inc., Baltimore MD; containing 71.4% monounsaturated fat, 14.3% polyunsaturated fat, and 14.3% of saturated fat), corn oil (Mazola 100% pure; CPC International, Englewood Cliffs NJ; containing 57.1% of polyunsaturated fat and 14.3% of saturated fat).

The phospholipids used to form the surrounding layer of lipospheres were lecithin (mainly egg phosphatidylcholine, Coatsome NC-10S; Nippon Oil &

Fats Co. Ltd., Japan) and dimyristoyl phosphatidylglycerol (DMPG; Avanti Polar Lipids, Alabaster AL).

The polymers used in the preparation of polymeric biodegradable lipospheres, described in Section V, are polylactide (Resomer L 104, MW = 2000; Boehringer Ingelheim, Germany), and polycaprolactone-diol (MW = 2000; Aldrich Chemical Co., Milwaukee WI).

B. Antigens

The feasibility of developing a human malaria sporozoite vaccine was demonstrated in a clinical trial by using irradiated sporozoites as antigens. Protection against sporozoite infection apparently can be achieved by inducing a high titer of antisporozoite antibodies. It is currently presumed that it is only during the brief period (a few minutes or hours), when the sporozoite resides in the blood, that antibodies can gain access to it and prevent continuation of the malarial infection. Therefore, a major goal of a sporozoite immunization scheme is to maintain a high titer of antibodies at the time of transfer of the organism from the mosquito to the host. A major challenge is to induce a high antisporozoite antibody titer that is also long-lived, with a protective duration of several months to a year or more.

The major sporozoite antigen that is responsible for inducing protective immunity is a protein, the circumsporozoite protein (CS), that covers the sporozoite's outer surface. A region containing repeating tetrapeptides in the middle of the CS protein is also thought to be capable of inducing protective immunity. It is widely believed that high titers of antibodies to the CS protein can interrupt the life cycle at the sporozoite stage and provide protection against infection [1].

In the past decade, the Walter Reed Army Institute of Research in Washington, D.C., has undertaken a major program to develop an effective vaccine to sporozoite-induced malaria. As part of this effort, in collaboration with the US National Institutes of Health, identification of the complete structure of the gene encoding the immunodominant protein, the CS protein, on the surface of the sporozoite form of the human malaria parasite *P. falciparum* was achieved in 1984 [9]. This development has provided an opportunity for the use of synthetic antigens as the basis for vaccines against malaria. Two types of sporozoite antigens are being explored; namely, synthetic peptide-carrier protein conjugates and recombinant expression proteins.

The two malaria antigens used in this study, R32NS1 and R32LR, were supplied under a Cooperative Research and Development Agreement by SmithKline Beecham Pharmaceuticals (King of Prussia PA). R32NS1 is a fusion protein with the following amino acid sequence: $[MDP(NANP)_{15}NVDP(NANP)_{15}NVDP]NS1_{81}$. The R32 refers to the 30 repeats of the tetrapeptide

NANP interspersed with two tetrapeptide NVDP repeats from the immunodominant repeat region of the CS protein of *P. falciparum*, and $NS1_{81}$ refers to 81 amino acids from the nonstructural protein of influenza virus. $NS1_{81}$ is added because it is thought to include human T-helper–cell epitopes and to function as a carrier protein [10]. For R32LR, R32 is linked to the first two amino acids, leucine and arginine (LR), from a nonsense reading of the tetracycline gene of the vector [11]. The R32LR was used as capture antigen in the enzyme-linked immunosorbent assay (ELISA), because it contains the same repeating units as the R32NS1 antigen used for immunization [12].

C. Adjuvants

It is widely believed that optimal methods for immunization against certain synthetic antigens may require the use of adjuvants, and this has stimulated a considerable amount of research aimed at developing new or improved adjuvants. Those most widely used consist of aluminum compounds, particularly aluminum hydroxide (alum), which is used in diphtheria and tetanus toxoid vaccines [13].

A variety of lipid adjuvants and protein mediators also influence the immune response to antigens encapsulated in liposomes. The most widely used examples of such adjuvants for practical immunization procedures are endotoxin (including lipid A and lipopolysaccharide), and numerous types of lipophilic derivatives of muramyl dipeptide (MDP).

Lipid A is the terminal portion of gram-negative bacterial lipopolysaccharide (LPS). In addition to containing nearly all of the endotoxic activity of LPS, lipid A is responsible for numerous other biological activities that are ordinarily associated with LPS [14]. Because of its potent endotoxic activities, lipid A, by itself, has had limited applicability as an adjuvant for use in human vaccines.

Lipid A isolated from *Salmonella minnesota* R595 (obtained from List Biological Laboratories, Campbell CA) was used as adjuvant in some of liposphere–vaccine formulations.

Alum (aluminum hydroxide, Rehsorptar Adsorptive Gel, obtained from Armour Pharmaceutical Co., Kankakee IL) was also used as additional adjuvant in some of the liposphere–R32NS1 formulations.

D. Method of Preparation

Manufacture of lipospheres was accomplished by gently melting the neutral fat, in the presence of phospholipid, and dispersing the mixture in an aqueous solution of the antigen by vigorous shaking, which results on cooling, in the formation of a phospholipid-stabilized, solid hydrophobic fat core containing the antigen [7].

The lipid components (neutral fat, phospholipid, and lipid A) at 1:1 molar ratio of fat to phospholipid were dissolved in chloroform in a round-bottomed flask. Lipid A was added to the lipid mixture at a weight ratio of 1 mg lipid A per 500 mg neutral fat. For the preparation of polymeric lipospheres, the lipid components and the biodegradable polymers (polylactide or polycaprolactone) were codissolved in chloroform in a round-bottomed flask at 1:1 weight ratio. Lipid A was added to the lipid mixture at a weight ratio of 1 mg lipid A per 200 mg polymer. The organic solvent was evaporated, using a rotary evaporator (Buchi, Switzerland), and the flask was placed in a desiccator under reduced pressure for 1 h to remove traces of residual solvent. The dry lipid mixture was then heated from 40° to 80°C to melt the fat, depending on the melting point of the specific type of fat. Warm phosphate-buffered saline (Dulbecco's PBS, Gibco Labs, NY) containing the R32NS1 malaria antigen (3.3 mg/ml) was added to give fat and phospholipid concentrations of 10–20 mg/ml, and the formulations were vigorously mixed for 1 min using a multiwrist shaker (Lab-Line Instruments Inc., Melrose Park IL) until a homogeneous dispersion was obtained. The uniform milky-appearing suspensions were immediately cooled below 20°C by immersing the flask in a dry ice–acetone bath for several seconds with continued shaking. Unencapsulated antigen was removed by centrifugation at $12,000 \times g$ for 30 min at 20°C and then washing with fresh PBS solution. Antigen encapsulation was more than 80%, as determined by a modified Lowry method for protein determination [15]. The polymeric, biodegradable lipospheres containing polylactide (PLD) or polycaprolactone (PCL) were prepared by the same melt technique as the standard liposphere formulations.

III. PHYSICAL CHARACTERIZATION OF LIPOSPHERES

A. Morphology

Figure 1 shows a light microscope picture of a typical preparation of antigen-free lipospheres composed of tristearin and lecithin, characterized by particles having a uniform spherical shape.

B. Particle Size Determination

Analysis of particle size distribution of lipospheres was performed using a LS 100 Coulter Counter Particle Size Analyzer (Coulter Corp., Hialeah FL). This instrument can measure particles from 0.4 to 800 μm by particle size-dependent, light diffraction patterns. The individual diffraction patterns from the many moving particles in the sample cell are superimposed, creating a single, integrated composite diffraction pattern that reflects the contribution from each particle and allows the determination of particle size distribution. The computer attached to the instrument uses an LS 100 software program (an integrated set of Microsoft Windows-based application programs) that controls the LS series instruments, processes the data, analyzes test results, and prints test reports. In

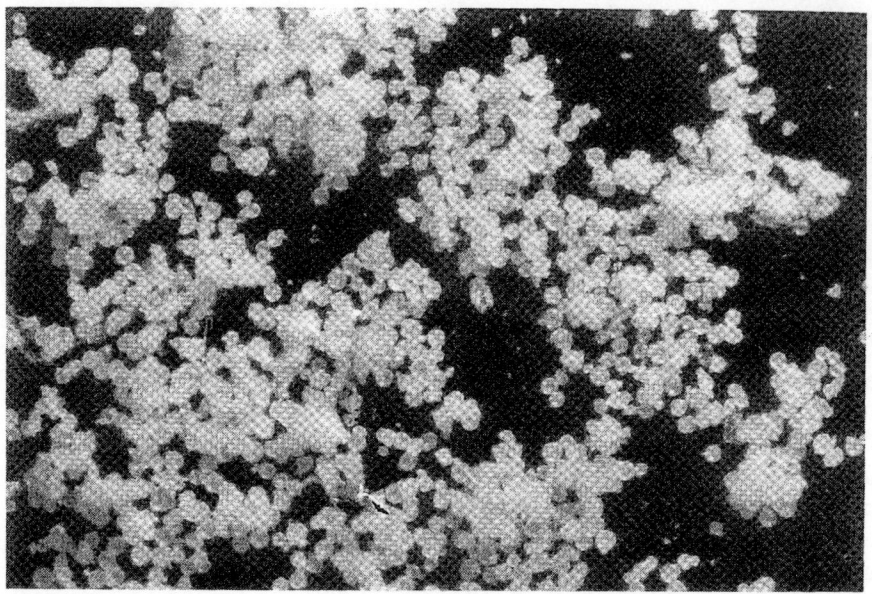

Fig. 1 Light microscopy of plain lipospheres composed of 4% tristearin (w/v) and 2% egg phosphatidylcholine (w/v) in phosphate-buffered saline.

the analysis mode, the particle size distribution plots was computed as volume-weighted distributions of particle diameter. The volume-weighted distribution (weighted by diameter cubed), which takes into consideration the actual mass of the particles, is a more accurate approach to determine the total particle size distribution of a given sample.

The particle size distribution of lipospheres containing the malaria R32NS1 antigen, but differing in their fat composition, was determined. Three groups of neutral fats were used: (a) solid fats, such as tristearin and stearic acid, with melting points in the range of 65°–70°C; (b) semisolid fats, such as tricaprin and ethylstearate, with melting points in the range of 30°–35°C; and (c) liquid fats, such as olive oil and corn oil. Two populations of liposphere particles usually coexisted, one in the size range of 1–10 μm in diameter (population A), and a second population with a diameter between 10 and 100 μm (population B; Table 1). No correlation was found between the fats' physical state (solid, semisolid, or liquid) or the melting point and particle size distribution of lipospheres formed. Lipospheres made of tristearin were the most homogeneous formulations, with 100% of the particles with an average diameter of about 7 ± 3 μm (see Table 1).

Table 1 Particle Size Distribution of R32NS1–Liospheres as a Function of Their Fat Composition

Type of fat	m.p. (°C)	Appearance	Population A		Population B	
			Size ($\mu \pm$ SD)	V%	Size ($\mu \pm$ SD)	V%
Tristearin	65	Solid	7 ± 3	100	—	—
Stearic acid	69	Solid	8 ± 4	46	81 ± 52	54
Tricaprin	30	Semisolid	7 ± 3	71	51 ± 18	29
Ethylstearate	35	Semisolid	8 ± 4	45	96 ± 51	55
Olive Oil	—	Liquid	8 ± 3	86	46 ± 14	14
Corn oil	—	Liquid	8 ± 3	91	44 ± 10	10

Biodegradable polymeric liospheres made of polylactide and lecithin showed a very broad particle size distribution from 2 to 100 μm in diameter, with mean average size of 30.6 ± 25.9 μm and median (percentage of particles > 50 μm) equal to 21.6 [8]. Polycaprolactone liospheres showed a similar range of particle size distribution, but with 1.5-fold mean average size (45.6 ± 29.5 μm) and double median value (43.6), compared with polylactide liospheres [8]. Inclusion of lipid A in the composition of the polymeric liospheres reduced their mean average particle size by a factor of 0.25, regardless of the polymer type [8].

All the polymeric liosphere formulations prepared remained stable during the 3-month period of the study, and no phase separation or appearance of aggregates were observed. The difference between polymeric liospheres and the standard liosphere formulations is the composition of the internal core of the particles. Standard liospheres, such as those previously described, consist of a solid hydrophobic fat core composed of neutral fats such as tristearin, whereas, in the polymeric liospheres, biodegradable polymers, such as polylactide or polycaprolactone substituted for the tryglycerides. Both types of liospheres are thought to be stabilized by one layer of phospholipid molecules embedded in their surface.

IV. IMMUNIZATION PROCEDURES

Mice (Balb/c, four to five per group) were immunized by intramuscular injections in the legs at 0 and 4 weeks with 0.1 ml of liosphere formulations. Antigen and lipid A doses were 2.5 μg/mouse.

Rabbits (White New Zealand, four to five animals per group) were immunized intramuscularly at 0 and 4 weeks with 0.5–1.0 ml of liosphere formulations. The antigen and lipid A dose were 100 μg/rabbit. Some of the formulations were adsorbed on alum (0.8 mg/ml final concentration) before injection. The animals were bled before the primary immunization and every 2 weeks thereafter. The sera were collected and stored at −20°C until tested for antibody production.

In the animal protocol for testing the effect of route of administration on liposphere immunogenicity, the liposphere–vaccine formulations were given to rabbits in the absence of alum by oral or parenteral routes (subcutaneous, intraperitoneal, intravenous, and intramuscular injections).

All the liposphere formulations injected containing lipid A were nonpyrogenic and nontoxic, and they were generally well tolerated. No local inflammation or other side effects were observed at the site of administration during the 3-month period of the experiment.

A. Enzyme-Linked Immunosorbent Assay

Solid-phase ELISAs were carried out to evaluate the levels of IgG antibody activity against a capture antigen (R32LR) containing the same repeating units as the antigen that was used for immunization [12]. Assays were performed at room temperature in 96-well, U-bottom Immulon-2 polystyrene microplates (Dynatech Laboratories, Alexandria VA). The wells were coated with 0.1 μg of R32LR antigen dissolved in PBS. Approximately 18 h later the contents of the wells were aspirated, filled with blocking buffer (0.5% casein, 0.01% thimerosal, 0.005% phenol red, and 1% Tween-20 in PBS) and held for 1 h at room temperature. Rabbit sera to be tested were diluted in 0.5% blocking buffer containing 0.025% Tween-20, and aliquots of each dilution were added in triplicate wells. After a 2-h incubation at room temperature, the contents of the wells were aspirated, washed three times with PBS–Tween-20 (0.05%), and 50 μg of horseradish peroxidase-conjugated goat antihuman IgG (Bio-Rad, Richmond CA; diluted 1:1000 in 0.5% blocking buffer containing 0.025% Tween-20) was added to each well. After 1 h the contents of the wells were aspirated, the wells were washed three times with the PBS-Tween-20 washing solution, and 100 μl of ABTS-peroxidase substrate (Kirkgaard and Perry Laboratories, Inc., Gaithersburg MD) was then added to each well. Absorbance was read at 405 nm 1 h after addition of peroxidase substrate using an automatic ELISA plate reader (Skatron, Norway).

V. IMMUNOGENICITY OF LIPOSPHERES
A. Effect of Liposphere Fat Composition

The effect of type of fat used in the preparation of liposphere on their immune response to encapsulated antigen was tested. Mice were immunized twice at

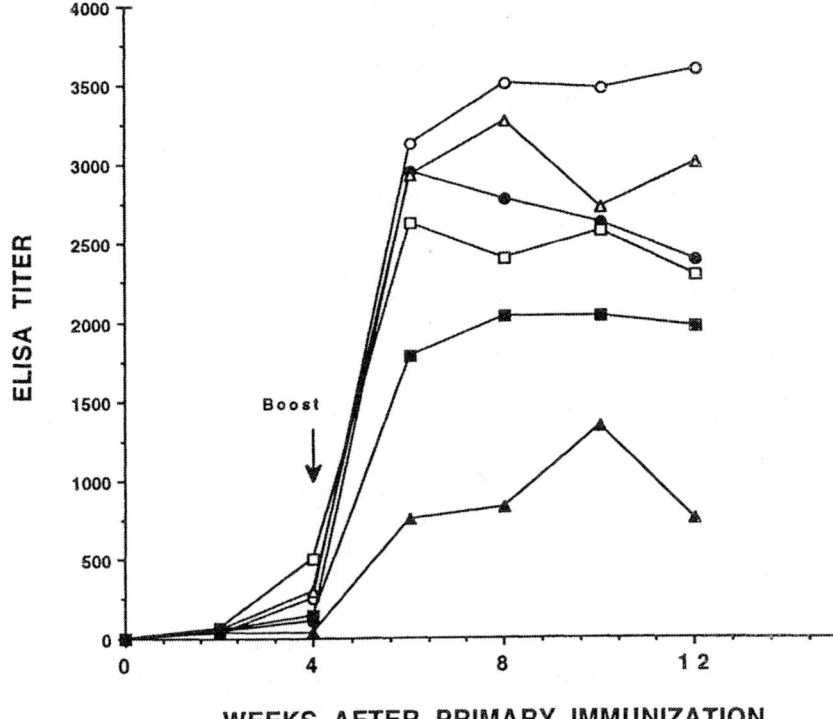

Fig. 2 Effect of liposphere fat composition on IgG antibody activity in Balb/c mice immunized with R32NS1 malaria antigen encapsulated in lipospheres. Each point represents the mean ELISA antibody response (four to five mice per group) after subtraction of the preimmune value at a serum dilution of 1:1600. Each mouse was immunized intramuscularly at weeks 0 and 4 (boost). Antigen dose = 2.5 μg/mouse. Open circles, ethylstearate; closed circles, tristearin; open squares, tricaprin; closed squares, corn oil; open triangles, olive oil; closed triangles, stearic acid.

weeks 0 and 4 with lipospheres containing R32NS1 malaria antigen. For all liposphere formulations, the first immunization at week 0 caused a very small immune response (Fig. 2). However, after the booster injection, a very marked increased of mean IgG antibody levels was observed. For most of the six liposphere–vaccine formulations tested, the immune response obtained remained at very high levels of IgG antibody titers, even after the 12-week period of the experiment (see Fig. 2). The most immunogenic liposphere formulation was the one made of ethylstearate, whereas lipospheres made of stearic acid showed the lowest IgG ELISA titers. The complete order of immunogenic activity of the six liposphere formulations tested was:

ethylstearate > olive oil > tristearin > tricaprin > corn oil stearic acid

No correlation between liposphere particle size or fat chemical characteristics and immunogenicity was found (see Table 1 and Fig. 2).

The IgG antibody ELISA titers obtained on immunizing rabbits with liposphere (R32NS1) were superior to those obtained following similar immunizations with the free antigen absorbed to alum, which showed no antibody activity at the same antigen concentrations (R. L. Richards and C. R. Alving, unpublished results). It was previously shown that this antigen was also poorly immunogenic in humans when injected alone as an aqueous solution, or when adsorbed on alum [10].

B. Influence of Phospholipid Composition

Incorporation of a negatively charged phospholipid, dimyristoyl phosphatidylglycerol (DMPG) in the liposphere lipid phase caused a significant increase in the antibody response to the encapsulated R32NS1 antigen [7]. Enhancement of immunogenicity by inclusion of charged lipids has also been observed with certain antigens in liposomes. Negatively charged liposomes produced a better immune response to diphtheria toxoid than positively charged liposomes [16]. However, when liposomes were prepared with other antigens, positively charged liposomes worked equally as well as those bearing a negative charge [17,18]. Further studies are needed to determine whether negative charges in liposphere have general abilities to enhance immunogenicity, or whether, as with liposomes, charge effects are dependent on individual antigen composition.

C. Influence of Fat/Phospholipid Molar Ratio

An interesting correlation was observed between the liposphere fat/phospholipid (F/PL) molar ratio, particle size, and immunogenicity. Low F/PL ratios (≤ 0.75) induced the formation of lipospheres of small particle size (70% smaller than 10 μm in diameter), and this apparently resulted in increased antibody titers [7]. Among the ratios tested, a maximal level of IgG antibody production was obtained at a F/PL ratio of 0.75, whereas at higher ratios decreased antibody production was observed. Although the reason for this phenomenon is unknown, a possible explanation may be the occurrence of better antigen orientation and epitope exposures in the small lipospheres because of higher surface curvature. It may also be relevant that small liposomes have been reported to generate higher antibody titers against encapsulated antigen than large liposomes do [19].

D. Effect of Particle Size

Two populations of particles usually coexist in liposphere formulations: one in the size range of 1–10 μm in diameter (population A), and a second population with a diameter between 10 and 80 μm (population B). The particle size distribution of liposphere depends on the fat/phospholipid (fat/PL) molar ratio, and the immune response to liposphere-encapsulated R32NS1 was also dependent on the F/PL ratio. The average size of the particles increases with increasing fat/PL molar ratio. Under conditions in which the fat/PL ratio is high (≥ 2.5), the large-particle population is predominant (approximately 80% of the particles had an average size of 73 μm); whereas for fat/PL ratios of 0.75 or less, most of the liposphere have a diameter of less than 10 μm [7].

E. Effect of Route of Administration

To examine the influence of different routes of liposphere administration on their immunogenicity, rabbits were immunized orally or parenterally (by subcutaneous, intraperitoneal, intramuscular, or intravenous routes) with liposphere made of tristearin and lecithin (1:1 molar ratio) containing the malaria antigen. The immune response obtained was followed with time for 12 weeks after immunization.

There was no antibody activity after oral immunization in any of the individual rabbits immunized with liposphere–R32NS1 vaccine formulation (Fig. 3). However, rabbit immunization by all parenteral routes tested resulted in enhanced immunogenicity, with increased antibody IgG levels over the entire postimmunization period. The individual rabbit immune response in Fig. 3 shows that immunization by subcutaneous injection was the most effective vaccination route among all parenteral administration routes tested.

VI. LIPOSPHERES AS CARRIERS OF ADJUVANTS

The adjuvant activity of lipid A on the immunogenicity of liposphere was investigated. Lipid A was included in the lipid phase of liposphere, as it has been used effectively by many laboratories to enhance humoral immunity to a wide range of antigens owing to its adjuvant properties [2]. The adjuvant activity of liposomal lipid A has been investigated [2]: Liposomes can serve as a vehicle that allows expression of the adjuvant activity of lipid A, and can simultaneously reduce certain unwanted side effects of lipid A. Incorporation of lipid A into liposomes greatly reduces many of the toxic effects normally associated with endotoxin, such as pyrogenicity and neutropenia, with no substantial reduction of its adjuvant activity [2].

A successful human trial of alum-adsorbed liposomes containing monophosphoryl lipid A demonstrated that a formulation consisting of a combination of

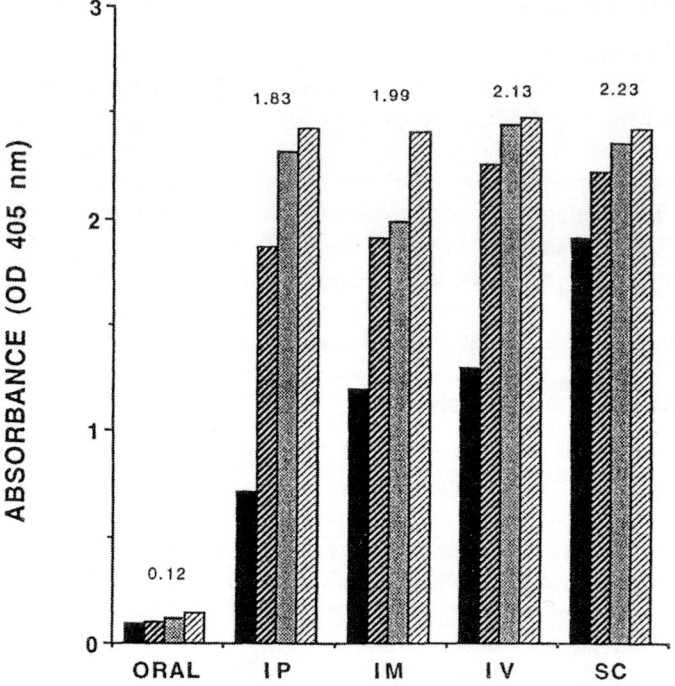

Fig. 3 Effect of administration route on immunogenicity of lipospheres–R32NS1. A liposphere-R32NS1 formulation made of tristearin and egg lecithin was administered orally, intraperitoneally, subcutaneously, intramuscularly, or intravenously to rabbits (four per group) at 0 and 4 weeks, and the IgG antibody responses in their sera were measured with time. Each bar represents the individual rabbit immune response at week 6 after subtraction of the preimmune value at a serum dilution of 1:200. Antigen dose = 100 μg/rabbit. The values above the bar groups indicate the mean ELISA antibody response for each group.

oil-in-water (O/W) and adsorbent adjuvants can have considerable safety and efficacy and may be useful in the development of a potential vaccine against the human malaria parasite (*P. falciparum*) [20].

A comparison was made between a 100-μg injection of R32NS1 malaria antigen incorporated in lipospheres lacking lipid A and R32NS1 entrapped in lipospheres containing lipid A (Fig. 4), both formulations administered in the absence of alum. Incorporation of lipid A in lipospheres significantly increased the immune response to R32NS1 malaria antigens resulting in double IgG lev-

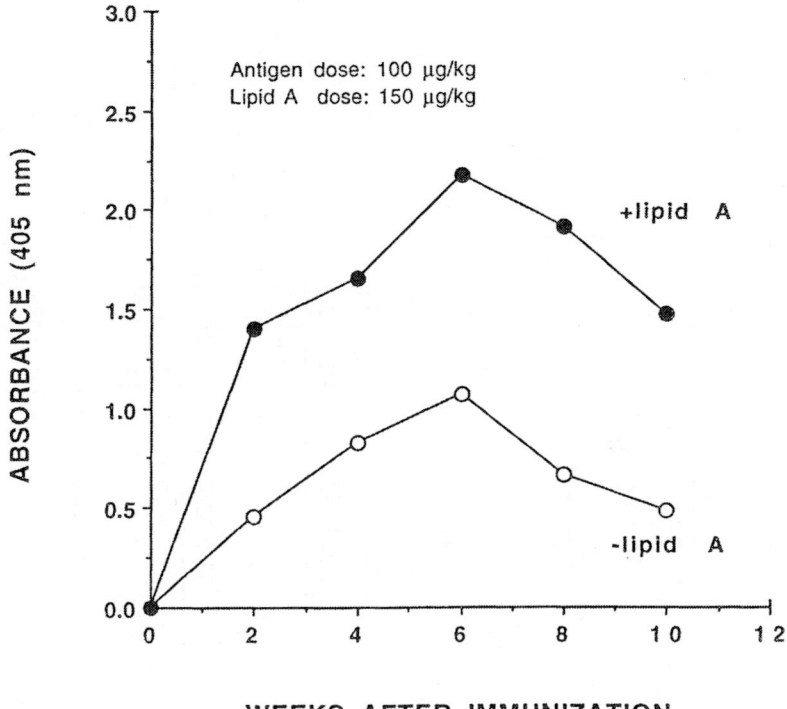

Fig. 4 Time course of the adjuvant effect of lipid A on the immunogenicity of lipospheres containing R32NS1 antigen. Rabbits were immunized intramuscularly twice (at 0 and 4 weeks) with 100-μg doses of R32NS1 malaria antigen entrapped in lipospheres containing or lacking lipid A (150 μg/kg). Both formulations were administered in the absence of alum. Each point represents the mean ELISA absorbance value (four rabbits per group) after subtraction of the preimmune value at a serum dilution of 1:200.

els, compared with R32NS1 lipospheres lacking the lipid A. The adjuvant effect of lipid A incorporated in lipospheres was observed even after 1600-fold dilution of the rabbit sera.

The adjuvant effect of different doses of lipid A in lipospheres was also examined by immunizing rabbits with lipospheres containing R32NS1 that was prepared at a different final concentrations of lipid A. The ELISA titers of the individual rabbit groups immunized, as determined by dilution of serum obtained at 6 weeks after primary immunization, are illustrated in Fig. 5. A gradual increase in IgG antibody titer with increasing lipid A dose was observed. The strongest antibody activity was obtained with lipospheres containing 150 μg of lipid A per rabbit. At a higher lipid A dose (200 μg/rabbit), a decrease

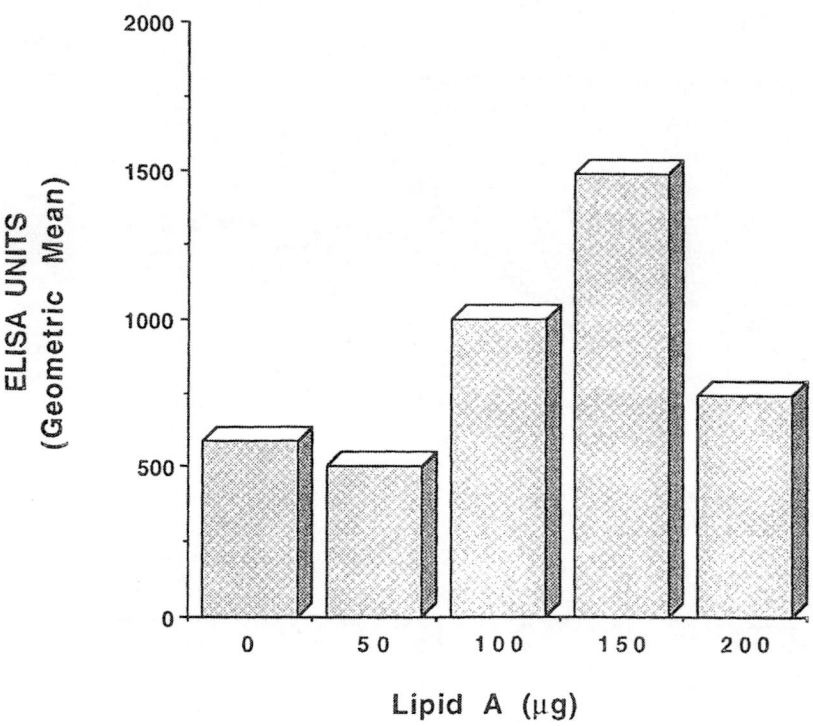

Fig. 5 ELISA IgG antibody activity at 6 weeks in rabbits immunized twice (at 0 and 4 weeks) with a 100-μg dos dose of R32NS1 malaria antigen and different doses of lipid A (0–200 μg/kg) both encapsulated in lipospheres. Each point represents the mean ELISA antibody response (four rabbits per group) at the indicated serum dilution after subtraction of the preimmune value.

in ELISA units was observed. Figure 4 also shows that in the presence of lipid A, an enhanced immune responsed is obtained, even in the absence of alum. This observation is very important, for there is an increasing concern about the long-term toxic side effects of alum [13].

VII. POLYMERIC BIODEGRADABLE LIPOSPHERE VACCINES

Over the past decade the use of polymeric materials for the administration of pharmaceuticals and as biomedical devices has increased dramatically. The most important biomedical applications of biodegradable polymers are in the form of implants and devices for surgical dressings, and in the area of controlled drug-delivery systems. Several articles have recently been published describing the adjuvant effect achieved by the association of antigens with biodegradable polymeric microparticulate-delivery systems, showing controlled-release of several immunogens [21–25].

The improvement of efficiency of essential vaccines by the combination of new immunological adjuvants and advanced delivery systems based on controlled-release technology is actually one of the major priorities of the World Health Organization Programme for Vaccine Development, as recently announced by the WHO [26]. The general objective is to improve vaccine immunogenicity and simplify delivery through conversion of multiple-dose vaccines to single-dose vaccines, with emphasis on controlled-release systems to induce a protective immune response as soon as possible after first immunization, with a delayed boost of immunity.

The preparation and use of polymeric biodegradable lipospheres as a potential vehicle for the controlled-release of vaccines was studied. The recombinant R32NS1 malaria antigen was incorporated into biodegradable polymeric lipospheres in the absence or presence of lipid A as an adjuvant.

The immunogenicity of polymeric lipospheres composed of polylactide (PLD) or polycaprolactone (PCL) was tested in rabbits after intramuscular injection of the formulations [8]. High levels of specific IgG antibodies were observed in the sera of the immunized rabbits for up to 12 weeks after primary immunization, using solid-phase ELISA (Fig. 6). The PCL lipospheres containing the malaria antigen were able to induce sustained antibody activity after one single injection, in the absence of immunomodulators. The PCL lipospheres showed superior immunogenicity when compared with PLD lipospheres, the difference being attributed to the different biodegradation rates of the polymers.

The biodegradation of lactide polymers has been reviewed by several groups [27–29]. Important factors in polymer biodegradation are the molecular weight and polydispersity, as well as the crystallinity and morphology of the polymers. Others factors that may affect polylactide degradation include chemical and configurational structure, fabrication conditions, site of implantation, and degradation conditions [30]. Biodegradation of the aliphatic polyesters occurs by bulk erosion. The lactide–glycolide polymer chains are cleaved by random nonenzymatic hydrolysis to the monomeric lactic and glycolic acids and are eliminated from the body through the Krebs cycle, primarily as carbon dioxide, and in urine.

Given the differences in the biodegradation profiles of PLD and PCL, it can be assumed that the higher degradation rate of PLD results in faster release of R32NS1 malaria antigen from the lipospheres, causing the observed temporary increase in antibody activity followed by a gradual time-dependent decrease in IgG ELISA titers. In contrast, for PCL, which biodegrades at a slower rate,

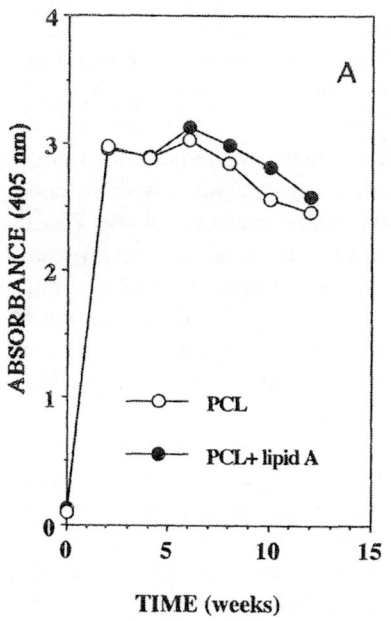
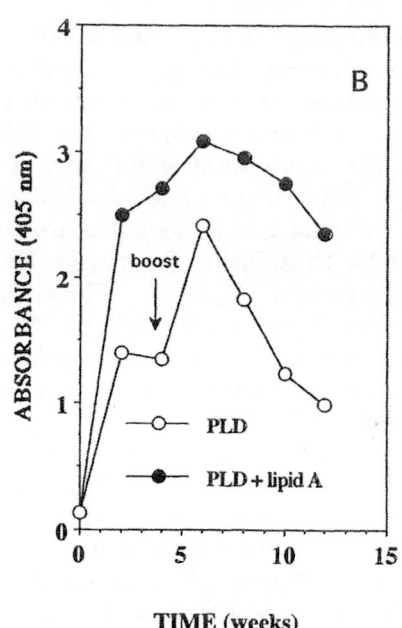

Fig. 6 Time course of the immune response of polymeric liposheres containing R32NS1 malaria antigen composed of (A) polycaprolactone (PCL), or (B) polylactide (PLD). Each point represents the mean ELISA absorbance value (four rabbits per group) after subtraction of the preimmune value at a serum dilution of 1:200. Each rabbit was immunized at 0 and 4 weeks with 100-μg doses of antigen.

the antigen is probably released from the liposheres in a more sustained manner over a longer time, resulting in prolonged immunogenicity.

The adjuvant effect of lipid A on the immunogenicity of polymeric liposheres was also tested [8]. Incorporation of lipid A in PCL liposheres had no effect on the IgG ELISA titers. However, for PLD liposheres, lipid A significantly increased the immune response to R32NS1 malaria antigen, resulting in IgG levels similar to those obtained with PCL liposheres. The adjuvant effect of lipid A incorporated in PLD liposheres was observed even after 1600-fold dilution of the rabbit sera [8].

Most vaccines require a primary immunization, followed by two or more boosters, for optimum immune response. If one injection of the immunization schedule is missed, it leads to manifold loss of effective antibody titers. According to WHO statistics, more than 30% of the patients do not return for the next injection at each time point of the immunization schedule. The effect of noncompliance is most severe in Third World countries, where more than a million children die each year from vaccine-preventable diseases.

The ideal way to substantially improve current vaccines is to develop formulations that will provide time-released doses of immunogens that could replace the need for multiple visits and booster shots. Controlled-release vaccines would be particularly advantageous in the Third World, where repeated immunizations of the vaccinee by the health care personnel are difficult to achieve [31].

The data described herein shows sustained high levels of IgG antibody production following one single immunization of rabbits immunized with biodegradable lipospheres containing malaria antigen. These results are very promising, with the expectation that biodegradable polymeric lipospheres might be very useful in the conversion of multiple-dose vaccines to single-dose vaccination, avoiding the need for repeated immunizations.

VIII. CONCLUSIONS

The results presented in this study demonstrate that enhanced immunogenic efficacy can be achieved by using liposphere-based formulations, indicating the potential usefulness of lipospheres in the formulation of human and veterinary vaccines. The liposphere approach employs a fat–lipid environment to achieve several goals: to serve as a carrier to protect the antigen; to serve as "depot"; and to provide a surface interphase necessary for adjuvant activity.

It is reasonable to presume that the immunogenic and adjuvant activity of lipospheres may be due to a combination of factors. These factors may include a focused and enhanced delivery of the antigen to an antigen-presenting cell (macrophage) and protection of the antigen from metabolic destruction at other sites in the body that do not participate in the immune response.

The binding of an antigen to a surface or presentation of a special type of surface for antigen adsorption appears to be critical for the biological activities of many agents that are reported to have adjuvant activities [32]. The data obtained with the liposphere-encapsulated antigen in the present study confirm the proposed relation that exists between physicochemical properties of surface-active systems and their ability to serve as adjuvants. It has been proposed that the abilities of surfactants to act as adjuvants is dependent on their capability of concentrating adjuvant and immunogen on hydrophobic surfaces, where they are more effectively presented to cells of the immune system [33].

The liposphere-delivery system, as a fat-based adjuvant formulation, may provide both the surface interphase necessary for solubilization and proper orientation of the adjuvant-active material, and as potential carriers for vaccines, which may allow better position for processing and presentation of the incorporated antigens, resulting in enhanced immunogenicity.

The feasibility of polymeric biodegradable lipospheres as carriers for the controlled release of a recombinant malaria antigen was also demonstrated.

The microencapsulated peptide induced efficient and potent primary immune responses. Polymeric liposheres containing R32NS1 malaria antigen were able to induce very high levels of antibody activity after one single injection in the absence of immunomodulators.

Polymeric liposheres prepared with a copolymer mixture of poly(caprolactone-*co*-lactide); and other biodegradable polymers can be also prepared using the same procedure described here. An advantage of the copolymer liposhere-delivery system could be the ability to control the time or rate at which the incorporated immunogen is released. For vaccines, this permits better scheduling of the antigen release in a manner such as to maximize the antibody response following a single administration, thereby avoiding the need for repeated vaccinations.

REFERENCES

1. C. R. Alving, Liposomes as carriers of vaccines, *Liposomes: From Biophysics To Therapeutics* (M. J. Ostro, ed.), Marcel Dekker, New York, 1987, pp. 195–218.
2. C. R. Alving, Liposomes as carriers of antigens and adjuvants, *J. Immunol. Methods 140*:1–13 (1991).
3. J. H. Eldridge, J. K. Staas, J. A. Meulbroek, J. R. McGhee, T. R. Tice, and R. Gilley, Biodegradable microspheres as a vaccine delivery system, *Mol. Immunol. 28*:287 (1991).
4. M. Zahirul, I. Kan, J. P. Opdebeeck, and I. G. Tucker, Immunopotentiation and delivery systems for antigens for single-step immunization: Recent trends and progress, *Pharm. Res. 11*:2–11 (1994).
5. M. Maniar, D. Hannibal, S. Amselem, X. Xie, R. Burch, and A. J. Domb, Characterization of liposheres: Effect of carrier and phospholipid on the loading of drug into the liposheres, *Pharm. Res. 8*:S185 (1991).
6. M. Maniar, T. Hylton, D. Hannibal, M. Green, M. Rock, S. Amselem, R. Burch, and A. J. Domb, In vitro and in vivo evaluation of a sustained release local anesthetic formulation, *Pharm. Res. 8*:S196 (1991).
7. S. Amselem, A. J. Domb, and C. R. Alving, Liposheres as a vaccine carrier system: Effects of size, charge, and phospholipid composition, *Vaccine Res. 1*:383–395 (1992).
8. S. Amselem, C. R. Alving, and A. J. Domb, Polymeric biodegradable liposheres as vaccine delivery systems, *Polym. Adv. Technol. 3*:351–357 (1992).
9. J. B. Dame, J. L. Williams, T. F. McCutchan, J. L. Weber, R. A. Wirtz, W. T. Hockmeyer, W. L. Maloy, J. D. Haynes, I. Schneider, D. Roberts, S. Sanders, E. P. Reddy, C. L. Diggs, and L. H. Miller, Structure of the gene encoding the immunodominant surface antigen on the sporozoite of the human malaria parasite *Plasmodium falciparum*, *Science 225*:93 (1984).
10. L. S. Rickman, D. M. Gordon, R. Wistar, Jr., U. Krzych, M. Gross, M. Hollingdale, J. E. Egan, J. D. Chulay, and S. L. Hoffman, Use of adjuvant containing

mycobacterial cell-wall skeleton, monophosphoryl lipid A, and squalene in malaria circumsporozoite protein vaccine, *Lancet* 337:998–1001 (1991).

11. S. L. Hoffman, L. T. Cannon, Sr., J. A. Berzofsky, W. R. Majarian, J. F. Young, W. L. Maloy, and W. T. Hockmeyer, *Plasmodium falciparum*: Sporozoite boosting of immunity due to a T-cell epitope on a sporozoite vaccine, *Exp. Parasitol.* 64:64–70 (1987).

12. J. V. Verma, N. M. Wassef, R. A. Wirtz, C. T. Atkinson, M. Aikawa, L. D. Loomis, and C. R. Alving, Phagocytosis of liposomes by macrophages: Intracellular fate of liposomal malaria antigen. *Biochim. Biophys. Acta* 1066:229–238 (1991).

13. R. Edelman, Vaccine adjuvants, *Rev. Infect. Dis.* 2:370–383 (1980).

14. H. Takada and S. Kotani, Structural requirements of lipid A for endotoxicity and other biological activities, *CRC Crit. Rev. Microbiol.* 16:477 (1989).

15. O. H. Lowry, N. J. Rosebrough, A. L. Farr, and R. J. Randall, Protein measurement with the folin phenol reagent, *J. Biol. Chem.* 193:265 (1951).

16. A. C. Allison and G. Gregoriadis, Liposomes as immunological adjuvants, *Nature* 252:252 (1974).

17. T. D. Heath, D. C. Edwards, and B. E. Ryman, The adjuvant properties of liposomes, *Biochem. Soc. Trans.* 4:129 (1976).

18. N. Van Rooijen and R. Van Nieuwmegen, Endotoxin enhanced adjuvant effect of liposomes, particularly when antigen and endotoxin are incorporated within the same liposome, *Immunol. Commun.* 9:747 (1980).

19. P. N. Shek, B. Y. K. Yung, and N. S. Stanacev, Comparison between multilamellar and unilamellar liposomes in enhancing antibody formation, *Immunology* 49:37 (1983).

20. L. F. Fries, D. M. Gordon, R. L. Richards, J. E. Egan, M. R. Hollingdale, M. Gross, C. Silverman, and C. R. Alving, Liposomal malaria vaccine in humans: A safe and potent adjuvant strategy, *Proc. Natl. Acad. Sci. USA* 89:358–362 (1992).

21. M. Singh, A. Singh, and G. P. Talwar, Controlled delivery of diphtheria toxoid using biodegradable poly (D,L,-lactide) microcapsules, *Pharm. Res.* 8:958 (1991).

22. D. T. O'Hagan, D. Rahman, J. P. McGee, H. Jeffery, M. C. Davies, P. Williams, S. S. Davis, and S. J. Challacombe, Biodegradable microparticles as controlled release antigen delivery systems, *Immunology* 73:239 (1991).

23. J. Kreuter, U. Berg, E. Liehl, M. Soliva, and P. P. Speiser, *Vaccine* 4:125–129 (1986).

24. P. Artursson, I. L. Martensson, and I. Sjoholm, Biodegradable microspheres III. Some immunological properties of polyacryl starch microspheres, *J. Pharm. Sci.* 75:697 (1986).

25. M. E. D. Martin, J. B. Dewar, and J. F. E. Newman, Polymerised serum albumin beads possessing slow release properties for use in vaccines, *Vaccine* 6:33 (1988).

26. Classified announcements, *Nature* 23:353 (1991).

27. A. Domb, S. Amselem, and M. Maniar, Biodegradable polymers as drug carrier systems, *Polymeric Biomaterials* (S. Dumitriu, ed.), Marcel Dekker, 1994, pp. 399–433.

28. A. Domb, S. Amselem, J. Shah, and M. Maniar, Degradable polymers for site-specific drug delivery, *Polym. Adv. Technol.* 3:279–292 (1992).

29. D. H. Lewis, Controlled release of bioactive agents from lactide/glycolide polymers, *Biodegradable Polymers as Drug Delivery Systems* (M. Chasin and R. Langer, eds.), Marcel Dekker, New York, 1990, pp. 1–41.

30. S. M. Li, H. Garreau, and M. Vert, Structure–property relationships in the case of the degradation of massive aliphatic poly(α-hydroxy acids) in aqueous media, *J. Mater. Sci. 1*:198 (1990).

31. B. R. Bloom, Vaccines for the Third World, *Nature 342*:115 (1989).

32. L. F. Woodard, Surface chemistry and classification of vaccine adjuvants and vehicles, *Bacterial Vaccines* (A. Mizrahi, ed.), Alan R. Liss, New York, 1990, pp. 281–306.

33. R. Hunter, F. Strickland, and F. Kezdy, The adjuvant activity of nonionic block polymer surfactants: I. The role of hydrophile–lipophile balance, *J. Immunol. 127*:1244–1250 (1981).

6

The Characterization of Polyanhydride Microspheres

Anette Brunner and Achim Göpferich

University of Erlangen-Nürnberg
Erlangen, Germany

I. INTRODUCTION

During the last two decades, degradable polymers have steadily gained importance in the field of drug delivery. Degradable polymers have the advantage of being biocompatibile and biodegradable into nontoxic monomers; therefore, they need no postapplication removal. This paved the way to sophisticated applications, such as tissue engineering [1,2] and drug delivery [3]. The first degradable polymers that were used for drug delivery originally served other purposes. Poly(lactic acid) and poly(glycolic acid), for example, were originally used for the manufacture of resorbable sutures [4,5]. When such "off-the-shelf" materials were screened for use in drug delivery, it became obvious that some of their properties had to be improved. Through the manufacture of copolymers from racemic lactic acid, the copolymerization of lactic acid with glycolic acid, and the synthesis of oligomers, it was possible to obtain materials that had better degradation and erosion control and could be processed easier to dosage forms. In the following years, research focused on the synthesis of new degradable polymers and on the development of technologies needed for the manufacture of new degradable, controlled-release dosage forms. During this period, microspheres and nanoparticles emerged as injectable drug-delivery systems. They overcame some of the problems with large implants, such as painful implantation procedures, and opened the way to new applications, such as drug targeting [6], and the improvement of existing applications, such as vaccination [7,8]. Since then, microspheres have become one of the most popular controlled-release–dosage forms and are commercialized as preparations for the treatment of prostate cancer with gonadorelin agonists (lutenizing hormone-releasing hormone; LH-RH agonists). Despite tremendous progress in the formulation of pharmaceutical microsphere preparations, they still remain a complicated dosage form, bearing many difficulties and problems. The incorporation of sensitive drugs into such systems may serve as a good example. Proteins and peptides often fail to yield a desired drug-release behavior because of poor drug-release control or stability problems [9]. Many problems with microspheres are a direct result of their microstructure and its changes during erosion. Additional problems arise from complicated drug–polymer interactions. One of the keys to a better understanding of drug release from microspheres is the detailed knowledge of the degradable polymer properties and of degradation and erosion, as well as the detailed characterization of the microsphere microstructure. It is the intention of this chapter to review the properties of polyanhydrides before and during erosion and to show physicochemical ways of polyanhydride microsphere characterization.

II. POLYANHYDRIDES

A. Historical Development and Presently Used Polyanhydrides

Polyanhydrides were first synthesized in 1909, by Bucher and Slade, and were made of aromatic monomers [10]. In 1930, the first aliphatic polyanhydrides were synthesized as prospective raw material for the manufacture of textile fibers [11,12], a goal also pursued in vain during the 1950s [13,14], as polyanhydrides are not enough hydrolysis-resistant to serve as long-lasting materials. In the early 1980s, they were rediscovered in the search for fast-degrading polymers that could be used for erosion-controlled drug delivery [15]. Today, polyanhydrides can be regarded as "designer polymers" for many reasons. They can be synthesized from a large pool of monomers, they can be manufactured with various degrees of crystallinity [16], they allow control of degradation rates and water uptake [17], they can be manufactured with a branched structure [18], or they may be cross-linked [19]. Probably, their most important advantage, though, is their biocompatibility [20,21] in combination with excellent drug-release control [22]. Numerous polyanhydrides have now been synthesized, most of them by melt polycondensation, which is the standard way of synthesis [23], although other methods might be also used [24]. The general formula of polyanhydrides is shown in Fig. 1A. The monomers are bifunctional carboxylic acids, which differ by the chemical groups R_1 and R_2, separating the carboxylic acid ends. Polyanhydrides can be synthesized as homopolymers ($R_1 = R_2$), or as copolymers ($R_1 \neq R_2$). Some of the numerous monomers that have been used for the manufacture of polyanhydrides are shown in Fig. 1B. To simplify nomenclature, the monomer names are abbreviated as indicated in parenthesis. From these short forms, the polymer names are derived. The co-polymer poly[1,3-bis(p-carboxyphenoxy)propane-co-sebacic acid], with a monomer ratio of 20:80, for example, is abbreviated p(CPP–SA) 20:80. This terminology will be used throughout this chapter.

Compared with the relatively short period during which they have been synthesized as drug carriers, polyanhydrides have been very successful. At present, p(CPP–SA) polymers are used for the treatment of brain cancer in humans after promising clinical trials [25]. The p(FAD–SA) polymers are evaluated for the same type of therapy, with microspheres because they allow stereotactic injection [26].

B. The Properties of Polyanhydrides

To understand the characteristics of polyanhydride microspheres, it is necessary to have a detailed knowledge of the properties of the polymers. Polyanhydrides have been investigated thoroughly during the last couple of years, which makes much physicochemical data available [27–30]. Not all polyanhydrides made of the monomers shown in Fig. 1b are ideal materials for the manufacture of microspheres. For example, p(SA), is highly crystalline, has poor mechanical properties, and erodes too fast, whereas p(CPP) erodes too slowly. Through the synthesis of copolymers, these properties can be tremendously improved. Copolymers were first investigated for the randomness of the monomer distribution in the polymer backbone. In a copolymer with monomers A and B, there are three possible types of bonds AA, BB, and AB. The relative number of these bonds was determined experimentally by using nuclear magnetic resonance (NMR) [31] and compared with predictions based on the random distribution of monomers [16]. Figure 2 shows the results for p(CPP–SA) and p(CPH–SA). There is reasonable agreement between predictions and experimental values, which allows one to conclude that p(CPP–SA) and p(CPH–SA) are random copolymers.

Polyanhydrides are partially crystalline and, therefore, have a typical appearance under polarized light (Fig. 3). The images are taken from melt-processed polyanhydrides and show typical Maltese crosses, which indicate that polyanhydrides consist of spherulitic structures [32]. This microstructure is confirmed by scanning electron microscopy (SEM) investigations of melt-processed polyanhydrides. In SEM images of the p(CPP–SA) 20:80 polymer matrix cross sections (Fig. 4A), the spherulites appear as concentric structures. In Fig. 4B, SEM images of p(FAD–SA) 30:70 cross sections show the crystalline skeleton of spherulites.

The crystallinity of polyanhydrides can be adjusted through the copolymer composition (Fig. 5). The degrees of crystallinity were determined by wide-

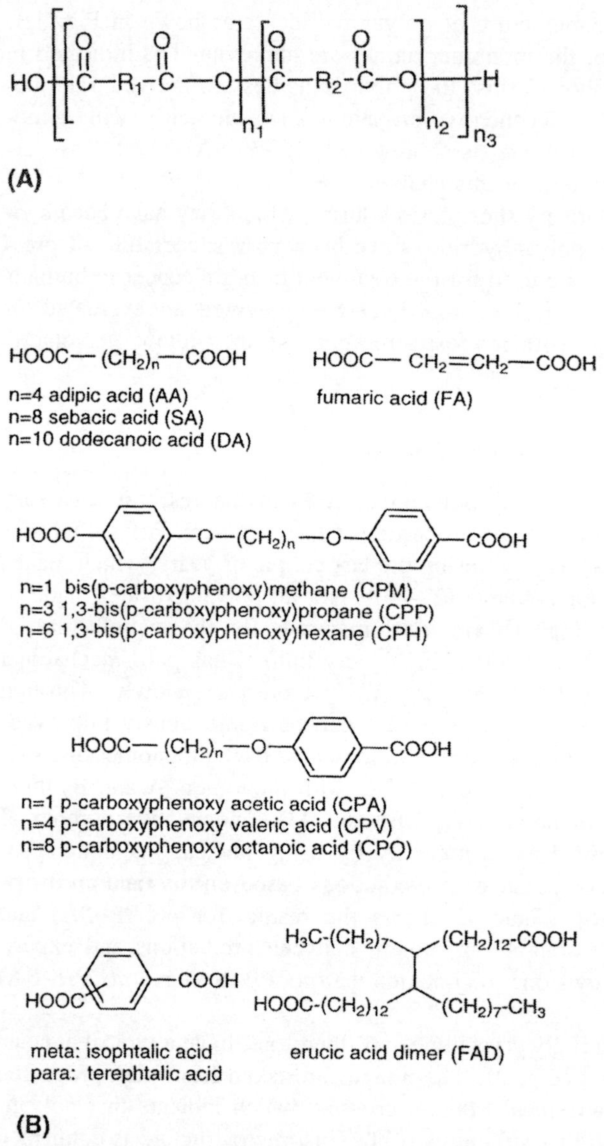

(A)

HOOC— (CH$_2$)$_n$—COOH

n=4 adipic acid (AA)
n=8 sebacic acid (SA)
n=10 dodecanoic acid (DA)

HOOC— CH$_2$=CH$_2$—COOH

fumaric acid (FA)

HOOC—⟨benzene⟩—O—(CH$_2$)$_n$—O—⟨benzene⟩—COOH

n=1 bis(p-carboxyphenoxy)methane (CPM)
n=3 1,3-bis(p-carboxyphenoxy)propane (CPP)
n=6 1,3-bis(p-carboxyphenoxy)hexane (CPH)

HOOC— (CH$_2$)$_n$—O—⟨benzene⟩—COOH

n=1 p-carboxyphenoxy acetic acid (CPA)
n=4 p-carboxyphenoxy valeric acid (CPV)
n=8 p-carboxyphenoxy octanoic acid (CPO)

HOOC—⟨benzene⟩—COOH

meta: isophtalic acid
para: terephtalic acid

H$_3$C-(CH$_2$)$_7$ (CH$_2$)$_{12}$-COOH

HOOC-(CH$_2$)$_{12}$ (CH$_2$)$_7$-CH$_3$

erucic acid dimer (FAD)

(B)

Fig. 1 (A) General formula of polyanhydride polymers; (B) monomers used for the synthesis of polyanhydrides.

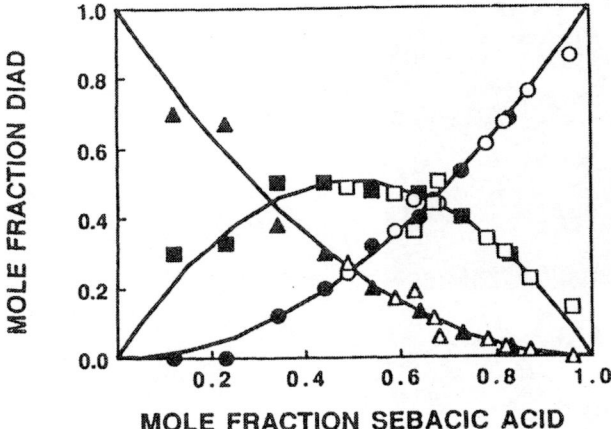

Fig. 2 Comparison of copolymer sequence distribution with that predicted for a random copolymer for p(CPH–SA) and p(CPH–SA). For p(CPH–SA): (●) SA–SA, (■) SA–CPH, (▲) CPH–CPH. For p(CPP–SA): (○) SA–SA, (□) SA–CPP, (△) CPP–CPP. (From Ref. 16.)

angle X-ray diffraction (WAXD) [33] and by differential-scanning calorimetry (DSC) [34] for these copolymers. The values are confirmed qualitatively by polarized light microscopy. For example, p(CPP–SA) 50:50 of 15% crystallinity shows no spherulites under polarized light.

Important parameters for processing polyanhydrides to dosage forms are the glass transition temperature and the melting point. Both have been measured by DSC [16] and are listed in Table 1. The figures, again, indicate a strong dependence on the copolymer composition. More data, including heat of fusion values, can be found in Ref. [16].

III. THE CHARACTERISTICS OF POLYANHYDRIDE EROSION

An understanding of polyanhydride erosion and degradation is essential to understand the properties of eroding polyanhydride microspheres. The techniques used for the characterization of polymer bulk erosion are also suited for the investigation of microspheres.

A. The Definition of Degradation and Erosion

The erosion of degradable polymers is a complicated process, in which various reaction and transport processes are involved. Erosion starts with the intrusion

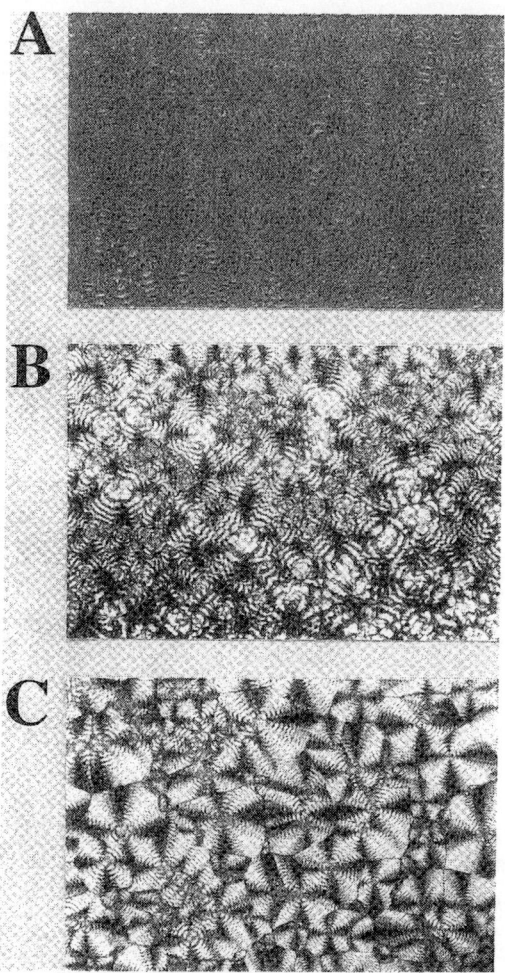

Fig. 3 Polarized light microscopic images of polyanhydride films: (A) pSA; (B) p-(FAD–SA) 20:80; (C) p-(FAD–SA) 50:50. (From Ref. 34.)

of water into the polymer bulk and triggers degradation. *Degradation* is the polymer chain scission process and is the most important part of erosion. Through degradation, oligomers and monomers are created that finally diffuse to the polymer surface, where they are released from the polymer bulk. *Erosion* is the sum of all these processes that finally lead to the loss of mass from the polymer bulk.

Table 1 Glass Transition (T_g) and Melting Point (T_m) of Polyanhydrides

Polymer	T_m(°C)	T_g(°C)
p(SA)	86	60.1
p(CPP–SA) 4:96	76	41.7
p(CPP–SA) 9:91	78	—
p(CPP–SA) 13:87	75	47
p(CPP–SA) 17:83	72	47
p(CPP–SA) 27:73	66	44
p(CPP–SA) 41:59	178	4.2
p(CPP–SA) 60:40	200	0.2
p(CPP–SA) 80:20	205	15
p(CPP)	240	96
p(CPH–SA) 27:73	57.2	45
p(CPH–SA) 55:45	49.5	11.5
p(CPH–SA) 64:36	43.3	11.8
p(CPH–SA) 70:30	110.5	14.5
p(CPH)	143	47

Source: Ref. 42.

B. The Importance of Erosion for Drug Release

Polyanhydrides differ from other polymers by the reactivity of the anhydride bond [35]. Table 2 gives a survey on the half-lives of functional groups that are typical for degradable polymers. Carboxylic acid anhydrides and *ortho*-esters are the most reactive bonds, which makes polyanhydrides and poly(*ortho*-esters) fast-degrading polymers.

Table 2 Half-Lives of Degradable Polymers

Polymer class	Hydrolysis rate
Polyanhydride	0.1 h
Polyketal	3 h
Poly(*ortho*-ester)	4 h
Polyacetal	0.8 yr
Polyester	3.3 yr
Polyurea	33 yr
Polycarbonate	42,000 yr
Polyurethane	42,000 yr
Polyamide	83,000 yr

Source: Ref. 35.

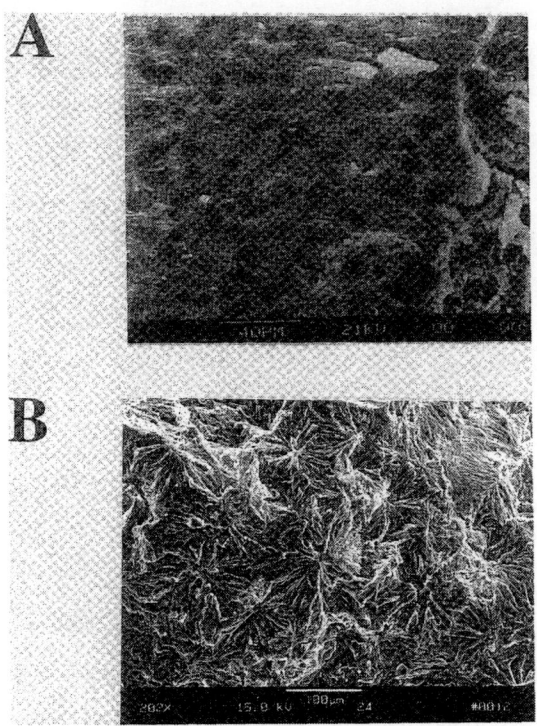

Fig. 4 An SEM image of polyanhydride cross sections containing spherulites: (A) *p*(CPP–SA) 20:80; (B) *p*(FAD–SA) 30:70.

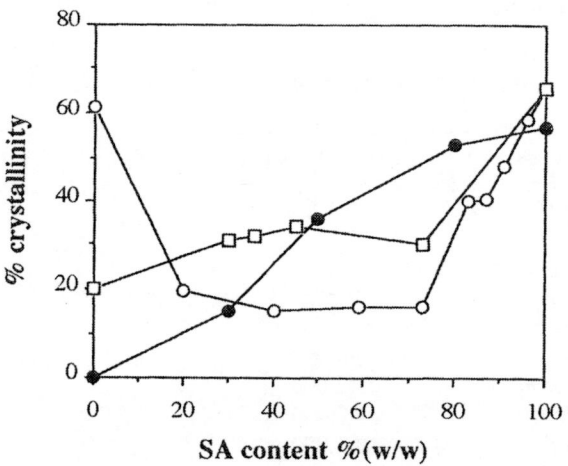

Fig. 5 Crystallinity of polyanhydride copolymers depending on copolymer composition: (●) *p*(FAD–SA); (○) *p*(CPP–SA); (□) *p*(CPH–SA). (Data from Refs. 33 and 34.)

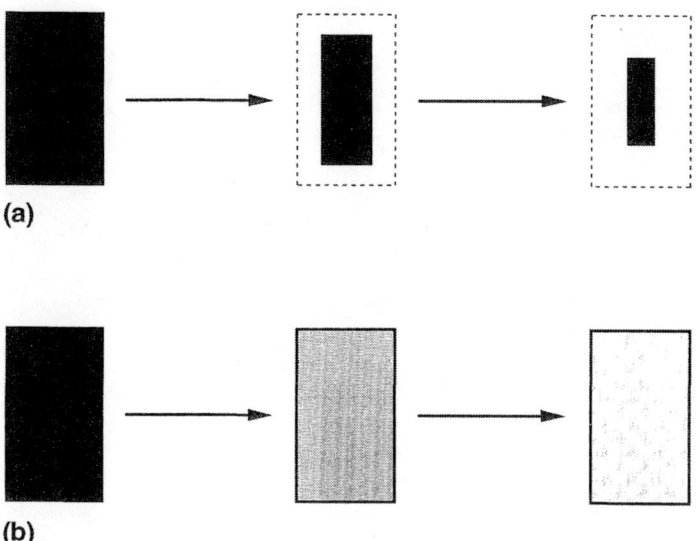

Fig. 6 Schematic illustration of (a) surface and (b) bulk erosion.

The fast degradation of polyanhydrides has consequences for the erosion mechanism. For degradable polymers, two different erosion mechanisms have been proposed: surface or heterogeneous, and bulk or homogeneous erosion [36]. The difference is illustrated in Fig. 6. In the surface-eroding polymers, degradation is faster than the intrusion of water into the polymer bulk and, therefore, is confined to the polymer surface. Consequently, erosion also affects only the outermost polymer layers. Bulk-eroding polymers, in contrast, degrade slowly and, because of the fast intrusion of water into the bulk, throughout their cross section. Here, therefore, erosion is not limited to the polymer surface. Polymers containing reactive functional groups tend to degrade fast and to be surface-eroding, whereas polymers containing less reactive functional groups tend to be bulk-eroding.

The erosion mechanism has consequences for the release of drugs from degradable polymers. Drug release has been classified into diffusion-, swelling-, and erosion-controlled release [3]. A degradable polymer might release drugs by all three mechanisms. The quickest mechanism, however, will dominate (Fig. 7). The faster a polymer erodes, the greater its chances that drug release might be erosion-controlled. Polyanhydrides are, therefore, an ideal material for the manufacture of erosion-controlled drug-delivery systems.

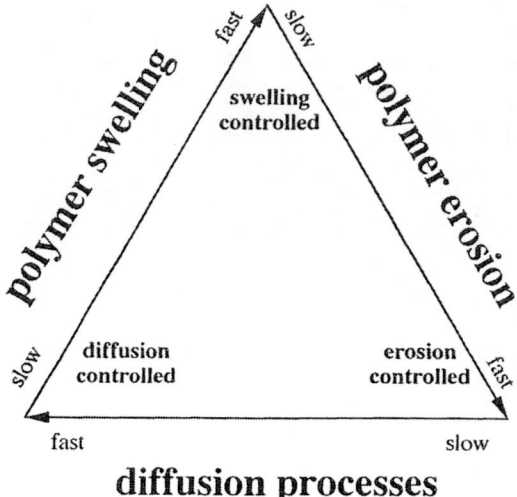

Fig. 7 Possible mechanisms of drug release from degradable polymers.

C. Changes of Polyanhydride Properties During Erosion

1. Morphological Changes During Erosion

Surface and bulk erosion are ideal cases, and most polymers cannot be unequivocally assigned to one of them. The same is true for polyanhydrides. This can be seen clearly from SEM images of partially eroded polyanhydride cylinder cross sections. The SEM image of such a cross section is shown in Fig. 8. Characteristic for p(CPP–SA) polyanhydrides is the formation of erosion fronts, such as the one shown in Fig. 8A. Such erosion fronts separate eroded from noneroded polymer. In Fig. 8B, noneroded polymer is shown that consists of the concentric banding of intact spherulites. In the eroded part of the disk, amorphous polymer has disappeared, whereas the crystalline skeleton of the spherulites is still in place, as shown in Fig. 8C. It can be concluded, that the crystalline parts of polyanhydrides degrade and erode too slowly to allow perfect surface erosion. Crystalline erosion zones remain on the polymer surface. Because of the high porosity in the erosion zone, however, p(CPP–SA) comes close to a perfect surface-eroding polymer. Erosion zones in p(FAD–SA) are different. This is because of the physical state of FAD, which is an oily liquid. Rather than building erosion zones, the FAD monomer created during erosion sticks to the surface of the polymer, leading to the steady accumulation of an FAD film. Such FAD films might act as diffusion barriers

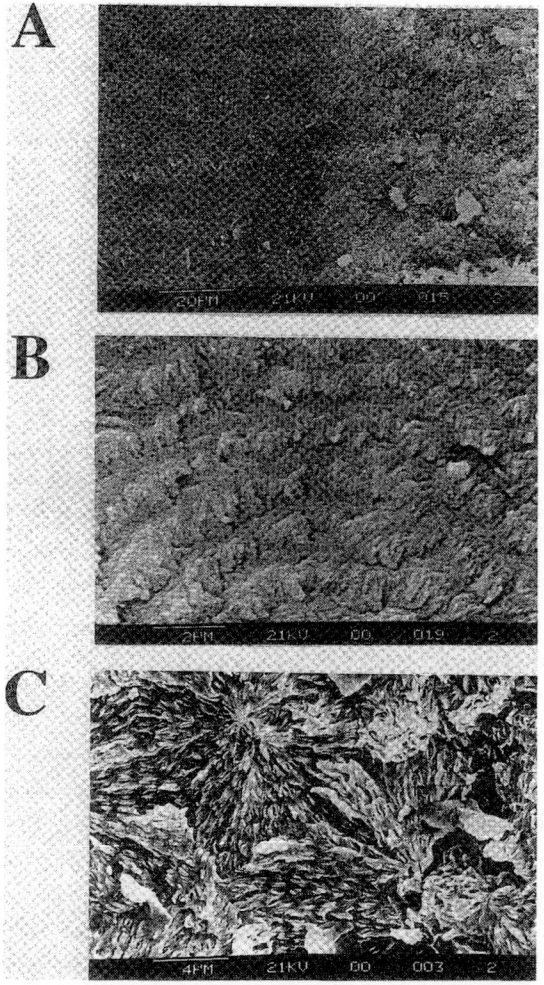

Fig. 8 SEM images of p(CPP–SA) 20:80 cross sections after 2 days of erosion: (A) erosion front separating eroded (right) from noneroded polymer (left); (B) noneroded polymer in detail; (C) eroded polymer in detail. (From Ref. 39.)

and contribute substantially to the control of drug release from an eroding polymer.

The porosity of eroding p(CPP–SA) disks has been investigated by mercury intrusion porosimetry. The pore size distributions were estimated from the Washburn equation [37] and are shown in Fig. 9 [38]. This method can distinguish between macro- and micropores. Macropores are created immediately

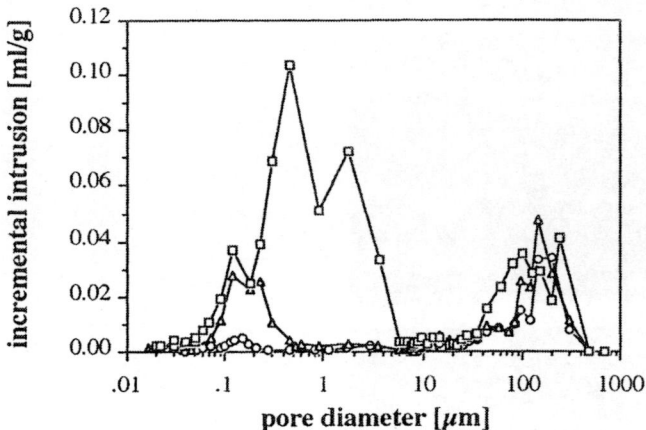

Fig. 9 Pore size distribution of eroding *p*(CPP–SA) 20:80 polymer matrix disks determined by mercury intrusion porosimetry: (○) day 1; (△) day 2; (□) day 4. (From Ref. 38.)

after contact with the erosion medium and have a size of approximately 100 *μ*m. They result from cracks on the surface of polyanhydride matrix disks [39]. The micropores are gradually created by erosion and result from the faster erosion of amorphous polymer areas compared with crystalline ones; they are approximately 100 nm. The total macropore volume is almost constant during erosion, which indicates that there is no crack propagation, whereas the volume of small pores increases steadily through the advancement of the erosion front.

2.   Crystallinity Changes During Erosion

The faster erosion of amorphous polymer, compared with the crystalline one, changes the overall crystallinity of polymer matrices during erosion. Changes in crystallinity can be exactly followed by wide-angle X-ray diffraction (WAXD) and differential-scanning calorimetry (DSC). The spectra obtained by both methods show very characteristic peak patterns. Time series of DSC spectra are shown in Fig. 10a. All three polymers, *p*(SA), *p*(CPP–SA) 20:80, and *p*(CPP–SA) 50:50, show a melting peak at approximately 75°C. When they erode, one new endothermal peak appears in the spectrum for the homopolymer *p*(SA), at approximately 100°C, and two new endothermal peaks appear in the spectra of the copolymers, at approximately 100° and 250°C. This indicates

Fig. 10 Changing DSC thermograms during erosion: (A) *p*(SA); (B) *p*(CPP–SA) 20:80; (C) *p*(CPP–SA) 50:50. (From Ref. 39.)

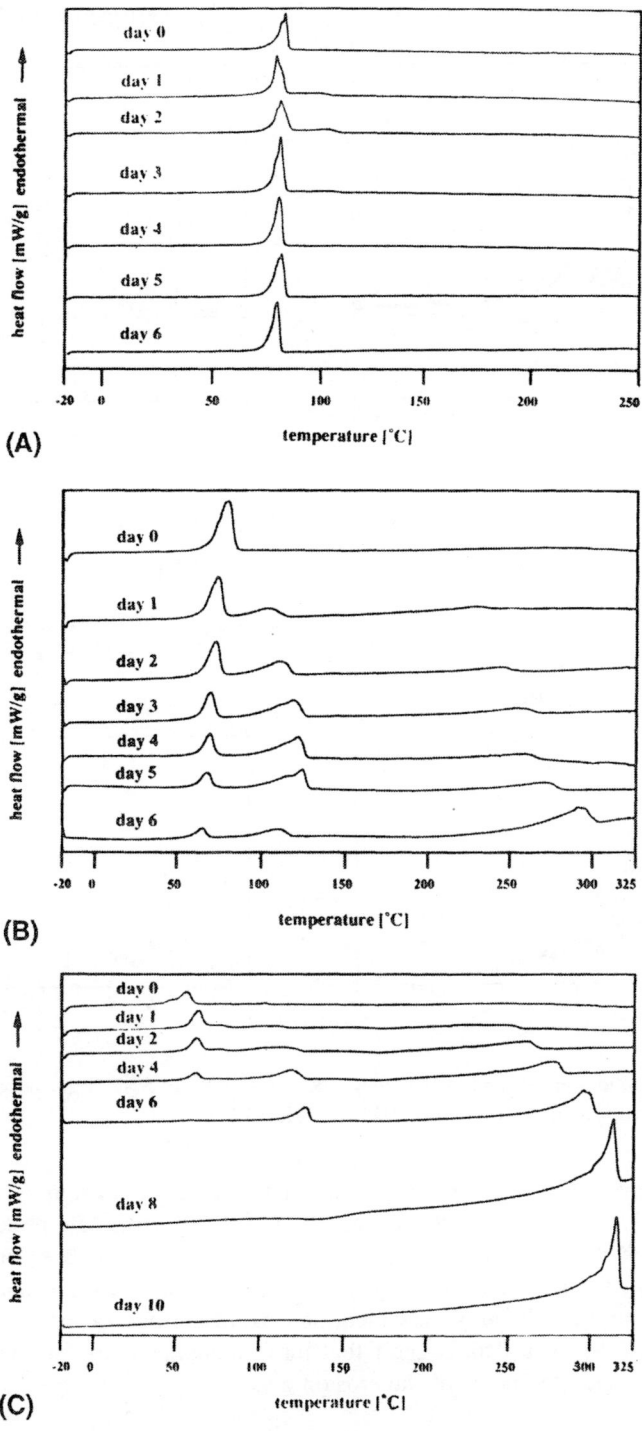

(A)

(B)

(C)

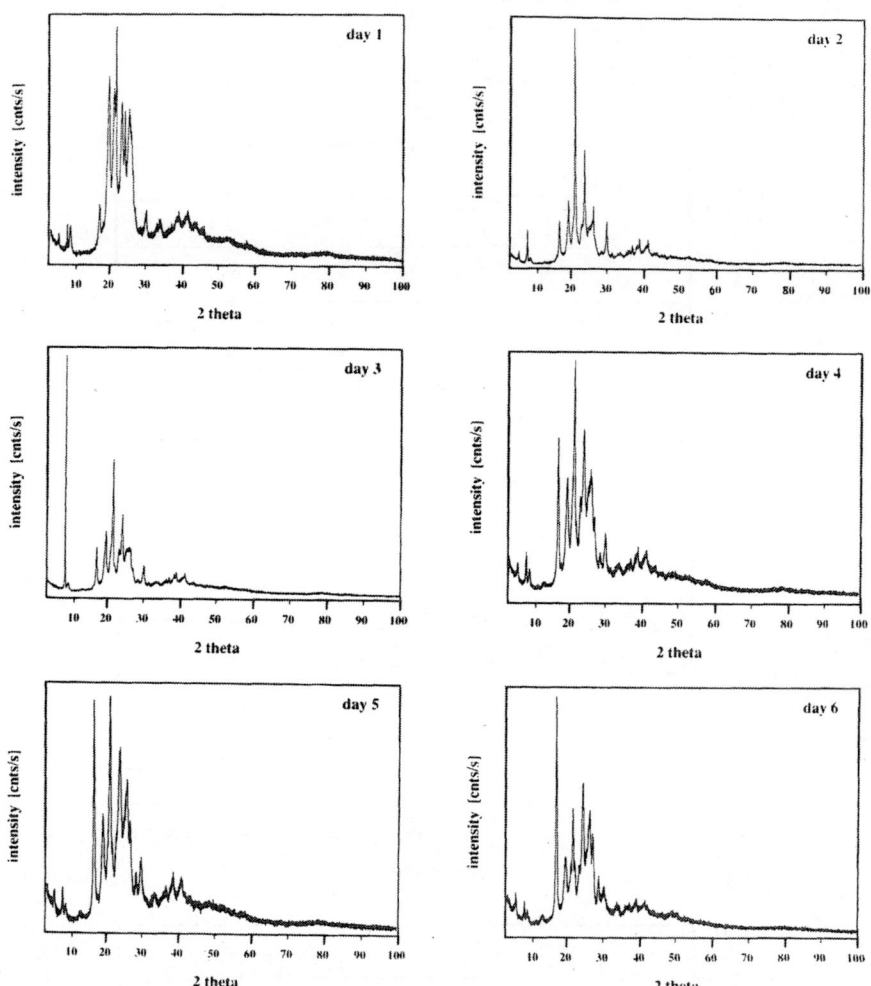

Fig. 11   Wide-angle X-ray diffraction spectra of *p*(CPP–SA) 20:80 polymer matrix disks after various times of erosion. (From Ref. 39.)

that new crystalline matter may have been created by erosion. The WAXD spectra (Fig. 11) show similar changes. In these spectra, new peaks also appear, which clearly indicate that new crystalline matter has been created. By determining the DSC and WAXD spectra of the monomers SA and CPP, the source of new crystallinity was identified to arise from the crystallization of monomers [39]. The data suggest that these monomers have the tendency to crystallize inside the pores of the erosion zone.

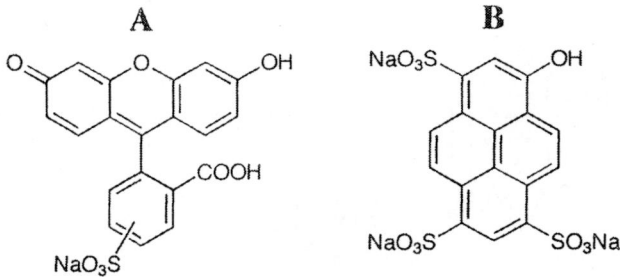

Fig. 12 pH sensitive fluorescent dyes: (A) fluorescein-5 (and -6) sulfonic acid; (B) pyranine.

3. pH Changes During Erosion

An issue of special interest is the question of pH within the eroding polyanhydrides, as the pH inside eroding microspheres might affect the solubility and the stability of incorporated drugs. As already seen by SEM, p(CPP–SA) 20:80 exhibits a highly tortuous network of pores in the erosion zone. The exact measurement of pH inside these pores has so far not been possible. The pH on the surface of eroding polyanhydride matrix disks, however, could be measured using scanning confocal fluorescence microscopy. For pH measurements, pH-sensitive fluorescent dyes, such as those shown in Fig. 12, were added to the degradation medium. The pH changes alter the emission spectra of these sub-

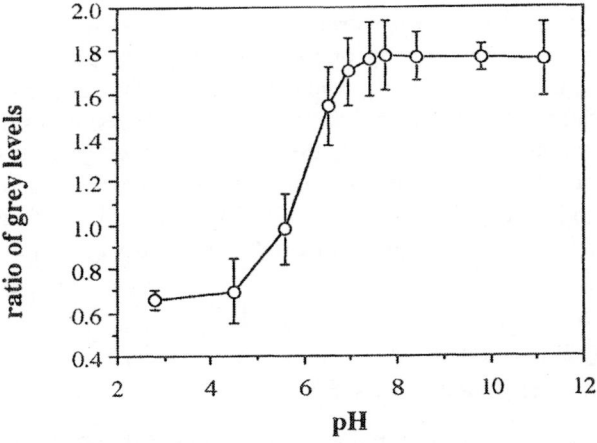

Fig. 13 Calibration curve for the determination of the surface pH by scanning confocal microscopy. Ratio of gray levels in channel 1 (515 nm) and channel 2 (600 nm). (From Ref. 39.)

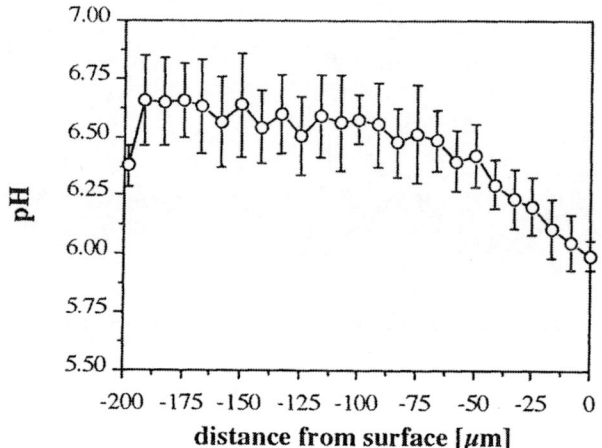

Fig. 14 pH profile of the buffer next to the surface of p(CPP–SA) 20:80 disks after 18.5 h of erosion (n = 6). (From Ref. 39.)

stances, which can be measured microscopically by simultaneously assessing confocal images at 515 and 600 nm. The ratio of the gray values obtained at both wavelengths is strongly pH-dependent (Fig. 13) for fluorescein-5 (and -6) sulfonic acid. From such calibration curves, pH profiles can be calculated from the gray values of images obtained at two wavelengths by scanning confocal fluorescence microscopy. The pH was measured as a function of the distance from the surface of the eroding polymer matrix disk [39]. These values are obtained by averaging the z-sections on top of the implants. The results are shown in Fig. 14. The pH drops significantly when approaching the surface of the device. A clear indication that the pH within polymer pores must be even lower to permit compensation of the buffer on the matrix surface. From the pK_a of monomers and their solubility, the pH inside the monomer-saturated pores can be calculated and is in a range of 4–5 [39]. Such values have been confirmed experimentally by embedding glass electrodes into eroding polyanhydride cylinders [40].

IV. THE MANUFACTURE OF POLYANHYDRIDE MICROSPHERES

Polyanhydride microspheres have been manufactured by four different methods: solvent evaporation, solvent removal, hot-melt encapsulation, and spray-drying. In addition, two methods for the manufacture of dual-walled microspheres have been developed.

A. Solvent Evaporation

For the preparation of microspheres by solvent evaporation, the polymer is first dissolved in organic solvents, such as methylene chloride. This polymer solution is processed to an oil-in-water emulsion by dispersion into an aqueous solution containing a surfactant, such as hydrolyzed poly(vinyl acetate). The emulsion is stirred for a couple of hours, during which the organic solvent evaporates, leaving the hardened microspheres [41]. With slight modifications, the solvent evaporation technique is suited to the encapsulation of hydrophilic substances. First a small amount of aqueous phase is dispersed in the organic polymer solution to form an oil-in-water emulsion, which is then processed to microspheres as just described. According to the multiple (w/o/w) emulsion that is created, the method is termed double-emulsion technique. Polyanhydride microspheres that are manufactured by solvent evaporation tend to be porous. The porosity, which increases drug release from microspheres, depends on the process parameters [42]. The disadvantages of any kind of solvent evaporation technique include solvent residues in the polymer, the potential instability of proteins during microsphere preparation [43], and the risk of polymer degradation.

B. Solvent-Removal Technique

The solvent-removal technique uses only organic phases for the manufacture of microspheres, which has the advantage of preventing hydrolysis during microsphere preparation [42]. The polymer is again dissolved in an organic solvent, such as methylene chloride, and is dispersed in a mixture of silicone oil, methylene chloride, and a surfactant, such as Span 85. The microspheres are hardened by adding a nonsolvent, such as petroleum ether, to the suspension. The microspheres obtained by solvent extraction are porous. A problem might be the use of organic solvents and the danger of silicone oil residues in the microspheres.

C. Hot-Melt Encapsulation

An interesting approach to the reduction of organic solvent residues in polyanhydride microspheres is the formation of microspheres from melted polymer. For this hot-melt encapsulation procedure polyanhydrides were melted, and drugs were dispersed in the melt as solid particles. This suspension was then formed to microspheres by dispersion into a nonsolvent, such as silicon or olive oil, at 5°C above the melting point of the polymer. The spheres solidify on cooling and are washed with petroleum ether. Microspheres made by hot-melt encapsulation have smooth surfaces and are less porous [44]. However, the temperatures to which polymer and drug are exposed limit the broad applicability of the method.

D. Spray-Drying

For the manufacture by spray-drying, polyanhydride polymers were dissolved in methylene chloride and were spray-dried with the drug suspended therein [45]. Microspheres made by spray-drying tend to have an irregular shape and a high porosities that causes the fast release of drugs [45].

E. Double-Walled Microspheres

Double-walled microspheres consist of two different polymer layers that allow polyanhydrides to combine with other degradable polymers, such as poly(lactic acid). They might be useful for suppressing the burst release of drugs, or for generating pulsatile-release profiles [46]. There are two methods by which such microspheres might be prepared. The first takes advantage of the partial or complete insolubility of polymers in one another. The cosolution of such polymers in organic solvents is dripped into aqueous solutions of poly(vinyl alcohol). On solvent evaporation the two polymers begin to separate. In their final state, they consist of an inner core made of one polymer and an outer wall that consists of the second polymer [47]. An alternative method by which double-walled microspheres can be prepared is a modified double-emulsion technique that is shown schematically in Fig. 15. The polymer is dissolved in an organic solvent, such as methylene chloride or ethyl acetate, into which a small volume

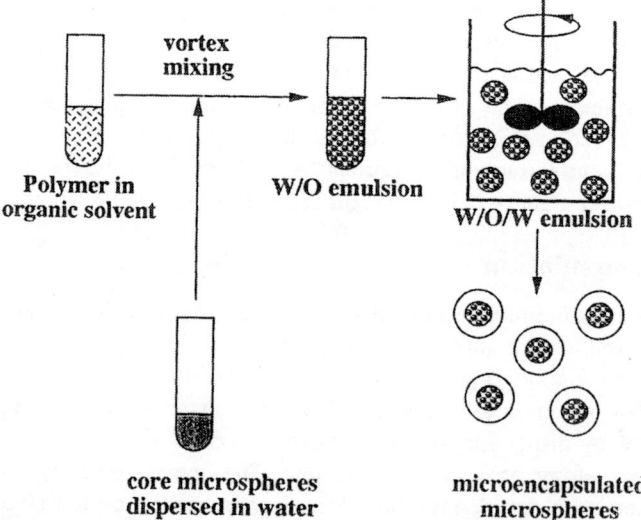

Fig. 15   Scheme for the manufacture of multiple-walled microspheres by a modified double-emulsion technique.

of aqueous phase containing microspheres is dispersed to form a water-in-oil emulsion. This emulsion is then dispersed into an aqueous solution of poly(vinyl alcohol) on which new microspheres are spontaneously formed, which contain a core that consists of one type of polymer and a coating that consists of a second type of polymer [48].

V. THE CHARACTERIZATION OF MICROSPHERES

Microspheres are very delicate systems. They have a distinct microstructure that depends strongly on the manufacturing conditions [49]. The microstructure, in return, affects the stability of drugs and drug release. The careful microstructural characterization of microspheres, therefore, will enable us to solve problems in which drug-release and drug-stability issues are involved. There are numerous physicochemical methods by which microspheres might be characterized, such as wide-angle X-ray diffraction [50], differential scanning calorimetry [41], scanning electron microscopy [51], transmission electron spectroscopy [47], gel-permeation chromatography [52], and many others. Because of the multitude of methods, in the following section only standard techniques are reviewed, and some recently used approaches to microsphere characterization are presented that could have the potential for better characterization of degradable microspheres.

A. Light Microscopy

Light microscopy has the advantage that it requires little sample preparation and that microscopes are widely accessible. Light microscopy is a valuable method for microsphere characterization. For example, it allows successful control of the coating procedure for the manufacture of double-walled microspheres. When viewed under polarized light, as shown in Fig. 16A, polyanhydride microspheres shine brightly owing to their partially crystalline nature. When such microspheres are coated, they change their appearance (see Fig. 16B). The Maltese crosses that appear result from passing the polarized light through the spherical interface between core and coating [53] and thereby prove the success of the coating procedure.

B. Scanning Electron Microscopy

Use of SEM is one of the standard techniques for microsphere characterization, as it offers, compared with light microscopy, a much higher resolution. In contrast with scanning transmission electron microscopy (STEM), the sample preparation is simple, as particles do not have to be cut with a microtome, which cannot be easily achieved for brittle polymers. SEM allows investigation of microsphere surfaces and, after particles are cut, their cross sections (Fig.

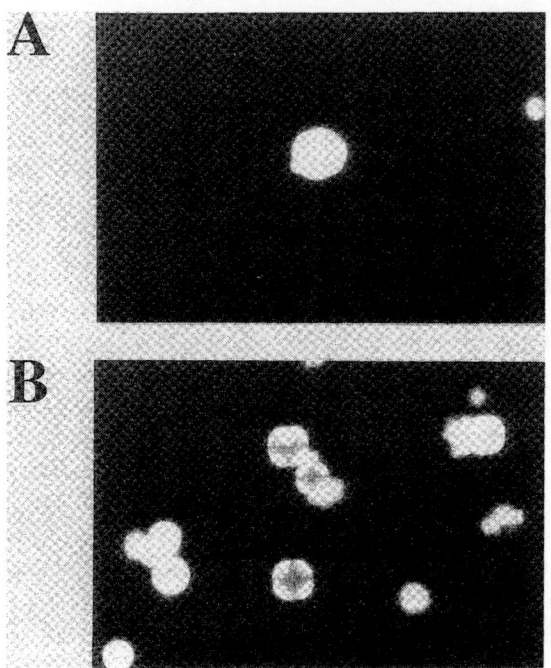

Fig. 16 Appearance of microspheres as viewed by polarized light microscopy: (A) Noncoated particle made of *p*(CPP–SA) 50:50; (B) after coating with poly(lactic-*co*-glycolic acid). (From Ref. 48.)

17). It has also proved useful for the investigation of multiple-walled microspheres made of polyanhydrides and poly(lactic-*co*-glycolic acid) microspheres. After cutting such systems, the internal structure can be revealed. Such cross sections are then viewed best at two angles. Figures 18A and B show the cross section of the same particle at two different angles. It can clearly be seen that the core of these microspheres consists of one polymer, whereas the coating consists of a second polymer. Figure 18C shows that the phase boundary between core and coating can be seen in detail with SEM.

C. Confocal Fluorescence Microscopy

Scanning confocal fluorescence microscopy can be used to characterize the structure of multiple-walled microspheres. So far this technique has been used

Fig. 17 SEM images of *p*(CPP–SA) 50:50 microspheres: (a) External surface of polyanhydride microspheres made by solvent evaporation; (b) cross section of polyanhydride microsphere made with low polymer concentration; (c) cross section of microsphere made by using high concentration of polymer. (From Ref. 48.)

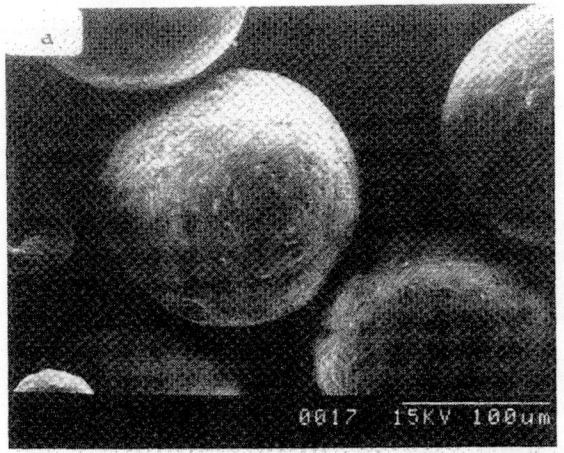

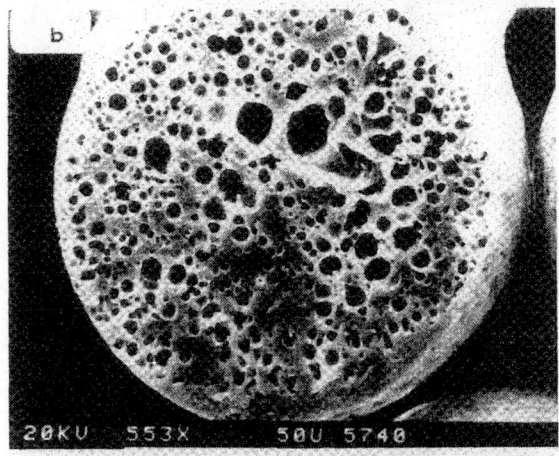

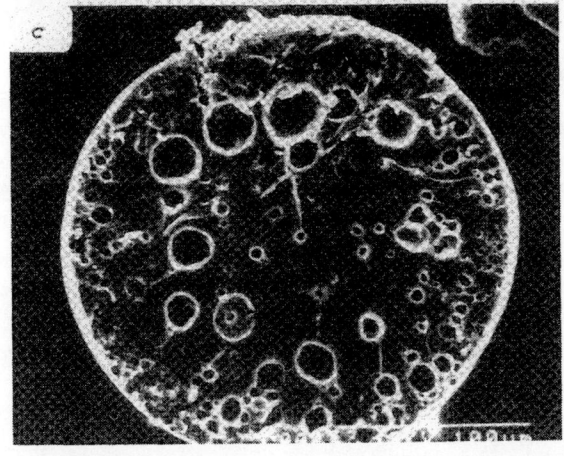

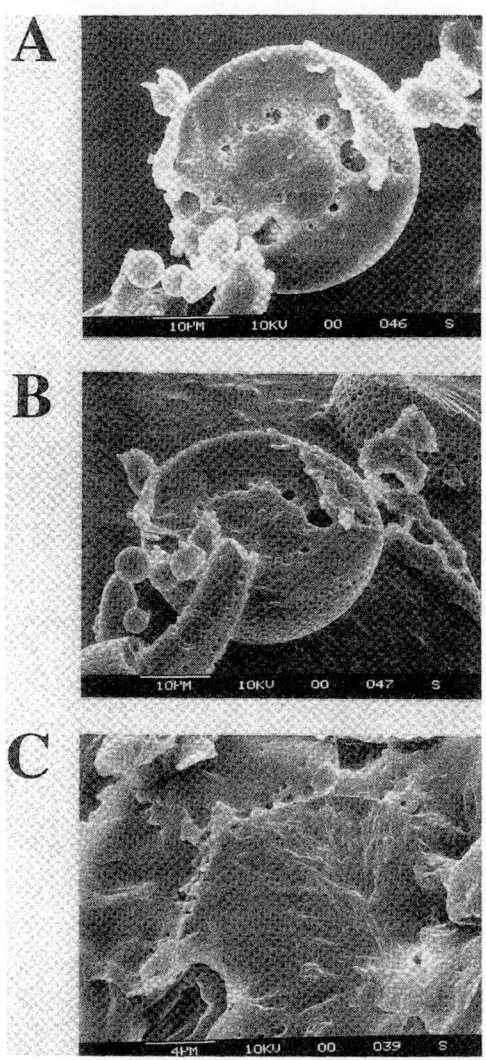

Fig. 18 Cross section through a PLA–GA-coated p(FAD–SA) 50:50 microsphere: (A) view from top (180°); (B) side view of the same particle (135°); (C) boundary between core and coating polymer. (From Ref. 48.)

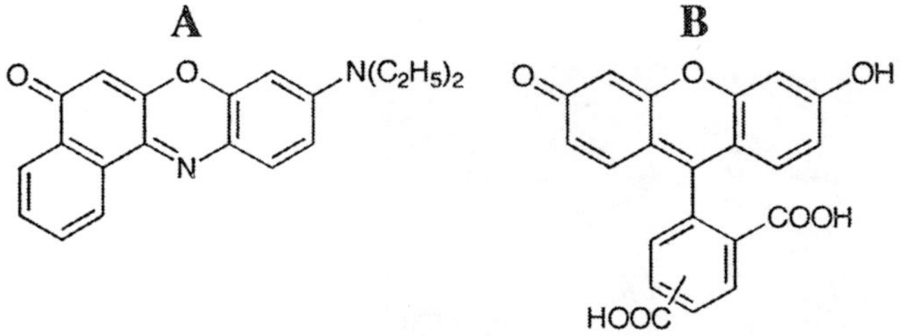

Fig. 19 Fluorescent dyes for the labeling of multiple-walled microspheres: (A) nile red; (B) 5- (and 6)-carboxy fluorescein.

only for nondegradable polymers, but it might also have some potential for the investigation of degradable microspheres. The core and the coating of double-walled microspheres are stained with two different dyes that have two different absorption spectra, such as carboxyfluorescein and nile red, which are shown in Fig. 19. Microspheres were manufactured from poly(ethylene-*co*-vinyl acetate), which served as a model polymer using the double-emulsion technique to allow the encapsulation of an aqueous carboxyfluorescein solution. The coating of these microspheres was stained with nile red. Confocal images were obtained simultaneously at 550 and 600 nm for both coated an noncoated microspheres. By the use of appropriate filters, carboxyfluorescein shows up at 550 nm and nile red at 600 nm, which makes it possible to locate the dyes inside the microspheres. Figure 20A shows an image of the microspheres before coating. At 550 nm, the irregularly dispersed carboxyfluorescein inside the particle shines brightly, whereas it is not visible at 600 nm. Pictures at 600 nm are obtained only after coating these spheres. Figure 20B shows that, in addition to carboxyfluorescein, nile red is present on the surface of the coated particles, which is a clear indication that the coating procedure was successful.

D. Electron Spectroscopy for Chemical Analysis

Electron spectroscopy for chemical analysis (ESCA) allows investigation of the surface chemistry of microspheres. It is possible to determine the atomic composition, within a few nanometers of a material surface. This can be used effectively for the investigation of degradable polymer surfaces, such as the surfaces of nanospheres made of degradable block-copolymers [54]. The ESCA spectra have been used to determine the atomic surface composition of microspheres. This was useful for determination of the surface composition of multiple-walled microspheres. To prove the principle, polyanhydride microspheres were encapsulated into PLA microspheres. Figure 21 shows the ESCA spectra

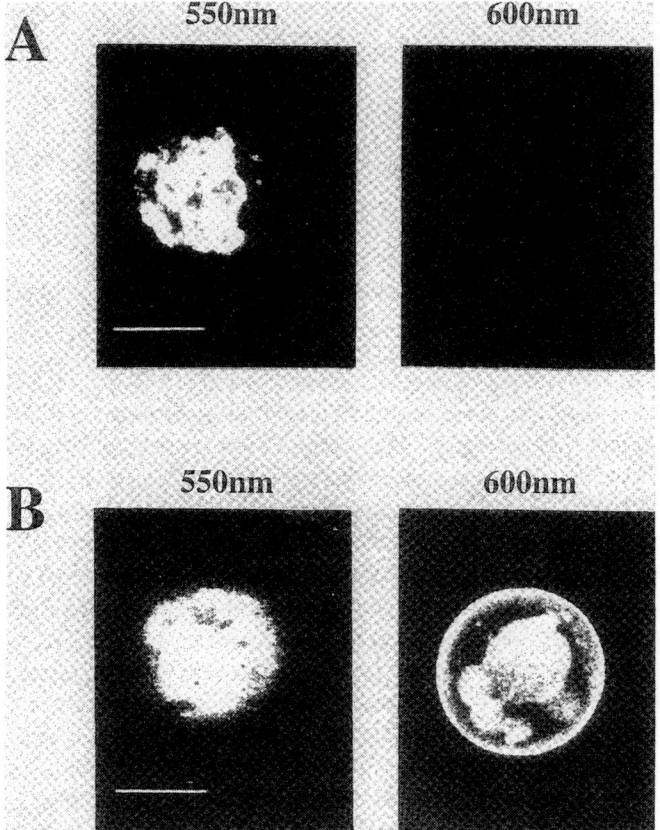

Fig. 20 (A) Images of an EVAC microsphere, assessed at 550 and 600 nm (scale bar = 100μm); (B) images of an EVAC microsphere coated with the same material assessed at 550 and 600 nm (scale bar = 100 μm). (From Ref. 48.)

of three types of microspheres: p(FAD–SA)-, poly(D,L-lactic acid), and p(FAD–SA) microspheres encapsulated into PLA. It is obvious, that the spectrum for the coated microspheres is the same as for the PLA microspheres, which shows that poly(D,L-lactic acid) is the surface polymer of the double-walled system. These spectra can also be evaluated quantitatively, by determining the carbon/oxygen ratio of the spheres, which is achieved by integration of the carbon signal between 283 and 293 eV and the oxygen signal between 530 and 538 eV. The results are shown in Table 3. It is confirmed by investigating the carbon peak at higher resolution. Because of the chemical neighborhood of

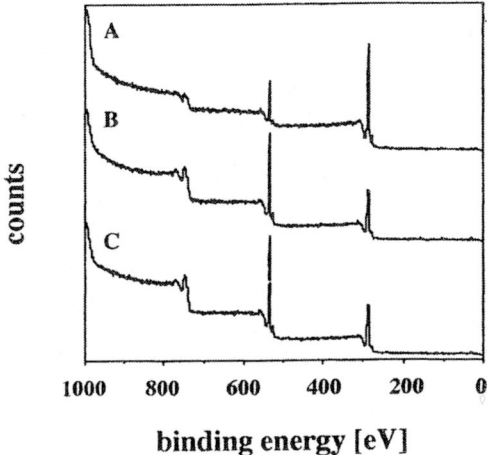

counts

1000　800　600　400　200　0

binding energy [eV]

Fig. 21　XPS spectra of (A) *p*(FAD–SA) 50:50 microspheres; (B) D,L-PLA-microspheres; (C) D,L-PLA-coated *p*(FAD–SA) 50:50 microspheres. (From Ref. 48.)

an atom, the energy at which electrons are lost might vary; therefore, these carbon peaks consist of several individual overlapping peaks (Fig. 22). Again the peak pattern of PLA is identical with the peak pattern of the multiple-walled microspheres, which proves that PLA is the surface polymer.

E. Attenuated Total Reflectance–Fourier-Transformed Infrared Spectroscopy

The degradation of polymers can be investigated using Fourier-transformed infrared (FTIR) spectroscopy by following the changing IR signals that are obtained from degrading films. Microsphere surfaces can be investigated by using

Table 3　C/O Ratio of Coated and Noncoated Microspheres

Polymer	C/O ratio
p(FAD–SA) 50:50 microspheres	5.54
poly(D,L-lactic acid) microspheres	1.61
p(FAD–SA) 50:50 microspheres coated with poly(D,L-lactic acid)	1.57

Source: Ref. 48.

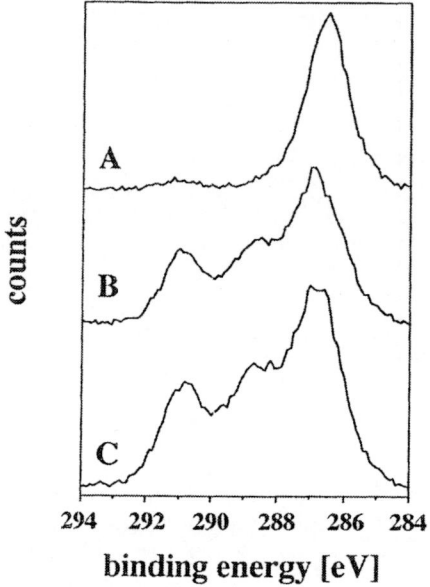

Fig. 22 Shape of the carbon peaks taken by XPS: (A) *p*(FAD–SA) 50:50 micro-spheres; (B) D,L-PLA microspheres; (C) D,L-PLA-coated *p*(FAD–SA) 50:50 micro-spheres. (From Ref. 48.)

a horizontal attenuated total reflectance (ATR) cell. The principle of the ATR–FTIR setup is shown in Fig. 23. The IR beam enters the ATR crystal (e.g., zinc selenide) and is reflected several times along its way from the bottom to the top. At the top of the crystal, it interacts with the material on the plate. The method permits obtaining IR spectra mainly of the material surface, with typical depths of penetration of 0.2–0.4 μm [55]. To demonstrate its potential, ATR–FTIR investigations of double-walled microspheres might serve as an

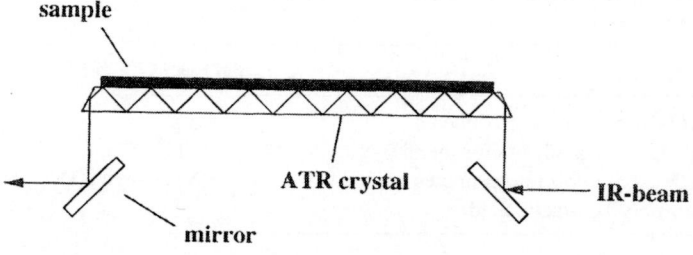

Fig. 23 ATR–FTIR measuring setup.

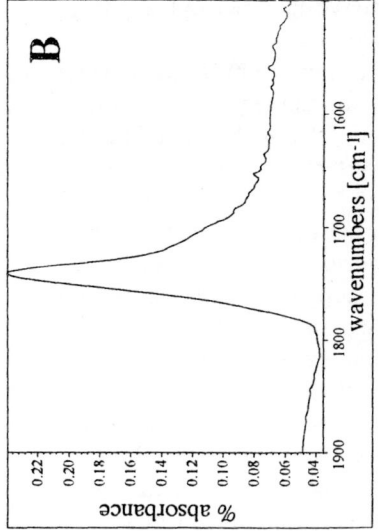

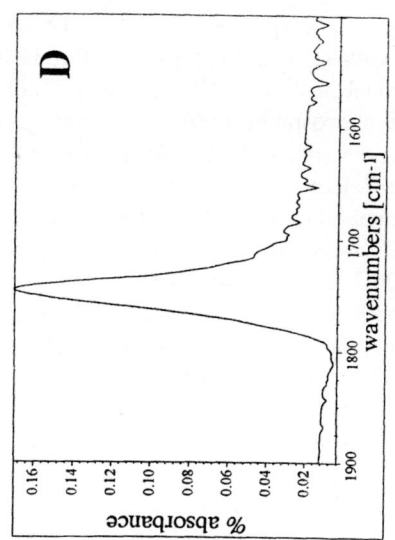

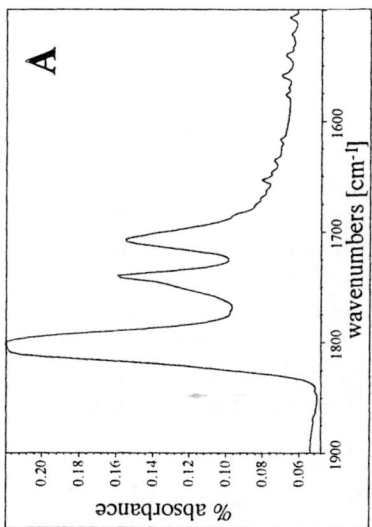

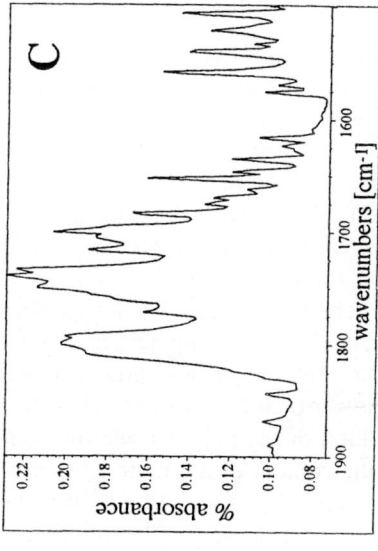

Fig. 24 ATR–FTIR spectra of microspheres made by solvent evaporation: (A) *p*(FAD–SA) 50:50; (B) poly(D,L-lactic acid); (C) microspheres made of a 1:1 mixture of *p*(FAD–SA) 50:50 and poly(D,L-lactic acid); (D) *p*(FAD–SA) 50:50 microspheres coated with poly(D,L-lactic acid).

example. The spectra of p(FAD–SA) 50:50 polyanhydride microspheres (Fig. 24A) consist of a typical doublet at 1740 cm^{-1} and 1810 cm^{-1}. There is one additional peak at 1710 cm^{-1} that is not visible in the spectrum of the polymer before microsphere preparation. This peak may be due to the partial hydrolysis of the reactive polymer during solvent evaporation. The spectra obtained for poly(D,L-lactic acid) microspheres are distinctly different. Figure 24B shows one characteristic peak at approximately 1760 cm^{-1}. When microspheres are manufactured from a mixture of both polymers, spectra such as that shown in Fig. 24C result, which contains the peaks of anhydrides and esters, indicating that both polymers are present in the microsphere surface. Figure 24D shows the spectra of dual-walled microspheres, consisting of a p(FAD–SA) 50:50 core and a poly(D,L-lactic acid) hull. They show the same carbonyl signals as poly(D,L-lactic acid) microspheres, which proves that poly (D,L-lactic acid) is the surface polymer. In summary, ATR–FTIR spectra allow one to determine the surface composition of microspheres and could yield valuable information on their surface composition depending on the manufacturing procedure. ATR–FTIR might also allow one to investigate degradation processes, such as the degree of degradation of particles made by solvent evaporation techniques.

F. Drug Release

Drug release from degradable microspheres might give useful hints on the microstructure of particles and the mechanism of erosion. The release of low molecular weight compounds from polyanhydride microspheres has been studied extensively and has revealed some of the polymer properties. The release of acid orange from p(SA), for example, has been observed to be very rapid (Fig. 25). This reflects that aliphatic polyanhydrides erode faster than aromatic ones. In addition, surface cracking after contact with water is much more pronounced for p(SA) than it is for p(CPP–SA) copolymers, for example. The slower release of drugs from copolymers containing aromatic monomers has been proved by releasing acid orange from p(CPH–SA) copolymers. Figure 26 shows the release of acid orange from p(CPH–SA) 50:50, which is substantially slower than from p(SA). The decreased release rates are due to the slower degradation of the polymer and the increased hydrophobicity of the matrix owing to the content of aromatic CPH monomer.

An interesting observation is the release of peptides from p(FAD–SA). Figure 27 shows release profiles of BSA from p(FAD–SA) 25:75 at various protein loadings. No initial burst is observed, irrespective of the protein loading. The protein is released for up to 3 weeks at a near-constant rate. The release rate of protein depends on the monomer composition of polyanhydrides used,

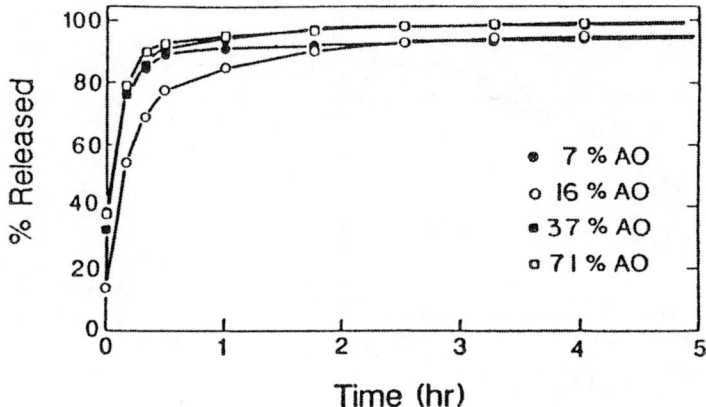

Fig. 25 Acid orange released from p(SA) microspheres for various acid orange loadings. (From Ref. 50.)

as can be seen from Fig. 28. Release periods, ranging from days to weeks, are possible for polyanhydride microspheres by changing their monomer composition. The deposition of FAD monomer on the surface of particles during erosion might be responsible for the changing release profiles.

Although release profiles can yield valuable information on microspheres, often they can be interpreted only if information on the polymer properties and the erosion mechanism is available.

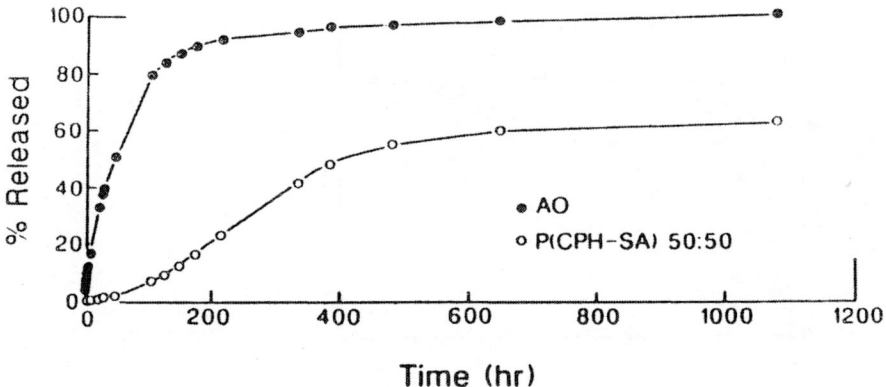

Fig. 26 Acid orange (AO) released from p(CPH–SA) 50:50 microspheres and erosion of the polymer [p(CPH–SA) 50:50]measured by sebacic acid release. (From Ref. 50.)

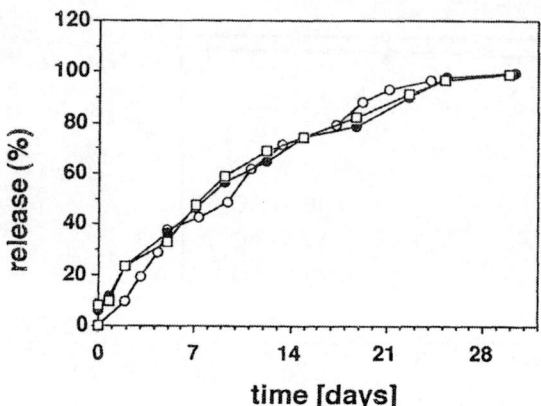

Fig. 27   Release of BSA from *p*(FAD–SA) 25:75 microspheres depending on load-ing: (○) 1% BSA; (●) 2% BSA; (□) 4% BSA. (From Ref. 43.)

VI.   OUTLOOK ON FUTURE DEVELOPMENTS

The future research on microspheres will focus on a more precise characteriza-tion of microspheres' properties. For example, there is no data available on the pH, the surface tension, or the osmotic pressure inside the pores of eroding microspheres that might account for the instability of proteins and peptides inside them. Apart from better ways of characterization, there is still a need for improvement of microsphere-manufacturing technologies. Such technolo-

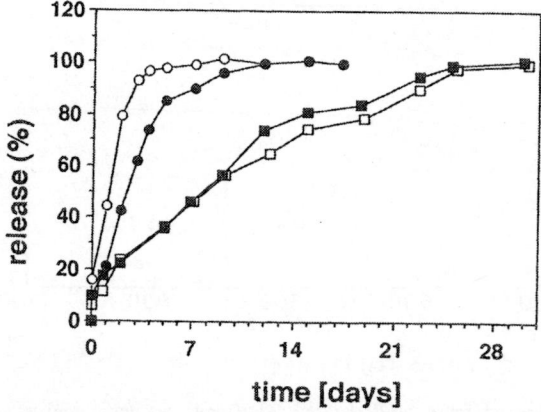

Fig. 28   Release of BSA from polyanhydride microspheres depending on polymer type: (○) *p*(SA); (●) *p*(FAD–SA) 8:92; (□) *p*(FAD–SA) 25:75; (■) *p*(FAD–SA) 44:56. (From Ref. 43.)

gies should allow the manufacture of microspheres with better erosion and drug-release control, should guarantee the stability of drugs during microsphere preparation and erosion, should reduce organic solvent residues, and should minimize polymer degradation during manufacturing. The advancement of analytical technology and the better characterization of microspheres will pave the way to such progress.

ACKNOWLEDGMENTS

The authors thank the Deutsche Forschungsgemeinschaft (DFG) for sponsoring parts of the presented research by grants GO 565/1 and GO 565/3-1.

REFERENCES

1. R. Langer and J. P. Vacanti, Tissue engineering, *Science 260*:920–926 (1993).
2. C. A. Vacanti, J. P. Vacanti, and R. Langer, Tissue engineering using synthetic biodegradable polymers, *Protein Formulations and Delivery*, (J. Cleland and R. Langer, eds.), American Chemical Society, Washington DC, 1994, p. 16–35.
3. R. Langer, New methods of drug delivery, *Science 249*:1527–1532 (1990).
4. N. D. Miller and D. F. Williams, The in vivo and in vitro degradation of PGA suture material as a function of applied strain, *Biomaterials 5*:365–368 (1984).
5. J. B. Herrmann, R. J. Kelly, and G. A. Higgins, Polyglycolic acid sutures, *Arch. Surg. 100*:486–490 (1970).
6. R. H. Müller, *Colloidal Carriers for Controlled Drug Delivery and Targeting*, Wissenschaftliche Verlagsgesellschaft, Stuttgart, 1990.
7. I. Preis and R. Langer, A single-step immunization by sustained antigen release, *J. Immunol. Methods 28*:193–197 (1979).
8. I. Esparza and T. Kissel, Parameters affecting the immunogenicity of microencapsulated tetanus toxoid, *Vaccine 10*:714–720 (1992).
9. R. E. Johnson, L. A. Lanaski, V. Gupta, M. J. Griffin, H. T. Gaud, T. E. Needham, and H. Zia, Stability of atriopeptin III in poly(D,L-lactide-*co*-glycolide) microspheres, *J. Controlled Release 17*:61–68 (1991).
10. J. E. Bucher and W. C. Slade, The anhydrides of isophthalic and terephthalic acids, *J. Am. Chem. Soc. 31*:1319–1321 (1909).
11. J. Hill, Studies on polymerization and ring formation. VI. Adipic anhydride, *J. Am. Chem. Soc. 52*:4110 (1930).
12. J. Hill and W. H. Carothers, Studies on polymerization and ring formation XIV: A linear superpolyanhydride and a cyclic dimeric anhydride from sebacic acid, *J. Am. Chem. Soc. 54*:1569 (1932).
13. A. Conix, New high-melting fiber-forming polymers, *Makromol. Chem. 24*:76–78 (1957).
14. A. Conix, Aromatic polyanhydrides, A new class of high-melting fiber-forming polymers, *J. Poly. Sci. 29*:343–353 (1958).

15. H. B. Rosen, J. Chang, G. E. Wnek, R. J. Linhardt, and R. Langer, Biodegradable polyanhydrides for controlled drug delivery, *Biomaterials 4*:131 (1983).

16. J. Tamada and R. Langer, The development of polyanhydrides for drug delivery applications, *J. Biomater. Sci. Polym. Ed. 3*:315–353 (1992).

17. A. J. Domb and M. Maniar, Absorbable biopolymers derived from dimer fatty acids, *J. Polym. Sci. [A] Polym. Chem. 31*:1275–1285 (1993).

18. M. Maniar, X. Xie, and A. J. Domb, Polyanhydrides V. Branched polyanhydrides, *Biomaterials 11*:690–694 (1990).

19. A. J. Domb, E. Mathiowitz, E. Ron, S. Giannos, and R. Langer, Polyanhydrides. IV. Unsaturated and crosslinked polyanhydrides, *J. Polym. Sci. [A] Polym. Chem. 29*:571–579 (1991).

20. H. Brem, A. Kader, J. I. Epstein, R. J. Tamargo, A. Domb, R. Langer, and K. Leong, Biocompatibility of a biodegradable controlled-release polymer in the rabbit brain, *Select. Cancer Ther. 5*:55–65 (1989).

21. H. Brem, A. Domb, D. Lenartz, C. Dureza, A. Olivi, and J. I. Epstein, Brain biocompatibility of biodegradable controlled release polymer consisting of anhydride copolymer of fatty acid dimer and sebacic acid, *J. Controlled Release 19*:325–330 (1992).

22. E. Ron, T. Turek, E. Mathiowitz, M. Chasin, M. Hageman, and R. Langer, Controlled release of polypeptides from polyanhydrides, *Proc. Natl. Acad. Sci. USA 90*:4176–4180 (1993).

23. K. W. Leong, V. Simonte, and R. Langer, Synthesis of polyanhydrides: Melt-polycondensation, dechlorination and dehydrative coupling, *Macromolecules 20*:705–712 (1987).

24. K. Leong, A. Domb, E. Ron, and R. Langer, Polyanhydrides, Encyclopedia of Polymer Science and Engineering, Suppl. Volume, 2nd edition, John Wiley & Sons Inc., 1989, pp. 648–665.

25. H. Brem, K. A. Walter, and R. Langer, Polymers as controlled drug delivery devices for the treatment of malignant brain tumors, *Eur. J. Pharmacol. Biopharmacol. 39*:2–7 (1993).

26. T. Painbeni, P. Menei, M. C. Venier, M. Boisdron-Celle, and J. P. Benoit, Characterization and toxicity of biodegradable BCNU-loaded microspheres. Efficacy on C6 rat malignant glioma after intraperitoneal and stereotactic implantation, Proceedings of the 13th Pharmaceutical Technology Conference, Strassbourg, France, 1994, pp. 65–47.

27. J. Tamada and R. Langer, Erosion mechanism of hydrolytically degradable polymers, *Proc. Natl. Acad. Sci. USA 90*:552–556 (1993).

28. E. Mathiowitz, J. Jacob, K. Pekarek, and D. Chickering III, Morphological characterization of bioerodible polymers. 3. Characterization of the erosion and intact zones in polyanhydrides using scanning electron microscopy, *Macromolecules 26*:6756–6765 (1993).

29. A.-C. Albertsson and S. Lundmark, Synthesis, characterization, and degradation of aliphatic polyanhydrides, *Br. Polym. J. 23*:205–212 (1990).

30. M. Chasin, A. Domb, E. Ron, E. Mathiowitz, R. Langer, K. Leong, C. Laurencin, H. Brem, and S. Grossman, Polyanhydrides as drug delivery systems, *Biode-*

gradable Polymers as Drug Delivery Systems (M. Chasin and R. Langer, eds.), Marcel Dekker, New York, 1990, pp. 43–70.

31. E. Ron, E. Mathiowitz, G. Mathiowitz, A. Domb, and R. Langer, NMR-characterization of erodible copolymers, *Macromolecules* 24:2278–2282 (1991).

32. D. C. Bassett, *Principles of Polymer Morphology*, Cambridge University Press, Cambridge, 1981, pp. 22–28.

33. E. Mathiowitz, E. Ron, G. Mathiowitz, C. Amato, and R. Langer, Morphological characterization of bioerodible polymers. 1. Crystallinity of polyanhydride copolymers, *Macromolecules* 23:3212–3218 (1990).

34. A. Göpferich, R. Gref, Y. Minamitake, L. Shieh, M.-J. Alonso, Y. Tabata, and R. Langer, Drug delivery from bioerodible polymers: Systemic and intravenous administration, *Protein Formulations and Delivery* (J. Cleland and R. Langer, eds.), American Chemical Society, Washington DC, 1994, pp. 242–277.

35. K. Park, W. S. W. Shalaby, and H. Park, *Biodegradable Hydrogels for Drug Delivery*, Technomic Publ., Lancaster, 1993.

36. R. Langer and N. Peppas, Chemical and physical structure of polymers as carriers for controlled release of bioactive agents: A review, *J. Macromol. Sci. Rev. Macromol. Chem. Phys. C23*:61–126 (1983), pp. 61–126.

37. E. W. Washburn, The dynamics of capillary flow, *Phys. Rev. 17*:273–283 (1921).

38. A. Göpferich, Mechanisms of polymer degradation and erosion, *Biomaterials 17*:103–114 (1996).

39. A. Göpferich and R. Langer, The influence of microstructure and monomer properties on the erosion mechanism of a class of polyanhydrides, *J. Polym. Sci. 31*:2445–2458, 1993.

40. C. Laurencin, Novel bioerodible polymers for controlled release analysis of in vitro/in vivo performance and characterizations of mechanism. PhD thesis, Massachusetts Institute of Technology, 1987.

41. Y. Tabata and R. Langer, Polyanhydride microspheres that display near-constant release of water-soluble model drug compounds, *Pharm. Res. 10*:391–399 (1993).

42. E. Mathiowitz and R. Langer, Polyanhydride microspheres as drug delivery systems, *Microcapsules and Nanoparticles in Medicine and Pharmacy* (M. Donbrow ed.), CRC Press, Boca Raton FL, 1992, pp. 100–122.

43. Y. Tabata, S. Gutta, and R. Langer, Controlled delivery systems for proteins using polyanhydride microspheres, *Pharm. Res. 10*:487–496 (1993).

44. E. Mathiowitz and R. Langer, Polyanhydride microspheres as drug carriers. I. Hot-melt microencapsulation, *J. Controlled Release 5*:13–22 (1987).

45. E. Mathiowitz, H. Bernstein, S. Giannos, P. Dor, T. Turek, and R. Langer, Polyanhydride microspheres. IV. Morphology and characterization of systems made by spray drying, *J. Appl. Polym. Sci. 45*:125–134 (1992).

46. N. A. Peppas, Fundamentals of pH and temperature-sensitive delivery systems, *Pulsatile Drug Delivery, Current Applications and Future Trends*, (R. Gurny, H. E. Junginger, and N. A. Pappas, eds.), APV Paperback Vol. 33, Stuttgart, 1993, pp. 41–56.

47. K. J. Pekarek, J. S. Jacob, and E. Mathiowitz, Double-walled polymer microspheres for controlled drug release, *Nature 367*:258–260 (1994).

48. A. Göpferich, M.-J. Alonso, and R. Langer, Development and characterization of microencapsulated microspheres, *Pharm. Res. 11*:1568–1574 (1994).
49. C. Schugens, N. Laruelle, N. Nihant, C. Grandfils, R. Jérome, and P. Teyssié, Effect of the emulsion stability on the morphology and porosity of semicrystalline poly(L-lactide) microparticles prepared by W/O/W double emulsion-evaporation, *J. Controlled Release 32*:161–176 (1994).
50. E. Mathiowitz, C. Amato, P. Dor, and R. Langer, Polyanhydride microspheres: 3. Morphology and characterization of systems made by solvent removal, *Polymer 31*:547–556 (1990).
51. E. Mathiowitz, D. Kline, and R. Langer, Morphology of polyanhydride microsphere delivery systems, *Scan. Microsc. 4*:329–340 (1990).
52. E. Mathiowitz, M. D. Cohen, and R. Langer, Novel microcapsules for delivery systems, *React. Polym. 6*:275–283 (1987).
53. H. Freund, *Handbuch der Mikroskopie in der Technik*, Vol. 1, Umschau Verlag, Frankfurt, 1957.
54. R. Gref, Y. Minamitake, M. T. Peracchia, V. Trubetskoy, V. Torchilin, and R. Langer, Biodegradable long-circulating polymeric nanospheres, *Science 363*:1600–1603 (1994).
55. J. Van Straaten and N. A. Peppas, ATR–FTIR analysis of protein adsorption on polymeric surfaces, *J. Biomater. Sci. 2*:113–121 (1991).

7

Nanoparticulate Drug Carrier Technology

Maria J. Alonso

University of Santiago de Compostela
Santiago de Compostela, Spain

I. INTRODUCTION

The technologies of producing colloidal dispersions from nonbiodegradable polymers have a relative short history in the pharmaceutical field. These colloidal dispersions have been otherwise defined as pseudolatexes. Pseudolatexes are now being used for coating pharmaceutical dosage forms, such as tablets or granulates, owing to their favorable rheological and film-forming properties [1]. A major advantage over polymer solutions is the concentration–viscosity relation for pseudolatex, thus allowing the coating process to occur in an aqueous environment very rich in polymer. Because of this, several companies manufacture cellulosic and acrylic polymers as aqueous polymeric colloidal dispersions. Reviews of these processes and their implications on pharmaceutical technology have been presented elsewhere [2].

A major effect in the pharmaceutical field has been caused by the use of colloidal dispersions of biodegradable polymers as nanoparticulate drug carriers. This idea emerged, more than 20 years ago, following the discovery that biodegradable nanoparticles can carry drug molecules to specific organs within the body [3]. Since the conception of this idea, a great deal of information about the potential and limitations of nanoparticles as drug carrier systems has been gathered. This improved understanding of the in vivo behavior of nanoparticulate drug carriers has motivated the development of new strategies to optimize their physicochemical properties (e.g., size and hydrophilicity) and, as a result, their in vivo fate after parenteral administration. Moreover, the demonstrated usefulness of nanoparticles as drug carriers for other routes of administration, such as the oral and ocular routes, has broadened their clinical potential.

Although drug delivery and targeting issues with nanoparticulates have been extensively reviewed [3–5], little attention has been paid to the critical analysis of the technologies available to prepare them. Therefore, the main purpose of this chapter is to provide an overview of the techniques to produce nanoparticles and nanocapsules, and to classify them based on the polymer properties and their ability to encapsulate different types of drugs. Mechanisms of incorporation and subsequent release of lipophilic and hydrophilic molecules from different types of polymeric carriers will be discussed. In addition, special emphasis will be placed on the technological variables that can be adapted to control the size of the nanospheres as well as their inner structure and surface properties. Finally, pharmaceutical aspects, such as sterilization and conservation, and biopharmaceutical issues, such as different routes of administration, will be addressed in this chapter.

II. POLYMER-BASED NANOPARTICULATE CARRIERS

While taking into account the high potential of polymeric colloidal suspensions as drug carriers, the purpose of the present chapter will be to review those made of polymers that are considered biodegradable and biocompatible. These polymers are either amphiphilic macromolecules, obtained from natural sources, or hydrophobic polymers, synthesized chemically. Some of these polymers were originally investigated for other biomedical applications; consequently, there is literature available concerning their safety and biodegradation.

A. Natural Hydrophilic Polymers

The natural hydrophilic group includes two categories of polymers: proteins, such as gelatin and albumin; and polysaccharides, such as alginate, dextran, and chitosan. In spite of the recognized interest on exploiting natural polymer sources, natural polymers have some disadvantages related to poor batch-to-batch reproducibility, the specific conditions for their degradation, and potential antigenicity. Among these drawbacks, the major concern with parenteral administration of proteins and polysaccharides is their antigenicity. The relevant information provided in the literature is deficient; thus, more in vivo toxicity studies will have to be performed to assess the safety of carriers based on these polymers. For example, alginate, which has been approved for oral and ophthalmic administration, was recently reported to have adequate hemocompatibility and not to elicit an immunogenic responses [6]. Consequently, it may represent a new candidate for parenteral administration. On the other hand, dextran, albumin, and gelatin are acceptable for parenteral administration; however, their aggregation and the introduction of a cross-linking agent during nanosphere preparation may render them immunogenic (see Section III. A). As a matter of fact, clinical experience with heat-denatured macroaggregated albumin systems has indicated a few severe problems, such as anaphylactic reactions [7]. On the other hand, the polysaccharide chitosan is not hemocompatible and requires the presence of some specific enzymes (i.e., lysozyme or bacteria) to undergo degradation [8]. Therefore, the present application of chitosan should be restricted to extraparenteral routes of administration.

B. Synthetic Hydrophobic Polymers

Most synthetic biodegradable polymers that have been used in the preparation of colloidal dispersions have been previously employed to prepare microspheres. As a consequence, the most relevant physicochemical and biological properties of these polymers have been extensively evaluated. Polymers can be synthesized before or during nanoparticle preparation. The first group includes polyesters such as poly(ϵ-caprolactone) (PECL) and the family of poly(lactic acid) (PLA) and poly(lactic–glycolic acid) (PLGA) copolymers [see extensive reviews of these polymer properties in Refs. 9, 10]. The second group is represented by the poly(alkylcyanoacrylates) (PACA) which have received the greatest attention as polymeric colloidal drug carriers. The safety of the polyesters is clearly illustrated by the fact that some formulations based on these polymers have been approved for human use. On the other hand, the safety of the PACAs has been a controversial issue because of the toxicity of the corresponding alkylcyanoacrylate monomers. Nevertheless, extensive studies carried out by Couvreur's group indicate that degradation of PACA nanoparticles does not produce significant amounts of monomers, but hydrophilic cyanoacrylic acid oligomers, which are easily eliminated from the body by glomerular filtration [11]. Furthermore, phase I clinical trials with doxorubicin-loaded PACA nanoparticles have shown the ability of these particles to reduce the cardiotoxicity of doxorubicin and have defined the recommended dose for phase II studies [12].

All these polymers will eventually be interesting, relative to their safety, for the design of nanoparticulate drug carriers. Other criteria, such as the preparation conditions of the nanospheres, drug–polymer compatibility, expected drug release behavior, and the final purpose of the formulation (i.e., route of administration) should be taken into account on the final choice of the polymer carrier.

III. PREPARATION TECHNIQUES

The selection of the appropriate method for preparing drug-loaded nanoparticles depends on the physicochemical properties of the polymer and the drug. On the other hand, the procedure and the formulation conditions will determine the inner structure of these polymeric colloidal systems. Two types of systems with different inner structures are possible: (a) a matrix-type system composed of an entanglement of oligomer or polymer units, defined here as a *nanoparticle* or *nanosphere*; and (b) a reservoir-type system, consisting of an oily core surrounded by a polymer wall, defined here as a *nanocapsule*. As shown in the following section, the in vitro release behaviors of these types are radically different.

The classification presented here is based on the nature of the materials used to prepare the nanospheres (Table 1). This list includes (a) amphiphilic macromolecules that undergo a cross-linking reaction during the preparation of the nanospheres; (b) monomers that polymerize during the formation of the nanospheres; and (c) hydrophobic polymers, which are initially dissolved in organic solvents and then precipitated under controlled conditions to produce nanospheres.

A. Cross-Linking of Amphiphilic Macromolecules

Nanoparticles can be prepared from amphiphilic macromolecules, polysaccharides, and proteins, by inducing their aggregation and further stabilization either by heat denaturation or chemical cross-linking. These processes may occur in a water-in-oil emulsion system or in an aqueous environment. In the latter, the cross-linking reaction may take place after a phase separation process (desolvation, in which no chemical reactions are involved) or, it may be the cause for the phase separation process (cross-linking reaction initiated by changes in pH or charged molecules).

1. Cross-Linking in a Water-in-Oil Emulsion

The cross-linking technique was first applied to the nanoencapsulation of drugs by Kramer et al. in 1974 [13] and, subsequently, optimized over the last two decades. The main steps of the procedure are schematized in Fig. 1. Formerly, an aqueous solution of the protein (human serum albumin [HSA] or bovine serum albumin [BSA]) was emulsified in an oil using a high-speed homogenizer [13] or a sonicator [14]. The water-in-oil emulsion was then poured into heated oil (temperature higher than 100°C) and held at that temperature for a specific time to denature the protein, thereby leading to the formation of proteinaceous submicroscopic particles. The particles were finally washed with organic solvents and subsequently collected by centrifugation. With this procedure, the crucial factors in nanosphere production include the emulsification energy and the stabilization temperature [15,16]. To overcome the limitation of high stabilization temperatures, a chemical cross-linking agent, most frequently glutaraldehyde, was incorporated into the system, thereby avoiding the thermal denaturation process [17]. However, other disadvantages, besides either heat

Table 1 Biodegradable Polymers Used for the Preparation of Nanoparticles and Nanocapsules

Polymer type	Nanoencapsulation technique	Candidate drug
Hydrophilic		
Albumin, gelatin	Heat denaturation and cross-linking in W/O emulsion	Hydrophilic
	Desolvation and cross-linking in aqueous medium	Hydrophilic and protein affinity
Alginate, Chitosan	Cross-linking in aqueous medium	Hydrophilic and protein affinity
Dextran	Polymer precipitation in an organic solvent	Hydrophilic, no soluble in polar solvent
Hydrophobic		
Poly(alkylcyanoacrylates)	Emulsion polymerization	Hydrophilic
	Interfacial O/W polymerization	Hydrophobic (soluble in oils)
Polyesters (PLA, PLGA, PECL)	Solvent extraction–evaporation	Hydrophilic or hydrophobic
	Solvent displacement	Soluble in polar solvent
	Salting-out	Soluble in polar solvent

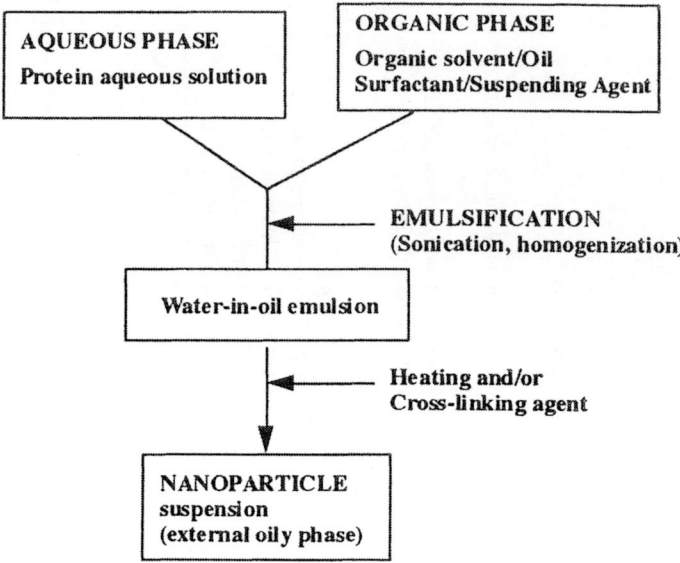

Fig. 1 Preparation of proteinaceous nanoparticles by the cross-linking in a water-in-oil (w/o) emulsion technique.

denaturation or chemical cross-linking, included the variable size of the nanospheres obtained (mean particle diameter between 0.4 and 1 μm). To optimize the nanosphere preparation method in terms of achieving a narrow-size distribution, Gallo et al. [18] analyzed the influence of several formulation variables on the particle size, thus providing conditions for the production of nanospheres with a mean size of 0.5 μm. New techniques using chemical derivatization of albumin in water-in-oil emulsions have recently been reported. Nakagawa et al. [19] have described a method to prepare bovine serum albumin (BSA) nanospheres with a diameter of 170 nm that relies on the steric stabilization of the albumin droplets during cross-linking through viscous polymer solutions in organic solvents. Namely, the aqueous BSA solution was emulsified in a solution of ethyl cellulose in chloroform by ultrasonication; then, the cross-linker glutaraldehyde was incorporated into the system and agitated for several hours. The resulting product was washed successively with toluene, isopropanol, and water, and finally, freeze-dried. This procedure was recently simplified and optimized by Roser et al. [20], who produced nanospheres of various sizes by using a hydroxypropylcellulose–chloroform solution as a continuous external phase. Chemical dehydration has also recently been reported, producing BSA nanospheres with a narrow-size distribution [21]. For example, the chemical dehydrating agent 2,2-dimethoxypropane was used to convert a water-in-oil emulsion (aqueous BSA solution in cottonseed oil) into a suspension, thereby avoiding the coalescence of the droplets and reducing the nanosphere size to 300 nm. With this technique, the sonication time required for the emulsification was substantially reduced.

2. Phase Separation in an Aqueous Medium

Proteins and polysaccharide nanoparticles can also be prepared by a phase-separation process in an aqueous medium. The phase separation can be induced by desolvation of the macromolecule, by changes in the pH or temperature, or by adding counterions to the aqueous medium. The cross-linking reaction may be subsequent to the phase separation, or it may occur simultaneously.

To determine the best conditions for the formation of nanospheres according to a desolvation process, a three-component diagram was generated that included the major components of the system (i.e., gelatin, water, and a solvent-competing solute, such as sodium sulfate or an alcohol). As illustrated in Fig. 2, three main steps may be considered in this method: First, protein (albumin or gelatin) dissolution in water; second, protein aggregation and precipitation by adding a desolvating or solvent-competing agent, such as sodium sulfate;

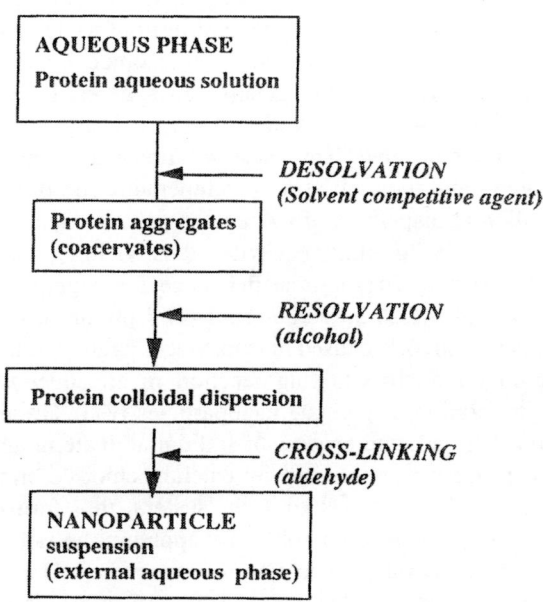

Fig. 2 Preparation of proteinaceous nanoparticles by the desolvation in an aqueous medium technique.

and, finally protein deaggregation by incorporation of a resolvating agent (e.g., a small amount of ethanol or isopropanol). If the procedure is stopped in the second step, the large aggregates of protein will lead to the formation of microspheres. However, because of the resolvation process, a milky suspension, consisting of colloidal particles, is obtained. Finally, these embryo nanoparticles are cross-linked with glutaraldehyde. A discussion of the key steps of this procedure was initially presented by Marty et al. [22] and later extensively reviewed by Oppenheim [23]. This similar procedure was adapted to the encapsulation of lipophilic compounds by Krause et al. [24]. Here, the drug triamcinolone acetonide was dissolved in a small volume of chloroform, which was then emulsified in an aqueous solution of gelatin by ultrasonication for 30 min. The system was then desolvated by incorporating sodium sulfate and isopropanol and hardened by adding a glutaraldehyde solution. However, this technique has not been further used, probably because of the necessity of using ultrasonication for long periods.

A phase separation may also occur, as a consequence of changes in the pH [25,26], or by the presence of counterions in the aqueous medium [27,28]. For example, in 1982, Oppenheim and co-workers succeeded in preparing a colloidal dispersion of insulin based on pH changes in the medium [25]. Insulin was initially precipitated and then partially redissolved to form nanodroplets, which were finally hardened by adding the cross-linker glutaraldehyde. In this particular example, the objective was not to encapsulate a specific drug, but to design a nanoparticulate form for insulin. A similar procedure was adopted by El-Samaligy and Rohdewald [26] for the preparation of gelatin nanospheres. First, the gelatin and the surfactant Tween-20 were dissolved in water, and the pH of the solution was adjusted to an optimal value. This clear solution was heated at 40°C, quenched at 4°C for 1 day, and then left at room temperature for 48 h, leading to the formation of a colloidal dispersion of aggregated gelatin. Gelatin aggregates were finally cross-linked with glutaraldehyde, thus yielding nanospheres with a mean size of 200 nm. The optimal pH range for aggregate formation in the nanometer range was 5.5–6.5. Values below 5.5 produced no aggregation, whereas pHs higher than 6.5 caused uncontrolled aggregation. More recently, a method, based on a cross-linking reaction in an aqueous phase, was developed for the preparation of alginate nanoparticles [27]. These particles were formed by a controlled gelation process of sodium alginate in an aqueous medium. The gelation process was induced by calcium chloride and then continued by the incorporation of poly(L-lysine) to the system, thus forming nanospheres composed of a polyelectrolyte complex. By applying the same principle we have succeeded in the preparation of chitosan nanospheres. Here, the precipitation of cationic polymer chitosan was induced by adding a highly negatively charged compound, such as tripolyphosphate, to the medium [28]. The size of these nanospheres was affected by the concentration of both the chitosan and tripolyphosphate in the medium. In addition, chitosan nanoparticles bear a positive net charge, a fact that makes them an interesting vehicle for negatively charged molecules.

To summarize the techniques that are based on the cross-linking of natural macromolecules, the water-in oil emulsion technique could be considered especially suitable for the encapsulation of hydrophilic drugs in terms of loading capacity. However, the usefulness of this technique is limited by the use of sonication or high-speed homogenization as well as the use of high volumes of organic solvents and oils. Phase-separation techniques, carried out in aqueous media, produce nanospheres of a narrow-size distribution (200–300 nm) and obviate the need for organic solvents; however, their efficiency in encapsulating hydrophilic drugs is strongly affected by the drug–polymer affinity (drug molecules are partitioned between the protein nanoparticles and the suspending aqueous medium). Nevertheless, the major general concern of the techniques included in this section lies in the necessity for using a cross-linker or hardening agent (most frequently glutaraldehyde) that may react with the encapsulated drug and may also confer toxicity on the nanosphere formulation.

B. Polymerization of Acrylic Monomers

The earliest nanoparticles prepared by the polymerization of a monomer were those proposed by Birrenbach and Speiser in the 1970s [29]. These nanoparticles, composed of polyacrylamide, were prepared by the polymerization of acrylamide and N,N'-methylene bisacrylamide, and they were intended for the encapsulation of antigens. Some years later, in the search for an adjuvant-carrier vaccine, Kreuter and Speiser, developed nanoparticles based on poly(methyl methacrylate) [30]. The main limitation of these polymeric carriers was related to their slow degradation rate and, therefore, their accumulation in the body. However, these initial nanoparticle formulations opened up new prospects in the area of drug targeting. Thus, Couvreur et al. [31] developed biodegradable poly(alkylcyanoacrylate) (PACA) nanoparticles using a simple polymerization reaction. These nanoparticles have been the object of the most extensive investigations in the area of polymeric colloidal drug carriers.

The PACA nanoparticles are prepared by an emulsion polymerization method in which droplets of water-insoluble monomers are emulsified in an external aqueous and acidic phase that contains a stabilizer [32] (Fig. 3). The monomers polymerize relatively fast by an anionic polymerization mechanism, the polymerization rate being dependent on the pH of the medium [33]. Thus, if the pH is neutral, the monomer polymerizes extremely fast, leading to the formation of aggregates. However, at acidic pH, between pH 2 and 4, the reaction is slowed, yielding nanospheres with a narrow-size distribution (frequently 200 nm). The system is maintained under magnetic agitation while the

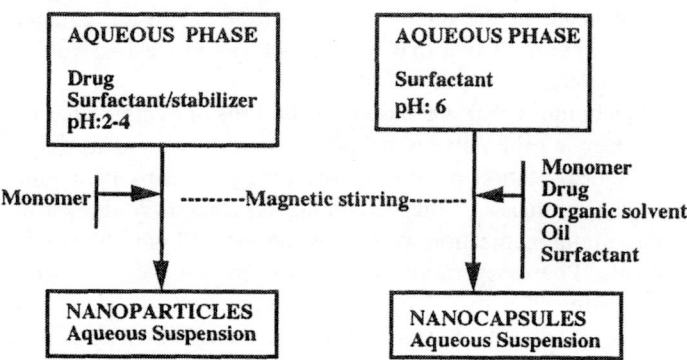

Fig. 3 The emulsion polymerization technique for the preparation of PACA nanoparticles and nanocapsules.

polymerization reaction takes place. The duration of the polymerization reaction is determined by the length of the alkyl chain (the longer the chain, the longer the polymerization time) varying from 2 h up to 12 h for ethyl and hexylcyanoacrylate, respectively. Finally, the colloidal suspension is neutralized and lyophilized following the incorporation of glucose as a cryoprotectant.

Water-soluble drugs may be associated with PACA nanospheres either by dissolving the drug in the aqueous polymerization medium or by incubating blank nanospheres in an aqueous solution of the drug. In the former, the drug molecules are entrapped within the polymer matrix and are also adsorbed onto the surface of the nanospheres. In the latter, the drug molecules are adsorbed onto the surface of the nanospheres. Logically, the loading efficiency is generally higher by incorporating the drug before the polymerization reaction, rather than adding it afterward. However, if the drug is incorporated into the polymerization medium, it may act as an initiator of the polymerization reaction, thus bonding covalently with the polymer [34,35]. The drug-loading efficiency is dependent on various factors, including the pK_a and polarity of the drug, size and surface charge of the nanospheres, and drug concentration in the aqueous medium. For example, the adsorption of the aminoglucoside amikacin onto PACA nanoparticles is dependent on the nanosphere size and surface charge [36]. In addition, the incorporation of the surfactant sodium lauryl sulfate to the polymerization medium induces the precipitation of amikacin and, as a result, increases the loading efficiency of the nanoparticles [37].

A new technique was developed, in 1986 by Al Khouri Fallouh et al. [38], for the encapsulation of lipophilic drugs into PACA polymers. In this technique (see Fig. 3), the monomers and the drug are dissolved in a mixture of a polar solvent (acetone or methanol), an oil (benzyl benzoate, coconut oil), and a lipophilic surfactant, such as lecithin. The organic phase is added to an aqueous phase containing a hydrophilic surfactant (e.g., Poloxamer 188) under magnetic agitation. Following dispersion of the organic phase into the aqueous phase, two processes occur simultaneously: the diffusion of the polar solvent into the aqueous phase, and the polymerization of the monomer at the water–oil interface. The polymerization is initiated by the hydroxy anions and leads to the formation of nanocapsules having an oily core surrounded by a polymer coat. The organic solvent is finally eliminated from the colloidal suspension under reduced pressure. Various lipophilic drugs have been entrapped in these nanocapsules. Lipophilic drugs entrapped in PACA nanocapsules, as well as hydrophilic drugs associated to PACA nanoparticles, have been extensively reviewed [39]. The selection of oil is normally based on the drug solubility criteria; however, it should be remembered that the type of oil has an influence on not only the size of the nanocapsules, but also on the molecular weight of the polymer coat and the stability of the suspension after storage [40].

A technique to prepare PACA nanocapsules with a hydrophilic core was also developed for the encapsulation of hydrophilic molecules [41–43]. Here, the interfacial polymerization of the monomer occurs in a water-in-oil emulsion. The drug to be encapsulated was incorporated into the internal aqueous phase, and the hydrophobic monomer was dissolved in the external organic medium (e.g., cyclohexane or chloroform). The main drawbacks are related to the use of organic solvents and, therefore, the necessity for washing and transferring to an aqueous medium, a process that causes the aggregation of the nanocapsules.

Briefly, from a technological perspective, PACA nanoparticles and nanocapsules may be considered interesting biodegradable carriers for hydrophilic and lipophilic drugs, respectively. Only some concerns, related to instability of some specific drugs in the acidic polymerization medium or to their reactivity with the monomer, limit the potential of these nanoencapsulation procedures.

C. Polymer Precipitation

The generic term *polymer precipitation* was chosen to designate the techniques based on the dissolution of the polymer in a particular solvent, followed by its dispersion in a continuous external phase, in which the polymer is insoluble. Usually, the inner phase is a solution of hydrophobic polymer (i.e., polyester) in an organic solvent. The external phase is an aqueous medium that contains a stabilizer. The main difference between the techniques included in this section is the miscibility of the organic and aqueous phases. Thus, the solvent extraction–evaporation technique is based on the use of solvents that have a limited solubility in water and form emulsions when dispersed in water (oil-in-water emulsions). The polymer precipitation occurs as a consequence of the solvent extraction or evaporation, whereas, when using a solvent that is completely soluble in the continuous aqueous phase (e.g., acetone), the immediate polymer precipitation occurs because of the complete miscibility of both phases.

Solvent precipitation techniques have been generally applied for hydrophobic polymers, except for dextran nanospheres. In this particular example, the polymer dextran was dissolved in water and then emulsified in an oily phase. The precipitation of dextran was induced by the incorporation of acetone. This technique was applied to the encapsulation of ovalbumin as a model antigen, yielding nanospheres with a mean size of 1 μm [44]. The main difference between this technique and those based on a cross-linked process is the absence of covalent bonds stabilizing the nanospheres, but other weak bonds, such as hydrogen bonds or Van der Waals forces, may apply.

Several techniques described in the literature are based on the mechanism of polymer precipitation:

1. Solvent Extraction–Evaporation

The solvent evaporation technique was first applied to the preparation of poly(D,L-lactic acid) (PLA) nanospheres in 1981 [45]. For this polymer, the organic phase containing the polymer and a lipophilic drug, testosterone, was emulsified into an aqueous phase that contained a surfactant. No specific details about the nature of the organic solvent or the emulsification procedure were reported in this initial work. A similar procedure was adopted some years later for the encapsulation of triamcinolone in PLA nanoparticles. The drug and polymer were dissolved in chloroform and emulsified in a gelatin solution by sonication; then, the solvent was eliminated by evaporation [46]. Bodmeier and Chen [47] proposed the replacement of sonication by a high-pressure homogenization to facilitate scaling-up. The homogenizer breaks the initial coarse emulsion in nanodroplets (microfluidization), yielding nanospheres with narrow-size distribution [47]. This procedure was later optimized for the production of biodegradable PLGA nanospheres, using a factorial experimental design [48]. The variables investigated were, the emulsification procedure (high-speed or high-pressure homogenizers), the polymer concentration in the organic phase (methylene chloride), the type and concentration of the stabilizer introduced in the external aqueous phase, and the phase volume ratio (O/W). Nanospheres of the smallest size (less than 100 nm) were obtained for the lowest phase volume ratio, the lowest PLGA concentration, and the highest poly(vinyl alcohol) (PVA) concentration, when acted as a stabilizer in the external aqueous phase. Other authors have prepared PLA and PECL using this technique, the emulsification step being the main difference between the procedures reported [49,50]. The emulsification procedure should be chosen on the basis of the

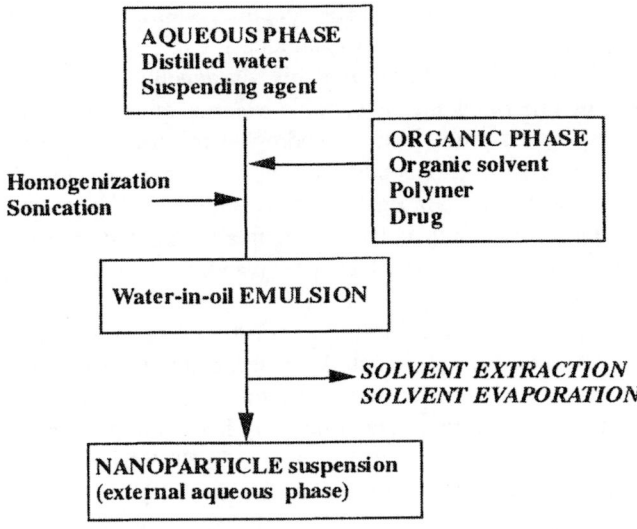

Fig. 4 The solvent evaporation and solvent extraction processes for the preparation of polyester nanoparticles.

drug sensitivity. However, the use of a microfluidizer or a high-speed homogenizer seems to be more suitable than the sonication for large-scale production.

The diagram in Fig. 4 summarizes the solvent extraction–evaporation technique carried out in an external aqueous medium. The hydrophobic polymer is dissolved in an organic solvent, such as chloroform, methylene chloride, or ethyl acetate. This organic phase is emulsified into an aqueous phase that contains a stabilizer, normally PVA. Once the nanoemulsion is formed, the solvent diffuses to the external phase until saturation. The solvent molecules that reach the water–air interface evaporate, thus driving the continuous diffusion of the solvent molecules from the inner droplets of the emulsion to the external phase. As a consequence of solvent elimination, the polymer precipitates, leading to the formation of nanospheres. The evaporation of the solvent may occur at room temperature or may be accelerated by using temperatures close to the solvent boiling point or vacuum. The extraction of the solvent from the nanodroplets to the external aqueous medium can be induced by adding an alcohol (e.g., isopropanol), thereby increasing the solubility of the organic solvent in the external phase. The extraction of the solvent can also be promoted by incorporating an additional amount of water into the emulsion.

Solvent extraction–evaporation techniques have been applied to mainly the encapsulation of lipophilic drugs, which can be dissolved in the polymer solution. For example, using the solvent evaporation procedure, we succeeded in encapsulating cyclosporin A, a very hydrophobic peptide, within PLGA nanospheres [51]. We have further modified this technique for the encapsulation of amphiphilic compounds, such as proteins. For this, an aqueous solution of the protein (25 μl) was incorporated into a variable volume of ethyl acetate (0.5–3 ml) that contained the polymer PLGA, leading to the formation of a water-in-oil emulsion or a suspension, depending on the volume of the organic phase. The resulting dispersion was then emulsified by sonication (10 s) in a variable volume of an aqueous solution of PVA, and finally, diluted in water or a 2% isopropanol solution, or introduced in a rotavapor to accelerate the evaporation of the solvent [52].

Most of the formulations developed according to solvent extraction–evaporation techniques used PVA as surfactant, which is not acceptable for parenteral administration. Consequently, a purification step is required to assure the elimination of the surfactant in the preparation. This limitation has been lately overcome by Verrecchia et al. [53], who succeeded in preparing PLA nanospheres (100-nm diameter) using human serum albumin as surfactant [53].

2. Solvent Displacement or Nanoprecipitation

The technique, schematized in Fig. 5, involves the use of an organic phase that is completely soluble in the external aqueous phase; consequently, neither evaporation nor extraction of the solvent is required for the polymer precipitation [54]. Polymer and drug are dissolved in acetone, ethanol, or methanol and

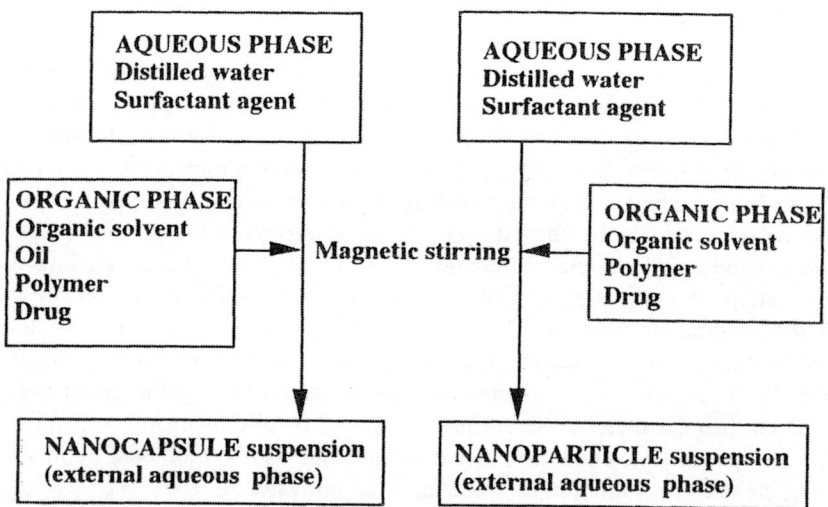

Fig. 5 The solvent displacement technique for the preparation of polyester nanoparticles and nanocapsules.

incorporated, under magnetic stirring, into an aqueous solution of a surfactant (Poloxamer 188). The organic solvent diffuses instantaneously to the external aqueous phase, inducing the immediate precipitation of the polymer and the drug. After nanoparticle formation, the solvent is eliminated and the suspension concentrated under reduced pressure. When using this procedure, further purification is not required because of the safety of the surfactants employed. However, the usefulness of this technique is limited to drugs that are highly soluble in a polar solvent, but only slightly soluble in water. If the drug is very hydrophilic, it will diffuse to the external aqueous phase, whereas if the drug is very hydrophobic, it may precipitate in the aqueous phase as nanocrystals, which will grow during storage. Indomethacin is a typical drug that is adequately encapsulated using this procedure, owing to its moderate lipophilic character [55]. However, with a very lipophilic drug, such as cyclosporine, the encapsulation efficiency observed in PECL nanospheres prepared by the nanoprecipitation technique was only 0.46 mg/g of polymer [56], whereas with a solvent evaporation method the encapsulation efficiency in PLGA nanoparticles reaches 60 mg/g of polymer [51].

The loading efficiency of lipophilic drugs can be improved by designing a nanocapsule system composed of an oily core surrounded by a polymer coat (see Fig. 5). In this system, the drug is dissolved in a small volume of an appropriate oil and then diluted in the polar solvent [57]. When the organic solution is dispersed in an aqueous phase, the polymer precipitates around the nanodroplets, forming a reservoir system. This technique has been used successfully for the encapsulation of lipophilic drugs, such as indomethacin [58], metipranolol [40], and betaxolol [59], using polyesters such as PLA, PLGA, and PECL.

More recently, Niwa et al. [60] proposed a spontaneous emulsification method for the production of nanospheres, based on the use of a mixture of a nonpolar and a semipolar solvent (methylene chloride and acetone) to dissolve the drug and the polymer. The acetone plays a role in improving the dissolution of certain drugs in the organic phase and favors the dispersion of the organic phase in the external aqueous phase.

3. Salting-out

A salting-out procedure has been developed by Bindschaedler et al. [61–64], the main steps of which are summarized in Fig. 6. This technique, which could be considered a combination of those previously mentioned, is based on the incorporation, under magnetic stirring, of a saturated aqueous solution of PVA into an acetonic solution of the polymer to form an emulsion of oil-in-water. In the nanoprecipitation technique, the polymer acetonic solution was completely miscible in the external aqueous medium; however, in this procedure the miscibility of both phases is prevented by the saturation of the external aqueous

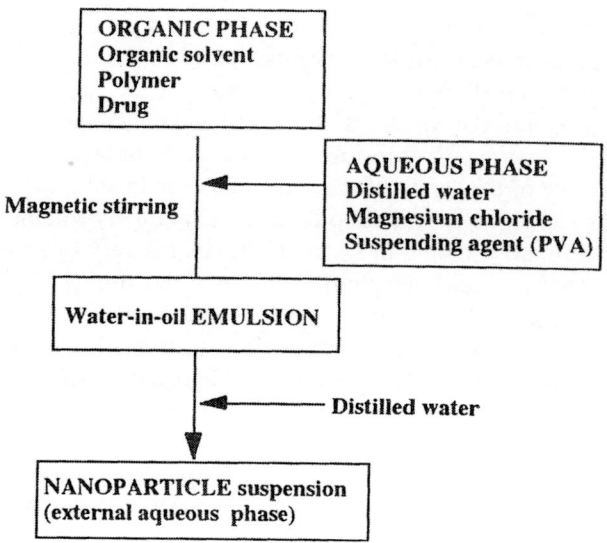

Fig. 6   Preparation of polyester nanoparticles by the salting-out method.

phase with PVA. The precipitation of the polymer occurs when a sufficient amount of water is added to allow complete diffusion of the acetone into the aqueous phase. Obviously, this technique is suitable for drugs and polymers that are soluble in polar solvents, such as acetone or ethanol.

   To summarize, techniques based on the precipitation of a hydrophobic polymer are useful for the encapsulation of either hydrophilic or hydrophobic drugs because a variety of solvents, including polar (e.g., acetone or methanol) and nonpolar (methylene chloride or chloroform) solvent can be chosen to dissolve the drug. Furthermore, a small amount of water (solvent extraction–evaporation) or an oil (solvent displacement) can be incorporated into the polymer solution, thus improving the encapsulation of very hydrophilic or very hydrophobic drugs, respectively. The main limitation, which is a common feature of the polymer precipitation techniques, is the necessity for using organic solvents to dissolve the hydrophobic polymers. In particular, the most important drawback of the solvent extraction–evaporation method is the requirement of a chlorinated solvent; however, it is possible to obtain nanospheres using other less toxic solvents, such as ethyl acetate. The requirements the solvent must have include: a certain solubility in the external aqueous phase (to promote the extraction); a relatively low boiling point; and a compatibility with the stabilizer. The advantages of these techniques over the polymerization and cross-linking techniques are that they avoid potential reactions between drug and monomer or the cross-linker, respectively. Furthermore, by using hydrophobic polymers, the potential toxicity of residual monomer or cross-linker molecules is obviated.

IV. DRUG-LOADING AND IN VITRO RELEASE PROCESSES

Guidelines concerning the suitability of the various nanoencapsulation techniques for the incorporation of drugs of different solubility were presented in the previous section. Although the type of polymer largely determines the nanoencapsulation procedure (see Table 1), the final selection is made according to the drug properties. In this section, the techniques available for the encapsulation of different types of drugs in terms of drug-loading efficiency and the drug controlled-release capacity, are compared.

A. Incorporation and Release of Lipophilic Drugs

Lipophilic drugs can be either incorporated in a nanocapsule (reservoir) or a nanoparticle (matrix) system using hydrophobic polymers (PACA, PLA, PLGA, or PECL). Encapsulation procedures involve the use of an inner organic phase that contains polymer or monomer, drug, oil (nanocapsules), and an external aqueous phase that contains a stabilizer. The selection of the organic phase and, therefore, the nanoencapsulation technique, should be based mainly on drug and polymer solubility criteria. Very lipophilic drugs may require dissolution in nonpolar solvents (such as chloroform or methylene chloride) or in oils. Therefore, the solvent extraction–evaporation or interfacial polymerization–polymer deposition techniques may be useful. One of the most important problems in the nanoencapsulation of very hydrophobic drugs is the migration of the drug during polymer precipitation to the external aqueous medium, forming nuclei of crystallization. These nanocrystals might be detectable only when they grow during storage. This problem could be eventually overcome by increasing the amount of polymer or modifying the external aqueous medium.

The in vitro release behavior of a lipophilic compound from a polymeric colloidal system is largely affected by its inner structure. It has been reported that the release process of drugs from nanocapsules (reservoir system) is dominated by the drug partition from the colloidal suspension to the external sink solution. This behavior, observed for various drugs, such as indomethacin [65], metipranolol [40], betaxolol [59], and cyclosporine [66], is, therefore, dependent on drug solubility in the oily core and the characteristics of the external receptor medium. For example, the in vitro release of cyclosporin A from PACA nanocapsules was clearly affected by the volume of the receptor aqueous medium [66]. Similarly, the release of indomethacin from PLA nanocapsules was incomplete when the study was performed in a dialysis system; however, most of the drug was very rapidly released when performed in a diffusion cell [67]. Furthermore, the release rate of indomethacin from PLA nanocapsules was significantly enhanced when albumin was incorporated into the receptor sink medium [68]. The authors attributed this increased drug release to the gradient established between the oily core and the external aqueous medium, because of the protein-binding affinity of the drug. These results confirm the partitioning of the drug into oily nanodroplets and aqueous receptor medium. In addition, the molecular weight of the polymer forming the coat of the nanocapsules as well as the nanocapsules' size remain unmodified during the release process [40]. Consequently, these colloidal nanocapsular drug carriers should not be considered as controlled-release drug-delivery systems, the main function of the polymer coat being to avoid the coalescence of the oily droplets.

The in vitro release behavior of lipophilic compounds from nanoparticles (matrix systems) may be markedly affected by polymer erosion, as the drug is dispersed within the polymer matrix. However, other factors, such as particle size and drug loading, may also, to a large extent, affect the release rate of the drug. Gurny et al. [45] observed that the release rate of testosterone from PLA nanospheres was a biphasic process characterized by an initial rapid release in the first day, followed by a slower release thereafter (90% released in 2 days).

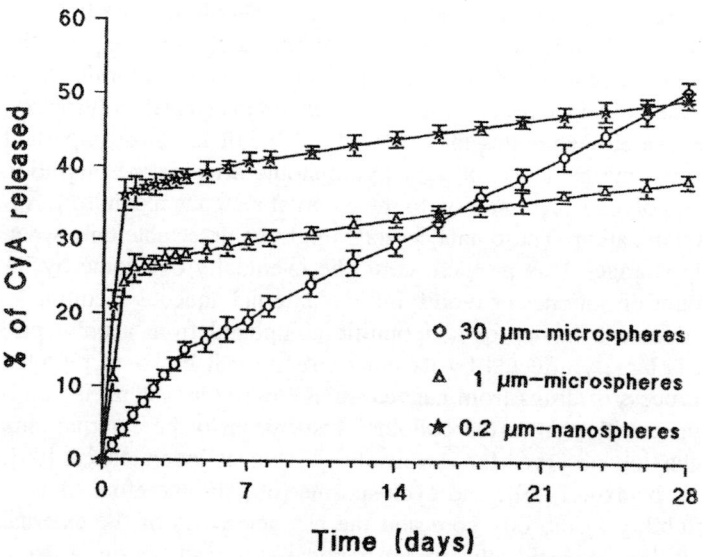

Fig. 7 In vitro release profiles of cyclosporine from PLGA microspheres and nanospheres prepared by the solvent evaporation technique.

The first phase is considered to be due to the release of the drug located on the particle surface, and the second part of the release profile being due to drug diffusion out of the polymer particles and the breakdown of polymeric material [45]. After taking into consideration that PLA degrades quite slowly (several months), it was concluded that polymer degradation rate does not control the release of testosterone. This may be attributed to the high loading of the nanospheres and the creation of large pores and channels, as a consequence of the drug-release process. A common feature of most of the nanoparticulate formulations thus far developed is the initial burst effect on the first day of release. Figure 7 shows the in vitro release profiles of cyclosporin A from PLGA micro- and nanospheres and indicates that the initial phase of release on the first day is related to the size of the microspheres, whereas the second phase is affected by the degradation rate of the polymer [69].

B. Association and Release of Hydrophilic Drugs and Bioactive Macromolecules

The association of hydrophilic compounds, especially macromolecules, to polymeric carriers remains an important issue under investigation. Basically, there are two main ways of association: one is based on the incorporation of the bioactive molecule during the formation process of the nanospheres, and the other on the drug adsorption onto preformed nanospheres (adsorbates).

1. Adsorption–Desorption of Hydrophilic Drugs

Several hydrophilic drugs and amphiphilic macromolecules (i.e., antibodies, antigens, and oligonucleotides) have been adsorbed onto preformed PACA nanospheres [36,70–72]. The mechanism and extent of adsorption process is dependent on the properties of the nanospheres (size and hydrophobicity) and the drug–polymer affinity. For example, the extent of the adsorption of amikacin, which is a polar molecule, onto PACA nanospheres was directly related to the nanospheres' surface charge. Thus, nanospheres with the highest negative charge adsorbed the highest number of positively charged amikacin molecules, suggesting an ionic interaction mechanism between the drug and the polymer nanospheres [37]. In other instances, the complexation of drugs with surfactants or salts may change their charge or solubility, thus leading to a larger attachment of the drugs to the nanoparticles. For example, the association of amikacin to PACA nanospheres was substantially enhanced by the formation of complexes with the surfactant sodium lauryl sulfate [36]. Likewise, the adsorption of oligonucleotides onto nanoparticles was mediated by the formation of ion pairs between the negatively charged phosphate groups of the nucleic acid chain and some hydrophobic cations (i.e., alkyltrimethylammonium salts) [71]. These data suggest that the interaction of hydrophilic molecules with the polymer nanoparticles is due to ionic forces or hydrophobic bonds. However, exceptions are illustrated by DNA and monoclonal antibodies, which were found to bind strongly to PACA nanospheres [72]. The mechanism of interaction of these bioactive molecules with the polymer was not elucidated; however, the binding of DNA molecules inhibited their transcription. On the other hand, the antibody fragments adsorbed to PACA nanospheres were released only after the degradation of the polymer [70], suggesting a strong and specific antibody–polymer interaction.

The mechanism of drug release from adsorbates is governed by the desorption process of the drug from the polymer carrier. The release rate being dependent on the strength of the drug–polymer interaction and the characteristics of the release medium. For example, the release rate of amikacin from PACA nanospheres occurred very rapidly when the system was maintained under sink conditions [40]. In general, with the aforementioned exceptions, the drug is desorbed before the polymer is significantly degraded.

2. Incorporation and Release of Hydrophilic Drugs

A different situation is observed when the hydrophilic molecule is incorporated during the nanoparticle's formation. In those instances when the encapsulation process is carried out in an aqueous environment, such as in the polymerization of cyanoacrylates, the desolvation of proteins, and the cross-linking of carbohydrates, the drug is dissolved in the aqueous medium, and its entrapment in the nanoparticles is very much dependent on the drug–polymer affinity; whereas, when the encapsulation process requires the use of an organic solvent (PLA or PECL), the drug-loading efficiency will be dependent on the formulation conditions.

In a study of the entrapment of two hydrophilic antibiotics, ampicillin and gentamicin, in PACA nanoparticles, it was observed that the level of association of gentamicin was considerably lower than that of ampicillin, a fact that was attributed to the higher polarity of gentamicin [73]. Another problem arises when the drug to be incorporated into the polymer nanoparticles reacts with the monomer, or it is unstable in the acidic polymerization conditions. For example, the antiviral agent vidarabine induces the polymerization of cyanoacrylic monomers through a zwitterionic pathway, resulting in covalent linkages between vidarabine and the PACA nanoparticles [74]. This problem may be eventually overcome by incorporating the drug at different stages during the polymerization process. In this sense, Grangier et al. noted [35] that the stage at which growth hormone-releasing factor was incorporated into the polymerization medium was critical in avoiding degradation of this peptide and its covalent linkage to the polymer. This indicates that the initial formation of the polymerization nuclei represents the critical stage in which drug–monomer reactions are favored. On the other hand, the initial acidic pH (2–4) of the polymerization medium can be gradually neutralized while the polymerization reaction takes place, thus minimizing degradation of pH-sensitive drugs.

The release of drugs entrapped within PACA nanoparticles normally results from the erosion of the polymer. Indeed, drug release was increased when esterases (enzymes that degrade PACA polymers) were added to the incubation medium [35,73]. Since the degradation rate of PACA varies from a few hours to more than 1 day, depending on the length of the alkyl chain, similar variation in release rates may be expected for drugs entrapped in these polymeric carriers.

The association of hydrophilic drugs with natural macromolecules, such as proteins and polysaccharides, using a continuous aqueous media procedure, is also dependent on the specific drug–macromolecule affinity. Doxorubicin has been frequently chosen as a model hydrophilic compound for investigating the loading capacity of nanospheres made of different materials, such as proteins (collagen, albumin) [15,16,23], polysaccharides (alginate) [27], and synthetic polymers (PACA) [43]. Among these polymer carriers, the nanospheres made of natural macromolecules, such as collagen and alginate, have shown the highest loading capacity, which was explained by the large affinity of doxorubicin for those macromolecules [23,27]. In summary, the nanoencapsulation techniques involving continuous aqueous media are particularly useful for hydrophilic drugs that have an affinity for proteins and polysaccharides.

Generally, the release of hydrophilic drugs from macromolecular carriers occurs relatively fast (1–2 days) and the release rate reflects the affinity of the drug for the polymer [23,27]. Nevertheless, the release rate can be adjusted by changing the formulation variables that have an effect on the protein cross-linkage [75,76].

Hydrophilic molecules can also be encapsulated within polyester nanospheres (i.e., PLGA) using the procedures indicated in Section III. C. These techniques are based on the displacement, diffusion, or extraction–evaporation of solvent or on the salting-out effect. The solvent displacement or diffusion techniques are limited to drugs that are soluble in a polar solvent, such as acetone. Consequently, they are not suitable for the encapsulation of macromolecules that are very sensitive to polar solvents (e.g., proteins). 5-Fluorouracil was used as a water-soluble model drug to investigate its encapsulation in PLGA nanospheres using the solvent diffusion method [60]. The drug content in the nanospheres was very low, ranging from 0.08 to 2.65%, probably because of the normal tendency of the hydrophilic drug to diffuse to the external aqueous phase. Furthermore, the drug was completely released in 1 day, suggesting that the release process is not controlled by the polymer degradation rate. The salting-out method can also be applied to the encapsulation of hydrophilic compounds that are soluble in acetone. In the encapsulation of savoxepin methanesulfonate in PLA nanospheres [77] the entrapment efficiency was improved by adjusting the pH of the external aqueous medium. However, the maximum drug loading obtained was very low, in the range of 0.20–1.12% (drug/polymer ratio). In contrast, with savoxepin base, which is less hydrophilic, but more soluble in acetone, drug loading increased up to 9%. The release of this basic compound was prolonged for more than a month, thus showing the ability of these nanospheres to control the release.

To achieve high loading of very hydrophilic compounds, in particular macromolecules, the solvent extraction–evaporation technique seems to be the most efficient approach. In this method, the drug is dissolved in water and then dispersed in an organic solvent that has a limited solubility in water, such as ethyl acetate. The resulting emulsion or suspension (depending on the volume of the inner water phase) is redispersed in an aqueous phase to allow solvent extraction–evaporation and, thereby, the formation of nanospheres. The diffusion of the drug into the external aqueous phase is prevented by the limited solubility in the nonpolar solvent, which results in an increased drug-loading efficiency. This technique was successfully applied to the encapsulation of model proteins, such as BSA and lysozyme, in PLGA nanospheres. The size of the nanospheres is largely affected by the molecular weight of the polymer. As shown in Fig. 8, nanospheres provide an extended release of BSA after an

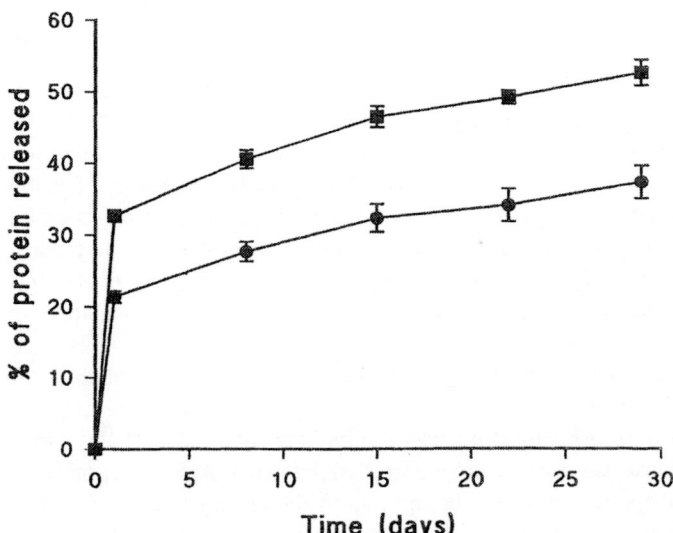

Fig. 8 In vitro release profiles of BSA from PLGA nanospheres prepared by the solvent extraction technique.

initial burst of release. The first phase, which corresponds to the release of protein molecules located at the nanospheres' surface, is logically affected by the size of the nanospheres; whereas, the second phase is related to the degradation rate of the polymer.

To summarize, lipophilic drugs can be encapsulated in nanocapsules or nanoparticles made of synthetic polymers (PACA, PECL, or PLGA). The release is markedly affected by the type of polymer and inner structure. Thus, the release of a lipophilic drug from PACA or PECL nanocapsules may occur in several hours, as the process is controlled by the partitioning of the drug between the oily core and the external release medium. As a contrast, the release of a lipophilic compound from PLGA nanoparticles is controlled by the degradation rate of the polymer and, therefore, it may be prolonged for more than a month. A hydrophilic compound can be adsorbed onto preformed nanospheres or entrapped within a polymer matrix composed of natural macromolecules or synthetic polymers. In the first instance, the release rate of the drug is dependent on mainly the drug–polymer affinity, whereas in the second example, the polymer degradation rate and nanoparticles' inner structure (porosity, degree of cross-linkage) may have an important effect on the drug-release process.

VI. PHYSICOCHEMICAL CHARACTERIZATION

In this section, the properties that have a special relevance for the colloidal carriers, namely particle size distribution, surface charge, and hydrophilicity will be examined. These physicochemical properties affect not only drug loading and release (as seen in the previous section), but also the interaction of these particulate carriers with biological membranes. Reviews focused on the in vivo fate of drug carriers with different physicochemical properties have been published elsewhere [3]; therefore, guidelines to modulate these properties and select the appropriate characterization techniques will be presented here.

A. Particle Size Analysis

Two main techniques are now being used to determine the particle size distribution of colloidal systems: photon correlation spectroscopy (PCS) and electron microscopy, including both scanning (SEM) and transmission (TEM) electronic microscopy. The usefulness of these techniques was previously discussed by Kreuter [78].

The PCS method, also called dynamic light scattering (DLS) or quasi-elastic light scattering (QELS), measures the Brownian motion of the particles and provides an accurate procedure for measuring the size distribution of nanoparticles. A description of the technique and the correct interpretation of the particle size distribution data have been discussed in detail [79]. The main advantages of PCS include its efficiency, its noninvasive character, and that this method does not require any particular sample preparation for analysis, such as particle isolation or sample drying, only an adequate dilution with pure water. However, because the measurements are based on the Brownian motion of the particles, the conditions of the suspending medium (e.g., adsorbed surfactants) may affect the particle size determinations.

Electron microscopy has the great advantage of providing an image of the particles to be measured. The main disadvantage is that the preparation of the sample is time-consuming. In particular, SEM involves the use of vacuum-dried nanoparticles that are coated with a conductive carbon–gold layer for analysis. A practical problem arises during the drying process because the nanoparticles are coated with the surfactants normally dissolved in the suspending medium. This coat hinders the observation of individual structures. This problem may be overcome by a previous purification of the nanoparticle suspension. However, occasionally, the total removal of surfactants may lead to particle aggregation. Use of TEM is an alternative and a powerful technique in the determination of particle size and shape and also of inner structures. Figure 9 shows TEM micrographs of PECL and chitosan nanoparticles prepared by a solvent displacement and phase separation, respectively. It may be observed that both types of nanoparticles are spherical, with homogeneous size and continuous inner structure. In combination with freeze-fracture procedures, TEM permits differentiation among nanocapsules, nanospheres, and emulsion droplets. The first observation of the internal structure of biodegradable nanospheres was performed by TEM after spray-freezing and cryofracture in 1979 [31]. This observation revealed that PACA nanospheres were composed of a continuous matrix. Conversely, as it was demonstrated some years later, the internal structure of nanocapsules appeared with a thin envelope around the oily nanodroplets [80]. As mentioned for SEM, the main limitation in TEM is the treatment of the samples before observation. Usually, the samples are prepared by placing a drop of the preparation on a cooper grid, followed by negative staining [38,58,80].

Atomic force microscopy (AFM) is an advanced microscopic technique that has been recently applied for the characterization of PLA nanospheres [81]. The AFM images can be obtained in an aqueous medium [82], which makes AFM a potentially powerful aid for the investigation of nanoparticle behavior in biological environments.

B. Surface Charge and Hydrophilicity

The nature and intensity of the surface charge of nanoparticulate drug carriers is very important because it determines their interaction with the biological environment as well as their electrostatic interactions with bioactive compounds. In this sense, we have observed that negative charges promote the

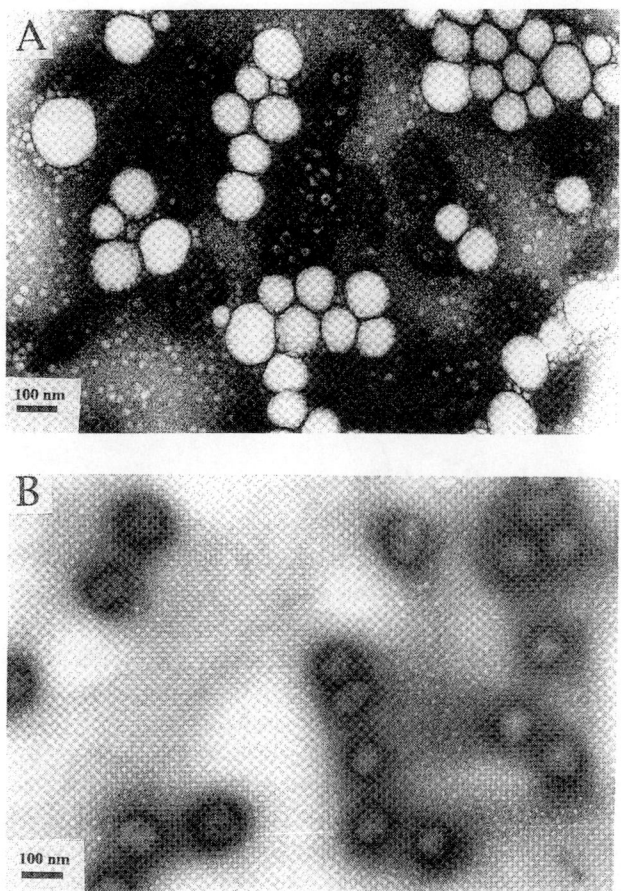

Fig. 9 A TEM micrograph of (A) PECL and (B) chitosan nanoparticles.

adsorption of positively charged drug molecules, such as aminoglucosides as well as enzymes and proteins [e.g., lysozyme and poly(L-lysine); 35,84]. The surface charge of colloidal particles can be determined by measuring the particle velocity in an electrical field. Laser light-scattering techniques, in particular laser Doppler anemometry, has become a fast and a high-resolution technique for the determination of nanoparticle velocities [83,85].

Although it is recognized that surface hydrophobicity of nanoparticles has an important influence on the interaction of colloidal particles with the biological environment (e.g., protein adsorption and cell adhesion), rigorous methods for the evaluation of appropriate parameters are presently being elucidated. Several approaches, including hydrophobic interaction chromatography, two-phase partition, adsorption of hydrophobic fluorescent or radiolabeled probes, and contact angle measurements have been previously reviewed [86]. These methods permit one to compare particles of different polarity, thereby giving an estimation of their comparative in vivo behavior. Recently, several sophisticated methods of surface chemical analysis have been developed, thus permitting the identification of specific chemical groups on the surface of nanospheres. Among these techniques x-ray photoelectron spectroscopy (XPS) seems to be the most accurate to analyze the nanospheres' surface, as it identifies chemical groups in the 5-Å–external nanosphere coat. For example, when this technique was applied to the characterization of PEG-coated PLGA nanospheres, it identified the PEG chemical elements that were concentrated on the nanospheres' surface [81].

C. Methods of Altering Particle Size and Surface Characteristics

There is extensive literature to support the notion that the fate of the colloidal particles in the body is determined by three main factors: particle size, particle charge, and surface hydrophobicity [87–89]. Particles with a very small size (less than 100 nm), low charge, and a hydrophilic surface are not recognized by the mononuclear phagocytic system (MPS) and, therefore, have a long half-life in the blood circulation. Consequently, it is not surprising that several authors have investigated approaches for the production of particles with the desired properties. We have used experimental factorial designs to deduce the optimal conditions to produce nanospheres of a particular size [48,90]. Obviously, the factors involved in the control of the nanospheres' size are dependent on the preparation technique; however, in general, the nature and the concentration of the surfactants used play an important role in determining the particle size, as well as the surface charge. For example, by increasing the concentration of Poloxamer 188, Seijo et al. [91] succeeded in producing nanospheres with a mean size of less than 50 nm.

The approaches for modifying surface charge and hydrophilicity were initially based on the adsorption of hydrophilic surfactants, such as block copolymers of the Poloxamer and Poloxamine series [92]. These studies provided valuable information on the in vivo behavior of hydrophilic nanospheres. However, the desorption of the hydrophilic coatings and their toxicity may limit the usefulness of these nanospheres for intravenous injection. Lately, the idea of using diblock copolymers made of PLA and poly(ethylene oxide) (PLA–PEO) is gaining widespread interest, owing to the safety and stability of the hydrophilic coat. For this purpose, different PLA–PEO copolymers have been synthesized and used either to produce nanospheres or to be adsorbed onto nanospheres [81,93]. In the first method, the copolymer is dissolved in an organic solvent, and then emulsified in an external aqueous phase, thereby orienting the PEO toward the aqueous surrounding medium [81]. In the latter method, the PLA–POE copolymer is adsorbed onto preformed PLGA nanoparticles [93]. Both approaches have been efficient in prolonging the nanospheres circulation time following intravenous administration. Nevertheless, it may be expected that a more stable hydrophilic coat is obtained when PLA–PEO is used to prepare the nanospheres, rather than when it is adsorbed onto the nanospheres. Another alternative investigated has been the adsorption of proteins, in particular, the adsorption of sialic acid-rich glycoprotein (orosomucoid), to mimic the sialic acid-rich surface of red blood cells as a way to reduce the uptake by the MPS [95]. This approach turned out to be inefficient, because the glycoprotein was rapidly desorbed from the nanospheres after incubation in serum. To prevent protein desorption, Verrechia et al. [53] proposed the incorporation of the protein during the nanospheres' formation process [53]. In this case, the protein attachment was more effective, as evidenced by the fact that 50% of the protein remains at the nanosphere surface after several hours' incubation in serum. Nevertheless, the fixed coat did not avert capture of the nanospheres by the MPS, since the distribution pattern of coated nanospheres was essentially identical with that of the uncoated nanospheres [96]. More information about the technology of surface modification of nanospheres can be obtained in Chapter 9.

VII. PHARMACEUTICAL CONSIDERATIONS

A. Purification and Isolation

Nanoparticles prepared in an external oily phase are difficult to isolate because they require successive centrifugation steps to eliminate the oil. As a consequence, the particle size distribution is generally broad and the mean size is close to 1 μm. Preferably, nanospheres need to be prepared using an aqueous phase containing safe stabilizers, thereby avoiding further purification steps. The necessity of purification is dependent on the final purpose of the formulation developed. For example, Poloxamers and dextrans are considered safe polymers (at relatively low concentrations) for parenteral administration and, therefore, nanospheres suspended in these stabilizers do not require further purification. However, the stabilizer PVA, which is frequently involved in the preparation of polyester nanospheres (i.e., solvent evaporation and salting-out procedures), is not acceptable for parenteral administration and, consequently, must be eliminated. This necessity is not so critical, however, if the nanospheres are intended for oral or ocular administration, as PVA has been consistently administered by these routes. Tangential filtration or cross-flow filtration is the most suitable procedure for the purification of nanoparticles. This technique avoids the irreversible aggregation that occurs when nanoparticles are isolated by centrifugation [69,97].

Nanospheres are normally isolated by freeze-drying, using a cryoprotective agent to assess the redispersability of the colloidal system. Sugars, such as glucose and trehalose, have been efficient in preventing the aggregation of nanoparticles during the freeze-drying process. This easy step becomes complicated for the isolation of nanocapsules composed of an oily core surrounded by a tiny polymeric wall. In this case, owing to the fragility of the polymer coat, nanocapsules tend to aggregate during the freeze-drying process. The nature of the internal oily droplets seems to be the most important factor in selecting the lyophilization conditions. In some particular cases, trehalose concentrations as high as 30% were required to prevent the aggregation of PLA and PECL nanocapsules [98]. This extremely high sugar concentration causes hypertonicity of the final suspension and, thus, is not acceptable for parenteral administration. Probably the surface characteristics of these colloidal systems will have to be substantially modified to finally achieve a stable or readily dispersible formulation.

B. Stability

Because most of the colloidal systems described in this chapter can be easily isolated in a solid freeze-dried form, little is known about the stability of these colloidal systems during storage. It is expected, however, that nanoparticles made of hydrolytic degradable polymers will degrade (although at a low rate if the temperature and pH are controlled) over time. For example, when PLA and PECL nanoparticles were stored for 350 days at 5°C, only minor changes in the molecular weight of the polymers and in size of the nanospheres were observed. However, when PLA nanoparticles were stored at 25°C, changes in the polymer molecular weight were detected after a 3-month storage period [99]. In addition, it was observed that the presence of anionic surfactants in the dispersion causes a more rapid degradation of PLA. Therefore, it could be expected that drugs with strong nucleophilic groups may catalyze the degradation of the polymer.

Nanocapsules are more delicate structures, and this must be taken into account in terms of stability during storage. Even though it was previously reported that nanocapsules are stable for 7 months [58], we have observed that metipranolol-loaded PIBCA and PECL nanocapsules aggregate and partially release their drug content after 3 months of storage at 4°C [40]. This process, however, was very much dependent on the nature of the encapsulated oil. Although a total phase separation was observed using the oil Labrafil, only slight increases in the nanocapsule diameter were detected using the oil Migliol. On the other hand, when studying the stability of colloidal carriers, it is important to analyze not only the particle size and polymer molecular weight, but also the eventual leaking of the drug from the carrier during storage. For example, during the encapsulation of strong hydrophobic drugs, precipitation in the external aqueous phase causes nanocrystals that may be misinterpreted as nanoparticles. The nanocrystals, however, will became evident if the preparation is stored for several days or weeks to allow the crystallization nuclei to grow.

From these few studies it may be concluded that the stability of the colloidal polymer carriers during storage is dependent on not only the storage conditions, but also on the exact composition of the formulations stored. Consequently, for each, specific formulation, the corresponding stability study will have to be performed to assess the quality of the product.

C. Sterilization

Most nanoparticles are expected to be used parenterally; therefore, they are required to be sterile. Sterilization may be achieved, as for any other microparticulate system, by using aseptic conditions throughout formulation, or by use of gamma-radiation at the end of the process. In addition, some authors have reported the use of autoclaving for sterilization of nanospheres and nanocapsules [22,38]. However, for some specific formulations, it was observed that autoclaving yields a substantial increase in the nanocapsules' size [80], a fact that could induce the release of the encapsulated drug or even lead to the total destruction of the system. Finally, the possibility of sterilizing the final suspension by filtration may represent a useful approach for those systems with a particle size smaller than 0.2 μm. When using this alternative, the loss of drug or polymer particles during the filtration process should be carefully tested.

VIII. BIOPHARMACEUTICAL ISSUES

A. Parenteral Administration

As stated at the beginning of this chapter, biodegradable nanoparticles were formed to target drugs to specific sites in the body, thus interfering with normal distribution patterns. The success of this approach has been demonstrated in particular for PACA nanospheres for intracellular antibiotherapy [100] and cancer therapy [101]. In addition, the capacity of nanospheres to transport drug molecules to some specific cell types has been extensively investigated in in vitro cell culture, thus providing evidence of their ability to penetrate cell membranes [5]. Nevertheless, it is now known that the ability to modulate the fate of drug-loaded nanoparticles administered intravenously is limited by their uptake by the MPS. Major efforts have been directed at changing the nanospheres' surface, to minimize their capture by MPS cells and prolong their residence time in the blood circulation. The achievement of this goal would not necessarily imply that the transported drug will reach the specific target. Nevertheless, modulation of the distribution patterns of a particular drug may lead to a reduction in its toxicity, or to a prolongation in its therapeutic effect (for more information see Refs. 3–5).

Nanospheres can also behave as sustained or controlled drug-delivery systems when injected subcutaneously or intramuscularly. The distribution pattern of the peptide cyclosporin A was essentially the same when encapsulated in nanospheres or in microspheres [102]. In both experiments, the residence time of the peptide in the bloodstream was considerably prolonged, whereas the blood concentration was related to the release rate of the peptide from the microspheres and nanospheres. Similarly, [^{14}C]savoxepin-loaded PLA nanoparticles injected intramuscularly remained at the injection site for at least 7 days [103]. In this sense, the advantage of nanoparticles over microparticles may be related to their injectability and, probably, to their improved tolerance. The subcutaneous route has been also exploited for the administration of vaccines adsorbed onto poly(methyl methacrylate) nanoparticles [see review in Ref. 104]. Kreuter and colleagues reported the adjuvant effect of these nanospheres for various vaccines. They indicated that particle size and hydrophobicity play an important role in the adjuvant effect: an increase in hydrophobicity and a decrease in particle size lead to an improved adjuvant effect. The same authors also studied the in vivo fate of these particles and noted that they degrade very slowly and remain in the subcutaneous tissue for extended periods [105]. Consequently, they may be unsuitable for the long-term administration of drugs, but they may be an acceptable material for vaccination applications, for which a continuing immunostimulation by the associated antigen may be required.

B. Nanoparticles as Drug Carriers to Overcome Epithelial Barriers

There are a few articles supporting the idea that nanoparticles may improve the bioavailability of drugs administered orally. For example, Maincent et al. observed an enhanced peroral bioavailability of vincamine [106] and dihydropirine [107], when associated with poly(hexylcyanoacrylate) and PECL nanoparticles, respectively, in comparison with a drug solution. Insulin was also incorporated into PACA, with the purpose of improving its bioavailability [108,109]. After oral administration of the nanoencapsulated insulin to diabetic rats, a prolonged hypoglycemic effect was observed. This was attributed to the retarded passage of intact nanocapsules through the rat mucosa and to the protection of encapsulated insulin against proteolitic enzymes in the gut. This information suggested that nanoparticulates overcome the gastrointestinal barrier, a fact that was further evidenced by the studies performed by by Jani et al. [110,111], in which they show the crucial role of the size of the polystyrene nanoparticles in their intestinal uptake. More specifically, these authors observed that the total uptake ranged from 33% (50 nm latex) to about 7% (1 μm latex). Consequently, the possibility of improving the oral bioavailability of drugs using nanoparticulate carriers exists. On the other hand, the fact that nanoparticles accumulate in the Peyer's patches of the intestinal mucosa [112] can be exploited for the oral administration of vaccines. Several studies have also been designed to address the importance of the particle size on the uptake of particles into the Peyer's patches and their application for oral vaccination [see review in Ref. 113]. Although most of these studies have been done with microparticles (larger than 1 μm), they have shown that small microspheres accumulate into the Peyer's patches to a higher extent than large microspheres. As a result of these interesting findings, it is expected that nanoencapsulated oral vaccine formulations may represent a new way to improve immunization regimens.

Nanoparticles have also proved to be useful in overcoming epithelia when applied topically to the eye. Following the observation that nanospheres adhere to the ocular epithelial surfaces [114], several drugs have been encapsulated in either PACA or PECL nanoparticles or nanocapsules, with the aim of improving their ocular bioavailability. For some specific formulations (i.e., β-adrenergic blockers) a reduction in the systemic side effects was also noted [115]. Confocal microscopy has been a useful tool to investigate the fate of nanoparticles and nanocapsules after topical instillation [116]. With this technique, we have observed that PECL nanocapsules are taken up by the corneal epithelial cells without causing damage or alteration of the cell membrane. Therefore, these nanocapsules may have application as carriers transporting drugs from the lacrymal fluid toward the inner part of the eye. To further assess the efficacy and extent of this passage, a quantitative evaluation of the diffusion of the polymer carrier is required. Finally, PECL nanoparticles and nanocapsules are well tolerated by the eye, a fact that makes them promising vehicles for the administration of ophthalmic drugs [117].

As nanoparticles can cross the eye corneal epithelium, it would be very interesting to investigate their ability to pass through the skin. In a recent review [118], Rolland reported that nanoparticulate carriers can cross the stratum corneum and act as microreservoirs of drug in the horny layer, thus providing a way for delivery of drugs to the lower layers of the skin and, eventually, to the blood circulation. Nevertheless, the work published until now has been oriented only to the targeting of drugs to the pilosebaceus unit [119], and no specific information on the uptake of polymeric nanospheres by corneocytes has yet been presented. Therefore, more work will have to be done to assess the efficacy of this new approach.

IX. CONCLUDING COMMENTS

Over the last two decades, considerable progress has been made in the development of new procedures to prepare well-characterized polymeric particulate drug carriers. Efforts have been mainly addressed to the simplification and optimization of the nanoencapsulation procedures in terms of safety of the ingredients involved and mildness of the procedures. As a matter of fact, a long road has been paved from the development of albumin nanospheres by thermal denaturation in the 1970s, to current technology, such as that disclosed in the patent of the salting-out procedure. This evolution has also been promoted by the introduction of new polymers with optimized mechanical and biomedical properties. The properties of the drug to be associated, such as solubility and stability, and the final application of the formulation to be designed play important roles in the choice of the polymer and nanoencapsulation technique. Great advances on the characterization techniques have led to an improved knowledge of the inner structure and surface properties of the nanoparticulate carrier and, therefore, to a better comprehension of their in vivo behavior.

Significant efforts in this field are now being focused on the surface modification of nanoparticulate carriers. The main objective is the improvement of their targeting properties in terms of preventing their uptake by the MPS, or favoring their interaction with specific epithelial cells, as a way to overcome biological barriers. From the considerable amount of work described in the literature, it may be stated that nanoparticles hold promise as drug-delivery systems not only for parenteral, but also for oral, ocular, and transdermal administration. This increased potential of the nanoparticulate carriers demands the efficient encapsulation of new macromolecular drugs and antigens. Therefore, the successful exploitation of nanoparticulate carriers will be possible only if there is improved understanding of their biopharmaceutical behavior and, concurrently, if there are improvements in the technologies of nanoencapsulation.

ACKNOWLEDGMENTS

The author would like to thank Professor R. Langer and Professor P. Couvreur for thorough review of the manuscript and suggestions. In addition, the author is very grateful for the help from Drs. M. Cardamone, P. Calvo, and H. Eldering.

REFERENCES

1. G. S. Banker and G. E. Peck, The new water-based colloidal dispersions, *Pharm. Technol. 5*:55 (1981).

2. J. W. McGinity, ed., *Aqueous Polymeric Coatings for Pharmaceutical Dosage Forms*, Marcel Dekker, New York, 1989.

3. S. J. Douglas, S. S. Davis, and L. Illum, Nanoparticles in drug delivery, *CRC Crit. Rev. 3*:233 (1987).

4. J. Kreuter, Nanoparticle-based drug delivery systems, *J. Controlled Release 16*:169 (1991).

5. P. Couvreur and C. Vautier, Polyalkylcyanoacrylate nanoparticles as drug carrier: Present state and perspectives, *J. Controlled Release 17*:187 (1991).

6. A. Shamkhani, M. Bhakoo, A. Tuboku-Metzger, and R. Duncan, Evaluation of the biological properties of alginates and gellan and xanthan gums, Proceedings of International Symposium on Controlled Release and Bioactive Materials, Amsterdam, 1991, pp. 213–214.

7. R. L. Littenberg, Anaphylactoid reaction to human albumin microspheres, *J. Nucl. Med. 16*:236 (1975).

8. S. H. Pangburn, P. V. Trecony, and J. Heller, Partially deacetylated chitin: Its use in self-regulated drug delivery systems, *Chitin, Chitosan and Related Enzymes* (J. P. ed.), Academic Press, New York, 1984, p. 3.

9. D. H. Lewis, Controlled release of bioactive agents from lactide/glycolide polymers, *Biodegradable polymers as Drug Delivery Systems* (M. Chasin and R. Langer eds.), Marcel Dekker, New York, 1990, p. 1.

10. C. G. Pitt, Poly-ε-caprolactone and its copolymers, *Biodegradable Polymers as Drug Delivery Systems* (M. Chasin and R. Langer eds.), Marcel Dekker, New York, 1990, p. 71.

11. P. Couvreur, L. Grislain, V. Linaerts, P. Brasseur, P. Guiot, and A. Biernacki, Biodegradable polymeric nanoparticles as drug carriers for antitumor agents, *Polymeric Nanoparticles as Drug Carriers for Antitumor Agents* (P. Guiot and P. Couvreur eds.), CRC Press, Boca Raton, FL, 1986, p. 27.

12. J. Kattan, J. P. Droz, P. Couvreur, J. P. Marino, A. Boutan-Laroze, P. Rougier, P. Brauls, H. Vranks, J. Grognet, X. Morge, and H. Sancho, Phase I clinical trial and pharmacokinetic evaluation of doxorubicin carried by poly(isohexylacyanoacrylate) nanoparticles, *Invest. New Drugs 10*:191 (1992).

13. P. A. Kramer, Albumin microspheres as vehicles for achieving specificity in drug delivery, *J. Pharm. Sci. 63*:1647 (1974).

14. K. Sugibasayashi, Y. Morimoto, T. Nada, Y. Kato, A. Hasegawa, and T. Arita, Drug-carrier property of albumin microspheres in chemotherapy. Preparation and tissue distribution in mice of microsphere-entrapped 5-fluorouracil, *Chem. Pharm. Bull. 27*:204 (1979).

15. P. K. Gupta, J. M. Gallo, C. T. Hung, and D. G. Perrier, Influence of stabilization temperature on the entrapment of Adriamycin in albumin microspheres, *Drug Dev. Ind. Pharm. 13*:1471 (1987).

16. P. K. Gupta, C. T. Hung, F. C. Lam, and D. G. Perrier, Albumin microspheres. Synthesis and characterization of microspheres containing Adriamycin and magnetite, *Int. J. Pharm. 43*:167 (1987).

17. K. Widder, G. Flouret, and A. Senylei, Magnetic microspheres: Synthesis of a novel parenteral drug carrier, *J. Pharm. Sci. 68*:79 (1979).

18. J. M. Gallo, C. T. Hung, and D. G. Perrier, Analysis of albumin microsphere preparation, *Int. J. Pharm. 22*:63 (1984).
19. Y. Nakagawa, K. Takayama, H. Ueda, Y. Machida, and T. Nagai, Preparation of bovine serum albumin nanospheres as drug targeting carriers, *Drug Design Deliv. 2*:99 (1987).
20. M. Roser and T. Kissel, Surface-modified biodegradable albumin nano- and microspheres, I. Preparation and characterization, *Eur. J. Pharmacol. Biopharm. 39*:8 (1993).
21. K. Bhargava and H. Y. Aindo, Immobilization of active urokinase on albumin microspheres: Use of a chemical dehydrant and process monitoring, *Pharm. Res. 9*:776 (1992).
22. J. J. Marty, R. G. Oppenheim, and P. Speiser, Nanoparticles—a new colloidal drug delivery system, *Pharm. Acta Helv. 53*:17 (1978).
23. R. C. Oppenheim, Nanoparticulate drug delivery systems based on gelatin and albumin, *Polymeric Nanoparticles as Drug Carriers for Antitumor Agents* (P. Guiot and P. Couvreur eds.), CRC Press, Boca Raton FL, 1986, p. 27.
24. H. J. Krause and P. Rohdewald, Preparation of gelatin nanocapsules and their pharmaceutical characterization, *Pharm. Res. 5*:239 (1985).
25. R. C. Oppenheim, N. F. Stewart, L. Gordon, and H. M. Patel, The production and evaluation of orally administered insulin nanoparticles, *Drug Dev. Ind. Pharm. 8*, 531 (1982).
26. M. S. El-Samaligy and P. Rohdewald, Reconstituted collagen nanoparticles, a novel drug carrier delivery system, *J. Pharm. Pharmacol. 35*:537 (1983).
27. M. J. Rajaonaryvony, C. Vauthier, G. Couarraze, F. Puisieux, and P. Couvreur, Development of a new drug carrier made from alginate, *J. Pharm. Sci. 82*:912 (1993).
28. P. Calvo, C. Remuñán-López, J. L. Vilo-Jeto, and M. J. Alonso, Novel hydrophilic chitosan-polyethylene oxide nanoparticles as protein carriers, *J. Appl. Polymer Sci.* (submitted).
29. G. Birrenbach and P. Speiser, Polymerized micelles and their use as adjuvants in immunology, *J. Pharm. Sci. 65*:1763 (1976).
30. J. Kreuter and P. Speiser, In vitro studies of poly(methylmethacrylate) adjuvants, *J. Pharm. Sci. 65*:1624 (1976).
31. P. Couvreur, B. Kante, M. Roland, P. Guiot, P. Bauding, and P. Speiser, Polycyanoacrylate nanocapsules as potential lysosomotropic carriers: Preparation, morphological and sorptive properties, *J. Pharm. Pharmacol. 31*:331 (1979).
32. P. Couvreur, M. Roland, and P. Speiser, Biodegradable submicroscopic particles containing a biologically active substance and compositions containing them, U.S. Patent 4,329,332 (1982).
33. E. F. Donnelly, D. S. Jonston, D. C. Peffer, and D. J. Dunn, Ionic and zwiterionic polymerization of *n*-alkyl 2-cyanoacrylates, *Polym. Lett. Ed. 15*:399 (1977).
34. V. Guise, J. Y. Drouin, J. Benoit, J. Mahuteau, I. Genin, and P. Couvreur, Vidarabine-loaded nanoparticles: A physico-chemical study, *Pharm. Res. 7*:736 (1990).
35. J. L. Grangier, M. Puygrenier, J. C. Gautier, and P. Couvreur, Nanoparticles as carriers for growth hormone releasing factor, *J. Controlled Release 15*:3 (1991).

36. C. Losa, P. Calvo, E. Castro, J. L. Vila Jato, and M. J. Alonso, Improvement of ocular penetration of amikacin sulfate by association to poly(butylcyanoacrylate) nanoparticles, *J. Pharm. Pharmacol. 43*:548 (1991).

37. M. J. Alonso, C. Losa, P. Calvo, and J. L. Vila Jato, Approaches to improve the association of amikacin sulfate to poly(alkylcyanoacrylate) nanoparticles, *Int. J. Pharm. 68*:69 (1991).

38. N. Al Khouri Fallouh, L. Roblot-Treupel, H. Fessi, J. P. Devissaguet, and F. Puisieux, Development of a new process for the manufacture of polyisobutylcyanoacrylate nanocapsules, *Int. J. Pharm. 28*:125 (1985).

39. E. Alléman, R. Gurny, and E. Doelker, Drug-loaded nanoparticles—preparation methods and drug targeting issues, *Eur. J. Pharmacol. Biopharm. 39*:173 (1993).

40. C. Losa, L. Marchal-Heussler, F. Orallo, J. L. Vila Jato, and M. J. Alonso, Design of new formulations for topical ocular administration: Polymeric nanocapsules containing metipranolol, *Pharm. Res. 10*:80 (1993).

41. M. S. El Samaligy, P. Rohdewald, and H. A. Mahmood, Polyalkyl cyanoacrylate nanocapsules, *J. Pharm. Pharmacol. 38*:216 (1986).

42. M. R. Gasco and M. Trotta, Nanoparticles from microemulsions, *Int. J. Pharm. 29*:267 (1986).

43. M. R. Gasco, S. Morel, M. Trotta, and I. Viano, Doxorubicine englobed in polybutylcyanoacrylate nanocapsules: Behavior in vitro and in vivo, *Pharm. Acta Helv. 66*:47 (1991).

44. U. Schröder and A. Stähl, Crystallized dextran nanospheres with entrapped antigen and their use as adjuvants, *J. Immunol. Methods 70*:127 (1984).

45. R. Gurny, N. A. Peppas, D. D. Harrington, and G. S. Banker, Development of biodegradable and injectable lattices for controlled release of potent drugs, *Drug Dev. Ind. Pharm. 7*:1 (1981).

46. H. J. Krause, A. Schwarz, and P. Rohvewald, Polylactic acid nanoparticles, a colloidal drug delivery system for lipophilic drugs, *Int. J. Pharm. 27*:145 (1985).

47. R. Bodmeier and H. Chen, Indomethacin polymeric nanosuspensions prepared by microfluidization, *J. Controlled Release 12*:223 (1990).

48. M. C. Julienne, M. J. Alonso, J. L. Gómez, and J. P. Benoit, Preparation of poly(D,L-lactide/glycolide) nanoparticles of controlled particle size distribution: Application of experimental designs, *Drug Dev. Ind. Pharm. 18*:1063 (1992).

49. M. D. Coffin and J. W. McGinity, Biodegradable pseudolatexes: The chemical stability of poly(D,L-lactide) and poly(ε-caprolactone) nanoparticles in aqueous media, *Pharm. Res. 9*:200 (1992).

50. P. D. Scholes, A. G. A., Coombes, L. Illum, S. S. Davis, M. Vert, and M. C. Davies, The preparation of sub-200 nm poly(lactide-*co*-glycolide) microspheres for site-specific drug delivery, *J. Controlled Release 25*:145 (1993).

51. A. Sánchez, J. L. Vila Jato, and M. J. Alonso, Development of biodegradable microspheres and nanospheres for the controlled release of cyclosporin A, *Int. J. Pharm. 99*:263–273, (1993).

52. D. Blanco and M. J. Alonso, Protein-loaded polylactic acid/glycolic acid nanospheres: protein stability and controlled release properties, *Int. J. Pharm.* (submitted).

53. T. Verrechia, P. Huve, M. Veillard, G. Spenlehauer, and P. Couvreur, Adsorp-

tion/desorption of human serum albumin at the surface of poly(lactic acid) nanoparticles prepared by a solvent evaporation process, *J. Biomed. Mater. Res.* 27:1019 (1993).

54. H. Fessi, J. P. Devissaget, F. Puisieux, and C. Thies, Procédé de preparation de systèmes colloidaux dispersibles d'une substance, sous forme de nanoparticules, *French Patent*, 2, 608, 988.

55. G. Reich, Preparation and characterization of indomethacin-loaded nanoparticles using poly(D,L-lactide)/synperonic blends, Sixth International Conference on Pharmaceutical Technology: II, 1992, p. 308.

56. M. Guzman, J. Molpeceres, F. García, M. R. Aberturas, and M. Rodriguez, Formation and characterization of cyclosporin-loaded nanoparticles, *J. Pharm. Sci.* 82:498 (1993).

57. H. Fessi, F. Puisieux, and J. P. Devissaguet, Procedé de preparation de systèmes colloidaux dispersibles d'une substance sous forme de nanocapsules, *European Patent*, 274 961 (1987).

58. H. Fessi, F. Puisieux, J. P. Devissaguet, N. Ammoury, and S. Benita, Nanocapsule formation by interfacial polymer deposition following solvent displacement, *Int. J. Pharm. 55*:R1–R4 (1989).

59. P. Maincent, L. Marchal-Heussler, D. Sirbat, P. Thouvenot, M. Hoffman, and J. A. Vallet, Polycaprolactone colloidal carriers for controlled ophthalmic delivery, Proceedings International Symposium on Controlled Release and Bioactive Materials, 1992, p. 226.

60. T. Niwa, H. Takeuchi, T. Hino, N. Kunou, and Y. Kawashima, Preparation of biodegradable nanospheres of water-soluble and insoluble drugs with D,L-lactide/glycolide copolymer by a novel spontaneous emulsification solvent diffusion method, and the drug release behavior, *J. Controlled Release 25*:89 (1993).

61. C. Bindschaedler, R. Gurny, and E. Doelker, Process for preparing a powder of water-insoluble polymer which can be redispersed in a liquid phase, the resulting powder and utilization thereof. *US Patent* 4,968,350, 1990.

62. H. Ibrahim, C. Bindschaedler, E. Doelker, P. Buri, and R. Gurny, Aqueous nanodispersions prepared by a salting-out process, *Int. J. Pharm. 87*:239 (1992).

63. E. Allémann, R. Gurny, and E. Doelker, Preparation of aqueous polymeric nanodispersions by a reversible salting-out process, influence of process parameters in particle size, *Int. J. Pharm. 87*:247 (1992).

64. E. Allémann, E. Doelker, and R. Gurny, Preparation of aqueous polymeric nanodispersions by a reversible salting-out process: Purification of an injectable dosage form, *Eur. J. Pharmacol. Biopharm. 39*:13 (1992).

65. N. Ammoury, H. Fessi, J. P. Devissaget, F. Puisieux, and S. Benita, In vitro release kinetic pattern of indomethacin from poly(D,L-lactide) nanocapsules, *J. Pharm. Sci. 79*:763 (1990).

66. S. Bonduelle, C. Foucher, J. C. Leroux, F. Chouinard, C. Cadieux, and V. Linaerts, Association of cyclosporin to isohexylcyanoacrylate nanospheres and subsequent release in human plasma in vitro, *J. Microencapsul. 9*:173 (1992).

67. B. Magenheim, M. Y. Levy, and S. Benita, A new in vitro technique for the evaluation of drug release profile from colloidal carriers—ultrafiltration technique at low pressure, *Int. J. Pharm. 94*:115 (1993).

68. N. Ammoury, H. Fessi, J. P. Devissaget, F. Puisieux, and S. Benita, In vitro release kinetic pattern of indomethacin from poly(D,L-lactide) nanocapsules, *J. Pharm. Sci. 79*:763 (1990).

69. A. Sánchez and M. J. Alonso, Poly(D,L-lactide-*co*-glycolide) micro and nanospheres as a way to prolong blood/plasma levels of subcutaneously injected cyclosporin A, *Eur. J. Pharmacol. Biopharm. 41*:31 (1995).

70. C. Kubiak, L. Manil, and P. Couvreur, Sorptive properties of antibodies onto cyanoacrylic nanoparticles, *Int. J. Pharm. 41*:181 (1988).

71. C. Chavany, T. Le Doan, P. Couvreur, F. Puisieux, and C. Hélène, Polyalkylcyanoacrylate nanoparticles as polymeric carriers for antisense oligonucleotides, *Pharm. Res. 9*:441 (1992).

72. W. M. Bertling, M. Gareis, V. Paspaleeva, A. Zimmer, J. Kreuter, E. Nürnberg, and P. Harrer, Use of liposomes, viral capsids and nanoparticles as DNA carriers, *Biotechnol. Appl. Biochem. 13*:390 (1991).

73. M. J. Henry-Micheland, M. J. Alonso, A. Andremont, J. Sauzieres, and P. Couvreur, Attachment of antibiotics to nanoparticles: Preparation, drug-release and antimicrobial activity in vitro, *Int. J. Pharm. 35*:121 (1987).

74. V. Guise, J. Y. Drouin, J. Benoit, J. Mahuteau, P. Dumont, and P. Couvreur, Vidarabine-loaded nanoparticles: A physico-chemical study, *Pharm. Res. 7*:737 (1990).

75. P. K. Gupta, C. T. Hung, D. G. Perrier, Albumin microspheres. II. Effect of stabilization temperature on the release of Adriamycin, *Int. J. Pharm. 33*:147 (1986).

76. P. K. Gupta, F. C. Lam, and C. T. Hung, Albumin microspheres. IV. Effect of protein concentration and stabilization time on the release of Adriamycin, *Int. J. Pharm. 51*:253 (1989).

77. E. Alléman, J. C. Leroux, R. Gurny, and E. Doelker, In vitro extended release properties of drug-loaded poly(DL-lactic acid) nanoparticles produced by a salting-out procedure, *Pharm. Res. 10*:1732 (1993).

78. J. Kreuter, Physico-chemical characterization of polyacrylic nanoparticles, *Int. J. Pharm. 14*:43 (1983).

79. S. E. Bott, Submicrometer particle sizing by photon correlation spectroscopy: Use of multiple-angle detection, *Particle Size Distribution Assessment and Characterization* (T. Provder ed.), *ACS Symp. Ser 332*:74–88 (1987).

80. J. M. Rollot, P. Couvreur, L. Roblot Treupel, and F. Puisieux, Physico-chemical and morphological characterization of polyisobutylcyanoacrylate nanocapsules, *J. Pharm. Sci. 75*:361 (1986).

81. R. Gref, Y. Minimitake, M. T. Peracchia, V. Trubetskoy, V. Torchilin, and R. Langer, Biodegradable long-circulating polymeric nanospheres, *Science 263* 1600 (1994).

82. B. Drake, C. B. Prater, A. L. Weisenhorn, S. A. C. Gould, T. R. Albrecht, C. F. Quate, D. S. Cannell, H. G. Hansma, and P. K. Hansma, Imaging crystals, polymers and processes in water with the atomic force microscope, *Science 243*:1586 (1989).

83. B. R. Ware, Electrophoretic light scattering, *Adv. Colloid Interface Sci. 4*:1 (1974).

84. P. Calvo, J. L. Vila Jato, and M. J. Alonso, Unstabilization and stabilization of PECL nanoparticles by protein adsorption, Proceedings International Symposium on Controlled Release and Bioactive Materials, 21, 1994, p. 1150.

85. R. H. Muller, S. S. Davis, L. Illum, and E. Mak, Particle charge and surface hydrophobicity of colloidal drug carriers, *Targeting of Drugs with Synthetic Systems* (G. Gregoriadis, J. Senior, and G. Poste, eds.), Plenum Press, New York, 1986, p. 239.

86. R. H. Muller, S. S. Davis, L. Illum, and E. Mak, Particle charge and surface hydrophobicity of colloidal drug carriers, *Targeting of Drugs with Synthetic Systems* (G. Gregoriadis, J. Senior, and G. Poste, eds.), Plenum Press, New York, 1986, p. 239.

87. S. S. Davis, S. J. Douglas, L. Illum, P. D. E. Jones, E. Mak, and R. H. Muller, Targeting of colloidal carriers and the role of surface properties, *Targeting of Drugs with Synthetic Systems* (G. Gregoriadis, J. Senior, and G. Poste, eds.), Plenum Press, New York, 1986, p. 123.

88. S. D. Troster, K. H. Wallis, H. Rainer, R. H. Muller, and J. Kreuter, In vivo characterization of poly(methylmetacrylate) nanoparticles and correlation to their in vivo fate, *J. Controlled Release 20*:237 (1992).

89. R. H. Muller, K. H. Wallis, S. D. Troster, and J. Kreuter, In vitro characterization of poly(methyl metacrylate) nanoparticles and correlation to their in vivo fate, *J. Controlled Release 20*:247 (1992).

90. M. J. Alonso, A. Sánchez, B. Seijo, D. Torres, and J. L. Vila Jato, Joint effects of monomer and stabilizer concentrations on physico-chemical characteristics of poly(butyl 2-cyanoacrylate) nanoparticles, *J. Microencapsul. 7*:517 (1990).

91. B. Seijo, E. Fattal, L. Roblot-Treupel, and P. Couvreur, Design of nanoparticles of less then 50 nm diameter: Preparation characterization and drug loading, *Int. J. Pharm. 62*:1 (1990).

92. L. Illum, S. S. Davis, R. H. Muller, E. Mak, and P. West, The organ distribution and circulation time of intravenously injected colloidal carriers sterically stabilized with a block copolymer—Poloxamine 908, *Life Sci. 40*:367 (1987).

93. S. E. Dunn, S. Stolnik, M. C. Garnett, M. C. Davies, A. G. A. Coombes, D. C. Taylor, M. P. Irving, S. C. Purkiss, T. F. Tadros, S. S. Davis, and L. Illum, Biodistribution studies investigating poly(lactide-*co*-glycolide) nanospheres modified by novel biodegradable copolymers, Proceedings International Symposium on Controlled Release and Bioactive Materials 21, 1994, p. 210.

94. J. C. Leroux, P. Gravel, L. Balant, B. Volet, B. M. Anner, E. Allemann, E. Doelker, and R. Gurny, Internalization of poly(D,L-lactic acid) nanoparticles by isolated human leukocytes and analysis of plasma proteins adsorbed onto the particles, *J. Biomed. Mater. Res. 28*:471 (1994).

95. J. C. Olivier, C. Vauthier, M. Taverna, D. Ferrier-Baylocq, F. Puisieux, and P. Couvreur, Comparative study of the stability of orosomucoid adsorbed on polyisobutylcyanoacrylate nanoparticles in the presence of serum and of a low albumin concentration, Proceedings International Symposium on Controlled Release and Bioactive Material 21, 1994, p. 307.

96. D. V. Bazile, C. Ropert, P. Huve, T. Verrechia, M. Marland, A. Frydman, M. Veillard, and G. Spenlehauer, Body distribution of fully biodegradable [^{14}C]poly

(lactic acid) nanoparticles coated with albumin after parenteral administration to rats, *Biomaterials 13*:1093 (1992).

97. E. Alléman, E. Doelker, and R. Gurny, Drug loaded poly(lactic acid) nanoparticles produced by a reversible salting out process: Purification of an injectable dosage form, *Eur. J. Pharmacol. Biopharm. 39*:13 (1993).

98. M. Auvillain, G. Cavé, H. Fessi, and J. P. Devissaget, Lyophilisation de vecteurs colloidaux submicroniques, *STP Pharma 5*:738 (1989).

99. M. D. Coffin and J. McGinity, Biodegradable pseudolatexes: The chemical stability of poly(D,L-lactide) and poly(ε-caprolactone) nanoparticles in aqueous media, *Pharm. Res. 9*:200 (1992).

100. M. Youssef, E. Fattal, M. J. Alonso, L. Roblot-Treupel, J. Sauzieres, C. Tancrede, A. Ommes, P. Couvreur, and A. Andremont, Effectiveness of nanoparticles-bound ampicillin in treatment of *Listeria monocytogenes* infection in nude mice, *Antimicrob. Agents Chemother. 32*:1204 (1988).

101. C. Verdun, F. Brasseur, H. Vranckx, P. Couvreur, and M. Roland, Tissue distribution of doxorubicin associated with polyisohexylcyanoacrylate nanoparticles, *Cancer Chemother. Pharmacol. 26*:13 (1990).

102. A. Sanchez, R. Seoane, O. Quireza, and M. J. Alonso, In vivo study of the tissue distribution and immunosuppressive response of cyclosporin A-loaded polyester micro- and nanospheres, *Drug Delivery 2*:21 (1995).

103. E. Alleman, R. Gurny, E. Doelker, F. S. Skinner, and H. Schutz, Distribution, kinetics and elimination of radioactivity after intravenous and intramuscular injection of ^{14}C-savoxepina-loaded PLA nanospheres to rats, *J. Controlled Release 29*:97 (1994).

104. J. Kreuter, Possibilities of using nanoparticles as drug carriers for drugs and vaccines, *J. Microencapsul. 5*:115 (1988).

105. J. Kreuter, M. Nefzger, E. Liehl, R. Czok, and R. Voges, Distribution and elimination of poly(methyl metacrylate) nanoparticles after subcutaneous administration to rats, *J. Pharm. Sci. 72*:1146 (1983).

106. P. Maincent, R. Le Verge, P. Sado, P. Couvreur, and J. P. Devissaget, Disposition kinetics and oral bioavailability of vincamin-loaded polyalkyl cyanoacrylate nanoparticles, *J. Pharm. Sci. 75*:955 (1986).

107. P. Maincent, L. Fluckiger, M. Leroueil, M. Hoffman, and J. Atkinson, Interest of colloidal carriers as vectors of antihypertensive dihydropyrine calcium-entry blocker to decrease blood pressure in hypertensive rats, Proceedings International Symposium on Controlled Release and Bioactive Materials 21, 1994, p. 33.

108. C. Damgé, C. Michel, M. Aprahamian, P. Couvreur, and J. P. Devissaguet, Nanocapsules as carriers for oral peptide delivery, *J. Controlled Release 13*:133 (1990).

109. C. Damgé, C. Michel, M. Aprahamian, and P. Couvreur, New approach for the oral administration of insulin with polycyanoacrylate nanocapsules, *Diabetes 37*:246 (1988).

110. P. Jani, G. W. Halbert, J. Langridge, and A. T. Florence, Nanoparticle uptake by the rat gastrointestinal mucosa: Quantification and particle size dependency, *J. Pharm. Pharmacol. 42*:821 (1990).

111. P. U. Jani, A. T. Florence, and D. E. McCarthy, Further histological evidence

of the gastrointestinal absorption of polystyrene nanospheres in the rat, *Int. J. Pharm. 84*:245 (1992).

112. P. U. Jani, D. E. McCarthy, and A. T. Florence, Nanosphere and microsphere uptake via Peyer's patches: Observation of the rate of uptake in the rat after a single oral dose, *Int. J. Pharm. 86*:239 (1992).

113. D. T. O'Hagan, Microparticles as oral vaccines, *Novel Delivery Systems for Oral Vaccines* (D. T. O'Hagan, ed.), CRC Press, Boca Raton FL, 1994, p. 175.

114. R. W. Wood, V. H. K. Li, J. Kreuter, and J. R. Robinson, Ocular disposition of poly-hexyl-2-cyano(3-^{14}C)acrylate nanoparticles in the albino rabbit, *Int. J. Pharm. 23*:175 (1985).

115. C. Losa, M. J. Alonso, J. L. Vila, F. Orallo, J. Martinez, J. A. Saavedra, and J. C. Pastor, Reduction of cardiovascular side effects associated with ocular administration of metipranolol by inclusion in polymeric nanocapsules, *J. Ocul. Pharmacol. 8*:191 (1992).

116. P. Calvo, M. J. Thomas, Alonso, J. L. Vila Jato, and J. Robinson, Study of the mechanisms of interaction of poly-ϵ-caprolactone nanocapsules with the cornea by confocal laser scanning microscopy, *Int. J. Pharm. 103*:283 (1994).

117. Personal communication.

118. A. Rolland, Particulate carriers in dermal and transdermal drug delivery: Myth or reality, *Pharmaceutical Particulate Carriers, Therapeutic Applications* (A. Rolland, ed.), Marcel Dekker, New York, 1993, p. 367.

119. A. Rolland, N. Wagner, A. Chatelus, B. Shroot, and H. Schaefer, Site-specific drug delivery to pilosebaceous structures using polymeric microspheres, *Pharm. Res. 10*:1738 (1993).

8

Biodistribution of Surface-Modified Liposomes and Particles

Vladimir P. Torchilin and Vladimir S. Trubetskoy

Massachusetts General Hospital and Harvard Medical School
Charlestown, Massachusetts

I. INTRODUCTION

Nanospheres, nanocapsules, liposomes, and micelles are widely used for delivery of diagnostic agents and therapeuticals [1–3]. Surface modification of these carriers is often used to control their properties in a desirable fashion: (a) to increase their longevity and stability in the circulation; (b) to change their biodistribution; (c) to achieve targeting effect; and (d) to impart some "unusual" properties (such as pH- or thermosensitivity). Among the most important and well-studied modifiers are (a) antibodies and their fragments; (b) proteins; (c) mono-, oligo-, and polysaccharides; (d) chelating compounds (such as EDTA, DTPA, or deferoxamine); and (e) soluble synthetic polymers.

Targeting phenomena, usually resulting from the modification of nanocarriers with antibodies and other specific ligands (such as sugar moieties), lead to drastic changes in their biodistribution: specific recognition between a carrier-immobilized substance and the appropriate target within the body results in greatly enhanced carrier accumulation in the area of interest. Similar changes in biodistribution can be found for different carriers if they are modified with an identical ligand. Properties of specific ligand-targeted drug carriers have already been discussed in detail elsewhere [e.g., see Ref. 4] and will not be specifically discussed here. Carrier surface modification, with chelating compounds or moieties capable of firmly binding heavy metals, is a special application of metal-containing carriers as diagnostic (imaging) agents and will be

243

discussed, to some extent, later. We will begin, however, with the discussion on some problems arising as a result of carrier surface modification with certain polymers.

Independently of the carrier type and designation, its modification with soluble synthetic polymers results in interesting theoretical and experimental (practical) phenomena connected with the accessibility of the carrier surface for carrier-interacting substances. Moreover, the presence of some polymers on the carrier surface can influence physicochemical properties (e.g., spectral) of the carrier constituents or some other carrier-bound or carrier-associated moieties.

II. POLYMER-GRAFTED LIPOSOMES: POLYMER FLEXIBILITY AND ACCESSIBILITY OF THE LIPOSOMAL SURFACE

A. Theoretical Model

Liposomes (phospholipid bilayered vesicles) may serve as a good model to understand the influence of a grafted polymer on carrier properties, and many regularities found for liposomes might be successfully applied to any microparticulate drug carrier.

One of the most popular and successful methods to prepare biologically stable and long-circulating liposomes for drug-delivery purposes is their coating with poly(ethylene glycol) (PEG) [5,6]. To enhance PEG's capability to incorporate into the liposomal membrane, the reactive derivative of hydrophilic PEG is terminus-modified with a hydrophobic moiety, according to the scheme presented in Fig. 1. The explanation of the mechanism of PEG's protective effect on liposomes is based on the participation of PEG in the repulsive interactions between PEG-grafted membranes and other particles [7]. It imposes a special importance on the role of surface charge and hydrophilicity of PEG-coated liposomes [8] and, generally speaking, the decreased rate of plasma proteins (opsonins) adsorption on the hydrophilic surface of PEG-incorporated liposomes [9].

To further understand what peculiarities of PEG behavior on the liposome surface underlie its ability to prevent liposomal opsonization and blood clearance, we considered the phenomenon from a physicochemical perspective. We hypothesized that the molecular mechanism of PEG's protective action is determined by the properties of a flexible polymer molecule (free rotation of individual units around interunit linkages) in solution and includes the formation of an impermeable polymeric layer over the liposome's surface even at low polymer concentrations [10,11].

Liposomal elimination from the blood proceeds mainly by the recognition of liposomes by phagocytic cells that is mediated by opsonins. The most evi-

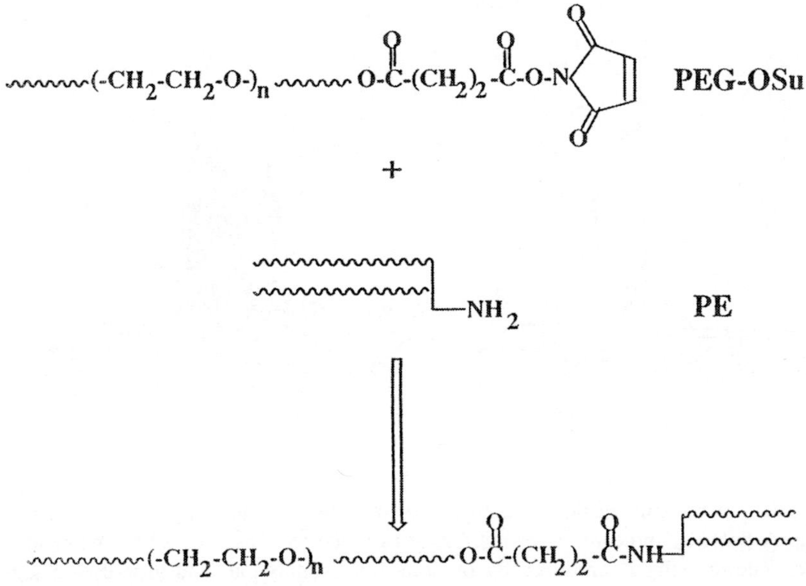

Fig. 1 Scheme of poly(ethylene glycol) (PEG) derivatization to give a terminal hydrophobic moiety for incorporation into liposomes.

dent approach to slow down the liposomal clearance and change biodistribution, therefore, is to prevent the contact of protein with the membrane. We assume that (a) protecting polymer contacts with plasma proteins does not result in the opsonization, and (2) the polymer itself does not contain any cell-specific moieties. Evidently, the protective layer of a polymer over the liposomal surface must combine an ability to escape opsonization and recognition by cells (it is best for it to look like water from outside) and to prevent the penetration of opsonizing proteins into the liposomal surface.

Some very simple quantitative criteria may help us characterize liposomal protection with a polymer [10]. If one considers the diffusional movement of "foreign" macromolecule toward the liposomal surface as the initial step of a protein–liposome interaction, we can express the degree of liposomal protection as the probability for the protein to collide with a polymer (P_{pol}) instead of a liposome (P_{lip}). Thus, if the protein is unable to reach the liposomal surface, $P_{lip} = 0$, so $P_{pol} = 1 - P_{lip} = 1$. When $P_{pol} = 0$, no interaction between protein and polymer occurs, and no protection is achieved.

To forecast possible P values, we can describe the behavior of liposome-grafted polymer molecule in terms of statistical physics (e.g., apply a simplified model of a polymer solution [12]. For example, we can consider it as a

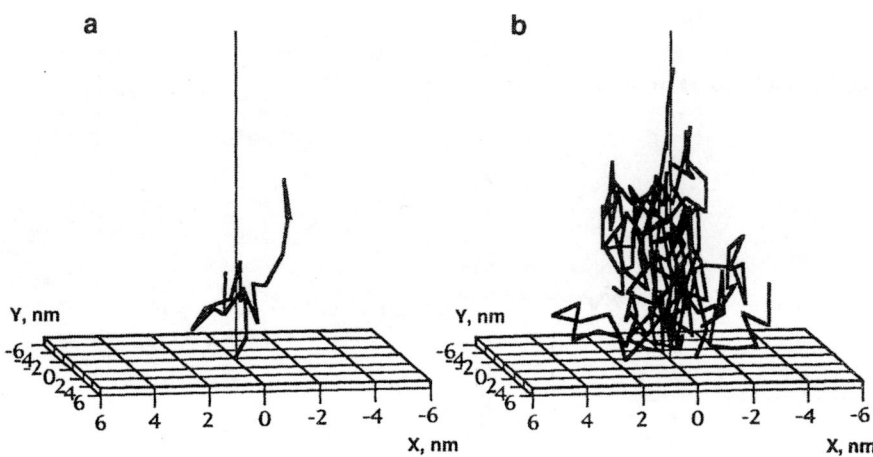

Fig. 2 Computer simulation of conformational cloud formation by surface-attached polymer. A polymer molecule is conditionally assumed to consist of 20 segments of 1 nm each. The drawing is produced by random-flight simulation, and unrestricted segment motion is assumed, with a space restriction equal to $Z > 0$. (a) One random conformation; (b) superposition of 11 random conformations. (From Ref. 10.)

three-dimensional network, in which each cell may be occupied either with a polymer unit or with a solvent (water) molecule. From this perspective, the more flexible the polymer is (i.e., the more independent the motion of any polymeric unit is relative to the neighboring one), the larger is the total number of its possible conformations and the higher is the transition rate from one conformation to another. This means, that a water-soluble, flexible polymer statistically exists as a distribution ("cloud") of probable conformations. Figure 2 shows how this cloud is formed (under some simple assumptions), and how its density and uniformity increase with the increase in the number of possible conformations. The polymer flexibility correlates with its ability to occupy, with high frequency, many cells in solution, temporarily squeezing water molecules out (i.e., making them impermeable for other solutes). To reach the liposomal surface, protein molecules must penetrate the whole cloud, formed by the liposome-attached polymer molecule. Thus, a relatively small number of water-soluble and very flexible polymer molecules can create a sufficient number of high-density conformational clouds over the liposomal surface to protect the latter from being opsonized and recognized by the reticuloendothelial system (RES) cells. From the kinetic analysis, we can easily obtain the following equation:

$$P_{pol} = \gamma \frac{S_P}{S_1} P*$$

where γ is the molar polymer/lipid ratio in the outer monolayer, and S_1 is the average area occupied by a single lipid molecule (for the given liposome size and composition). At high γ values, the polymer will be "stretched" out of the liposome. Therefore, $\gamma(S_P/S_1)$ is always less than 1. The maximal protection can be achieved when $\gamma S_P \approx S_1$ (the polymer clouds are practically fused) and $P*$ value is close to 1.

For a rigid-chain polymer (in which unit motion is hindered), even good water-solubility and hydrophilicity may not provide sufficient protection for the liposomal surface. The number of possible conformations for such polymers is lower; also, the conformational transitions proceed at a slower rate than those of a flexible polymer. This means that the density of the conformational cloud for a rigid polymer will be very uneven during a single collision act, and the number of water molecules disturbed will be much fewer. Within the "cellular" model, it appears that there should exist a sufficient water space through which the normal diffusion of plasma proteins toward the liposomal surface is still possible. Thus, to protect the liposome, one has to bind a much larger number of rigid polymer molecules on the liposomal surface. Only when the same polymer combines both hydrophilicity and flexibility, can it serve as an effective liposomal protector, even at relatively low concentration of the surface-immobilized macromolecules (the most obvious example of such polymers is PEG). Other possible candidates might be, for example, poly(acrylamide) and poly(vinylpyrrolidone).

The size of the area on the liposomal surface protected with a single polymer molecule of a given molecular weight, and the number of polymer molecules required for the effective protection of the liposome of a given size might be determined with the use of such parameters as the average end-to-end distance of a polymer random coil in solution, $R_{0,sol}$. A polymer molecule is located mainly in the volume "between the ends" (inside the appropriate sphere for the molecule in solution and inside the hemisphere for the surface-immobilized polymer). So, we can assume that $R_{0,sol}$ value closely corresponds to the radius of the impermeable dense cloud (R). The end-to-end distance for the surface-attached molecule might be about two times longer than $R_{0,sol}$. Thus, we can assume the radius of the protected area being from $R_{0,sol}$ to $2R_{0,sol}$. From this simple approach and using $R_{0,sol}$ values for PEG of different molecular weight [as published in Ref. 13], we can estimate the area of the liposomal surface that can be protected by a single PEG molecule. Assuming about 4.25×10^4 lipid molecules in the outer monolayer of single-bilayer, 100-nm liposome [14,15], we can calculate the molar PEG/lipid ratio required for 100% protec-

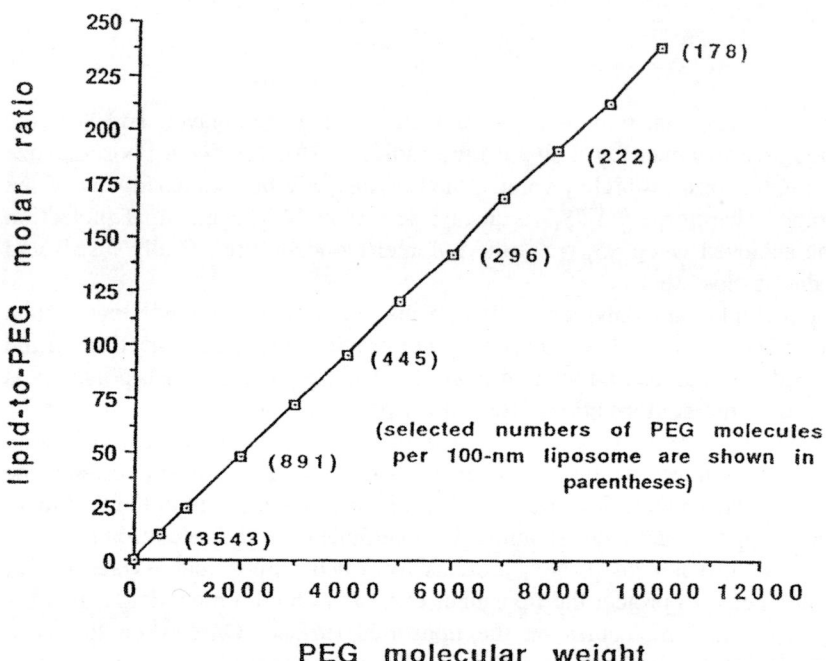

Fig. 3 The dependence on PEG molecular weight of the number of PEG molecules required for complete protection of a 100-nm liposome.

tion of the liposomal surface (surface area, S, 3.14×10^4 nm^2; Fig. 3). The figures calculated match well with published experimental data [16,17]; thus, the end-to-end distance may be used as an estimate for the protected square radius.

To further develop our model, we applied computer simulation [11]. Spatial distribution of short polymers grafted onto the liposomal surface was simulated using a three-dimensional random-flight model. The polymer was assumed to consist of absolutely rigid segments, with free segmental rotation around an intersegmental conjunction. The grafting point was assumed to be located at $x = y = z = 0$. At step 1, the random direction of the first segment was assumed by generation of two random angles: between the segment axis and axis z; and between the segment axis projection onto the xy plain and axis x. At step 2, the coordinates of the remote end of the first segment were calculated, conditionally assuming the segment length $l = 1$ nm for a flexible polymer and $l = 5$ nm for a rigid one. The results for a segment location below the liposomal surface were omitted. From the computer analysis it follows (Fig. 4) that the

Z, nm

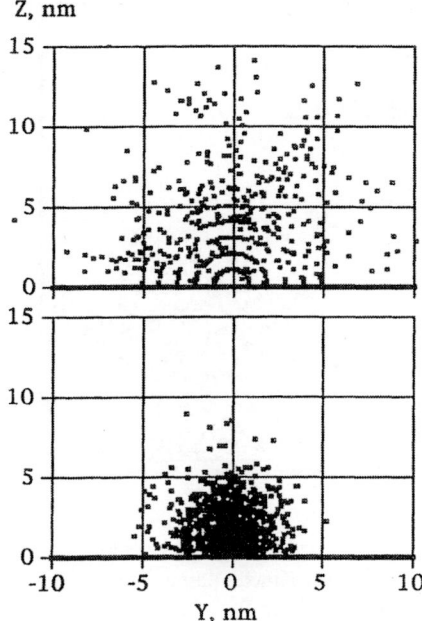

Fig. 4 Distribution of polymer conformations in space; slice $X = 0 \pm .25$ nm. Produced by random-flight simulation ($Z > 0$, polymer length 20 nm, 440 conformations). (Upper panel) segment length is 5 nm (rigid polymer); (lower panel) segment length is 1 nm (flexible polymer). Calculations for the rigid polymer were performed under the assumption that only every fifth 1-nm segment may change the direction. (From Ref. 11.)

flexible polymer forms a conformational cloud with a very high density in its central part. A rigid polymer of the same length (its segment was conditionally assumed to be five times longer than for the flexible polymer) forms a broad, but loose and thus permeable, cloud.

One interesting consequence suggested by the model is that, if there are any reactive centers on the liposomal surface (e.g., binding sites for certain macromolecules or other ligands), the presence of PEG in concentrations lower than necessary for complete coating of the surface, should lead to the appearance of two populations of those centers. The first population will consist of reactive moieties excluded from the volume occupied by PEG. Such reactive centers should possess all of the properties (including the reactivity) of similar centers on "plain" liposomes. Other centers remaining within the polymer cloud form the second population, with a sharply decreased ability to participate in

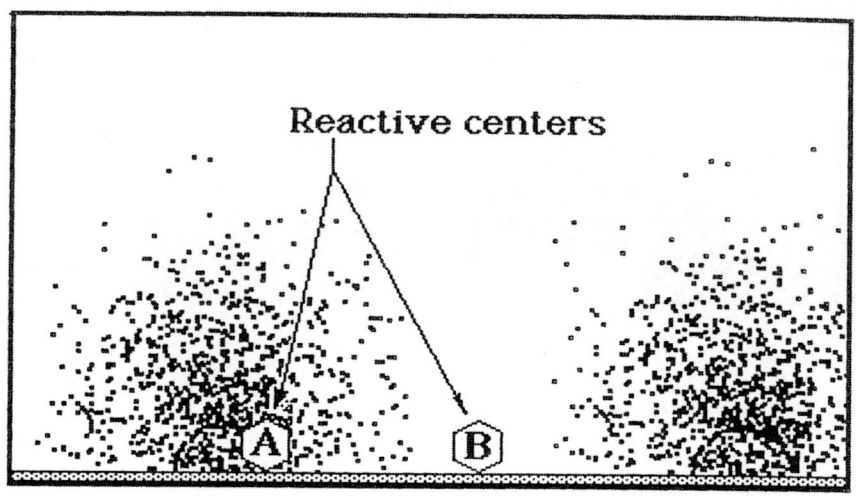

Fig. 5 The scheme of reactive site location on the liposome surface at low PEG concentration. The site can be either (A) sterically hindered by a polymer cloud, or (B) exposed and readily available for a variety of interactions. Kinetic parameters of chemically identical, but differently located, sites may differ. (From Ref. 11.)

"normal" interactions. Figure 5 illustrates this phenomenon, and permits the supposition that kinetic parameters of the reactive site will depend on its location on the liposomal surface that is partially coated with PEG.

More detailed analysis within the framework of the suggested model allows us to answer several important questions (V. Torchilin and B. Hoop, in preparation). How are PEG molecules distributed in the space above the liposomal surface (what is the mean thickness of PEG sheath)? How does the space available for water molecules depend on the distance above the liposomal surface? What is the thickness of the spherical volume around a liposome that can be entirely filled with PEG molecules? The proper answers for these questions might contribute to development of optimized, long-circulating, microparticulate drug carriers.

B. Experimental Evidence of a Polymer's Flexibility Importance

1. *Fluorescence Quenching*

To confirm the model experimentally, we have studied the efficacy of liposome-incorporated fluorescent marker quenching by a macromolecular quencher from the solution, depending on the polymer presence on the liposomal membrane [11]. To study the interaction of polymer-modified and *N*-[7-

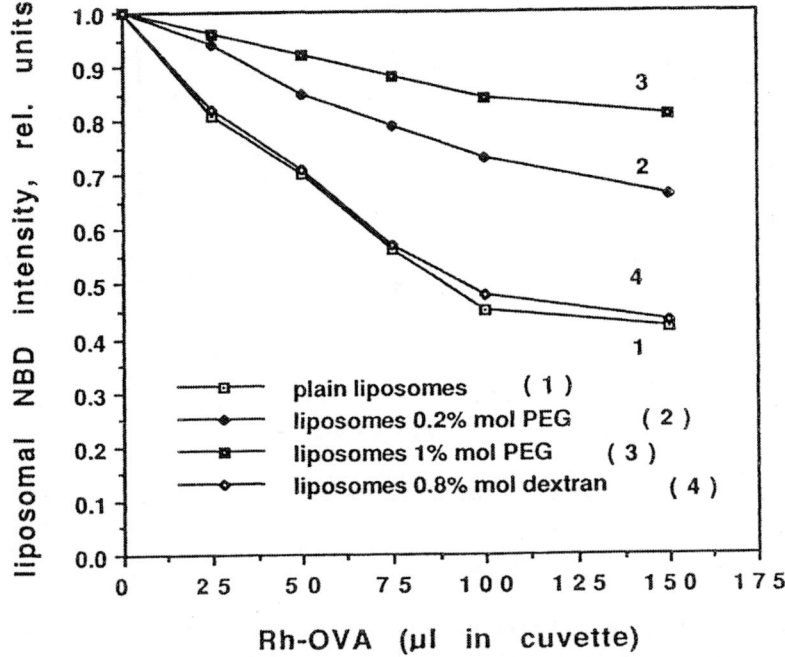

Fig. 6 Quenching of liposomal NBD fluorescence with Rh-OVA from solutions at different Rh-OVA concentrations. (1) Plain liposomes; (2) liposomes with 0.2 mol% PEG-PE; (3) liposomes with 1 mol% PEG-PE; (4) liposomes with 0.8 mol% dextran-stearylamine. (From Ref. 11.)

nitrobenz-2-oxa-1,3-diazol-4-yl]-dioleoylphosphatidyl ethanolamine (NBD)-containing liposomes with soluble rhodamine-modified ovalbumin (Rh-OVA), plain NBD-liposomes and NBD-liposomes, containing a certain quantity of either lipid-conjugated PEG (flexible polymer) or dextran, were treated with increasing quantities of Rh-OVA. The NBD fluorescence diminished identically for Rh-OVA-treated "plain" NBD-liposomes and NBD-liposomes with 0.8 mol% of dextran (Fig. 6). About 50% of the initial fluorescence can be quenched, which is evidence for even distribution of liposomal NBD between outer and inner monolayers of the membrane (only "outer" NBD is susceptible to quenching). However, NBD fluorescence quenching is drastically hindered in NBD–liposomes containing 1 mol% of PEG or even as few as 0.2 mol% PEG. Even at maximal Rh-OVA concentration, about 80% of the initial fluorescence can still be observed. Because the whole quenching process is limited only by Rh-OVA diffusion from the solution to the liposomal surface, it is evident that the presence of PEG (protective clouds) on the surface creates diffusional hindrance for this process.

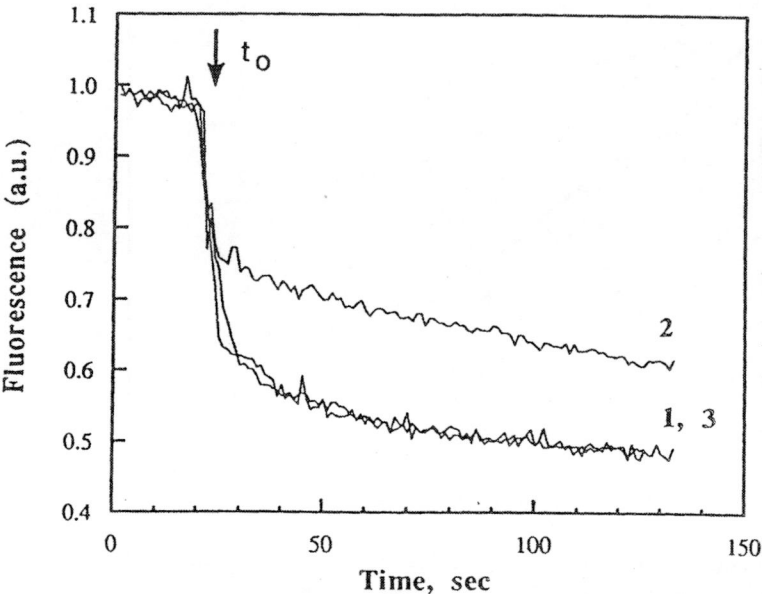

Fig. 7 Typical time courses of liposomal fluorescein quenching with antifluorescein antibody. (1) Plain fluorescein–liposomes; (2) fluorescein–liposomes with 1 mol% of PEG; (3) fluorescein–liposomes with 1 mol% of dextran. t_0, the point of antibody addition to liposomes; this parameter can vary slightly in different runs. (From Ref. 11.)

Another experimental proof was obtained using the system involving the liposomal surface-incorporated fluorescein derivative *N*-[glutarylamido-(5-aminoacetamido-fluorescein)]-*sn*-glycero-3-phosphoethanolamine-1,2-dioleoyl (GFl-PE) and antifluorescein antibody [11] (Fig. 7). In this example, fluorescence quenching with antibody was equal for plain GFl-liposomes and for GFl-liposomes with 1 mol% of dextran. This indicates that the presence of dextran in the quantities used does not create any diffusional limitations for antibody–fluorescein interactions. The presence of 1 mol% PEG in liposomes noticeably decreased both the rate of fluorescein quenching and the quantity of fluorescein residues accessible for the interaction with antibody (see Fig. 7) because of the diffusional limitations for antibody penetration imposed by PEG.

Incorporation of a smaller quantity of PEG (0.2 mol%) into liposomes indicated the existence of two different fluorescein pools on the liposomal surface (Fig. 8). One of them was quenched at the same time rate as fluorescein on PEG-free liposomes, whereas the quenching kinetics for the other was close to that for fluorescein on PEG–liposomes with a high PEG content. This can reflect different location of fluorescein molecules on the liposomal surface—

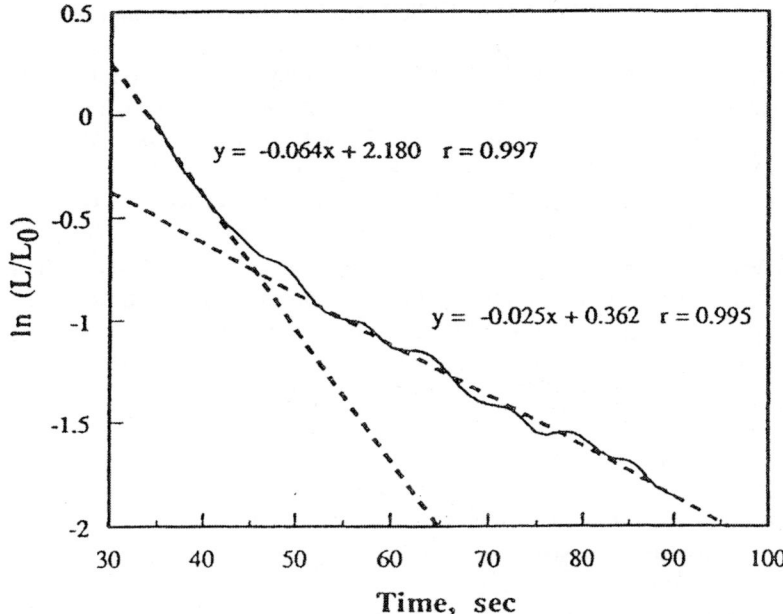

Fig. 8 Linearization (dotted lines) of liposomal fluorescein quenching with anti-fluorescein antibody. Typical experiment with 0.2 mol% PEG-PE. L_0 and L, initial and current fluorescence, respectively. Two phases on the kinetic curve can be seen: initial fast phase, reflecting the quenching of exposed antigen; and subsequent slow phase, reflecting the quenching of PEG-protected antigen. (Adapted from Ref. 11.)

between and inside PEG clouds — as we predicted on the basis of computer simulation data (see Fig. 5).

2. *Alternative Polymers on the Liposomal Surface*

The model of flexible polymer behavior on the liposomal surface enabled us to formulate some general requirements for polymers that can be used for liposomal protection: (a) polymers should be soluble and hydrophilic; and (b) they should have a highly flexible main chain. Polymer biocompatibility must be added to the list, if polymer–liposomes are intended for medical use. Poly (acrylamide) (PAA), poly(vinylpyrrolidone) (PVP), and poly(vinyl alcohol) were named as the most appropriate candidates among alternative liposome protectors.

We prepared PAA and PVP by free-radical polymerization of monomers in dioxane. By using different quantities of a growing chain terminator, both polymers were prepared with MW values of 6,000–8,000 (low molecular weight polymers; PAA-L and PVP-L) and 12,000–15,000 (high molecular weight polymers; PAA-H and PVP-H). The prepared polymers were made amphiphilic and membranotropic by chemical attachment of hydrophobic acyl groups of different lengths to a single terminus of a polymer molecule. The following products were used in further experiments: PAA-L with a terminal dodecyl group (PAA-L-D); PAA-L with a terminal palmityl group (PAA-L-P); PAA-H with a terminal palmityl group (PAA-H-P); PVP-L and PVP-H with terminal palmityl groups (PVP-L-P and PVP-H-P, respectively) [18].

Liposomes were prepared by the detergent (octyl glycoside) dialysis method from a mixture of egg phosphatidylcholine and cholesterol (7:3 molar ratio) with the addition, when necessary, of 2.5 or 6.5 mol% of the corresponding amphiphilic polymer. Liposomes were labeled with ^{111}In-pentetate(diethylene-triaminepentaacetic acid; DTPA)-stearylamine (prepared as in Ref. 19). The liposomes obtained were sized by passing through 0.6, 0.4, and 0.2-μm polycarbonate filters. The final liposomal size in all preparations was between 165 and 190 nm, with a narrow size distribution. Biodistribution of PAA- and PVP-coated liposomes was studied in BALB/C mice injected through the tail vein with 150 μl of liposomal suspension in HBS (ca. 700 μg of total lipid and 1–2 μCi of ^{111}In radioactivity).

The data presented in Figs. 9 and 10 clearly demonstrate that liposomes that are modified with amphiphilic derivatives of PAA and PVP change their biodistribution in a fashion similar to that for PEG-liposomes. Figure 9A shows that when modified with the same palmityl residue, PAA, PVP, and PEG of similar molecular weight (ca. 6000–8000), when used in similar concentrations, all provide steric protection and sharply increase the residence time of liposomes in the circulation. Half-clearance times for PVP-L-P-, PAA-L-P-, and PEG-liposomes with 2.5 mol% of protective polymer are about 45, 80, and 80 min, and for PVP-L-P-, PAA-L-P-, and PEG-liposomes with 6.5 mol% of protective polymer, about 120, 140, and 170 min, respectively, whereas half-clearance time for plain liposomes of the same size is only about 10 min [18].

The protective activity of PAA-L-D, and both PAA-H-P and PVP-H-P, however, is much lower (see Fig. 10). Despite a definite increase in the circulation time, these polymers are still much less effective steric protectors than are polymers of similar molecular weight, but with a longer acyl anchor (compare

---

Fig. 9 (A) Liposome clearance from the blood of experimental mice; (B) liposome accumulation in the liver: (1) plain liposomes; (2) PVP-L-P–liposomes (2.5 mol% PVP); (3) PAA-A-L-P–liposomes (2.5 mol% PAA); (4) PEG–liposomes (2.5 mol% PEG); (5) PVP-L-P–liposomes (6.5 mol% PVP); (6) PAA-L-P–liposomes (6.5 mol% PAA); (7) PEG–liposomes (6.5 mol% PEG). (From Ref. 18.)

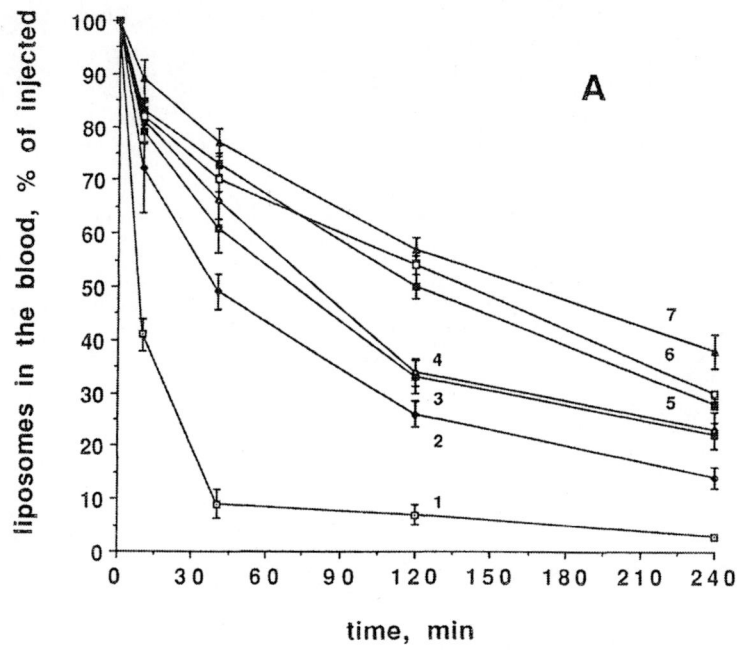

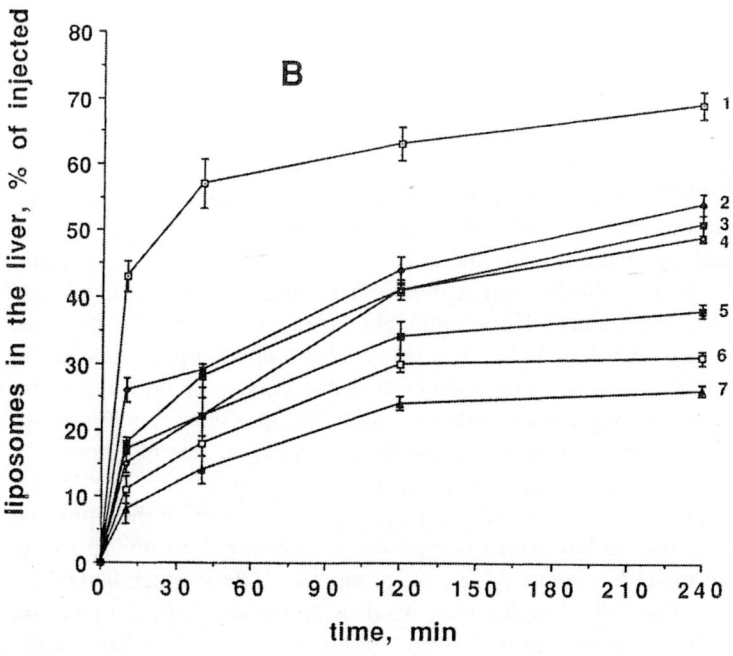

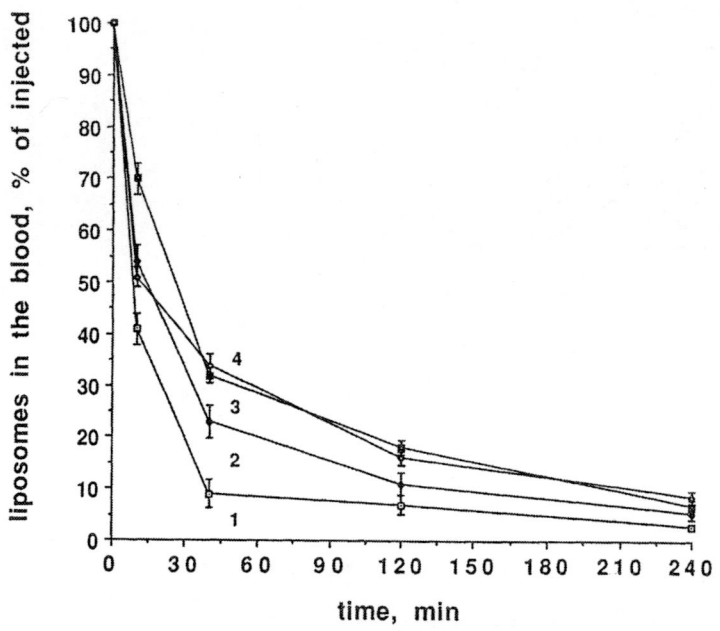

Fig. 10 Liposomal clearance from the blood of experimental mice. (1) Plain liposomes (reference curve); (2) PAA-L-D–liposomes (2.5 mol% PAA); (3) PVP-H-P–liposomes (2.5 mol% PVP); (4) PAA-H-P–liposomes (2.5 mol% PAA). (From Ref. 18.)

PAA-L-D- and PAA-L-P-liposomes), or for polymers with the same long acyl
anchor, but with a lower molecular weight of the hydrophilic moiety (compare
PAA-H-P- and PVP-H-P-liposomes with PAA-L-P, PVP-L-P, and PEG-liposomes). This can be explained by considering the energy of interaction between
the fatty acyl anchor, which keeps the polymer on the liposome surface, and
the hydrophobic area of the liposomal membrane. From a thermodynamic aspect, a relatively short dodecyl group is unable to keep a 6- to 8-kDa–polymer
molecule on the liposomal surface: the energy of the polymeric chain motion
is, probably, comparable (or even higher) with the energy of dodecyl group
interaction with phospholipid surroundings within the liposomal membrane. As
a result, PAA-L-D might be removed from the liposomal membrane; therefore,
it would demonstrate only a slight and transient protective effect. The longer
palmityl anchor provides a much more firm polymer binding with the liposome
(higher energy of interaction with hydrophobic membrane core owing to the
larger number of membrane-embedded CH_2-groups) and, thus, much better steric protection (see Fig. 9A). On the other hand, even the length of the palmityl
anchor might be insufficient to provide firm fixation of a 12- to 15-kDa polymer

on the liposomal surface, because of the much higher energy of polymer chain motion in solution compared with that for the shorter polymers [12]. So, the liposomal surface gradually loses its protective polymer coat, becomes opsonized, and ends in the RES.

Similar regularities have been found following liposomal accumulation in the liver (see Fig. 9B). Plain liposomes are captured by the liver very quickly (more than 50% in 45 min and close to 70% in 240 min). The longest circulating liposomes that contain 6.5 mol% of PAA-L-P, PVP-L-P, or PEG, demonstrate much slower liver uptake: less than 20% of these liposomes are captured in 45 min, and less than 40%, in 240 min. Liposomes with 2.5 mol% of protective polymer demonstrate intermediate liver uptake.

The general biodistribution pattern for PAA- and PVP-liposomes does not reveal any specific peculiarities when compared with PEG-liposomes. Spleen accumulation for all coated and noncoated liposomes depends slightly on the nature of the protecting polymer, but after 4h, always stay within 3–10% of the injected dose limit. If expressed as percentage injected dose per gram of tissue, the spleen demonstrates the maximal liposomal capture for all types of liposomes and, in 4 h after injection, may vary from 50 to 150% of dose per gram of tissue. Other investigated organs—kidneys, lungs, bone marrow, and brain—demonstrated minor liposomal accumulation, and insignificant variations in biodistribution of different liposomes.

Thus, hydrophilic and flexible synthetic polymers other than PEG (such as PAA and PVP), when made amphiphilic by modification at one terminus with a long-chain fatty acyl, can incorporate into the liposomal surface and make the liposomes long-circulating. The protective effects observed depend on both the length of the hydrophobic anchor and the length of hydrophilic polymer chain, and might be interpreted in terms of the balance between the energy of the hydrophobic anchor interaction with membrane core and the energy of the polymer chain motion in the water solution. These data are in good agreement with our theoretical considerations discussed earlier.

III. POLYMER MODIFICATION OF LIPOSOMES FOR ENHANCED MAGNETIC RESONANCE IMAGING

A good example of a practical use of the particle surface modification with synthetic polymers is a design of liposome-based contrast agents for lymph node magnetic resonance imaging (MRI). The imaging of lymph nodes plays a major role during the early detection of neoplastic involvement in cancer patients [20] because the lymphatic system is the major route for spreading metastases throughout the body. The replacement of the lymphatic tissue within the lymph node with tumor masses can restrict the contrast medium from entering, thereby revealing internal abnormalities by poor visualization.

Nanoparticulates of a different nature are taken up in local lymph nodes (more exactly, by the lymph node macrophages) after subcutaneous administration [21]. Liposomes, as typical nanoparticulates, fully retain the property of being accumulated in the lymph node after the subcutaneous, intraperitoneal, and intramuscular administration. Gadoliniuim (Gd)-containing liposomes have already been used as a contrast agent for liver and spleen MRI signal enhancement after intravenous administration [22,23]. For the latter purpose, liposomes are loaded with corresponding imaging agents, such as a Gd complex with the strong chelating agent pentetic acid (DTPA) [23]. The agent can be entrapped into the interior water of the liposome or attached to the liposome's surface. For better incorporation into the liposomal membrane a chelator can be preliminary modified with fatty acid or phospholipid residue [24]. We attempted to enhance the imaging potential of Gd-containing liposomes by simultaneously modifying the liposomal surface with PEG to design the MRI contrast agent for lymph node visualization with improved signal intensity.

The coating of MR-active Gd-liposomes with PEG might permit a change in the Gd water surroundings owing to the presence of water molecules tightly associated with the PEG molecule and, thus, to increase the possible signal. The Gd–DTPA-loaded PEG-modified liposomes (PEG MW 5000) were used for MR-imaging of lymph nodes in rabbits and compared with nonmodified Gd–liposomes: 200-nm liposomes were prepared from egg phosphatidylcholine, cholesterol, Gd–DTPA-PE (lipophilic paramagnetic probe), and PEG-PE in a 60:25:10:5 molar ratio by extruding the lipid suspension through polycarbonate filters. Liposome relaxation parameters were measured using a Praxis II Proton Spin Analyzer operated at 10.7 MHz. In vivo imaging of axillary–subscapular lymph node area in rabbits was performed using a 1.5-tesla GE Signa MRI scanner (T_1-weighed pulse sequence, fat suppression mode) during 2 h after the subcutaneous administration of a liposomal preparation into the paw of the anesthetized rabbit. Transverse T_1-weighed MR images of the axillary–subscapular lymph node area were acquired for 2 h immediately after the contrast agent was administration [25]. The raw data from the MRI instrument were analyzed by image-processing software to determine the relative target/nontarget (lymph node/muscle) pixel intensity.

The PEG–Gd–liposomes (20 mg of total lipid in 0.5 ml of saline) are able to outline axillar–subscapular lymph nodes within minutes (see signal accumulation kinetics in Fig. 11). The signal from plain (nonmodified) liposomes develops only weakly in both lymph nodes; node/muscle intensity ratio being at about 1.5, even after 80 min of observation. The PEG-coated Gd-containing vesicles rapidly and substantially increased the lymph node signal: the node/muscle ratio reached about 3.0 within 5–10 min. Figure 12 illustrates this remarkable ability of PEG–Gd–liposomes to delineate lymph nodes within min-

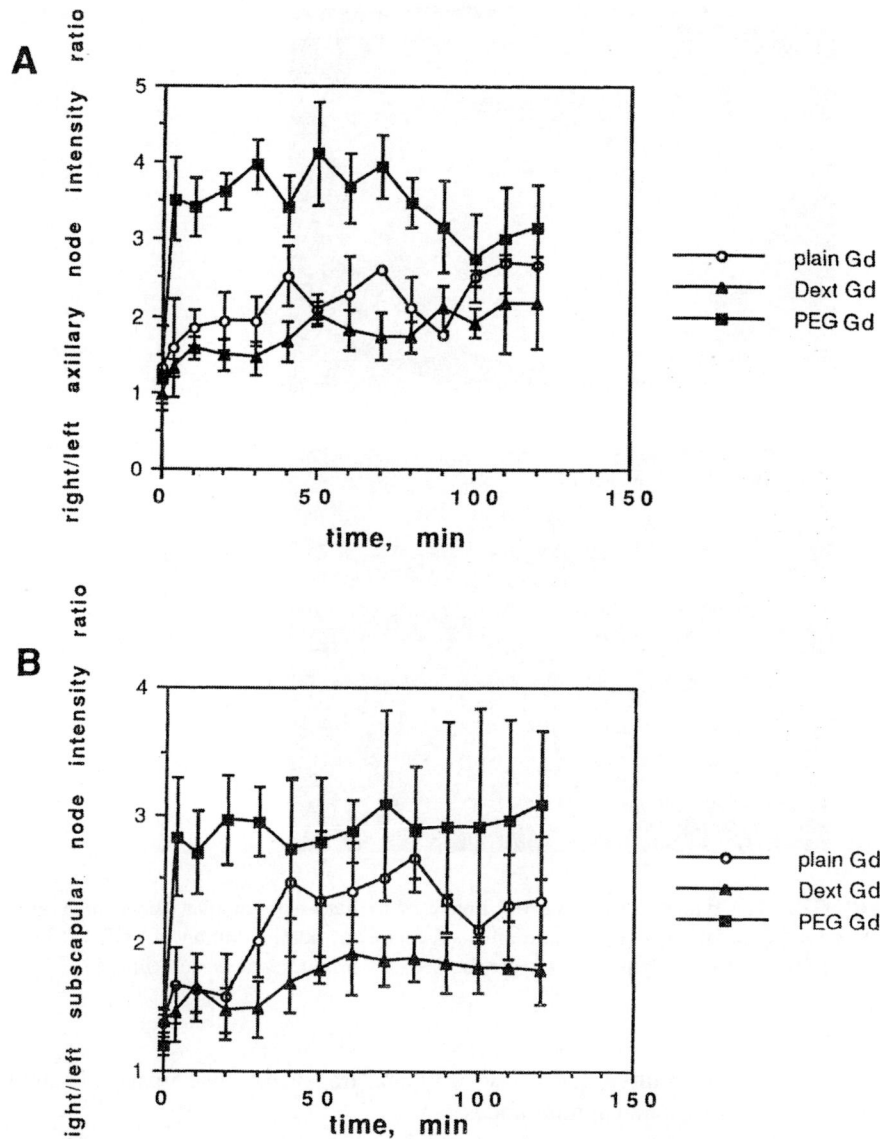

Fig. 11 The kinetics of experiment/control lymph node MR signal intensity ratio after subcutaneous administration of plain and PEG-modified Gd-containing liposomes for (A) axillary and (B) subscapular lymph nodes in rabbit. (Adapted from Ref. 46).

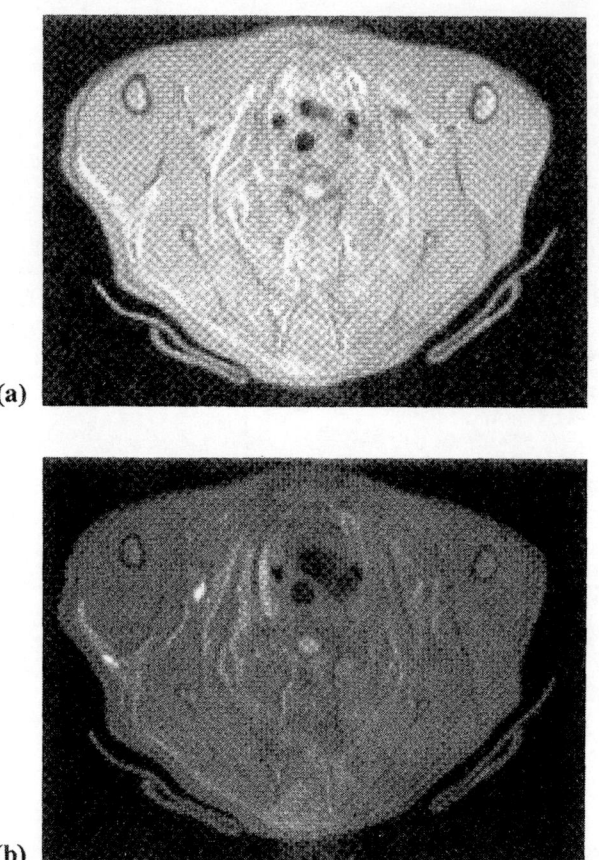

(a)

(b)

Fig. 12 The typical transverse MR image of axillary–subscapular lymph node area in rabbit **(a)** before and **(b)** 30 min after subcutaneous administration of PEG-Gd-liposomes (contrast-filled lymph nodes can be seen as bright spots on late image). (From Ref. 25.)

utes after administration, and shows a typical transverse slice image of rabbit axillary and subscapular lymph nodes.

However, measurements of the actual delivery of liposomes (using the surface-bound ^{111}In-radiolabel) demonstrated a decreased accumulation of PEG–Gd–liposomes (5 mol% PEG) in the lymph nodes (Fig. 13). This finding may reflect the macrophage-evading properties of PEG-modified liposomes. Yet, the smaller amount of the contrast material caused greater MRI signal enhancement. The mechanism of better lymph node visualization with PEG-modified Gd-liposomes might be explained by relaxivity (R_1) measurements of our lipo-

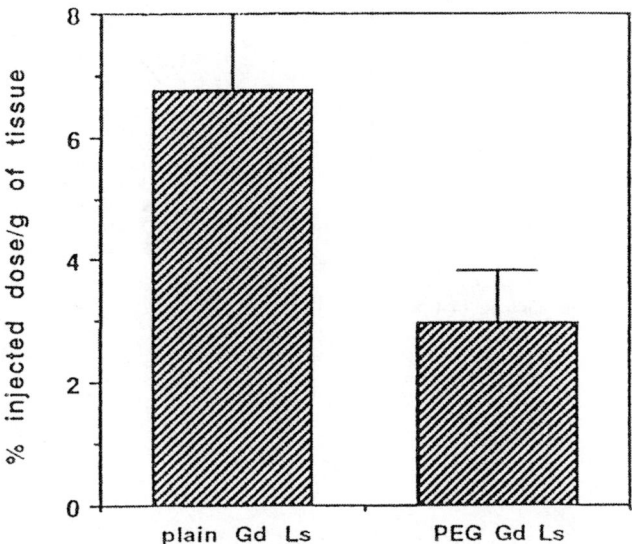

Fig. 13 Axillary lymph node accumulation of ^{111}In-labeled plain Gd–liposomes and PEG-coated Gd–liposomes after subcutaneous injection in rabbit; 2 h postinjection. (Adapted from Ref. 46).

some preparations. Figure 14 demonstrates that R_1 values of PEG–Gd–liposomes are approximately two times higher than the corresponding parameters for the plain Gd–liposomes. The relaxivity of paramagnetic ions is strongly dependent on the rate of exchange of Gd-coordinated water molecules with the bulk water. Because PEG binds a large number of water molecules by hydrogen bonding to the oxygen atoms in the main chain, one can conclude that there is an increased amount of water, in close vicinity to Gd atoms, available for the interactions with paramagnetic centers. After lymph node uptake, Gd–liposomes are compartmentalized within phagocytic cell lysosomes. As pointed out by Lauffer [26], this phenomenon, theoretically, might lead to a decrease of liposomal relaxivity in the intranodal tissue, because not all tissue water is available for interaction with paramagnetic ions. This effect should be less noticeable for PEG–Gd–liposomes because these vesicles "bring" their own water to the organelles, thereby reducing the tissue relaxivity decline.

The ability of PEG–Gd–liposomes to delineate lymph nodes very rapidly (within minutes after subcutaneous injection) is in a sharp contrast with other imaging modalities, for which it takes substantially longer to obtain a good lymph node image (24–48 h). Evidently, this is just a minor example of the polymer-modified liposomes' (and probably other nanocarriers') applicability

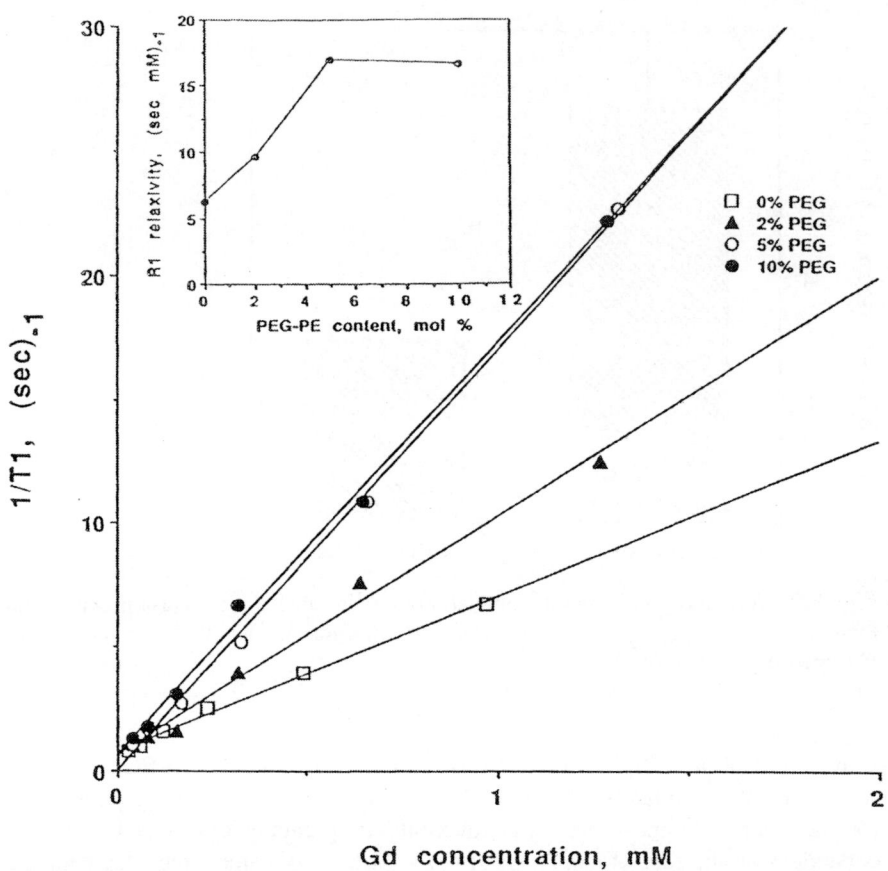

Fig. 14   Molar relaxivities of Gd–liposomes with different amounts of membrane-incorporated PEG-PE. (From Ref. 46).

for the delivery of diagnostic agents, including those for MRI. The approach could be generalized by using PEG–liposomes targeted to different areas of interest. Such sterically protected, targeted liposomes can be prepared by coimmobilization of PEG and monoclonal antibody (MAb) on the liposomal surface.

IV. POLYMER AND ANTIBODY ON THE LIPOSOMAL SURFACE

Is it possible to combine the unique properties of long-circulating (PEG-coated) nanocarriers (liposomes and, probably, others) and targeted antibody-modified carriers in one preparation? If yes, it can be very advantageous for the targeted delivery of imaging and therapeutic agents into affected areas with decreased blood flow or with low concentration of the target antigen. In areas with limited blood supply, even good carrier (liposome) binding to the target could not provide its high accumulation because of the small quantity of particles passing through the target with the blood during the time period when they are still present in the circulation. The same lack of targeting can occur if the concentration of the target antigen is very low, and even sufficient blood flow (and, consequently, carrier passage) through the target still does not result in good accumulation, owing to the few productive collisions between antigens and immunocarriers. Under such circumstances the combination of carrier longevity and specificity should lead to the increase of productive collisions between the targeted nanoparticles and the target with time, resulting in better drug accumulation in the target.

Let us consider this possibility, using liposomes as an example. The concern about using PEG for immunoliposomal protection is that it can create steric hindrances for normal antibody–target interaction. However, we can consider different cases for the coimmobilization of antibody and water-soluble flexible polymer on the same liposome [10], (Fig. 15; the scheme can be applied to any microparticulate carrier). In the first case (see Fig. 15a), the polymer does not cover the liposome completely. Antibody molecules are excluded from the volume occupied with PEG, and two separate zones appear on the surface— clouds of PEG with antibodies in between. Such liposomes should successfully interact with target antigens. At the same time, the clearance time of such liposome should differ little from that for normal immunoliposomes, because there is still PEG-free surface area available for opsonization.

In the second case (see Fig. 15b), the liposome surface might be completely coated with overlapping clouds of PEG, but there are still some areas of a more loose, conformational density into which free antibodies (capable of the lateral diffusion) can be squeezed from the more dense PEG areas, and be rather "free" there. Here, opsonins have very limited opportunity for the interaction with liposome, whereas antibodies are still able to recognize and to bind the target. Thus, we can obtain long-circulated, targeted liposomes.

The third case (see Fig. 15c) involves a very high degree of surface modification, and PEG is forming conformationally stretched "brushes" [27] on the liposome surface. Here, surface-immobilized antibody is trapped between polymer molecules and cannot overcome the steric hindrance and bind to the target. On the other hand, opsonins also cannot interact with the liposomal surface, providing very long circulation times for such liposomes; the half-clearance time in rabbits for immunoliposomes containing 10 mol% of PEG was about 10 h [16]. Brushes of this type, which should provide an extremely long circulation time for liposomes, also can be used for the long-circulating immunolipo-

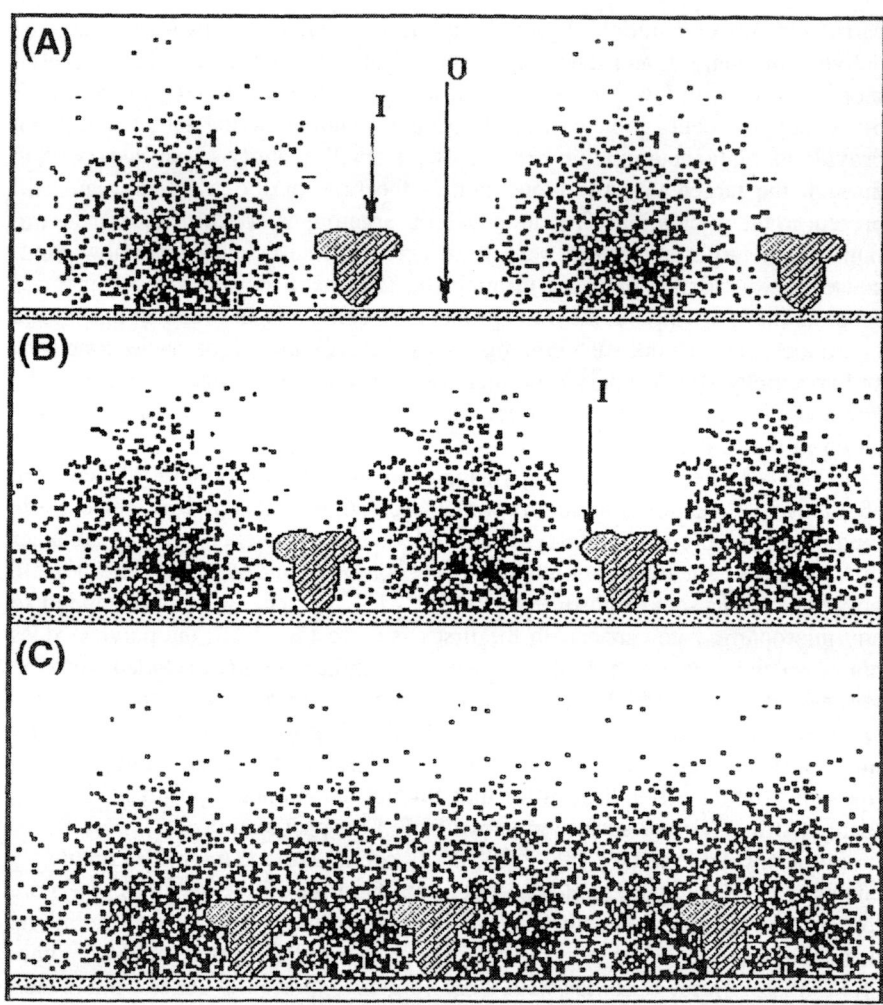

Fig. 15 Possible patterns for antibody and PEG coimmobilization on the liposome surface. Antibody is located between "umbrellas" of protective polymer molecules. **(A)** Low PEG concentration, partially nonprotected liposome surface can be reached by opsonins; **(B)** intermediate (optimal) PEG concentration, no free surface available for opsonins, whereas immunoglobulin can still interact with antigen; **(C)** high PEG concentration, polymer molecules form dense brushes, sterically hindering antibody mobility. (Adapted from Ref.10.)

somal preparation, if antibodies might be immobilized on the liposomal surface by the long-spacer group or even directly on termini of some PEG molecules [28].

To prove that long-circulating PEG-coated liposomes can be targeted by coincorporation of an antibody onto the liposomal surface, we have studied in vivo liposomes with anticardiac myosin antibody [16]. This antibody effectively binds myosin inside ischemic or necrotic cardiomyocytes that have affected or destroyed cellular membranes, but does not interact with normal cells, as it is unable to penetrate the intact plasmic membrane [29,30]. This forms the basis for the targeted delivery of radiolabeled PEG-coated long-circulating liposomes in the region of an ischemically compromised myocardium.

Infarcts in rabbits were generated [see Ref. 16]. Different radiolabeled liposomal preparations were injected intravenously after 30–60 min of reperfusion. Blood samples were taken at specified time intervals to measure liposomal radioactivity. Five to six hours after liposome injection, animals were killed by an overdose of pentobarbital. The heart was excised and cut into 5-mm slices, stained with 2% triphenyltetrazolium chloride to discover necrotic areas, and each slice was further divided into smaller samples. Samples of normal and infarcted myocardium were weighed and counted in a gamma-counter. The data on the liposomal accumulation in the heart were expressed as the infarct/normal myocardium radioactivity ratio. Biodistribution of liposomes in sacrificed animals at 5 h postinjection was studied by following the liposome-associated radioactivity accumulation in different organs.

Liposomes (ca. 150 nm) were prepared by detergent dialysis method from a mixture of phosphatidylcholine and cholesterol in 3:2 molar ratio. Liposomes were labeled with ^{111}In-DTPA-stearylamine, and also, when necessary, they contained 4 or 10 mol% of PEG-PE [16]. For incorporation into the liposomal membrane, antimyosin antibody was preliminary modified with a hydrophobic anchor, *N*-glutarylphosphatidylethanolamine [see Ref. 31].

The half-life values of different immunoliposomes in rabbits are shown in the Table 1, together with the biodistribution and target (infarct) accumulation data [16]. We can conclude that the increasing quantity of PEG increasingly protects liposomes from clearance, and coimmobilization of an antibody and PEG decreases the half-life of liposomes only at the lower PEG concentration, whereas a high PEG concentration blocks the recognition of antibody by liver cells. The results agree well with our scheme (see Fig. 15).

Table 1 also presents the typical patterns of liposome-associated ^{111}In-radioactivity accumulation in the infarcted heart in rabbits (expressed as infarct/normal ratios). Monoclonal antibody (MAb)-free PEG-coated liposomes showed only slight and, probably, nonspecific accumulation in the infarct with an uptake ratio of about 3:1–4:1. This observation might reflect the known phenomenon of plain liposomal accumulation in the infarcted heart region [de-

Table 1   Biodistribution of Different Liposomes Preparations in Infarcted
Rabbit Myocardium

Liposome type	Half-life (min)	Infarct/ normal ratio	Infarct	Liver	Spleen	Lung
			(% dose/g)			
Plain	10	2–3:1	0.018	0.80	0.85	0.018
plus 4 mol% PEG	300	3–4:1	0.13	0.16	0.42	0.15
plus antimyosin	40	14:1	0.145	0.37	0.55	0.03
plus antimyosin plus 4 mol% PEG	200	20:1	0.245	0.13	0.26	0.15
plus 10 mol% PEG	1000	12:1	0.18	0.07	0.19	0.20
plus antimyosin plus 10 mol% PEG	1000	12:1	0.19	0.08	0.20	0.18

Liposome concentration is expressed as percentage dose per gram tissue, following liposome-associated[111] In radioactivity. Data are given as an average of from three to five animals (without errors). One can see that coating of 150-nm liposomes with antimyosin and PEG results in high infarct accumulation, longer circulation, low liver and spleen uptake. Use of PEG itself imparts long circulation time and low liver and spleen uptake, but not enough accumulates in the infarct. Antimyosin alone does not give an infarct accumulation much better than PEG (both are far above plain liposomes), but the accumulation in liver and spleen is high. A high PEG concentration eliminates any antibody effects on liposome biodistribution and target accumulation.
Source: V.P. Torchilin, J. Narula, and B. A. Khaw, unpublished observations.

scribed in Ref. 32]. The uptake ratio with antimyosin liposomes without PEG was 14:1. The highest uptake ratio was achieved for PEG–immunoliposomes and reached 20:1. A very high PEG concentration (10 mol%) on immunoliposomes diminished the uptake ratio back to 12:1.

It is interesting to follow absolute figures of liposomal accumulation in the infarct. The PEG-coated antimyosin–liposomes effectively accumulate in the infarct zone — the percentage dose per gram of tissue for such liposomes is almost twice as high as for antibody-free PEG–liposomes or for PEG-free immunoliposomes, and 12-fold higher than for plain liposomes. Interestingly, liposomes modified either with a single MAb or with single PEG demonstrate about the same percentage dose per gram infarct accumulation (infarct/normal ratio is still much higher for antimyosin-modified liposomes — ca. 15 against < 5 — because of low, nonspecific accumulation of the targeted and fast-cleared antibody–liposomes in the normal myocardium and high nonspecific accumulation of long-circulating PEG-liposomes). This proves that target accumulation, in this example, can proceed both by specific recognition and nonspecifically by decreased filtration rate (liposomes can enter a necrotic area through permeabilized capillaries, but cannot leave it because of poorer drainage). The latter mechanism requires prolonged accumulation times and is realized only for long-circulating liposomes. For PEG–antibody–liposomes both mechanisms work in an additive fashion, so these liposomes demonstrate maximal "absolute" accumulation in the target. For liposomes with 10 mol% PEG, infarct accumulation does not differ for either preparation — PEG-liposomes or antibody–PEG-liposomes. This can be easily explained by the lack of antibody participation in the targeting in the latter instance (steric hindrances depicted in Fig. 13c). All acute myocardial infarctions have been confirmed by histochemical staining with triphenyltetrazolium chloride.

Thus, PEG-coated antimyosin–immunoliposomes can be used for the specific delivery of pharmaceuticals into the necrotic areas of the reperfused infarcted myocardium (areas with diminished blood supply). The data presented prove that liposomes (and, evidently, any other nanoparticulate carrier) can be targeting and long-circulating at the same time, if a proper polymer/monoclonal antibody ratio is maintained on the carrier surface.

V. POLYMERS ON THE SURFACE OF NANOPARTICLES AND LATEXES

In addition to liposomes, synthetic amphiphilic polymers also have been used for steric stabilization of particles with a hydrophobic surface to prolong their circulation in the blood and to alter their biodistribution. On one hand, these polymers demonstrate the ability to be easily adsorbed on the surface of the particulate carrier owing to hydrophobic interactions; on the other hand, their hydrophilic portion is exposed to the solution and effectively protects those particulates from interactions with plasma proteins in the blood after intravenous administration. Polystyrene latex particles, commercially available in various sizes in the range of $0.05-1$ μm, are the carriers of choice for these surface modification studies. Polystyrene latex particles have a more narrow size distribution compared with liposomes; hence, they might be more convenient model to study the phenomena connected with polymer modification of surfaces. The mechanism of protection is essentially the same as for PEG-containing liposomes — a conformational cloud of the polymer's flexible chain protects the particle's hydrophobic core from contact with opsonizing proteins [10,11].

Modification of the liposomal surface with amphiphilic polymer usually requires the incorporation of the polymer's hydrophobic moiety into the liposomes' phospholipid bilayer, which is impossible in solid particle modification. Consequently, a surface modification of the particle can be obtained by one of the following two methods: (a) absorption of a polymer on a particle surface; or (b) chemical grafting of polymer chains onto a particle. Possible examples of (a) include the absorption of a series of poly(ethylene oxide) and poly(propylene oxide) copolymers (Pluronic/Tetronic or Poloxamer/Poloxamine surfactants) on the surface of polystyrene latex particles by a hydrophobic interaction mechanism. Interestingly, the absorption of Poloxamer-type copolymers takes place on only those solid particles with a clearly hydrophobic surface: no interaction is detected, for example, between such copolymers and the liposomal surface [33]. There have been numerous studies on blood clearance and biodistribution of these particles after their coating with surfactants, including the use of surfactants for particle protection from uptake by the RES after intravenous injection [34], and enhanced delivery to lymph nodes after subcutaneous administration [35].

After intravenous administration, hydrophobic particles of submicron size became opsonized with macrophage-recognizable, but not yet identified, serum proteins [36]. Similar to liposomes, protecting the particle surface with hydrophilic flexible polymeric chains results in substantial decline in phagocytosis by liver macrophages with a subsequent prolongation of circulation time. Porter et al. [37] have demonstrated that the absorption of the foregoing copolymers leads to not only the decrease of particle uptake by resident macrophages in liver but, after coating with some specific copolymers, can redirect the injected nanoparticles to other organs. For example, the coating of 60-nm–polystyrene latex with Poloxamer 407 results in increased particle accumulation in bone marrow [37].

The same group have demonstrated that an analogous procedure also substantially helps to alter the biodistribution of subcutaneously injected nanospheres. The coating of 60-nm–diameter polystyrene nanospheres with certain Poloxamer/Poloxamine copolymers results in their increased accumulation in regional lymph nodes. The optimal length of the copolymer poly(oxyethylene) block is 5–15 oxyethylene units. Noncoated particles have a tendency to stay at the injection site, whereas particles coated with longer poly(oxyethylene)-containing copolymers are not retained in the nodes and eventually appear in the systemic circulation [35].

The hydrophobic surface of commercially available polystyrene nanospheres can be coated with other polymers of an amphiphilic nature. We have described the use of an amphiphilic polymer that has been developed specifically for the liposomal membrane incorporation: a hydrophilic linear polymer with a terminal lipid or fatty acyl group. The amphiphilic polymer can be synthesized either directly, by linking the polymer to lipid (such as for PEG–phosphatidylethanolamine; PEG-PE; MW 5 kDa) [5], or by using a fatty acyl moiety as a chain-terminating agent during the radical polymerization of some vinyl monomers

Fig. 16 (A) Blood clearance and (B) liver and (C) spleen accumulation of ^{111}In-labeled 100-nm polystyrene latex nanospheres with surface-adsorbed PEG and PAA in mice.

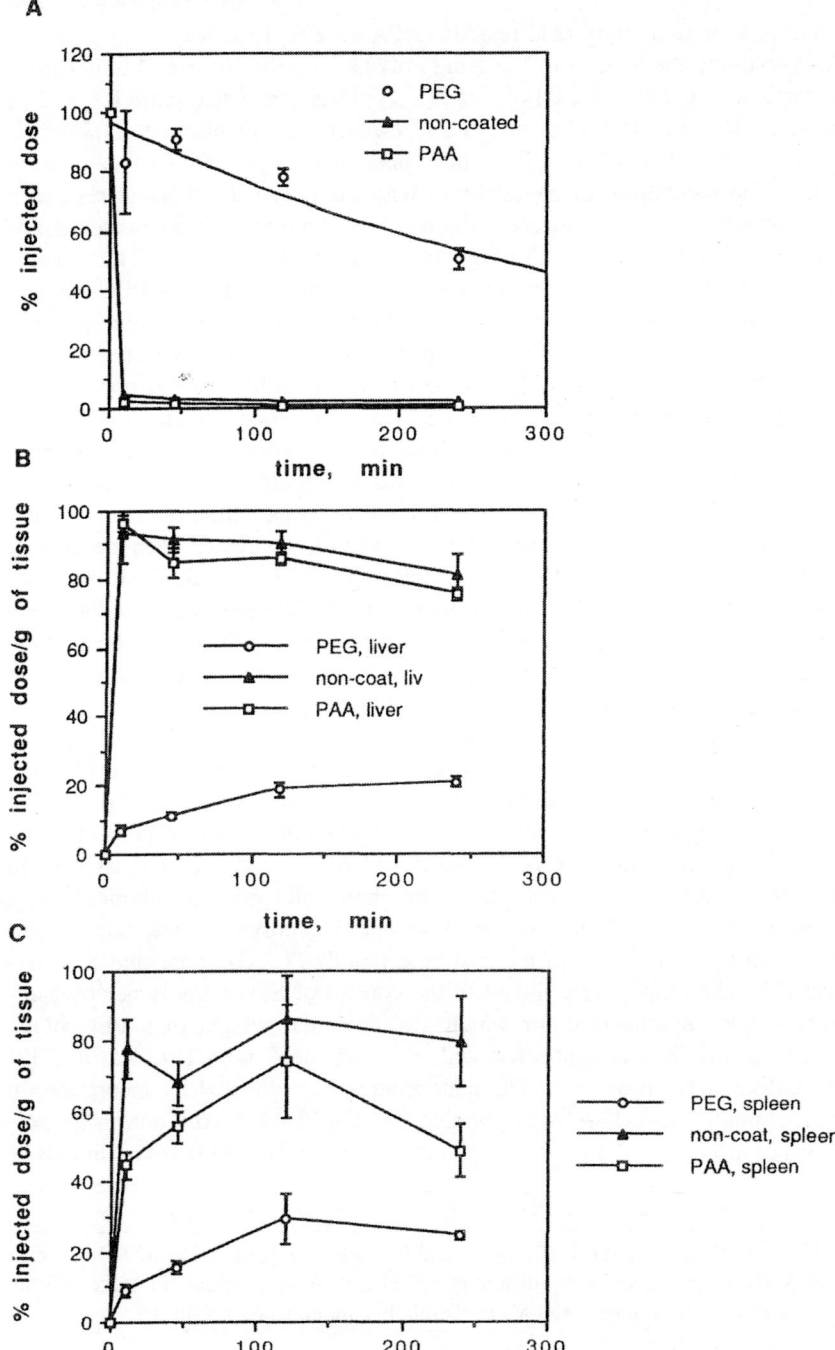

[18]. By the latter method we have synthesized acylpolyacrylamide capped with a C_{16}, saturated, fatty acid residue (APAA; MW 12 kDa).

To investigate the behavior of coating polymers on the surface of nanoparticulate carriers, we have used Polybeads — polystyrene latex particles with a diameter of 100 nm (Polysciences, Inc.). Two already mentioned amphiphilic polymers — APAA and PEG-PE — have been used to coat the surface of latex particles. The incubation of nanospheres with the polymers in water results in polymer attachment to the surface, which can be confirmed by measurement of the particle size before and after incubation with the polymer. The particle diameter increase after polymer adsorption was about 5 nm for PEG-PE and about 20 nm for APAA. The results of biodistribution studies in mice are presented in Fig. 16. The PEG-PE-coated particles, as one can expect, stayed in the circulation for a long time ($t_{1/2} = 4$ h). However, unlike APAA-coated liposomes, APAA-coated nanospheres cleared from the blood as fast as noncoated particles, with the similar pattern of liver accumulation. At the same time, spleen accumulation of APAA particles was substantially reduced compared with noncoated ones. Different effects of the same amphiphilic polymer coating on the behavior of different particulates in vivo has already been described. Poloxamer 407 can protect latex nanospheres from RES uptake, but not liposomes of similar size [33]. This difference in the in vivo behavior has been explained by the different orientation of polyacrylamide chains on the particle surface compared with the surface of a liposome, which results in different degree of carrier protection from absorbing plasma proteins.

Another important type of amphiphilic polymer includes copolymers in which, in a water medium, the hydrophobic block is able to form a solid phase (particle), whereas the hydrophilic part remains as a surface-exposed protective cloud. The example of such structure is the block-copolymer of poly(ethylene glycol) and poly(lactide–glycolide) (PEG–PLAGA), which is discussed in Chapter 9. To give a brief example of the chemically grafted polymer onto a solid protecting core, if one uses PLAGA–PEG copolymer, one can prepare long-circulating particles with an insoluble (solid) PLAGA core and a water-soluble PEG shell covalently linked to the core [38]. Several polymer preparations have been synthesized for which the molecular weight of a PEG block was 20 kDa, and it was connected with PLAGA block with 1:9, 1:5, and 1:4 PLAGA/PEG w/w ratios [39]. The nanospheres were labeled by incorporation of hydrophobic ^{111}In-DTPA–stearylamide into the PLAGA core during the solvent evaporation procedure. Blood clearance and biodistribution experiments in

---

Fig. 17 (A) Blood clearance (B) liver, and (C) spleen accumulation of ^{111}In-labeled PEG-PLAGA nanospheres with different PEG/PLAGA w/w ratios in mice. (From V. S. Trubetskoy, R. Langer, and V. P. Torchilin, unpublished results.)

A

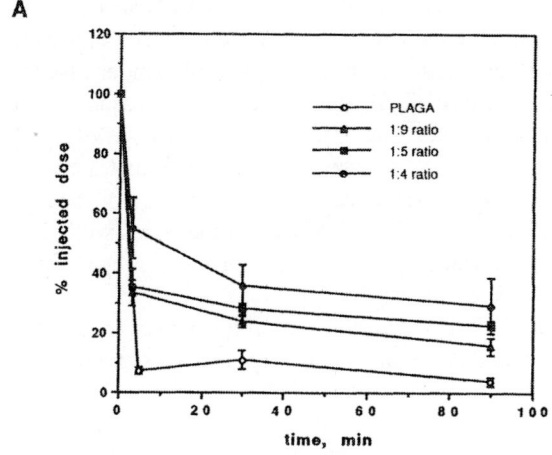

B

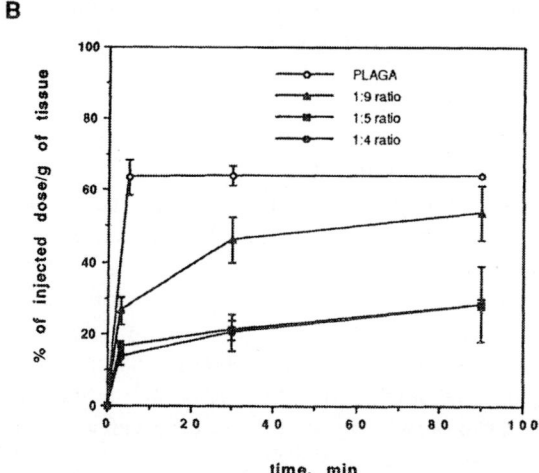

C

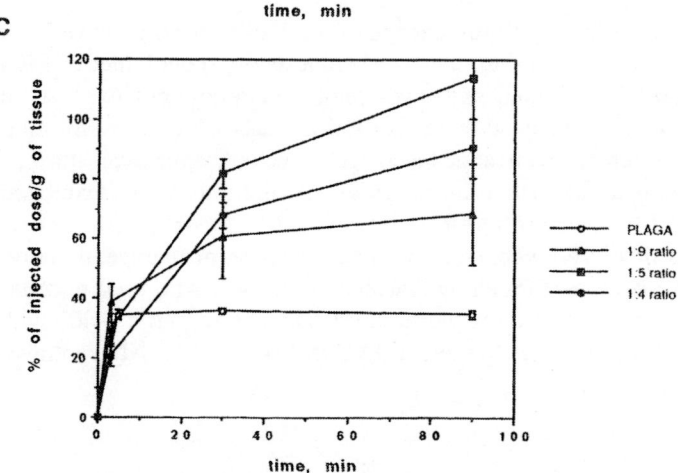

BALB/c mice have demonstrated that the protective effect of PEG in this system depends on the content of its block (Fig. 17). The clearance and liver accumulation patterns reveal one basic feature of the preparations under discussion: the higher the content of the PEG block, the slower is its clearance and the better its protection from liver uptake. The splenic uptake also reflects the particle size–dependent-filtering effect. As has been already demonstrated for liposomes, the vesicles with diameters exceeding 250 nm can be nonspecifically detained in spleen [40]: 1:5 and 1:4 PEG–PLAGA nanoparticles were 265 and 315 nm, respectively, so the elevated splenic uptake for these preparations can be explained by their passive retention in the spleen.

Similar effects on longevity and biodistribution of microparticular carriers might be achieved by direct chemical attachment of protective polyethylene oxide chains onto the surface of preformed particles [41].

Thus, similarly to liposomes, polymers on the surface of polymeric particles can influence their biodistribution after both intravenous and subcutaneous administration, which presents wide opportunities for controlling the in vivo behavior of various macroparticulate, polymeric drug carriers.

VI. POLYMERIC MICELLES

Amphiphilic copolymers in aqueous environment spontaneously form *polymeric micelles*, particles with hydrophobic blocks forming a core protected with hydrophilic chains of the copolymer. So, micelles prepared from amphiphilic block copolymers might be considered as another important example of particulate drug carriers that are surface-modified with polymers. This type of micelle exhibits relatively low critical micelle concentrations in solutions, a thermodynamic property that assures their existence as a particle. Eventually, after indefinite dilution, micellar nanoparticulates dissociate onto individual polymeric chains, a property that might be extremely useful in vivo (a drug-carrying micelle can slowly dissociate in the body, releasing the drug).

Several chemical designs for drug-carrying, polymeric micelles have been proposed. Ringsdorf et al. [42] described the synthesis of a copolymer of PEG and polylysine, partially modified with hydrophobic palmitoyl chains, that, in the form of a micelle interacts with rat peritoneal macrophages. A micellar carrier of similar structure, with anticancer agent cyclophosphamide attached to the remaining nonpalmitoylated amino groups of the polylysine, exhibited prolonged release of the active drug [43].

Detailed studies of in vivo behavior and biodistribution of therapeutic polymeric micelles have been performed by Kataoka et al. [44], who used micelles formed by a copolymer of poly(ethyleneoxide) and poly(aspartic acid) with covalently bound doxorubicin [Adriamycin; PEO-p(Asp(ADR))]. After intrave-

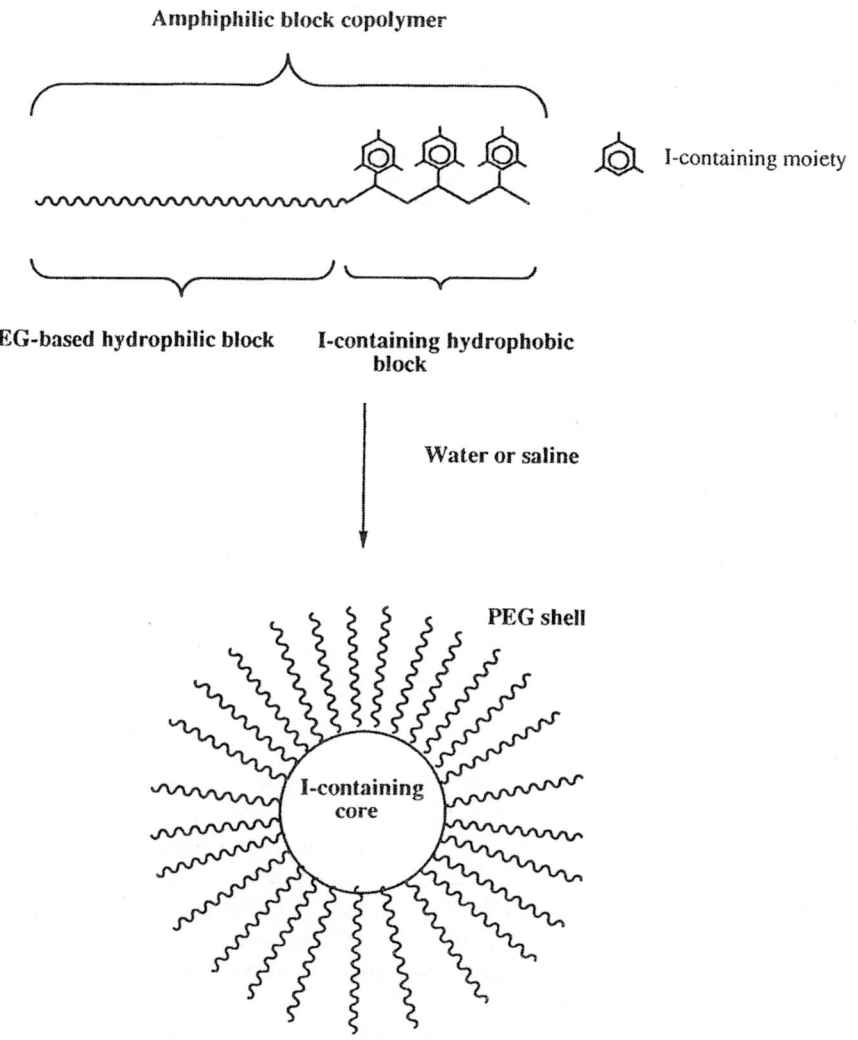

Fig. 18 Schematic representation of micelle formation from amphiphilic iodine-containing block-copolymer. (From Ref. 45.)

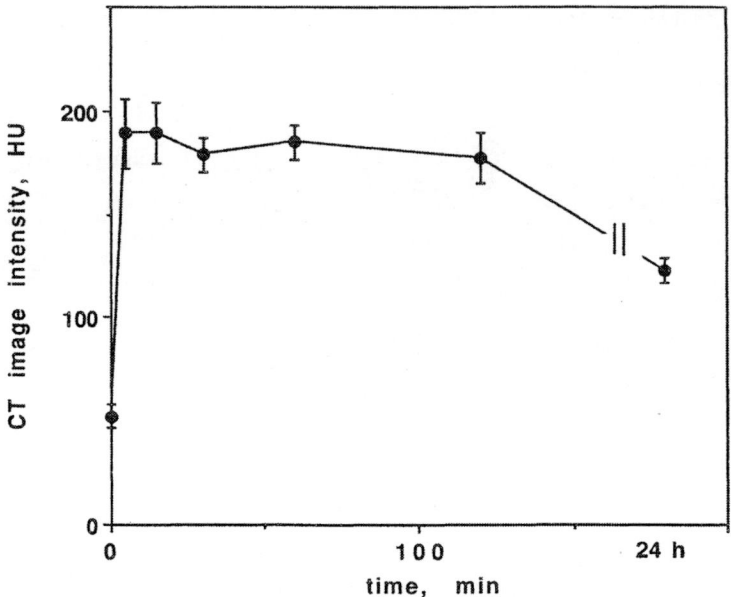

Fig. 19 Kinetics of the blood pool CT image intensity in rabbit after IV injection of iodine-containing micelles (iodine dose = 250 mg/kg; $n = 3$).

nous injection, the circulation time and biodistribution of micelles depend on the relative size of the copolymer blocks. For example, [PEO-p(Asp(ADR)] micelles with PEO block of 12 kDa in mice, has a $t_{1/2}$ of approximately 7 h, whereas micelles with a 5-kDa PEO block have a $t_{1/2}$ of about 1.5 h, and micelles with a 1-kDa block have a circulation $t_{1/2}$ of substantially less than 1 h. It seems that longer PEO blocks and shorter p(Asp) segments favor maintaining micellar structure in the blood and, consequently, longer circulation times and lower uptake by the RES [44].

Recently, we have proposed new polymeric micellar particles as a carrier for iodinated organic compounds, to be used as a contrast agent in x-ray imaging [45]. The copolymer, ϵ,N-(triiodobenzoyl)polylysine-monomethoxy PEG, can form particles with a hydrophobic iodine-containing nucleus that is sterically protected from the aqueous environment with flexible PEG chains (Fig. 18). In computed tomography (CT) experiments in rabbits, such micelles are long-circulating and can be used for CT-imaging of the blood pool: the blood remained opacified for more than 24 h after intravenous administration of the micellar agent (Fig. 19).

Thus, coating with polymers is a convenient way to control in vivo behavior and biodistribution of various microparticular carriers for drug and diagnostic agents.

REFERENCES

1. A. Rolland, ed., *Pharmaceutical Particulate Carriers*, Marcel Dekker, New York, 1993.
2. G. Gregoriadis, ed., *Liposomes as Drug Carriers*, John Wiley & Sons, New York, 1988.
3. R. Müller, *Carriers for Controlled Drug Delivery and Targeting*, CRC Press, Boca Raton FL, 1991.
4. V. V. Ranade, Drug delivery systems: Site-specific drug delivery utilizing monoclonal antibodies, *J. Clin. Pharmacol. 29*:873 (1989).
5. A. L. Klibanov, K. Maruyama, V. P. Torchilin, and L. Huang, Amphipathic polyethyleneglycols effectively prolong the circulation time of liposomes, *FEBS Lett. 268*:235 (1990).
6. A. Mori, A. L. Klibanov, V. P. Torchilin, and L. Huang, Influence of steric barrier activity of amphipathic poly(ethyleneglycol) and ganglioside GM_1 on the circulation time of liposomes and on the target binding of immunoliposomes in vivo, *FEBS Lett. 284*:263 (1991).
7. D. Needham, T. J. McIntosh, and D. D. Lasic, Repulsive interactions and mechanical stability of polymer-grafted lipid membranes, *Biochim. Biophys. Acta 1108*:40 (1992).
8. A. Gabizon and D. Papahadjopoulos, The role of surface charge and hydrophilic groups on liposome clearance in vivo, *Biochim. Biophys. Acta 1103*:94 (1992).
9. D. D. Lasic, F. G. Martin, A. Gabizon, S. K. Huang, and D. Papahadjopoulos, Sterically stabilized liposomes—a hypothesis on the molecular origin of the extended circulation times, *Biochim. Biophys. Acta 1070*:187 (1991).
10. V. P. Torchilin and M. I. Papisov, Why do polyethylene glycol-coated liposomes circulate so long? *J. Liposome Res. 4*:725 (1994).
11. V. P. Torchilin, V. G. Omelyanenko, M. I. Papisov, A. A. Bogdanov, Jr., V. S. Trubetskoy, J. N. Herron, and C. A. Gentry, Poly(ethylene glycol) on the liposome surface: On the mechanism of polymer-coated liposome longevity, *Biochim. Biophys. Acta 1195*:11 (1994).
12. J. des Cloizeaux and G. Jannink, *Polymers in Solution. Their Modelling and Structure*, Clarendon Press, Oxford, 1990, pp. 63, 280, 539.
13. M. Kurata, and Y. Tsunashima, Viscosity—molecular weight relationships and unperturbed dimensions of linear chain molecules, *Polymer Handbook* (J. Brandup and E. H. Himmelgut, eds.), John Wiley & Sons, New York, 1989, p. VII/1.
14. C. Huang and J. T. Mason, Geometric packing constraints in egg phosphatidyl-choline vesicles, *Proc. Natl. Acad. Sci. USA 75*:308 (1978).
15. H. G. Enoch and P. Strittmatter, Formation and properties of 1000-Å–diameter, single-bilayer phospholipid vesicles, *Proc. Natl. Acad. Sci. USA 76*:145 (1979).

16. V. P. Torchilin, A. L. Klibanov, L. Huang, S. O'Donnell, N. D. Nossiff, and B. A. Khaw, Targeted accumulation of polyethylene glycol-coated immunoliposomes in infarcted rabbit myocardium, *FASEB J. 6*:2716 (1992).

17. T. M. Allen and C. Hansen, Pharmacokinetics of stealth versus conventional liposomes: Effect of dose, *Biochim. Biophys. Acta 1068*:133 (1991).

18. V. P. Torchilin, M. I. Shtilman, V. S. Trubetskoy, K. Whiteman, and A. M. Milstein, Amphiphilic vinyl polymers effectively prolong liposome circulation time in vivo, *Biochim. Biophys. Acta 1195*:181–184 (1994).

19. C. Wood and E. Kabat, Immunochemical studies of conjugates of isomaltosyl oligosaccharides to lipid, *J. Exp. Med. 154*:432 (1981).

20. W. N. Charman and V. J. Stella, eds., *Lymphatic Transport of Drugs*, CRC Press, Boca Raton FL, 1992.

21. Y. Takakura, M. Hashida, and H. Sezaki, Lymphatic transport after parenteral drug administration, *Lymphatic Transport of Drugs*, (W. N. Charman and V. J. Stella, eds.), CRC Press, Boca Raton FL, 1992, p. 256.

22. G. Kabalka, E. Buonocore, K. Hubner, T. Moss, N. Norley, and L. Huang, Gadolinium-labeled liposomes: Targeted MR contrast agents for the liver and spleen, *Radiology 163*:255 (1987).

23. C. Tilcock, E. Unger, P. Cullis, and P. MacDougall, Liposomal Gd-DTPA: Preparation and characterization of relaxivity, *Radiology 171*:77 (1989).

24. G. W. Kabalka, M. A. Davis, T. H. Moss, E. Buonocore, K. Hubner, E. Holmberg, K. Maruyama, and L. Huang, Gadolinium-labeled liposomes containing various amphiphilic Gd-DTPA derivatives: Targeted MRI contrast enhancement agents for the liver, *Magn. Reson. Med. 19*:406 (1991).

25. V. S. Trubetskoy and V. P. Torchilin, New approaches in the chemical design of Gd-containing liposomes for use in magnetic resonance imaging of lymph nodes, *J. Liposome Res. 4*:961 (1994).

26. R. B. Lauffer, Magnetic resonance contrast media: Principles and progress, *Magn. Reson. Q. 6*:65 (1990).

27. S. T. Milner, Polymer brushes, *Science 251*:905 (1991).

28. G. Blume, G. Cevc, M. D. J. A. Crommelin, I. A. J. M. Bakker-Woundenberg, G. Kluft, and G. Storm, Specific targeting with poly(ethylene glycol)-modified liposomes: Coupling of homing devices to the ends of polymeric chains combines effective target binding with long circulation times, *Biochim. Biophys. Acta 1149*:180 (1993).

29. B. A. Khaw, J. A. Mattis, G. Melnicoff, H. W. Strauss, H. K. Gold, and E. Haber, Monoclonal antibody to cardiac myosin: Imaging of experimental myocardial infarction, *Hybridoma 3*:11 (1984).

30. B. A. Khaw, T. Yasuda, H. K. Gold, H. W. Strauss, and E. Haber, Acute myocardial infarct imaging with indium-111-labeled monoclonal antimyosin Fab, *J. Nucl. Med. 28*:1671 (1987).

31. V. Weissig, J. Lasch, A. L. Klibanov, and V. P. Torchilin, A new hydrophobic anchor for the attachment of proteins to liposomal membranes, *FEBS Lett. 202*:86 (1986).

32. V. J. Caride and B. L. Zaret, Liposome accumulation in regions of experimental myocardial infarction, *Science 198*:735 (1977).

33. S. M. Moghimi, C. J. H. Porter, L. Illum, and S. S. Davis, The effect of Poloxamer-407 on liposome stability and targeting to bone marrow: Comparison with polystyrene microspheres, *Int. J. Pharm. 68*:121 (1991).

34. L. Illum and S. S. Davis, Effect of nonionic surfactant Poloxamer 338 on the fate and deposition of polystyrene microspheres following intravenous administration, *J. Pharm. Sci. 72*:1086 (1983).

35. S. M. Moghimi, A. E. Hawley, N. M. Christy, T. Gray, L. Illum, and S. S. Davis, Surface engineered nanospheres with enhanced drainage into lymphatics and uptake by macrophages of the regional lymph node, *FEBS Lett. 344*:25 (1994).

36. H. Patel and S. Moghimi, Tissue specific opsonins and phagocytosis of liposomes, *Targeting of Drugs—Optimization Strategies* (G. Gregoriadis, A. Allison, and G. Poste, eds.), Plenum Press, New York, 1990.

37. C. J. H. Porter, S. M. Moghimi, L. Illum, and S. S. Davis, The polyethylene/polypropylene block copolymer Poloxamer-407 selectively redirects intravenously injected microspheres to sinusoidal endothelial cells of rabbit bone marrow, *FEBS Lett. 305*:62 (1992).

38. R. Gref, Y. Minamitake, M. T. Peracchia, V. S. Trubetskoy, V. P. Torchilin, and R. Langer, Biodegradable long-circulating polymeric nanospheres, *Science 263*:1600 (1994).

39. V. Trubetskoy, R. Langer, and V. Torchilin, unpublished results.

40. A. L. Klibanov, K. Maruyama, A. M. Beckerleg, V. P. Torchilin, and L. Huang, Activity of amphipathic poly(ethylene glycol) 5000 to prolong the circulation time of liposomes depends on the liposome size and is unfavorable for immunoliposome binding to target, *Biochim. Biophys. Acta 1062*:142 (1991).

41. G. R. Harper, M. C. Davies, S. S. Davis, T. F. Tadros, D. C. Taylor, M. P. Irving, and J. A. Waters, Steric stabilization of microspheres with grafted polyethylene oxide reduces phagocytosis by rat Kupffer cells in vitro, *Biomaterials 12*:695 (1991).

42. M. K. Pratten, J. B. Lloyd, G. Horpel, and H. Ringsdorf, Micelle-forming block copolymers: Pinocytosis by macrophages and interaction with model membranes, *Makromol. Chem. 186*:725 (1985).

43. H. Bader, H. Ringsdorf, and B. Schmidt, Water soluble polymers in medicine, *Angew. Makromol. Chem 123/124*:457 (1984).

44. G. S. Kwon, M. Yokoyama, T. Okano, Y. Sakurai, and K. Kataoka, Biodistribution of micelle forming polymer–drug conjugates, *Pharm. Res. 10*:970 (1993).

45. V. S. Trubetskoy, V. P. Torchilin, G. S. Gazelle, and G. L. Wolf, Amphiphilic radiopaque iodine-containing block–copolymer as micellar polymeric carrier with controlled in vivo performance, Proceedings of 21st International Symposium on Controlled Release of Bioactive Materials, Nice, France, 1994, pp. 676–677.

46. V. S. Trubetskoy, J. A. Cannillo, A. Milshtein, G. L. Wolf, and V. P. Torchilin, Controlled delivery of Gd-containing liposomes to lymph node: surface modification may enhance MRI contrast properties, *Magn. Res. Imaging 13*:31 (1995).

9

Poly(Ethylene Glycol)-Coated Biodegradable Nanospheres for Intravenous Drug Administration

Ruxandra Gref

Ecole Nationale Supérieure des Industries Chimiques
Nancy, France

Yoshiharu Minamitake

Suntory Limited
Ohra-Gun, Gunma-Ken, Japan

Maria Teresa Peracchia

Università degli Studi di Parma
Parma, Italy

Robert Langer

Massachusetts Institute of Technology
Cambridge, Massachusetts

I. INTRODUCTION

The development of injectable drug-loaded carriers with blood circulation times long enough to continuously deliver drugs, imaging agents, or other molecules, to specific sites in the body has been a major challenge. Such carriers could be used, for example, in the treatment of numerous illnesses that require intravenous administration of highly active compounds (such as those used in chemotherapy); by continuously delivering only the necessary amount of drug (without exceeding the toxicity level, or falling below the minimal therapeutic level), more favorable pharmacokinetics could be obtained. Another application could be the delivery of drugs with very short half-lives that are usually administered using pumps, which constrains the patient to remain in the hospital for long periods. In such cases, drug therapy could be improved by the design of long-circulating injectable particles that could encapsulate and, then, continuously liberate the drugs in the blood in a controlled manner. Patient compliance and convenience could thus be increased by avoiding the need of perfusion, and hospital costs could be reduced.

Some of the desired features of long-circulating systems are the ability to encapsulate, with good entrapment efficiency, a reasonably high-weight fraction (loading) of active compound. For example, more than 80% of the agent used in the first fabrication step of the encapsulation process should be incorporated into the final carrier, which should contain more than 30 wt% of active agent. Ideally, the carriers should be easy to prepare and store (for example, by freeze-drying). Their size should be small enough (less than 5 μm) to enable them to freely circulate through the smallest capillaries. One of the main feature of the carriers is their biodegradability; after drug liberation, the particles should decompose into nontoxic, eliminative fragments. Moreover, the carriers should have high stability to allow in vivo use.

Many systems (liposomes, niosomes, microemulsions, micelles, and nanoparticles) have been developed over the past decade to achieve drug delivery at optimal rates, or for the direct targeting of the particles to diseased tissues. Most of these systems are immediately recognized as foreign products and removed from blood circulation through phagocytosis by cells of the reticuloendothelial system (RES) [1–4]. For example, albumin or galactose microspheres disappear completely from blood within seconds [4,5]. Some progress has been made in reducing the rapid clearance of the carriers; for example, by chemical attachment or by adsorption of appropriate polymers or molecules to their surface [1,6–15]. However, it has been difficult to develop drug carriers that display all features presented in the previous paragraph. This chapter describes a recent approach to creating degradable polymeric nanospheres that, to a large extent, display the desired features of long-circulating drug carriers [16–23].

Nanoparticles (also called nanospheres) are defined as solid colloidal particles, less than 1 μm in size, that consist of macromolecular compounds [24]. Drugs or other biologically active materials are dissolved, entrapped, or encapsulated in the nanoparticles [25,26], or are chemically attached to the polymers or adsorbed to their surface [27].

A. Surface Properties of Long-Circulating Carriers

To achieve long blood circulation half-lives, the particles should be small enough to avoid mechanical clearance by filtration in the lungs, with particles larger than 5–7 μm [28], or in the spleen [29]. The RES uptake generally increases with increasing particle size, whatever the particles' nature: nanospheres [30], liposomes [31], or fat emulsion droplets [32,33]. A reduced carrier size (smaller than 200 nm), therefore, is desired for a long-circulating drug carrier.

Together with the small size, the particles' surfaces should be "stealthy" relative to RES cells, such as macrophages located in the liver and spleen. The adsorption of serum components (opsonins) on the particles' surfaces precedes their phagocytosis [34,35]. Serum complement is one of the most important components of the opsonic system that strongly activates macrophage phagocytosis [36,37]. The complement activation pathways have been investigated [38,39].

From the assumption that opsonin adsorption triggers the particles' uptake by the RES, the first attempts to prolong their circulation time were directed toward their surface charge [40] or hydrophilicity [41,42] modification, to reduce the interaction with opsonins as much as possible. Negatively charged particles are eliminated faster from blood than positively charged or neutral ones [43,44]. Neutral surfaces seem the most appropriate to reduce macrophage phagocytosis [45].

The creation of a hydrophilic coating on hydrophobic carriers significantly increases their circulation time. For example, a good theoretical and experimental basis exists on hydrophobic polystyrene particles coated by adsorption of Poloxamers [surfactants containing poly(ethylene glycol); PEG blocks], which significantly increases their blood half-life [8,41,46]. Polystyrene particles as small as 60 nm are completely removed from blood within minutes [1], whereas these same model carriers, when sterically stabilized by coating with Poloxamine 908, circulate in the blood over a few days [47]. The more hydrophobic blocks in the coating polymers adsorb on the surface, ensuring adhesion and, thereby, preventing polymer detachment, whereas the more hydrophilic PEG blocks project out and prevent protein adsorption [7,41]. However, these particles are nonbiodegradable and, therefore, cannot be administered intravenously. Moreover, the adsorption approach, based on specific interactions, is difficult to generalize from nondegradable polymers to biodegradable ones, such as poly(lactic acid) (PLA), poly(lactic-*co*-glycolic acid) (PLGA) or poly(ϵ-caprolactone) (PCL). Another problem is that the adsorbed polymers might desorb in vivo because of replacement by compounds from the blood with a higher affinity for the surface.

To ensure coating stability, an alternative to polymer adsorption is polymer chemical attachment. For example, macrophage phagocytosis was significantly reduced with albumin nanospheres on which PEG was covalently attached [9]. However, this approach sometimes encounters practical difficulties, owing to nonhomogeneous distribution of reactive sites on the surface [10] and the increasing steric hindrance as the coating reaction progresses.

B. Poly(ethylene glycol)-Based Carriers for Intravenous Administration

Hydrophilic PEG-based coatings enhance carrier blood circulation time [8,9,41,46]. Among the nonionic hydrophilic polymers, PEG has many advantages; it is considered nontoxic and is approved by the Food and Drug Administration (FDA) for internal use in the human body [48].

The protein resistance of PEG chains attached to a hydrophobic surface by one chain end, was studied theoretically [49,50]. The PEG dysopsonic effect depends on its molecular weight and surface density: the longer the PEG chains and the higher the surface density, the lower the blood protein adsorption. When attached to the liposomes' surface, PEG provides a steric barrier, avoiding opsonization [51]. PEG was used to increase the biocompatibility of surfaces [52,53] or to reduce the immunogenicity and antigenicity of the proteins to which it was attached [54]. The PEG-based micelles, with doxorubicin (Adriamycin) chemically attached to the polymers' backbone, circulate longer in blood [55].

Liposomes coated with PEG have significantly increased blood circulation times [11,56–58]. Liposomes have the advantage of being composed of substances that naturally exist in the body (such as cholesterol or lecithin). However, liposomes present some inconveniences, such as the strong dependence of the drug entrapment yield on the physicochemical properties of the drug, and a limited physical stability [59,60]. Because they are formed of a lipid membrane surrounding an aqueous phase, they are considered to be less stable than nanoparticles [61]. This is a disadvantage, because particle stability is important for their administration. Many coating materials, such as surfactants, can lead to disruption of the liposomes' membrane [62]. The higher stability of polymeric nanoparticles is an advantage for their longer shelf life after preparation.

C. Poly(ethylene glycol)-Coated Nanospheres

According to the foregoing considerations, stable nanoparticles, able to encapsulate drugs by physical entrapment and with a dysopsonic PEG coating should be appropriate candidates for the fabrication of long-circulating drug carriers. A recent approach was to form small-sized PEG-coated nanoparticles by a direct fabrication method, using specially tailored block copolymers [16–23]. These particles are schematized in Fig. 1; the steric effect produced by the PEG coating should avoid opsonin adsorption on the core. To obtain a coating that might prevent opsonization, PEG was covalently attached to the nanosphere core by the synthesis of amphiphilic diblock copolymers, PEG-R, where R is chosen among the family of hydrophobic polymers already known for their good bio-

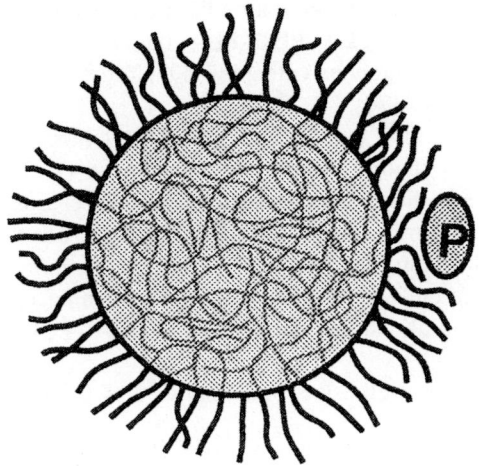

Fig. 1 Schematic representation of a PEG-coated nanosphere. The coating's steric hindrance should avoid blood protein (P) adsorption on the particle core, formed in a hydrophobic biodegradable polymer R (PLA, PLGA, PCL, PSA).

compatibility, such as PLA, PLGA, PCL [16–18]. Polyanhydrides, such as poly(sebacic acid) (PSA), were also used [19,20]. Moreover, by the adjustment of the molecular weight and chemical composition of R, the degradation time and the drug release kinetics from the core could be controlled.

This chapter is focused on PEG-coated nanospheres, including polymer synthesis, nanosphere fabrication, drug encapsulation, and in vivo applications.

II. BLOCK COPOLYMERS FOR THE PREPARATION OF POLY(ETHYLENE GLYCOL)-COATED NANOSPHERES

The PEG-R polymers were formed by reaction between monomethoxy-PEG (MPEG) or monomethoxy monoamine-PEG (MPEG-NH$_2$) and R composing a carboxyl end group [17]. The diblock PEG-R polymers were also formed by ring-opening polymerization of monomers (lactide, glycolide, caprolactone, or mixtures of them) on MPEG [17–19]. For example, PEG–PLGA polymers were formed by polymerization of D,L-lactide, glycolide, and MPEG, with molecular weights of 5, 12, and 20 kDa. PEG–PCL was formed by polymerization of ε-caprolactone and MPEG [19]. PEG–PLA was obtained by polymerization of lactide and MPEG (MW 2 and 5 kDa) [63]. In these cases, the ring-opening polymerization was initiated by the hydroxyl end groups of MPEG and was catalyzed by stannous octoate. This catalyst has been widely used to prepare PLA and PLGA [64–67] and is approved by the FDA as a food stabilizer [68]. It was also used for triblock PLA–PEG–PLA synthesis [69]. Moreover, among different catalysts used, stannous octoate was the most effective for triblock PLA–PEG–PLA synthesis [70].

Multiblock PEG_n-R_m polymers were also synthesized [19]. They have n MPEG blocks attached together at one chain end and continued by m (1–3) hydrophobic blocks R. For this synthesis, first, two or more MPEG-NH_2 chains were attached to the carboxyl groups of citric, mucic, or tartaric acids, or other natural polyfunctional molecules. The resulting polymers have one or more hydroxyl end groups, which were further used to initiate the ring opening or condensation polymerization of bioerodible R polymers (PLA, PLGA, PCL, PSA).

The chemical structure of some of the synthesized macromolecules is indicated on Scheme 1.

III. POLY(ETHYLENE GLYCOL)-COATED NANOSPHERE PREPARATION

Among the various methods to produce nanospheres listed in review articles [71], the emulsion–solvent evaporation and the nanoprecipitation are the most widely used. By using PEG-R copolymers, nanospheres were formed following an emulsion–evaporation procedure (Fig. 2). First, PEG-R was dissolved in a water-immiscible organic solvent (such as methylene chloride or ethyl acetate). Then, an oil-in-water emulsion was formed by vortexing in an aqueous phase containing a surfactant. The size of the emulsion was reduced either by sonication or by microfluidization. The two blocks, PEG and R, have different solubilities in the organic and the aqueous phases. For example, PEG is water-soluble, but is practically insoluble in ethyl acetate, whereas R is insoluble in water, but soluble in ethyl acetate or methylene chloride. The solubility difference should lead to a tendency of PEG migration inside the droplets, to locate at the water interface, whereas R chains remain and concentrate inside the droplets. Therefore, a phase-separated structure, as indicated in Fig. 1, is formed. After organic solvent removal, the R chains inside the core form an entangled structure, in which hydrophobic drugs can be entrapped during nanosphere fabrication. Finally, the solidified particles were recovered by centrifugation, washed to remove adsorbed surfactant, lyophilized, and stored anhydrously at 4°C. The lyophilized nanospheres can easily be redispersed in aqueous solutions.

Nanospheres were also prepared by nanoprecipitation [72]. In this technique, PEG-R is dissolved in a water-miscible organic solvent, such as acetone, which is poured into an aqueous phase, immediately leading to polymer precipitation. Compared with emulsion–solvent evaporation, nanoprecipitation forms smaller

MPEG-PLA

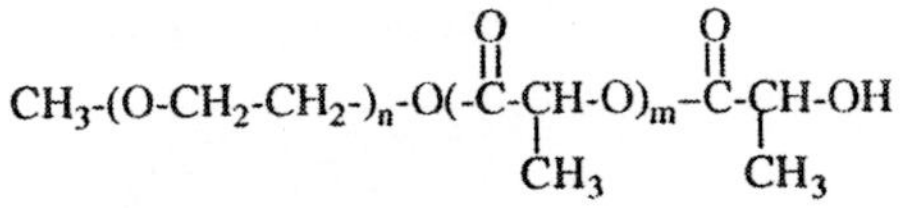

$$CH_3\text{-}(O\text{-}CH_2\text{-}CH_2\text{-})_n\text{-}O(\text{-}\overset{\overset{O}{\|}}{C}\text{-}CH\text{-}O)_m\text{-}\overset{\overset{O}{\|}}{C}\text{-}CH\text{-}OH$$
$$\qquad\qquad\qquad\qquad\qquad\quad CH_3\qquad\quad CH_3$$

MPEG$_3$-PLA

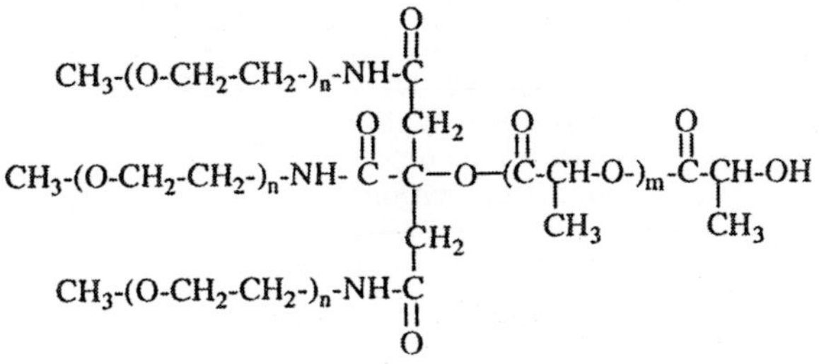

$$CH_3\text{-}(O\text{-}CH_2\text{-}CH_2\text{-})_n\text{-}NH\text{-}\overset{\overset{O}{\|}}{C}$$
$$CH_3\text{-}(O\text{-}CH_2\text{-}CH_2\text{-})_n\text{-}NH\text{-}\overset{\overset{O}{\|}}{C}\text{-}\underset{\underset{CH_2}{|}}{\overset{\overset{CH_2}{|}}{C}}\text{-}O\text{-}(\overset{\overset{O}{\|}}{C}\text{-}CH\text{-}O\text{-})_m\text{-}\overset{\overset{O}{\|}}{C}\text{-}CH\text{-}OH$$
$$\qquad\qquad\qquad\qquad\qquad\qquad\qquad\qquad\qquad CH_3\qquad\quad CH_3$$
$$CH_3\text{-}(O\text{-}CH_2\text{-}CH_2\text{-})_n\text{-}NH\text{-}\underset{\underset{O}{\|}}{C}$$

MPEG$_2$-PLA$_2$

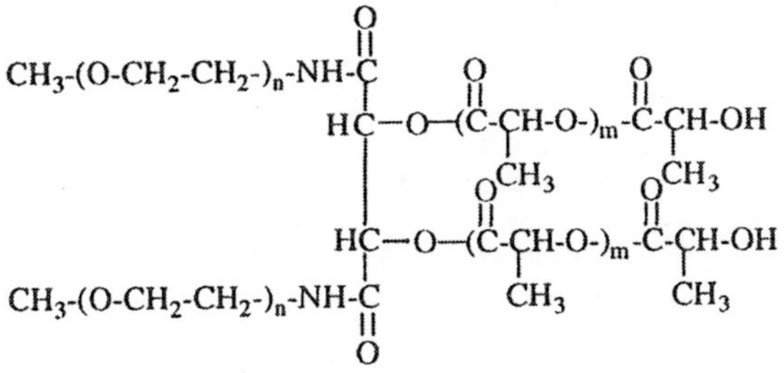

$$CH_3\text{-}(O\text{-}CH_2\text{-}CH_2\text{-})_n\text{-}NH\text{-}\overset{\overset{O}{\|}}{C}$$
$$H\overset{|}{C}\text{-}O\text{-}(\overset{\overset{O}{\|}}{C}\text{-}CH\text{-}O\text{-})_m\text{-}\overset{\overset{O}{\|}}{C}\text{-}CH\text{-}OH$$
$$\overset{|}{H}\overset{O^{CH_3}}{\underset{|}{C}}\text{-}O\text{-}(\overset{\overset{O}{\|}}{C}\text{-}CH\text{-}O\text{-})_m\text{-}\overset{\overset{O}{\|}}{C}\text{-}CH\text{-}OH$$
$$CH_3\text{-}(O\text{-}CH_2\text{-}CH_2\text{-})_n\text{-}NH\text{-}\underset{\underset{O}{\|}}{C}\qquad CH_3\qquad\quad CH_3$$

MPEG$_2$-PLA$_4$

$$CH_3\text{-}(O\text{-}CH_2\text{-}CH_2\text{-})_n\text{-}NH\text{-}\overset{\overset{O}{\|}}{C}$$
$$H\overset{|}{C}\text{-}O\text{-}(\overset{\overset{O}{\|}}{C}\text{-}CH\text{-}O\text{-})_m\text{-}\overset{\overset{O}{\|}}{C}\text{-}CH\text{-}OH$$
$$H\overset{|}{C}\text{-}O\text{-}(\overset{\overset{O}{\|}}{C}\text{-}CH\text{-}O\text{-})_m\text{-}\overset{\overset{O}{\|}}{C}\text{-}CH\text{-}OH$$
$$H\overset{|}{C}\text{-}O\text{-}(\overset{\overset{O}{\|}}{C}\text{-}CH\text{-}O\text{-})_m\text{-}\overset{\overset{O}{\|}}{C}\text{-}CH\text{-}OH$$
$$H\overset{|}{C}\text{-}O\text{-}(\overset{\overset{O}{\|}}{C}\text{-}CH\text{-}O\text{-})_m\text{-}\overset{\overset{O}{\|}}{C}\text{-}CH\text{-}OH$$
$$CH_3\text{-}(O\text{-}CH_2\text{-}CH_2\text{-})_n\text{-}NH\text{-}\underset{\underset{O}{\|}}{C}\qquad CH_3\qquad\quad CH_3$$

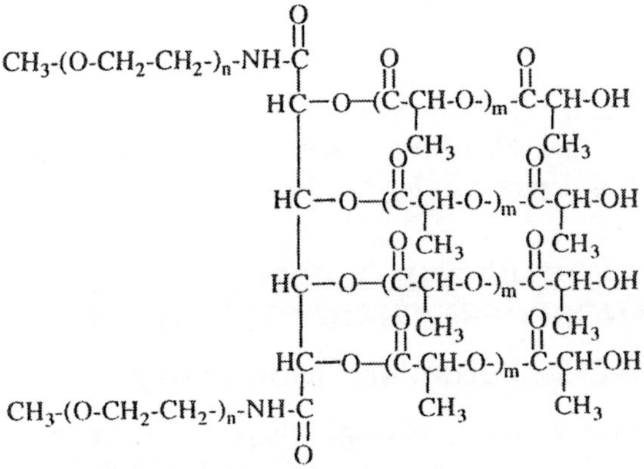

Scheme 1 Chemical structure of some synthesized diblock (PEG–PLA) and multiblock (PEG$_n$–PLA$_m$) copolymers.

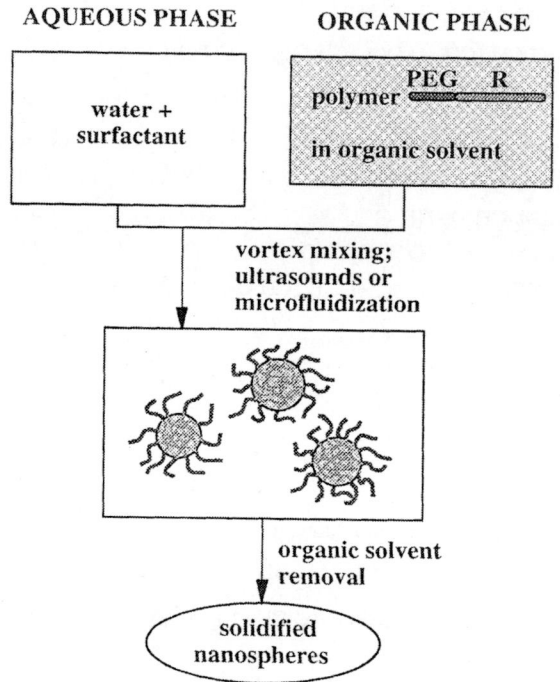

AQUEOUS PHASE   ORGANIC PHASE

Fig. 2   Schematic representation of the nanosphere preparation procedure.

particles, but it is questionable whether, during this fast preparation procedure, the polymer chains have the capability to rearrange, leading to phase-separated structures (as depicted in Fig. 1).

IV.   POLY(ETHYLENE GLYCOL)-COATED NANOSPHERE CHARACTERIZATION

A.   Size Distribution and Morphology

Size distribution and morphology (shape), two important characteristics of nanospheres, play a role in organ distribution after IV administration. Several techniques enable visualization of nanospheres' size distribution and morphology. Scanning electron microscopy (SEM) allows sample observation at high magnification, but requires previous coating of dried samples with a conductive material, usually gold or palladium. It was reported [73] that the gold-coating thickness can induce an important nanoparticle size overestimation from SEM pictures. Figure 3 shows typical SEM pictures of PLGA40K and PEG20K-PLGA180K nanospheres; the samples are spherical.

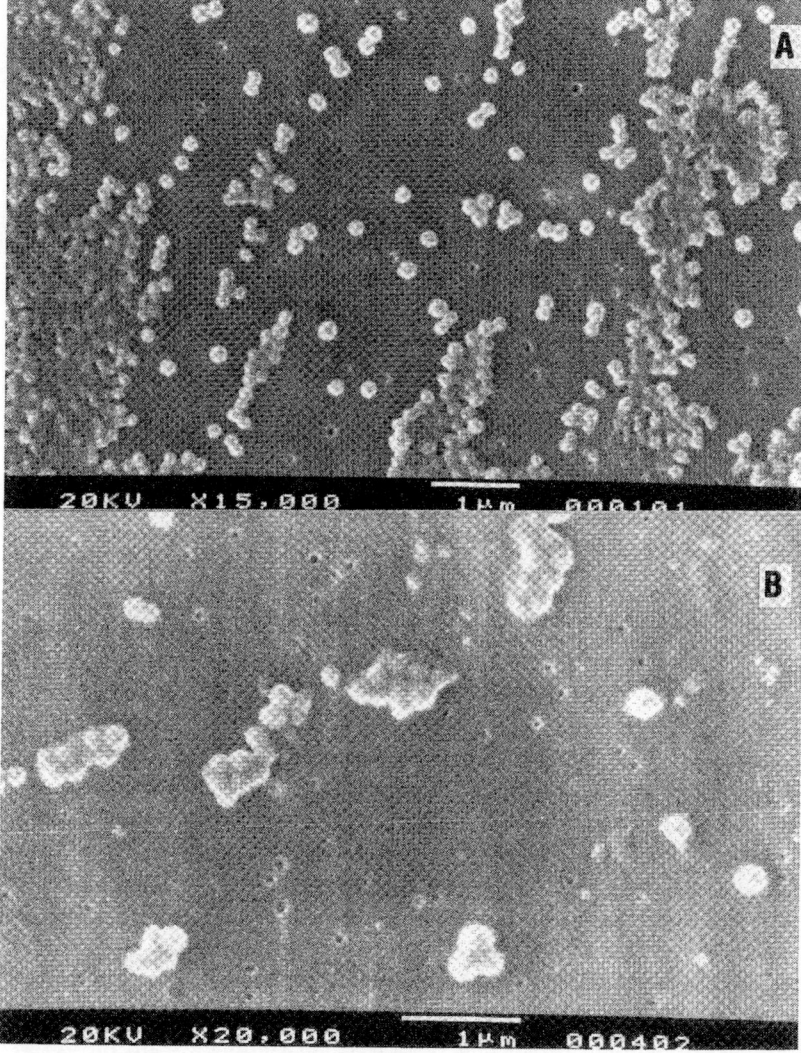

Fig. 3 Typical electromicrographs of (A) PLGA and (B) PEG20K–PLGA nanospheres taken by using scanning electron microscopy.

Other recently developed techniques allow sample observation at higher resolution than SEM and without the need of coating the sample. For example, atomic force microscopy (AFM) was used to observe PEG–PLGA and PLGA nanospheres [17] and confirmed their spherical shape. As with SEM, the samples were kept under vacuum during observation. Therefore, no information can be drawn concerning the PEG chain configuration at the surface.

A versatile technique used for PEG-coated nanospheres observation is freeze-fracturing [74]. The samples, dispersed in water, are cryoprotected in 30% glycerol. They are then frozen, fractured, and coated with carbon and platinum. The replicas are washed in water and then visualized by using transmission electron microscopy. With use of freeze-fracture, valuable information has been obtained on the nanospheres' surface morphology in water: Fig. 4 shows typical images of PLGA and PEG20K–PLGA nanospheres. Hemispheric replicas of nanospheres were found in all samples. The PLGA nanospheres showed a smooth surface, whereas PEG-coated ones present specific irregularities on their surface, with an average greatest diameter of about 12 nm, that were attributed to the PEG coating [74].

The nanospheres' hydrodynamic diameter and size distribution can be measured by quasi-elastic light scattering (QELS), also called photon correlation spectroscopy (PCS). This technique is based on scattered light fluctuation created by particles in movement. It was used to determine the mean diameter of PEG-R nanospheres and to follow their aggregation and deaggregation behavior [17].

B. Studies of the Poly(ethylene glycol) Coating

Because of the dysopsonic properties of PEG, PEG-coated nanospheres are expected to avoid opsonin adsorption, macrophage phagocytosis, prolong blood circulation time, and reduce liver and spleen accumulation. However, it is not obvious how to estimate the exact percentage of PEG chains located at the surface, or what percentage of PEG chains remain in the core. It is possible to measure other surface properties modified by the coating, such as chemical composition, hydrophobicity, and charge. These measurements are the result of PEG-coating density and thickness and do not distinguish between the respective effects of these two factors.

1. Surface Chemical Composition

Surface analysis techniques can be used to detect PEG on the nanospheres' surface by determining surface chemical composition changes. For example, x-ray photoelectron spectroscopy (XPS), also known as electron spectroscopy for chemical analysis (ESCA), allows nondestructive determination of surface chemical composition, at a depth in the typical range of 1–10 nm [75–77].

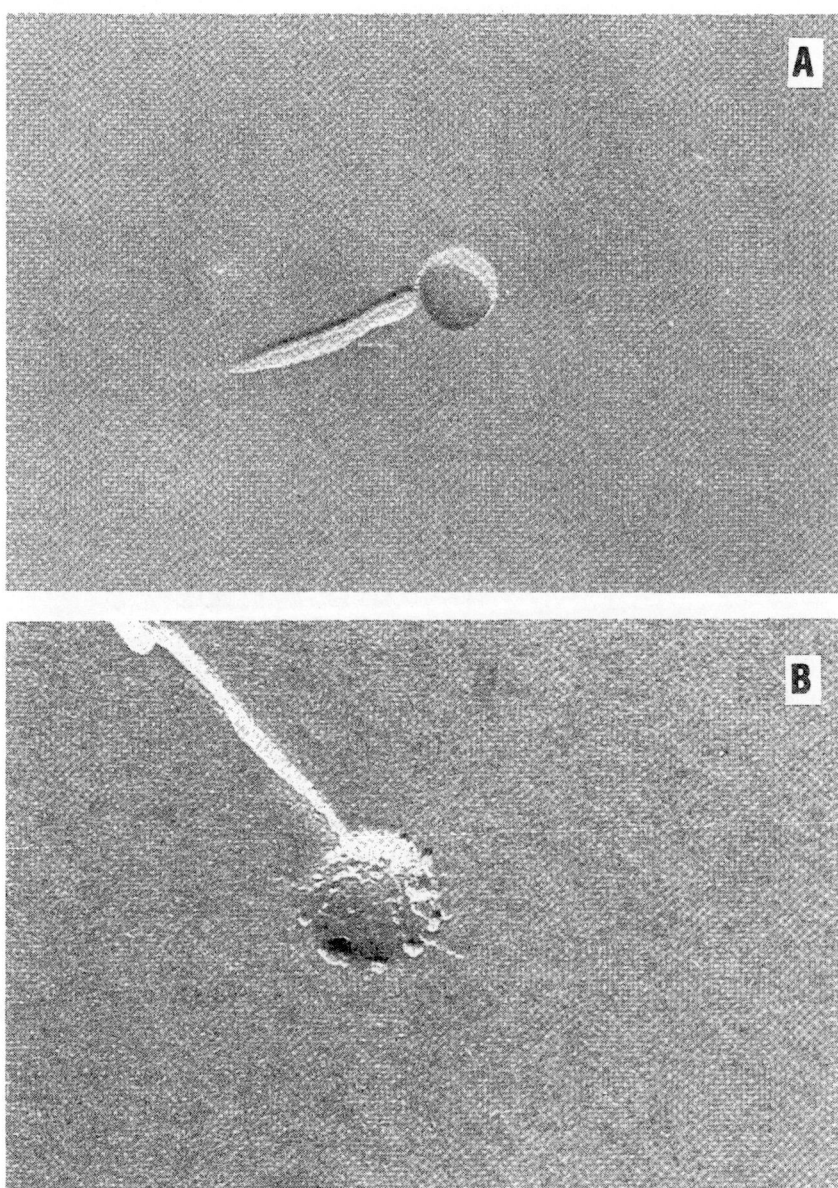

Fig. 4 Freeze-fracture studies: typical electromicrographs of (A) PLGA and (B) PEG20K–PLGA nanospheres taken by using transmission electron microscopy. The PLGA particles have a smooth surface, whereas the PEG chains adopt a specific conformation at the surface of the PEG20K–PLGA nanospheres.

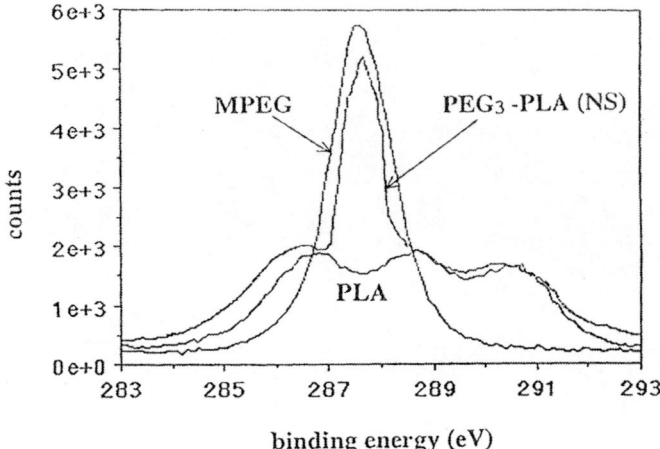

Fig. 5 Surface chemical analysis (ESCA) of PLA and PEG polymer powder and of (PEG20K)$_3$–PLA nanospheres. The binding energy spectrum of the nanospheres' surface is a superposition of the respective spectra of PLA and PEG.

Each polymer has a specific spectrum of binding energy. The analysis of a surface spectrum enables establishment of the nature and the relative amounts of the elements at the surface. ESCA was used [17,20] to comparatively study PEG–PLGA and PLGA nanospheres. The results thus obtained confirmed the presence of PEG chains on the first nanosphere layers. In nanospheres prepared without the use of any surfactant from multiblock PEG$_3$-PLA polymers (Fig. 5), it is possible to clearly distinguish that the spectrum of PEG$_3$–PLA particles on the surface is a superposition of the respective spectrum of PLA and PEG polymers.

To obtain maximum information, ESCA can be used in conjunction with another recently developed surface analysis technique, Secondary ion mass spectrometry (SIMS) [78]. This technique gives information on the average chemical composition on a surface of about 1-nm depth and is more molecularly specific than ESCA because the mass spectrum generated by SIMS can be analyzed by conventional mass spectrometry rules.

Both ESCA and SIMS experiments are conducted in a vacuum, so the PEG chain configuration at the analyzed surface is different from that in aqueous solutions. These techniques enable only semiquantitative conclusions: for example, they allow one to choose from among the different preparation techniques the one that will maximize the PEG relative density on the surface, or to study PEG detachment after various incubation times and, thereby, draw conclusions on coating stability.

2. Surface Hydrophobicity

Surface hydrophobicity can be determined by various techniques, reviewed recently [46], such as rose bengal binding methods, partitioning between two aqueous phases (dextran–PEG), or hydrophobic interaction chromatography (HIC). The use of HIC allows one to separate particles in a column on the basis of differences in their hydrophobic interactions with the gel matrix (usually agarose gels to which alkyl chains are coupled). The advantage of the chromatographic technique is that nonhomogeneous samples (on the basis of their surface hydrophobicity) can be separated. This was not true of any technique listed before (ESCA, SIMS, dye binding methods), from which only average data over the whole nanosphere population are obtained.

3. Surface Charge

Carrier charge is the main factor that mediates a carrier's interaction with different compounds present in living media. The surface zeta potential, proportional to the velocity of particles submitted to an electric field, gives an indication of surface charge. The zeta potential of PLA particles of a mean size of 1.7 μm, produced by grinding polymer slabs and dispersing them in water, was equal to -52.8 mV [79]. This negative value is attributed to the presence of carboxyl groups on the surface. The PEG–PLA nanospheres had a zeta potential nearly equal to zero [72], which is an expected result, as the polymers PEG–PLA are uncharged (i.e., they do not have carboxyl end groups, as in PLA).

4. Other Surface Characterization Techniques

Other techniques can be used to characterize the PEG-coated particles. For example, the measurement of the critical flocculation temperature (CFT)—the temperature at which the nanosphere solution becomes thermodynamically unstable and aggregates—gives information on the efficacy of the PEG coating to ensure the steric stabilization of the suspension [46].

The adsorption of plasma proteins on the surface of IV-administered particles is considered key to the explanation of their organ distribution. The efficacy of PEG-coated nanoparticles in avoiding opsonization can be determined by incubating samples with plasma proteins, then measuring the nature and the amount of the adsorbed proteins.

V. POLY(ETHYLENE GLYCOL)-COATED NANOSPHERE OPSONIZATION

Activation of the complement system, which is involved in natural protection against foreign compounds introduced into the body [39], has been studied in the presence of PLA70K and PEG2K–PLA30K particles [80]. In comparison with PLA nanospheres, the PEG-coated ones barely activate the complement system. Interestingly, PLA particles prepared by using a PEG-based surfactant (Pluronic F-68) strongly activate the complement system [80].

The aforecited study gives information only about complement activation. In vivo, the situation is more complex, as the particle's surface is exposed to numerous plasma proteins. Injected particles may also interact with fibronectin [81], or other compounds listed in reviews dealing with the topic [45,46,82]. Therefore, it is interesting to study the *competitive* adsorption of plasma proteins as they simultaneously come in contact with PEG-coated particles, as happens when the particles are suddenly injected. This study can be achieved by high-resolution, two-dimensional, polyacrylamide gel electrophoresis (2-D-PAGE). In the first dimension, proteins are separated according to the isoelectric point and, in the second dimension, the separation is based on the molecular weight. Thus, several proteins, including isoforms, can be detected semiquantitatively, after a desorption procedure from the surface of the particles previously incubated in serum. Use of 2D-PAGE was successfully [83] in establishing a correlation between the adsorbed proteins and the in vivo behavior of polystyrene carriers coated with Poloxamer 407 and Poloxamine 908. It was suggested that not only the composition, but also the conformation (exposing different binding sites), of the adsorbed proteins may affect their organ distribution. It is suggested that specific organ-targeting may be achieved by controlled adjustment of particle surface properties and related protein adsorption.

Two-dimensional PAGE was used to investigate PLGA and PEG–PLGA (PEG 5, 12, and 20 kDa) nanospheres [84]. The amount of plasma proteins adsorbed on the uncoated PLGA nanospheres is much higher than on the PEG–PLGA particles. On PLGA nanospheres, major proteins were apolipoproteins J, C-III, E and A-I, whereas their adsorption was practically eliminated on PEG-coated particles, whatever the molecular weight of PEG. Various other apolipoproteins still adsorb on PEG2OK–PLGA particles, but less than on PEG12K- and PEG5K-coated ones. These preliminary data indicate an effect of PEG chain length (or coating thickness) against opsonization of these proteins [84].

Interestingly, the biodistribution data obtained with these particles (reported in the following paragraphs) show that by increasing the PEG molecular weight, liver accumulation is reduced and blood circulation time is increased. This tends to prove that not only complement activation, but also various other opsonization processes, which are now under investigation, are responsible for the in vivo fate of IV-administered drug carriers.

VI. NANOSPHERE DEGRADATION AND DRUG RELEASE

One of the goals of PEG-coated nanospheres is to achieve drug-loaded carriers, able to liberate the active compound in a controlled manner directly into the bloodstream. Drug release and polymer degradation are two interconnected phenomena. Although few studies deal with biodegradable nanosphere degradation mechanisms, for larger systems, such as microspheres or polymer slabs, it is generally assumed that the drug is released by several processes:

Diffusion through the polymer matrix
Release by polymer degradation (either by surface or bulk erosion)
Solubilization and diffusion through microchannels that exist in the polymer matrix or are formed by erosion

Degradation of artificial polymers of the poly(α-hydroxy acid) type, some of which were used to form the core of PEG-coated nanospheres, have attracted much attention [64,85–87]. These polymers absorb water, which starts bulk degradation by polymer hydrolysis. Oligomers and monomers thus formed tend to diffuse out of the polymer matrix. If the rate of diffusion is smaller than the rate of reaction, a local concentration of oligomers can occur at the center of the sample and is responsible for an acidity increase which, in turn, catalyzes the chain scission by hydrolysis. Moreover, degradation may be complicated by the possibility that oligomers crystallize inside the matrix before diffusing out [85]. Slabs prepared from biodegradable PLA–PEG–PLA triblock copolymers showed faster degradation kinetics compared with PLA, and the overall biological response to implants was good and comparable in both experiments [88].

Nanospheres in PEG–PLGA degrade very slowly, keeping their mean diameter practically constant over several weeks [89]. In the first 60 h after incubation at 37°C, less than 8% of the total initial PEG amount was detached from nanospheres prepared from PEG5K–PLGA45K and less than 10% from PEG12K–PLGA100K. Much lower amounts of lactic acid were detected in the release medium; therefore, it was suggested that the ester bond between PEG and R is the most fragile one in the nanosphere structure [89].

VII. DRUG ENCAPSULATION IN POLY(ETHYLENE GLYCOL)-COATED NANOSPHERES

A. A Study Using a Model Drug: Lidocaine

Lidocaine, an antiarrhythmic drug, was used as a model lipophilic drug to study encapsulation properties of PEG-coated nanospheres [17,20]. Drug-loaded nanospheres were prepared by the procedure depicted in Fig. 2, in which lidocaine was dissolved together with the block copolymers in the organic solvent. The drug-loaded nanospheres were prepared from various diblock and multiblock PEG_n-R copolymers, where R was a polyester (PLA, PLGA, or PCL) or a polyanhydride (PSA). Drug-loading was determined spectrophotometrically by dissolving a weighed amount of lyophilized nanospheres in dichloromethane. In vitro release studies were performed under physiological conditions (phosphate-buffered solution 0.02 M; pH 7.2; 37°C), following a dialysis bag diffusion technique [90–92]. The same experimental procedure was used for all measurements: nanosphere suspensions (1 mg/ml) were placed inside dialysis bags (cutoff 50,000 Da) and the sink phase was totally replaced every hour. Lidocaine was assessed spectrophotometrically.

With all polymers studied, high loadings (up to 50% by nanosphere weight) and entrapment efficiencies (more than 95%) were achieved [17,20]. The entrapment efficiency and the nanosphere hydrodynamic mean diameter were both strongly dependent on the polymer structure. Table 1 summarizes some of the results obtained with diblock and multiblock polymers. Interestingly, the mean hydrodynamic nanosphere diameter practically does not change, whatever the drug loading. Conversely, the diameter is strongly dependent on PEG molecular weight for nanospheres prepared from both diblock PEG–PLGA and multiblock PEG_3–PLA polymers. The nanospheres' size increases with the molecular weight of PEG (see Table 1).

In PEG–PLGA polymers, for both loadings studied (20 and 33 wt%), the entrapment efficiency is constant, whatever the molecular weight of PEG. Conversely, for the multiblock PEG_3–PLA, the entrapment efficiency decreases with the molecular weight of the attached PEG blocks. A possible explanation could be a partial entanglement of PEG chains in the core caused by increased steric hindrance at the surface of nanospheres made of multiblock polymers, thereby forming a core structure with increased water affinity and lower ability to encapsulate drugs. Further studies would be necessary to establish those parameters related to polymer structure and which of these determine their ability to encapsulate drugs.

The cumulative amount of drug released from PEG–PLGA nanospheres is presented in Fig. 6A. Under identical experimental conditions, the higher the molecular weight of the coating PEG, the slower are the release kinetics. This suggests an effect of the PEG-coating layer on drug release; the PEG coating, which forms a steric barrier at the surface and prevents protein adsorption onto it, might also hinder drug release out of the core. The mean hydrodynamic diameter increased with increasing PEG molecular weight (see Table 1); the reduction of the surface area with the diameter increase would slow down release. As the diameter of the core and the thickness of the PEG coating were not evaluated, it is difficult to establish the exact influence of the PEG coating on lidocaine release.

Table 1 Lidocaine Encapsulation in a Series of PEG–PLGA and PEG$_3$–PLA Polymers; Influence of the Theoretical Loading on the Mean Diameter (Measured by QELS) and Real Loading

Polymer Structure		PLGA	PEG5–PLGA	PEG12–PLGA	PEG20–PLGA	PLA	(PEG5)$_3$–PLA	(PEG12)$_3$–PLA	(PEG20)$_3$–PLA
Theoretical loading: 20 wt%	Mean diameter (nm)	150	165	181	180	108	184	200	234
	Real loading (wt%)	18%	18%	18.4%	17.8%	20%	15%	13.6%	12.6%
Theoretical loading: 33 wt%	Mean diameter (nm)	135	170	175	181	114	200	206	238
	Real loading (wt%)	29.7%	28.1%	29.7%	25.7%	32%	25.7%	23.1%	22.3%

Figure 6B presents the cumulative amount of lidocaine released from PEG$_3$–PLA nanospheres. As in PEG–PLGA nanospheres, the drug-release kinetics are slowed down with increasing PEG molecular weight. It is difficult to compare these series of experiments, as the chemical structures of the cores were different (PLA and PLGA). Presumably, release kinetics are related to several properties of the polymer forming the core (such as water affinity, crystallinity, density, type, and strength of the interactions with the drug).

Drug-loading is also an important factor governing the release mechanism. Moreover, it has been reported (according to calorimetric and x-ray studies) that lidocaine could crystallize inside the nanosphere matrix at high loadings and that drug release was a function of nanosphere loading [17]. Surface analysis (ESCA) studies excluded the location of these crystals in the first 10 nm of the nanospheres' layers. SEM supported the fact that nanosphere samples were spherical and homogeneous, and no crystals were detected. It was thus hypothesized that nanocrystals of lidocaine were located at the nanospheres' center. Presumably, the type of drug dispersion inside the polymer matrix (homogeneous or nonhomogeneous, including nanocrystals) should also influence the release.

The results presented in the literature [17,19] gave an overview of different factors involved in drug release from PEG-coated polymeric nanospheres. It should be possible to obtain the desired drug-release patterns by appropriately choosing the diblock or multiblock copolymer composition and the drug-loading [20].

Because nanosphere degradation occurs relatively slowly compared with drug release, we might hypothesize that the polymer and the PEG coating act as a membrane barrier. Studies are underway to obtain a deeper understanding of the drug-release mechanism.

B. Encapsulation of Various Drugs into Poly(ethylene glycol)-Coated Nanospheres

Other drugs, such as prednisolone or carmustine, were entrapped in PEG-coated particles, yielding loadings and entrapment efficiencies similar to those obtained with lidocaine [19].

Carbon 14-labeled ibuprofen (IBP) was loaded in PLGA and PEG2K–PLA30K nanospheres [22]. The plasma half-life of this drug increased from a few minutes when encapsulated in nanospheres, to up to 2.5 h in the PEG–PLA nanoparticles. In the latter example, autoradiographs of IV-treated rats, 6 h after particle administration, showed that the drug-associated radioactivity was distributed all over the body. Conversely, in the PLGA nanospheres, [^{14}C]IBP was located essentially in the liver, the spleen, and the intestines [22].

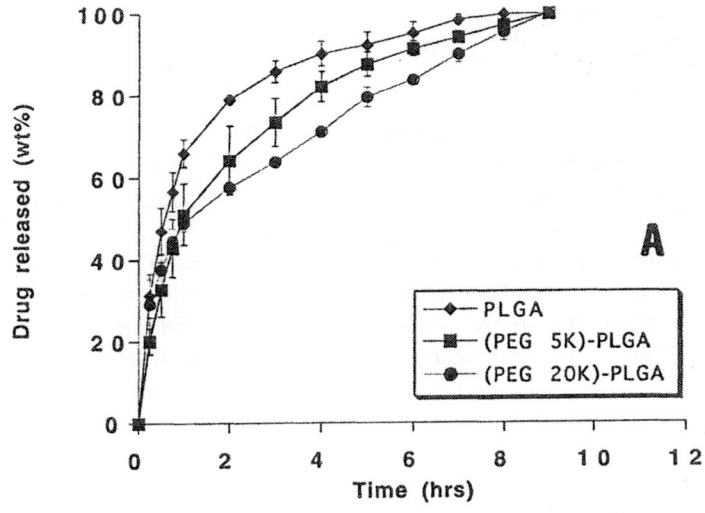

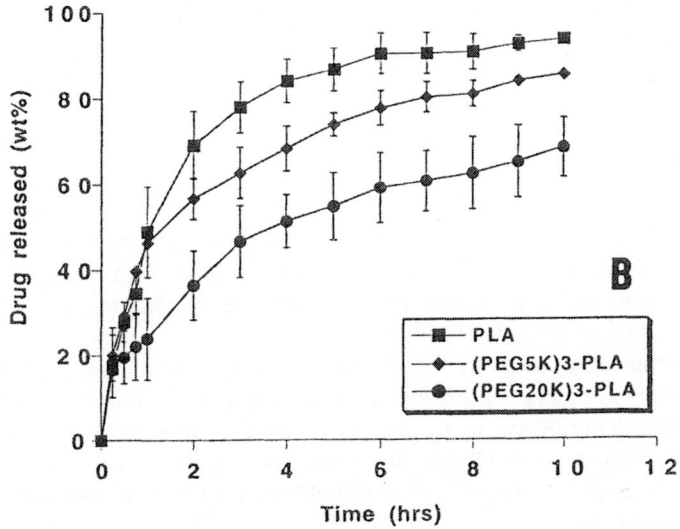

Fig. 6 Cumulative lidocaine release from nanospheres prepared from (A) PLA, PLGA, diblock PEG–PLGA and (B) multiblock PEG$_3$–PLA copolymers. The theoretical loading was in each case 33 wt%, the real loading is indicated in Table 1.

VIII. EX VIVO, IN VIVO, AND BIODISTRIBUTION STUDIES

Ex vivo studies with PEG-coated nanospheres proved their ability to prevent THP-1 macrophage cell line recognition [23]; after 2 h, PEG5K–PLA30K nanospheres were phagocytosed nine time less than noncoated PLA50K ones.

The in vivo behavior of PEG-coated nanospheres was studied [17]. For this, the particles were core-labeled with a radioactive gamma-emitting compound, indium 111 (^{111}In), which also allows study of the body distribution by using a gamma camera. The ^{111}In was chelated with pentetic acid (diethylenetri-aminepentaacetic acid) stearyl amide (DTPA-SA), and then encapsulated into the particles during the preparation procedure. The label is firmly incorporated, as proved by a practically negligible ^{111}In release in serum or buffer solutions at 37°C. A direct polymer chain labeling with ^{14}C during synthesis would be the most accurate way to study in vivo behavior. Indeed, labeling with ^{14}C does not alter polymer structure, or the resulting nanosphere structure. More-over, because of the high stability of the particles to degradation, a negligible label loss is expected after injection.

The ^{111}In-labeled PEG-coated and noncoated nanospheres were injected into BALB/c mice [17]. As expected, noncoated particles immediately accumulate in RES organs. Five minutes after injection, 66% of PLGA nanospheres accumulated in the liver. Conversely, after 5 h, the amount of 20K PEG-coated nanospheres in the liver still did not exceed 30%. This property was attributed to the presence of a protective PEG layer. Moreover, there is a correlation between the blood circulation time and the molecular weight of PEG. The higher the PEG chain length, the longer the blood persistence. An increase in blood circulation time and a reduction in liver and spleen accumulation was also found by increasing PEG surface density.

gamma-Scintigraphy studies with ^{111}In-labeled nanospheres support all these findings [17]. Pictures of mice were taken 15 min after particle injection; with PLGA nanospheres, only liver- and spleen-associated radioactivity was detected, whereas a high radioactivity in the blood pool (heart and lungs) was detected when PEG-coated particles were injected, showing their persistence in the vascular compartment.

However, the aforementioned results were obtained with particles prepared using PVA as a surfactant. It was recently found that a significant amount of this surfactant enters inside the particle, presumably by entanglement with the core chains [72] and might affect the PEG coating.

IX. CONCLUSION

Poly(ethylene glycol)-coated nanospheres are a promising new system that has potential for intravenous drug administration. The degradable core can be formed by using a variety of polymers, such as polyesters [poly(lactic acid), poly(lactic-*co*-glycolic acid), or polycaprolactone], or polyanhydride [poly(sebacic acid)] [16–23]. These nanospheres are relatively easy and inexpensive to prepare and store. They generally have a mean hydrodynamic diameter in solution between 90 and 200 nm. The PEG coating, evidenced by surface analysis, provides protection against human macrophage phagocytosis and confers on the particles a prolonged blood circulation time and reduced liver and spleen accumulation. To further improve the performance of PEG-coated nanospheres, studies are underway to design new polymers with tailored physicochemical properties. For example, multiblock PEG_n-R polymers—n PEG chains connected to one chain end of one biodegradable R block—could be used instead of diblock PEG-R polymers for the manufacture of PEG high-surface–density nanospheres [19].

Inside the biodegradable core, hydrophobic drugs (lidocaine, ibuprofen, or carmustine) were encapsulated by an emulsion–evaporation procedure. High loadings, up to 45 wt%, were obtained with lidocaine. The entrapment efficiency and the drug-release patterns strongly depended on the drug and polymers used. Drug release in vitro was controlled by choosing the appropriate polymer (polyester or polyanhydride) and the physicochemical properties of the drug or polymer (such as chemical composition, molecular weight, or crystallinity).

For each application, polymers of the PEG_n-R type could be tailored with appropriate degradation and release patterns. A variety of active substances could be efficiently administered if encapsulated into long-circulating PEG-coated nanospheres. For example, compounds with toxic side effects or very labile drugs, such as those used to fight cancer, for heart failure treatment, or for anesthesia prolongation, could be continuously released from nanospheres directly into the blood. We expect, with further study, this approach will reduce toxic side effects or will prolong the blood levels of drugs with a very short half-life. The observed tendency of nanoparticles to accumulate in inflamed tissues [93] may be used to deliver anti-inflammatory drugs (which often possess toxic side effects) or antibiotics to these sites; for example, in the treatment of rheumatic diseases.

Besides the treatment of diseases, other therapeutic applications could be envisaged, such as the encapsulation of contrast agents for magnetic resonance imaging. Magnetic particles could also be entrapped, offering a variety of medical applications for these types of nanospheres, such as cell separation. Magnetic nanospheres could eventually be directed toward a given site in the vascular compartment by using a magnetic field [94–96]. This would be a step forward toward local drug administration. With further studies, for example, after attachment of antibodies to their surface, the particles could potentially be used for active targeting to specific cells with corresponding surface epitopes in the diseased tissue [97]. Similarly, by the attachment of appropriate proteins, such as transferrin to the PEG end group on the surface of DNA-containing particles [98], nanosphere endocytosis for gene therapy could possibly be achieved.

ACKNOWLEDGMENTS

The authors would like to thank Professor R. Müller and Drs. T. Blunk and M. Lück (Freie Universität Berlin, Germany) for the 2D-PAGE studies. We are very grateful to Dr. J. M. Verbavatz (CEA Saclay, France), who carried out the freeze-fracture studies. We thank Professors A. Domb (The Hebrew University of Jerusalem, Israel), V. Torchilin (MGH East, Cambridge, MA), and N. Lothan (Technion City, Haifa, Israel) for helpful discussions.

REFERENCES

1. L. Illum, S. Davis, R. Müller, E. Mak, and P. West, The organ distribution and circulation time of intravenously injected colloidal carriers sterically stabilized with a block copolymer—Poloxamine 908, *Life Sci. 40*:367 (1987).
2. D. Bazile, C. Ropert, P. Huve, T. Verrecchia, M. Marlard, A. Frydman, M. Veillard, and G. Spenlehauer, Body distribution of fully biodegradable ^{14}C-poly (lactic acid) nanoparticles coated with albumin after parenteral administration to rats, *Biomaterials 13*:1093 (1992).
3. J. Kreuter, U. Täuber, and V. Illi, Distribution and elimination of poly(methyl-2-^{14}C-methacrylate) nanoparticles radioactivity after injection in rats and mice, *J. Pharm. Sci. 68*:1443 (1979).
4. S. Gottlleb, A. Ernst, L. Litt, K. Schwarz, and R. Meltzer, Effect of pressure on echocardiographic siderodensity from sonicated albumin: An in vitro model, *J. Am. Soc. Echocardiogr. 3*:238 (1990).
5. J. Saphiro, S. Reisner, G. Lichtenberg, and R. Meltzer, Intravenous contrast echocardiography with use of sonicated albumin in humans: Systolic disappearance of left ventricular contrast after transpulmonary transmission, *J. Am. Coll. Cardiol. 16*:1603 (1990).
6. T. Allen and D. Papahadjopoulos, Sterically stabilized ("stealth") liposomes: pharmacokinetics and therapeutic advantages, *Liposome Technology*, 2nd ed., Vol 3, (G. Gregoriadis, ed.), CRS Press, Boca Raton FL, 1993.
7. L. Illum and S. Davis, Effect of the nonionic surfactant poloxamer 338 on the fate and deposition of polystyrene microspheres following intravenous administration, *J. Pharm. Sci. 72*:1086 (1983).

8. S. Tröster and J. Kreuter, Contact angles of sufactants with a potential to alter the body distribution of colloidal drug carriers on poly(methyl methacrylate) surfaces, *Int. J. Pharm. 45*:91 (1988).
9. B. Müller and T. Kissel, Camouflage nanospheres: A new approach to bypassing phagocytic blood clearance by surface modified particulate carriers, *Pharm. Pharmacol. Lett. 3*:67 (1993).
10. P. Artursson, L. Brown, J. Dix, P. Goddard, and K. Petrak, Preparation of sterically stabilized nanoparticles by desolvation from graft copolymers, *J. Polym. Sci. [A] Polym. Chem. 28*:2651 (1990).
11. A. Klibanov, K. Maruyama, V. Torchilin, and L. Huang, Amphiphatic polyethylene glycols effectively prolong the circulation time of liposomes, *FEBS Lett. 268*:235 (1990).
12. T. Allen, C. Hansen, F. Martin, C. Redemann, and A. Yau-Young, Liposomes containing synthetic lipid derivatives of poly(ethylene glycol) show prolonged circulation half-lives in vivo, *Biochim. Biphys. Acta 1066*:29 (1991).
13. S. Dunn, A. Brindley, S. Davis, M. Davies, and E. Illum, Polystyrene–poly(ethylene glycol) (PS–PEG2000) particles as model systems for site specific drug delivery. 2. The effect of PEG surface density on the in vitro cell interaction and in vivo biodistribution, *Pharm. Res. 11*:1016 (1994).
14. T. Allen, C. Hansen, and J. Rutledge, Liposomes with prolonged circulation times: Factors affecting uptake by reticuloendothelial and other tissues, *Biochim. Biophys. Acta 981*:27 (1989).
15. B. Geho and J. Lau, US patent 4,501,728 (1983).
16. R. Gref, Y. Minamitake, M. Peracchia, V. Trubetskoy, A. Milshteyn, J. Sinkule, V. Torchilin, and R. Langer, Biodegradable PEG-coated stealth nanospheres, Proc. Int. Symp. Controlled Release Bioact. Mater., Controlled Release Soc., *20*:131, 1993.
17. R. Gref, Y. Minamitake, M. Peracchia, V. Trubetskoy, V. Torchilin, and R. Langer, Biodegradable long-circulating nanospheres, *Science 263*:1600 (1994).
18. Y. Minamitake, R. Gref, M. Peracchia, V. Trubetskoy, V. Torchilin, and R. Langer, Biodegradable injectable nanospheres composed of diblock polyethylene glycol–poly(lactic-*co*-glycolic) acid copolymers, *Macromolecules* submitted.
19. M. Peracchia, R. Gref, Y. Minamitake, N. Lotan, A. Domb, and R. Langer, PEG-coated long-circulating nanospheres prepared from amphiphilic diblock copolymers for intravenous drug delivery and targeting, *J. Controlled Release* submitted.
20. A. Domb, O. Elmalek, R. Gref, Y. Minamitake, M. Peracchia, and R. Langer, Synthesis and applications of non-linear block copolymers of PEG and poly(hydroxy acid) esters, in preparation.
21. A. Göpferich, R. Gref, Y. Minamitake, L. Shieh, M. Alonso, Y. Tabata, and R. Langer, Drug delivery from bioerodible polymers: Systemic and intravenous administration, *Formulation and Delivery of Proteins and Peptides*, (J. Cleland and R. Langer eds.), ACS Symposium Series, 567, Washington, DC, 1994.
22. T. Verrecchia, D. Bazile, Y. Archimbaud, M. Marlard, G. Spenlehauer, and M. Veillard, Compared bioavailability of IBP-5823 administered by the i.v. route in (i) stealth PLAPEG/cholate and (ii) non stealth PLAGA/albumin nanoparticles, Presented at the VIIIèmes Journées Scientifiques du GTRV, Nancy, France, 1993.

23. D. Bazile, T. Verrecchia, M. Bassoulet, M. Marlard, G. Spenlehauer, and M. Veillard, Ultradispersed polymer systems with rate and time control, *Yanuzaigaku* 53:10 (1993).

24. J. Kreuter, Evaluation of nanoparticles as drug delivery systems I. Preparation methods, *Pharm. Acta Helv.* 58:196 (1983).

25. J. Kreuter, Solid dispersion and solid solution, *Topics in Pharmaceutical Sciences* (D. Breimer and P. Speiser, eds.), Amsterdam, Elsevier, 1983.

26. L. Illum, M. Khan, E. Mak, and S. Davis, Evaluation of carrier capacity and release characteristics for poly(butyl-2-cyanoacrylate) nanoparticles, *Int. J. Pharm.* 30:17 (1986).

27. J. Kreuter, *Nanoparticles. Colloidal Drug Delivery Systems* (J. Kreuter, ed.), Marcel Dekker, New York, 1994.

28. M. Kanke, G. Simmons, D. Weiss, B. Bivins, P. DeLuca, Clearance of [141]Ce labelled microspheres from blood and distribution in specific organs following intravenous and intraarterial administration in beagle dogs, *J. Pharm. Sci.* 69:755 (1980).

29. S. Moghimi, H. Hedeman, N. Christy, L. Illum, and S. Davis, Enhanced hepatic clearance of intravenously administered sterically stabilized microspheres in zymosan-stimulated rats, *J. Leukoc. Biol.* 54:513 (1994).

30. S. Davis, Colloids as drug delivery systems, *Pharm. Technol.* 5:71 (1981).

31. G. Gregoriadis, D. Neerunjun, and R. Hunt, Fate of liposome-associated agent injected into normal and tumour-bearing rodents. Attempts to improve localization in tumour tissues, *Life Sci.* 21:357 (1977).

32. Y. Sato, H. Kiwada, and Y. Kato, Effects of dose and vehicle size on the pharmacokinetics of liposomes, *Chem. Pharm. Bull.* 34:4244 (1986).

33. A. Karino, H. Hayashi, K. Yamada, and Y. Ozawa, Effect of particle size and emulsifiers on the blood clearance and deposition of injected emulsions, *J. Pharm. Sci.* 76:273 (1987).

34. C. Van Oss, C. Gillman, and A. Neumann, *Phagocytic Engulfment and Cell Adhesiveness as Cellular Surface Phenomena*, Marcel Dekker, New York, 1975.

35. J. Kreuter, Evaluation of nanoparticles as drug delivery systems III. Materials, stability, toxicity, possibilities of targeting and use, preparation methods, *Pharm. Acta Helv.* 58:242 (1983).

36. J. Scieszka, L. Maggiora, S. Wright, and M. Cho, Role of complements C3 and C5 in the phagocytosis of liposomes by human neutrophils, *Pharm. Res.* 8:65 (1988).

37. N. Wassef, G. Matyas, and C. Alving, Complement-dependent phagocytosis of liposomes by macrophages: Suppressive effects of "stealth" lipids, *Biochem. Biophys. Res. Commun.* 176:866 (1991).

38. M. Kazatchine and M. Carreno, Activation of the complement system at the interface between blood and artificial surfaces, *Biomaterials* 9:30 (1988).

39. M. Frank and L. Fries, The role of complement in inflammation and phagocytosis, *Immunol. Today* 12:322 (1991).

40. D. Wilkins and P. Myers, Studies on the relationship between the electrophoretic properties of colloids and their blood clearance and organ distribution in the rat, *Br. J. Exp. Pathol.* 47:568 (1966).

41. L. Illum and S. Davis, Effects of the non ionic surfactant Poloxamer 338 on the fate and deposition of polystyrene microspheres following intravenous administration, *J. Pharm. Sci.* 72:1086 (1983).

42. D. Leu, B. Manthey, J. Kreuter, P. Speiser, and P. DeLuca, Distribution and elimination of coated polymethyl (2-^{14}C)methacrylate nanoparticles after intravenous administration in rats, *J. Pharm. Sci.* 73:1433 (1984).

43. K. Petrak, Design and properties of particulate carriers for intravascular administration, *Pharmaceutical Particulate Carriers* (A. Rolland, ed.), Marcel Dekker, New York, 1993.

44. Y. Tabata and Y. Ikada, Macrophage phagocytosis of biodegradable microspheres composed of L-lactic acid/glycolic acid homo- and copolymers, *Biomaterials* 9:356 (1988).

45. C. Porter, M. Davies, S. Davis, and L. Illum, Microparticulate systems for site-specific therapy—bone marrow targeting, *Site-Specific Pharmacotherapy* (A. Domb, ed.), John Wiley & Sons, London 1994.

46. R. Müller, ed., *Colloidal Carriers for Controlled Drug Delivery and Targeting*, CRC Press, Boca Raton FL, 1991.

47. L. Illum, S. Davis, R. Müller, E. Mak, and P. West, The organ distribution and circulation time of intravenously injected colloidal carriers sterically stabilized with a block copolymer—Poloxamine 908, *Life Sci.* 40:367 (1987).

48. J. Harris, Laboratory synthesis of polyethylene glycol derivatives, *J. Macromol. Sci. Rev. Macromol. Chem. Phys.* C25:325 (1985).

49. S. Jeon, J. Lee, J. Andrade, and P. De Gennes, Protein–surface interactions in the presence of polyethylene oxide. I. Simplified theory, *J. Colloid Interface Sci.* 142:149 (1991).

50. S. Jeon and J. Andrade, Protein–surface interactions in the presence of polyethylene oxide. II. Effect of protein size, *J. Colloid Interface Sci.* 142:159 (1991).

51. V. Torchilin and M. Papisov, Hypothesis: Why do polyethylene glycol-coated liposomes circulate so long? *J. Liposome Res.* 4:725 (1994).

52. A. Sawhney, C. Pathak, and J. Hubbell, Interfacial photopolymerization of poly (ethylene glycol)-based hydrogels upon alginate poly(L-lysine) microcapsules for enhanced biocompatibility, *Biomaterials* 14:1008 (1993).

53. D. Han, S. Jeong, K. Ahn, Y. Kim, and B. Min, Preparation and surface properties of POE-sulfonate grafted polyurethanes for enhanced blood compatibility, *J. Biomater. Sci. Polym. Ed.* 4:579 (1993).

54. A. Abuchowski, T. Van Es, N. Palczuk, and F. Davis, Alteration of immunological properties of bovine serum albumin by covalent attachment of polyethylene glycol, *J. Biol. Chem.* 252:3578 (1977).

55. K. Kataoka, G. Kwon, M. Yokoyama, T. Okano, and Y. Sakurai, Block copolymer micelles as vehicles for drug delivery, *J. Controlled Release* 24:119 (1993).

56. M. Woodle, M. Newman, and F. Martin, Liposome leakage and blood circulation: Comparison of adsorbed block copolymers with covalent attachment of PEG, *Int. J. Pharm.* 88:327 (1992).

57. K. Maruyama, T. Yuda, A. Okamoto, C. Ishikura, S. Kojima, and M. Iwatsuru, Effect of molecular weight in amphipathic polyethyleneglycol on prolonging the circulation time of large unilamellar liposomes, *Chem. Pharm. Bull.* 39:1620 (1991).

58. T. Allen, The use of glycolipids and hydrophilic polymers in avoiding rapid uptake of liposomes by the mononuclear phagocyte system, *Adv. Drug Deliv. Rev. 13*:285 (1994).

59. A. Kabanov, E. Batrakova, N. Melik-Nubarov, N. Fedoseev, T. Dorodnich, V. Alakhov, V. Chekhonin, I. Nazarova, and V. Kabanov, A new class of drug carriers: Micelles of poly(oxyethylene)–poly(oxypropylene) block copolymers as microcontainers for drug targeting from blood in brain, *J. Controlled Release 22*:141 (1992).

60. J. Delattre, P. Couvreur, F. Puisieux, J.-R. Philippot, and F. Schuber, *Les Liposomes. Aspects Technologiques, Biologiques et Pharmacologiques*, INSERM, Tec & Doc-Lavoisier, Paris 1993.

61. J. Kreuter, *Nanoparticles. Colloidal Drug Delivery Systems* (J. Kreuter, ed.), Marcel Dekker, New York, 1994, p. 220.

62. J. Kreuter, *Nanoparticles. Colloidal Drug Delivery Systems* (J. Kreuter, ed.) Marcel Dekker, New York, 1994, p. 314.

63. G. Spenlehauer, D. Bazile, M. Veillard, C. Prud'homme, and J. Michalon, EP 0520888A1 and EP 0520889A1 (1992).

64. D. Lewis, Controlled release of bioactive agents from lactide/glycolide polymers, in *Biodegradable Polymers as Drug Delivery Systems*, (M. Chasin and R. Langer eds.), 1990.

65. X. Zhang, U. Wyss, D. Pichora, and M. Goosen, Biodegradable polymers for orthopedic applications: Synthesis and processability of poly(L-lactide) and poly (lactide-*co*-caprolactone), *JMS Pure Appl. Chem. A30*:933 (1993).

66. H. Kricheldorf and R. Dunsing, Polylactones, 8: Mechanism of the cationic polymerization of L,L-dilactide, *Makromol. Chem. 187*:1611 (1986).

67. F. Kohn, J. Van Ommen, and J. Feijen, The mechanism of the ring-opening polymerization of lactide and glycolide, *Eur. Polym. J. 19*:1081 (1983).

68. D. Gilding and A. Reed, Biodegradable polymers for use in surgery—polyglycolic/poly(lactic acid) homo- and copolymers, *Polymer 20*:1459 (1979).

69. K. Zhu, L. Xiangzhou, and Y. Shilin, Preparation, characterization and properties of polylactide (PLA)–poly(ethylene glycol) (PEG) copolymers: A potential drug carrier, *J. Appl. Polym. Sci. 39*:1 (1990).

70. H. Kricheldorf and J. Meier-Haack, Polylactones, 22a) ABA triblock copolymers of L-lactide and poly(ethylene glycol), *Makromol. Chem. 194*:715 (1993).

71. J. Kreuter, Evaluation of nanoparticles as drug-delivery systems. I. Preparation methods, *Pharm. Acta Helv. 58*:196 (1983).

72. P. Quellec, R. Gref, and E. Dellacherie, Drug encapsulation in PEG-coated nanospheres in preparation.

73. J. Kreuter, *Nanoparticles. Colloidal Drug Delivery Systems* (J. Kreuter, ed.), Marcel Dekker, New York, 1994, p. 249.

74. R. Gref, M. T. Peracchia, J. M. Verbavatz, and R. Langer, unpublished results.

75. D. Clark, Advances in ESCA applied to polymer characterization, *J. Appl. Chem. 54*:415 (1982).

76. D. Clark and H. Thomas, Applications of ESCA to polymer chemistry. XVII. Systematic investigation of the core levels of simple homopolymers, *J. Polym. Sci. Polym. Chem. Ed. 16*:791 (1978).

77. J. Hayward, A. Durrani, Y. Lu, C. Clayton, and D. Chapman, Biomembranes as models for polymer surfaces. IV. ESCA analyses of a phosphorylcholine surface covalently bound to hydroxylated substrates, *Biomaterials* 7:252 (1986).
78. J. Gardella and D. Hercules, Static SIMS of polymer systems, *Anal. Chem.* 52:226 (1980).
79. R. Müller, ed., *Colloidal Carriers for Controlled Drug Delivery and Targeting*, CRC Press, Boca Raton FL, 1991, p. 94.
80. D. Labarre, M. Vittaz, G. Spenlehauer, D. Bazile, and M. Veillard, Complement activation by nanoparticulate carriers, Proc. Int. Symp. Controlled Release Bioact. Mater. Controlled Release Soc., 21:214, 1994.
81. H. Schreier, R. Abra, J. Kaplan, and C. Hunt, Murine plasma fibronectin depletion after intravenous injection of liposomes, *Int. J. Pharm.* 37:233 (1987).
82. M. Donbrow, ed., *Microcapsules and Nanoparticles in Medicine and Pharmacy*, CRC Press, Boca Raton, FL, 1992.
83. T. Blunk, D. Hochstrasser, J. Sanchez, B. Müller, and R. Müller, Colloidal carriers for intravenous drug targeting—determination of plasma protein adsorption patterns on surface-modified latex particles evaluated by two-dimensional polyacrylamide gel electrophoresis, *Electrophoresis* 14:1382 (1993).
84. R. Müller, T. Blunk, M. Lück, R. Gref, and R. Langer, unpublished results.
85. M. Vert, S. Li, and H. Garreau, More about the degradation of LA/GA-derived matrices in aqueous media, *J. Controlled Release* 16:15 (1991).
86. L. Pratt and C. Chu, Hydrolytic degradation of α-substituted polyglycolic acids: A semiempirical computational study, *J. Comput. Chem.* 14:809 (1993).
87. S. Shah, Y. Cha, and C. Pitt, Poly(glycolic acid-*co*-DL-lactic acid): Diffusion or degradation controlled drug delivery? *J. Controlled Release* 18:261 (1992).
88. H. Younes, P. Nataf, D. Cohn, Y. Appelbaum, G. Pizov, and G. Uretzky, Biodegradable PELA copolymers: In vitro degradation and tissue reaction, *Biomat. Artif. Cells, Artif. Organs* 16:705 (1988).
89. R. Gref, Y. Minamitake, M. Peracchia, and R. Langer, unpublished results.
90. M. Hashida, T. Yoshioka, S. Muranishi, and H. Sezaki, Dosage form characteristics of microspheres-in-oil emulsions 1: Stability and drug release, *Chem. Pharm. Bull.* 28:1009 (1980).
91. S. Benita, D. Friedman, and M. Weinstock, Pharmacological evaluation of an injectable prolonged release emulsion of physostigmine in rabbits, *J. Pharm. Pharmacol.* 38:653 (1986).
92. S. Miyazaki, N. Hashiguchi, W. Hou, C. Yokouchi, and M. Takada, Preparation and evaluation in vitro and in vivo of fibrinogen microspheres containing Adriamycin, *Chem. Pharm. Bull.* 34:3384 (1986).
93. J. Kreuter, *Nanoparticles, Colloidal Drug Delivery Systems*, (J. Kreuter, ed.) Marcel Dekker, New York, 1994, p. 315.
94. P. Gupta and C. Hung, Targeted delivery of low dose doxorubicin hydrochloride administered via magnetic albumin microspheres in rats, *J. Microencapsulation* 7:85 (1990).
95. J. Lalla and P. Ahuja, Drug targeting using non-magnetic and magnetic albumin–globulin mix microspheres of melefamic acid, *J. Microencapsulation* 8:37 (1991).
96. E. Hassan and J. Gallo, Targeting of anticancer drugs to the brain. I. Enhanced

brain delivery of oxantrazole following administration in magnetic cationic micro-spheres, *J. Drug Target.* *1*:7 (1993).

97. E. Wagner, M. Zenke, M. Cotten, H. Beug, and M. Birnstiel, Transferrin–polycation conjugates as carriers for DNA uptake into cells, *Proc. Natl. Acad. Sci. USA* *87*:3410 (1990).

98. L. Beck, V. Pope, D. Cowsar, D. Lewis, and T. Tice, *Adv. Contracep. Deliv. Syst.* *1*:79 (1986).

10

Polymerized and Microencapsulated Liposomes for Oral Delivery and Vaccination

Jun'ichi Okada

Sankyo Co., Ltd.
Shinagawa, Tokyo, Japan

I. INTRODUCTION

Liposomes have attracted much interest as drug carrier system. Advantages of the liposome system from the pharmaceutical aspect includes (a) the possibility to prepare them under mild conditions; (b) they are especially suitable for hydrophilic substances; (c) the ease of size control; and (d) the convenience for chemical modification of their surface. However, the major drawback of this system is instability, both physically and biologically. Physical instability, such as aggregation and relatively rapid leakage of entrapped substances, makes long storage difficult. Lipid exchange with cells [1], or extraction of lipids from liposomes by high-density lipoprotein [2], in vivo, is attributed to their physical instability. Biological instabilities, including quick removal of liposomes from the bloodstream by the reticuloendothelial system (RES) [3], often impair their ability as a drug-carrier system.

Improvements of physical stability are partially achieved with the addition of cholesterol into the bilayers [4], or by surface coating with polysaccharides [5,6]. Biological stability (i.e., avoiding recognition by the RES) is successfully improved with attachment of the ganglioside GM_1 [7] or poly(ethylene glycol (PEG) [3,8] on the liposomal surface.

The unique concepts for improving the physical stability of liposomes and to provide special characteristics on liposomes include polymerized liposomes and microencapsulated liposomes. In this chapter, a brief overview of these unique concepts is presented, and present research toward practical applications is introduced.

II. POLYMERIZED LIPOSOMES

A. Overview

The concepts of polymerized liposomes were first introduced by four separate research groups in 1980 [9–12]; detailed reviews have been written [13,14]. The fundamental objective or feature for improvement of polymerized liposomes is an increased physical stability over conventional (monomeric) liposomes. Introduction of covalent bonds between adjacent phospholipid molecules and formation of a polymer chain or network within a layer of the liposomal membrane can result in this improved physical stability.

The polymer backbone can run through both the hydrophilic head group and the hydrophobic hydrocarbon chain of each phospholipid molecule [15]. In addition, introduction of a polymerizable moiety at the tip of the hydrocarbon chain enables attachment across the two layers that constitute a bilayer. Liposome-forming polymerizable lipids that have been proposed are summarized by Regen [13]. Covalent bond formation between phospholipid molecules is induced by addition of a free-radical initiator, or by irradiation with an electromagnetic wave. Some of the often used free-radical initiators are azobis(isobutylonitrile), azobis(amidinopropane)hydrochloride [16], potassium sulfate [17], hydrogen peroxide [18], and others. The electromagnetic waves used for this purpose are represented by the ultraviolet [19] and gamma rays [20].

In the usual preparation procedure for polymerized liposomes, monomeric liposomes that entrap the substance of interest are first assembled from polymerizable phospholipid through a conventional method, and polymerization is carried out subsequently. In this procedure, the entrapped substances of interest are also exposed to electromagnetic waves, the radical initiator, or high temperature, which is often needed when using free-radical polymerization. Therefore, for pharmaceutical application, entrapped bioactive substances may be inactivated during the polymerization process. In addition, it is not easy to decrease the amount of radical initiator or its degradation products below an acceptable level while still keeping the liposomes intact.

To avoid these problems, a procedure with an alternative order is possible [21,22]. That is, empty liposomes without entrapment of the substance of interest are first assembled with a polymerizable phospholipid. Polymerization is carried out in the empty liposomes, and the phospholipid polymer prepared is recovered, for example, by freeze-drying, after complete washing out of the radical initiator and its degradation products, if needed. A phospholipid polymer was used to reconstitute the polymerized liposomes with the entrapped substance of interest. However, either a phospholipid polymer with very high molecular weight or with a cross-linked polymer network does not readily reconstitute polymerized liposomes, because of their low solubility in solvent [23] or their low flexibility [24]. Therefore, polymerized liposomes prepared in this procedure (polymerize first and reconstitution) also have the drawback that they are generally inferior to polymerized liposomes prepared in the former procedure (entraped first and then polymerized) in terms of physical stability.

When polymerized liposomes are administered into the body as a drug carrier, biodegradability and biocompatibility are required. Introduction of disulfide bonds as the polymer linkage was proposed for achieving biodegradability [25]. A peptide liposome was suggested by Ringsdorf et al. [26]. That is, monomeric liposomes were assembled with a lipid that has an amino acid as a head group, and addition of carbodiimide in the liposomal suspension leads to the formation of a peptide bond between the head groups of adjacent lipid molecules. Biodegradability is expected on the peptide bonds. Uptake of polymerized liposomes by macrophages in vitro was studied by Juliano et al. [27]. Mouse peritoneal macrophages took up polymerized liposomes more quickly than conventional monomeric liposomes. Their metabolic rate after uptake appeared to be similar for the polymerized and the conventional liposomes. According to Bonté et al. [28], polymerized liposomes do not seem to perturb hemolysis (i.e., polymerized liposomes had no substantial effect on platelet aggregation nor on fibrin clot formation). It is expected that polymerized liposomes can combine physical stability, biocompatibility, and biodegradability if the proper composition of polymerizable phospholipid is chosen. In addition, one can enhance the possibility of avoiding recognition by the RES by the attachment of PEG on surface.

B. Polymerized Liposomes for Oral Delivery

An interesting application of polymerized liposomes is as an oral drug carrier. Oral dosage is one of the most convenient routes for drug administration. However, many drugs, including peptides and proteins, are seldom delivered orally because of their low stability in, and poor absorption by, the gastrointestinal (GI) tract. On the other hand, lymphoid tissues, such as Peyer's patches, in the intestine are, by nature, responsible for the uptake of macromolecular antigens from the lumen and for processing them to confer immunity [29]. Macromolecules and fine particles can be taken up by the tissues [30], and this uptake is more efficient for the more hydrophobic particles [31]. Then, there is a possibility that liposomes are absorbed efficiently because they are composed of phospho"lipid," although the surface of liposomes is ionic. It is not clear whether all of the particles are processed by the immune system just after up-take, or whether some of the particles escape the process and are transferred into systemic circulation.

A preliminary examination was performed in vitro to investigate the feasibility of polymerized liposomes as an oral drug-carrier system [32]. Di(octadeca-dienoyl)phosphatidylcholine (DODPC) was the polymerizable phospholipid and azobis(isobutylonitrile) (AIBN) and azobis(amidinopropane)hydrochloride (AAPH) were the radical initiators used to prepare polymerized liposomes. Multilamellar vesicles (MLV) were formed, by Bangham's method [33], with an aqueous phase containing [^{14}C]sucrose as an aqueous marker. The AIBN was incorporated into the hydrophobic region of liposomal bilayers and AAPH was added in the aqueous phase. After freeze-thaw treatments, MLVs were kept at 60°C to carry out polymerization. The average diameter of polymerized liposomes thus prepared was 2 μm.

Laser light-scattering measurements were used to evaluate the physical stability of polymerized liposomes (Fig. 1). The addition of Triton X-100 to pre-polymerized liposomal suspension resulted in a complete decrease of scattered light intensity, indicating solubilization of the liposomes. In contrast, the intensity from a polymerized liposomal suspension was constant, irrespective of Triton concentration, showing that polymerized liposomes were stable against surface-active agents. Release profiles of the aqueous marker from polymerized liposomes in two types of release media that simulate GI fluids are shown in

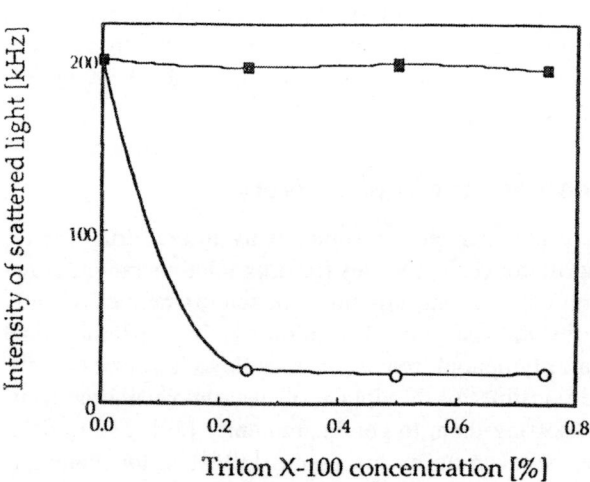

Fig. 1 Physical stability of polymerized (solid squares) and monomeric (prepolymerized) (open circles) DODPC liposomes against Triton X-100 measured with laser light-scattering intensities. (From Ref. 32.)

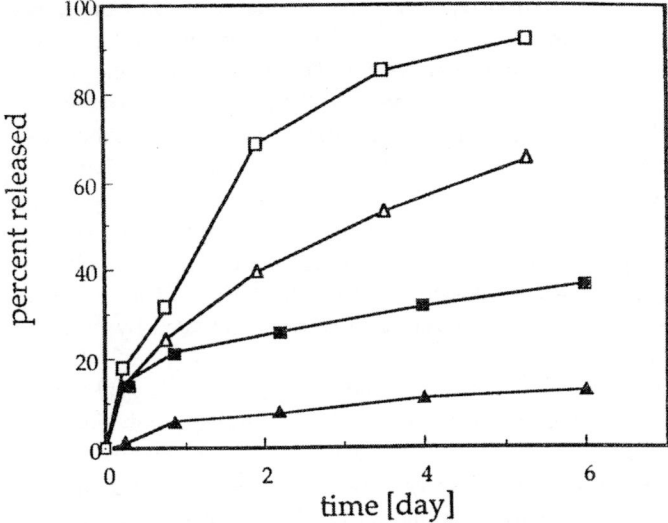

Fig. 2 *In vitro* release profiles of aqueous marker ([^{14}C]sucrose) from DODPC polymerized, or conventional (hydrogenated egg phosphatidylcholine plus cholesterol 1:1) liposomes in pH 2 isotonic saline or in pH 7.4 phosphate-buffered isotonic saline containing 10 mg/ml sodium taurocholate, 5 U/ml PLA$_2$, and 3 mM calcium chloride. (Solid squares) polymerized liposomes in pH 7.4; (solid triangles) polymerized liposomes in pH 2; (open squares) conventional liposomes in pH 7.4; (open triangles) conventional liposomes in pH 2. (From Ref. 32.)

Fig. 2. Polymerized liposomes showed slower release rates than conventional liposomes composed of hydrogenated egg phosphatidylcholine and cholesterol (1:1 mol). Conventional liposomes of this composition can be considered one of the most stable types of conventional liposomes [34,35]. Given the foregoing in vitro results, polymerized liposomes are expected to reach the intestine, keeping their liposomal structure, and retaining most of the aqueous content, when administered orally.

In vivo uptake of polymerized liposomes administered orally was observed in rats [36]. The DODPC polymerized liposomes and, as controls, monomeric liposomes (hydrogenated egg phosphatidylcholine and cholesterol 1:1) and a solution containing ^{125}I-bovine serum albumin (BSA), were gavaged into the stomach of fasted rats. Blood was sampled at appropriate intervals, and radioactivity was measured. Three routes for transfer of radioactivity from the digestive lumen into the bloodstream were considered: (a) liposomes are taken up and transferred into the bloodstream keeping ^{125}I-BSA inside (fraction 1); (b) liposomes are taken up and, subsequently, release ^{125}I-BSA: that is, ^{125}I-BSA or its degradation products are transferred into the bloodstream (fraction 2); and (c) ^{125}I-BSA is released from liposomes in the digestive lumen and is degraded to free ^{125}I or di- or tri-peptides attached to ^{125}I. These small molecules can be absorbed by the intestine [37] and transferred into the bloodstream (fraction 3). The presence of radioactive compounds of fractions 1 and 2 in the bloodstream were considered to be effective uptake of liposomes. Fraction 1, 2, and 3 in each blood sample were separated by centrifugation to pellet the liposomes, and then a trichloroacetic acid precipitation treatment, which precipitates peptides composed of more than three or four amino acid residues [38] was applied. A control experiment eliminated the possibility that free ^{125}I or ^{125}I-di- or tripeptides were adsorbed on the serum proteins and were precipitated by trichloroacetic acid treatment.

The sums of radioactivity of fractions 1 and 2 are shown in Fig. 3 for rats gavaged with polymerized liposomes, monomeric liposomes, and a solution containing the same amount of ^{125}I-BSA. If we assume that the sum of fractions 1 and 2 corresponds to the effective uptake of liposomes, polymerized liposome administration resulted in the highest uptake. Bioavailability of polymerized liposomes administered orally was 0.8% versus IV injection when compared with area under the concentration–time curve (AUC; 0–24–h).

In future studies, the exact amount of the polymerized liposomes' uptake by the intestine has to be measured using directly labeled polymerized liposomes. Studies are needed to examine the effects of changing size or surface character-

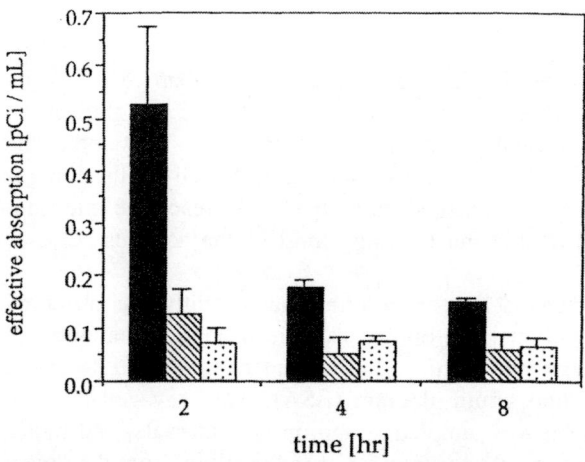

Fig. 3 Effective uptake by the intestine after oral administration of polymerized liposomes (solid square), conventional liposomes (hatched square), or solution (dotted square) containing ^{125}I-BSA. (From Ref. 36.)

istics, such as electrostatic charge and hydrophobicity, of the polymerized liposomes on their fate after oral delivery. Their fate after uptake may be changed by physical stability, which could be controlled by the extent of polymerization and surface characteristics of the polymerized liposomes.

III. MICROENCAPSULATED LIPOSOMES

A. Overview

Liposomes were combined with microcapsules to increase the physical stability of the liposomes. Microencapsulated liposomes (MELs) were first reported by Yeung and Nixon in 1988 [39]. They prepared microcapsules composed of a thin wall and an aqueous core that contained a liposomal suspension. Reverse-phase evaporation vesicles (REVs), from soy bean lecithin and cholesterol, were first prepared. Hexadiamine was dissolved in an aqueous phase of the liposomal suspension. The suspension was emulsified in a mixture of cyclohexane and chloroform (4:1). To this emulsion was added an organic base containing sebacoyl chloride, and polymerization occurred between the hexadiamine and sebacoyl chloride at the water–organic solvent interface to form the microcapsules' thin wall [40]. Organic solvents were washed away with ether that, because of its low aqueous solubility, did not degrade the liposomes. Release rates of aqueous markers were slower from MELs than from liposomes.

To obtain a free-flowing powder of MELs, gelatin or gelatin–acacia coacervate was added to the liposomal suspension [41]. This suspension was emulsified in the organic solvent and microencapsulated in the same manner as described earlier. A matrix of gelatin or gelatin–acacia coacervate inside the microcapsule allowed the resulting microcapsules to be isolated and dried.

B. Microencapsulated Liposomes with Pulsatile Release for Vaccination

One of the promising applications of MELs was as a vaccine carrier. Incorporation of vaccine antigen into liposomes that are physically stable can enhance antibody titers [42]. In addition, a vaccine carrier that will produce a pulsatile release of antigen could possibly eliminate booster shots. Microencapsulated liposomes that release their content in a pulsatile (i.e., lag-and-burst) manner were extensively studied by Langer and co-workers as shown below.

With large liposomes (2.8 $\pm$ 2 μm in diameter), absorption of phospholipase A_2 (PLA$_2$) was necessary to induce pulsatile release [43]. Reverse-phase evaporation vesicles (REVs) with an entrapped aqueous marker were incubated in PLA$_2$ solution. A centrifuged pellet of REVs was dispersed in sodium alginate solution. This suspension was sprayed into a CaCl$_2$ solution through a jet head equipped with a 25-gauge needle. On contact with the $CaCl_2$ solution, gelation occurred within the droplet of sodium alginate solution that contained the liposomes. Microcapsules thus prepared were immersed in poly(L-lysine) (PLL) solution to form a coating of the polyion complex between the alginate and PLL. An in vitro release study was performed in isotonic HEPES-buffered saline, containing 3 mM $CaCl_2$ and 0.02% gentamicin sulfate at pH 7.4, with gentle agitation at 37°C. Pulsatile release profiles of the entrapped aqueous marker were observed only when liposomes were adsorbed by PLA_2 before microencapsulation (Fig. 4). The aqueous marker that was entrapped directly in the microcapsule matrix (Ca alginate) was released quickly, suggesting that the rate-limiting step is diffusion through the liposomal membrane. Liposomes that were adsorbed by PLA_2, but not microencapsulated, showed the usual sustained release. Collectively, the mechanism of pulsatile release in this instance may be as follows; (a) PLA_2 adsorbs on liposomes and enzymatic degradation of liposome-forming phospholipids proceeds gradually, (b) liberation of PLA_2 from the liposomal surface is retarded by alginate matrix, and (c) the liposomal membrane degrades when the amount of lysolecithin reaches a level at which the bilayer structure can no longer be maintained. Those MELs prepared from a sodium alginate solution with a higher concentration showed faster release rates and a larger amount of released material during a particular period [44]; this may be related to a higher barrier ability for the alginate matrix

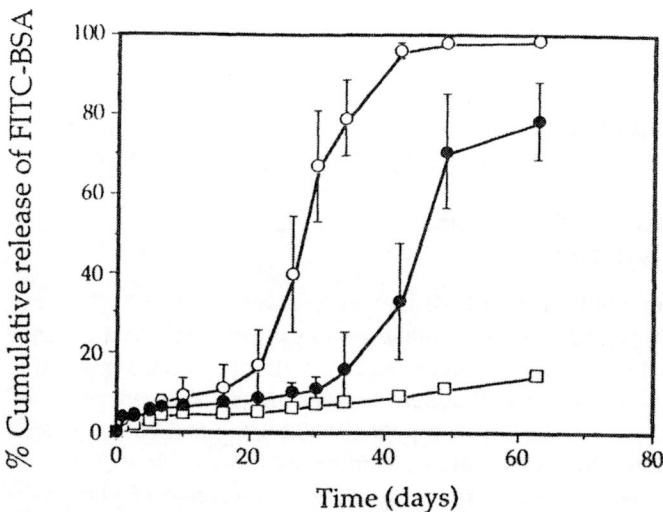

Fig. 4 In vitro-release profiles of aqueous marker (FITC-BSA) from MELs containing liposomes that were treated with different concentration of PLA_2. (open circle) 10 U PLA_2; (solid circle) 1 U PLA_2; (open square) not treated by PLA_2. (From Ref. 43.)

against PLA$_2$ liberation. By using a phospholipid that is a better substrate for PLA$_2$, a shorter lag time for the onset of pulsatile release results. Lower molecular weight and a longer coating time of PLL increased the lag time. The longest lag time reported was up to 30 days at 37°C in vitro [43].

Liposomes were extruded through a 0.6-μm–pore polycarbonate membrane to reduce their size, before their microencapsulation in the Ca-alginate matrix by the same procedure as that mentioned earlier [45]. In this experiment, pulsatile release profiles were observed even without PLA$_2$. The origin of this pulsatile release was related to the interaction of free alginate polyions with the phospholipid bilayers. Free alginate polyions before contact with Ca^{+2} ions in the microencapsulation process were inserted into phospholipid bilayers. This insertion caused destabilization of the bilayers, and an initial burst of release was induced (Fig. 5). During a release study, Ca^{+2} ions were gradually leached out from the Ca alginate matrix. When the amount of Ca^{+2} ion in the matrix decreased to a level at which the matrix structure could no longer be held, free alginate ions were increasingly liberated from the loosened matrix. They were inserted into phospholipid bilayers, and, as a result, the second burst was in-

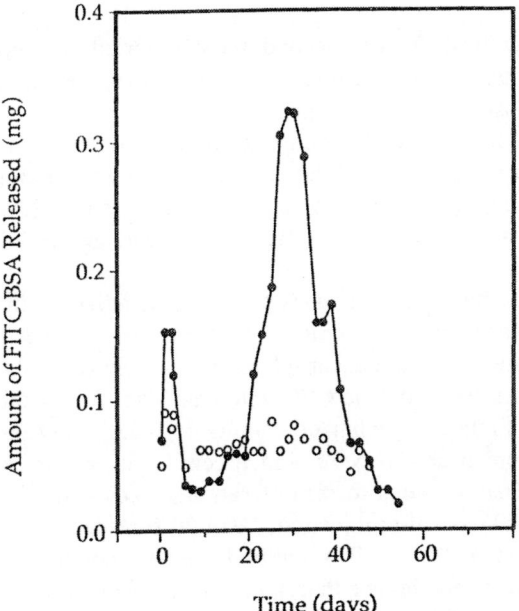

Fig. 5 In vitro release profiles of aqueous marker (FITC-BSA) from MELs (solid circle) or liposomes (open circle) composed of egg PC plus cholesterol (1:1 molar). (From Ref. 45.)

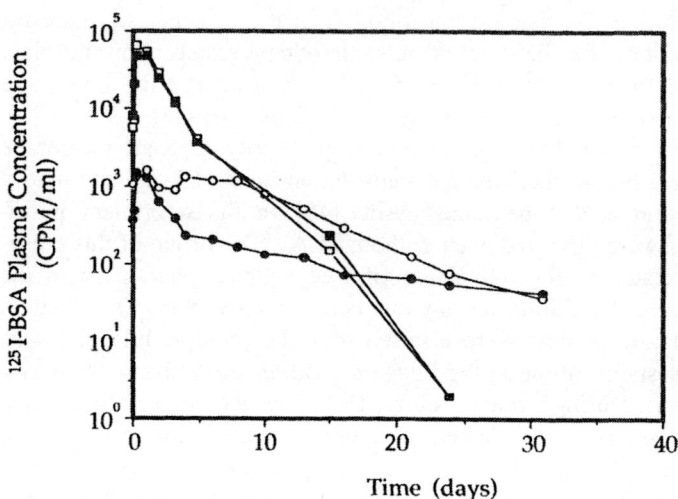

Fig. 6 Pharmacokinetics of ^{125}I-BSA appearance in rat plasma after SC injection of the protein in saline (open squares) or Freund adjuvant (solid squares), or encapsulated in liposomes (open circles) or MELs (solid circles). (From Ref. 46.)

duced (Fig. 5). The longest lag time for the second burst reported was up to 120 days at 37°C in vitro when a phospholipid with high-phase–transition temperature was used in the liposomes. When larger liposomes (2.8 $\pm$ 2 μm in diameter) were microencapsulated, pulsatile release was not observed without PLA$_2$ treatment. The difference may be attributed to the difficulty for a free alginate ion to be inserted between phospholipid molecules in bilayers with low curvature (i.e., in large liposomes), because the packing of head groups of phospholipid molecules is tighter.

The pharmacokinetics of, and humoral responses to, antigen delivered by MELs were studied [46]. Liposomes containing ^{125}I-BSA were extruded through a 0.6-μm–pore membrane and microencapsulated in Ca alginate without PLA$_2$ adsorption. The MELs were injected SC into rats. As a control, liposomes, saline, or complete Freund adjuvant containing ^{125}I-BSA was injected. Radioactivity derived from trichloroacetic acid precipitable protein in plasma was detected for more than 4 weeks in sera of rats injected with ^{125}I-BSA in MELs (Fig. 6). Fifty or 80 days after injection, 50 or 30% respectively, of the initial dose was recovered in MELs. The antibody levels induced by BSA in MELs were three to four times higher than those induced by BSA in liposomes and saline, and 1.3 times higher than those induced by BSA in complete Freund adjuvant at their maximum level (Fig. 7). In addition, antibody titers remained for a longer period in rats that were injected with MELs than in

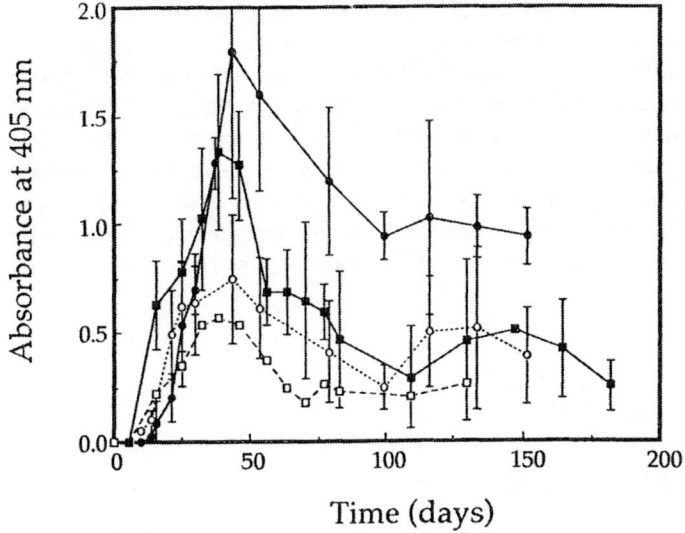

Fig. 7 Humoral response to BSA of rats immunized with the antigen in saline (open squares); Freund's adjuvant (solid squares); liposomes (open circles); or MELs (solid circles). Ordinate shows absorbance in ELISA of antibody. (From Ref. 46.)

those that were injected BSA in complete Freund's adjuvant. There was little histological change in the MELs injection sites and surrounding tissues. These results show microencapsulated liposomes (MELs) are not only an effective, but a safe system that could be applied to humans for vaccination.

IV. SUMMARY

Two approaches to increase the physical stability of liposomes are introduced.

Polymerized liposomes were prepared with introduction of covalent bonds between the phospholipid molecules that constitute a liposome. While keeping the original features of conventional liposomes, polymerized liposomes showed higher stability against solubilization by surface-active agents and slower release rates of entrapped substances than those of conventional liposomes. Development of biodegradable polymerized liposomes can lead to many possibilities for practical application, including a carrier system for oral delivery of vaccine antigen or peptidic drug substances. The key to development of a polymerized liposomal system is to achieve an appropriate polymerization process that does not impair the biological activity of the substance of interest, while keeping an acceptable level of trap efficiency.

Liposomes were microencapsulated to improve their physical stability. Microencapsulation can be performed with synthetic polymers as well as with natural hydrogels. Microencapsulated liposomes (MELs) with calcium alginate hydrogel as a capsule matrix showed a pulsatile release pattern of entrapped substances in the liposomes. This is an interesting characteristic that is useful, for example, in implantable vaccine-release systems, because it can eliminate the booster shots. The preparation process of a Ca alginate hydrogel microcapsule is simple, so that scale-up to a mass-production dose does not seem to be difficult. However, to impart sterility to MEL products, a sterilized manufacturing facility may be essential.

REFERENCES

1. R. Bittman and S. Clejans, Kinetics of cholesterol and phospholipid exchange between mycoplasma membranes and lipid vesicles, *Isr. J. Med. Sci. 23*:398 (1987).
2. P. Palatini, Disposition kinetics of phospholipid liposomes, *Adv. Exp. Med. Biol. 318*:374 (1992).
3. D. Papahadjopoulos, T. M. Allen, A. Gabizon, E. Mayhew, K. Matthay, S. K. Huang, K. D. Lee, M. C. Woodle, D. D. Lasic, C. Redemann, and F. J. Martin, Sterically stabilized liposomes: Improvements in pharmacokinetics and antitumor therapeutic efficacy, *Proc. Natl. Acad. Sci. USA 88*:11460 (1991).
4. K. Inoue, Permeability properties of liposomes prepared from dipalmitoyllecithin, dimyristoyllecithin, egg lecithin, rat liver lecithin and beef brain sphingomyelin, *Biochim. Biophys. Acta 339*:390 (1974).
5. J. Sunamoto, T. Sato, M. Hirota, K. Fukushima, K. Hiratani, and K. Hara, A newly developed immunoliposome: an egg phosphatidylcholine liposome coated with pullulan bearing both a cholesterol moiety and an IgM fragment, *Biochim. Biophys. Acta 898*:323 (1987).
6. Y. Atsuta, N. Muramatsu, and T. Kondo, Phagocytosis by guinea pig polymorphonuclear leukocytes of liposomes stabilized with polysaccharides, *Biomater. Artif. Cells Artif. Organs 17*:125 (1989).
7. N. M. Wassef, G. R. Matyas, and C. R. Alving, Complement-dependent phagocytosis of liposomes by macrophages: suppressive effects of "stealth" lipid, *Biochem. Biophys. Res. Commun. 176*:866 (1991).
8. A. L. Klivanov, K. Maruyama, A. M. Beckerleg, V. P. Torchillin, and L. Huang, Activity of amphiphatic poly(ethylene glycol) 5000 to prolong the circulation time of liposomes depends on the liposome size and its being unfavorable for immunoliposome binding to target, *Biochim. Biophys. Acta 1062*:142 (1991).
9. S. L. Regen, B. Czech, and A. Singh, Polymerized vesicles, *J. Am. Chem. Soc. 102*:6638 (1980).
10. H. H. Hub, B. Hupfer, H. Koch, and H. Ringsdorf, Polymerizable phospholipid analogues—new stable biomembrane and cell models, *Angew. Chem. Int. Ed. Engl. 19*:938 (1980).

11. D. S. Johnston, S. Sanghera, M. Pons, and D. Chapman, Phospholipid polymers—synthesis and spectral characteristics, *Biochim. Biophys. Acta 602*:57 (1980).

12. D. F. O'Brien, T. H. Whitesides, and R. T. Klingbiel, The photopolymerization of lipid-diacetylenes in bimolecular-layer membranes, J. Polym. Sci. Polym. Lett. Ed. *19*:95 (1981).

13. S. L. Regen, Polymerized liposomes, *Liposomes* (M. J. Ostro, ed.), Marcel Dekker, New York, 1987, p. 73.

14. D. A. Tirell, Polymer chemistry and liposome technology, *ACS Symp. Ser. 362*:152 (1988).

15. A. Akimoto, K. Dorn, L. Gros, H. Ringsdorf, and H. Schupp, Polymer model membranes, *Angew. Chem. Int. Ed. Engl. 20*:90 (1981).

16. H. Ohno, S. Takeoka, and E. Tsuchida, Unequivalent chemical environment of diene groups in 1- and 2-acyl chains of polymerizable lipids analyzed by radical polymerization, *Bull. Chem. Soc. Jpn. 60*:2945 (1987).

17. J. H. Fendler, Polymerized surfactant aggregates: Characterization and utilization, *Acc. Chem. Res. 17*:3 (1984).

18. S. L. Regen, K. Yamaguchi, N. K. P. Samuel, and M. Singh, Polymerized–depolymerized vesicles. A reversible phosphatidylcholine-based membrane, *J. Am. Chem. Soc. 105*:6354 (1983).

19. H. Kitano, N. Kato, and N. Ise, Mutual recognition between polymerized liposomes. III. Association processes between avidin and biotin on polymerized liposome surfaces, *Biotechnol. Appl. Biochem. 14*:192 (1991).

20. H. Ohno, Y. Ogata, and E. Tsuchida, gamma-Ray polymerization of phospholipid having diene or triene groups as liposomes, *J. Polym. Sci. Polym. Chem. Ed. 24*:2959 (1986).

21. S. Takeoka, H. Ohno, H. Iwai, and E. Tsuchida, Liposome formation of selectively-polymerized diene-containing phospholipids and their postpolymerization, *J. Polym. Sci. Polym. Chem. Ed. 28*:717 (1990).

22. T. Kunitake, N. Nakashima, K. Takarabe, M. Nagai, A. Tsuge, and H. Yanagi, Vesicles of polymeric bilayer and monolayer membranes, *J. Am. Chem. Soc. 103*:5945 (1981).

23. H. Ohno, Y. Ogata., and E. Tsuchida, Polymerization of liposomes composed of diene-containing lipids by UV and radical initiators: Evidence for the different chemical environment of diene groups on 1- and 2-acyl chain. *Macromolecules 20*:929 (1987).

24. R. Elbert, A. Laschewsky, and H. Ringsdorf, Hydrophilic spacer groups in polymerizable lipids: Formation of biomembrane models from bulk polymerized lipids, *J. Am. Chem. Soc. 107*:4143 (1985).

25. S. L. Regen, Polymerized phosphatidylcholine vesicles as drug carriers, *Ann. N.Y. Acad. Sci. 446*:296 (1985).

26. N. Nuemann and H. Ringsdorf, Peptide liposomes from amphiphilic amino acids, *J. Am. Chem. Soc. 108*:487 (1986).

27. R. L. Juliano, M. J. Hsu, and S. L. Regen, Interactions of polymerized vesicles with cells. Uptake, processing and toxicity in macrophages, *Biochim. Biophys. Acta 812*:42 (1985).

28. F. Bonté, M. J. Hsu, A. Papp, K. Wu, S. L. Regen, and R. L. Juliano, Interactions of polymerizable phosphatidylcholine vesicles with blood components: Relevance to biocompatibility, *Biochim. Biophys. Acta 900*:1 (1987).

29. S. B. Hanauer and S. C. Kraft, Immunology in the intestine, *Gastroenterology* (J. E. Berk, ed.), W. B. Saunders, Philadelphia, 1985, p. 1611.

30. J. P. Ebel, A method for quantifying particle absorption from the small intestine of the mouse, *Pharm. Res. 7*:848 (1990).

31. J. H. Eldridge, C. J. Hammond, J. A. Meulbroek, J. K. Staas, R. M. Gilley, and T. R. Tice, Controlled vaccine release in the gut-associated lymphoid tissues. 1. Orally administered biodegradable microspheres target the Peyer's patches, *J. Controlled Release 11*:205 (1990).

32. J. Okada, S. Cohen, and R. Langer, In vitro evaluation of polymerized liposomes as an oral drug delivery system, *Pharm. Res. 12*: 576 (1995).

33. A. D. Bangham, M. W. Hill, and N. G. A. Miller, *Methods in Membrane Biology* (E. D. Korn, ed.), Plenum Press, New York, 1974, p. 1.

34. T. Yotsuyanagi, H. Hashimoto, M. Iwata, and K. Ikeda, Effect of cholesterol on liposome stability to ultrasonic disintegration and sodium cholate solubilization, *Chem. Pharm. Bull. 35*:1228 (1987).

35. B. P. Gaber and J. P. Sheridan, Kinetic and thermodynamic studies of the fusion of small unilamellar phospholipid vesicles, *Biochim. Biophys. Acta 685*:87 (1982).

36. J. Okada, S. Cohen, and R. Langer, in preparation.

37. K. E. Webb, Jr, Intestinal absorption of protein hydrolysis products: A review, *J. Anim. Sci. 68*:3011 (1990).

38. N. A. Greenberg and W. F. Shipe, Comparison of the abilities of trichloroacetic acid, picric, sulfosalicylic, and tungstic acids to precipitate protein hydrolysates and proteins, *J. Food Sci. 44*:735 (1979).

39. V. W. Yeung and J. R. Nixon, Preparation of microencapsulated liposomes, *J. Microencapsulation 5*:331 (1988).

40. T. M. S. Chang, Semipermeable microcapsules, *Science 146*:524 (1964).

41. J. R. Nixon and V. W. Yeung, Preparation of microencapsulated liposomes, II. Systems containing nylon-gelatin and nylon-gelatin-acacia walling material, *J. Microencapsulation 6*:43 (1989).

42. G. Gregoriadis, Immunological adjuvants: A role for liposomes, *Immunol. Today 11*:89 (1990).

43. P. G. Kibat, Y. Igari, M. A. Wheatley, H. N. Eisen, and R. Langer, Enzymatically activated microencapsulated liposomes can provide pulsatile drug release, *FASEB J. 4*:2533 (1990).

44. Y Igari, P. G. Kibat, and R. Langer, Optimization of a microencapsulated liposome system for enzymatically controlled release of macromolecules, *J. Controlled Release 14*:263 (1990).

45. S. Cohen, M. C. Bañó, M. Chow, and R. Langer, Lipid–alginate interactions render changes in phospholipid bilayer permeability, *Biochim. Biophys. Acta 1063*:95 (1991).

46. S. Cohen, H. Bernstein, C. Hewes, M. Chow, and R. Langer, The pharmacokinetics of, and humoral responses to, antigen delivered by microencapsulated liposomes, *Proc. Natl. Acad. Sci. USA 88*:10440 (1991).

11

Pharmacokinetics of Microparticulate Systems

Pravin R. Chaturvedi

Vertex Pharmaceuticals Incorporated
Cambridge, Massachusetts

I. INTRODUCTION

Most drugs produce a pharmacological response as a result of a concentration-dependent, reversible interaction at specific receptor sites, leading to an alteration of some aspect of the cell structure or function. Therefore, it is imperative that a correct amount of drug should be absorbed and transported to the active site, and the concentration should be maintained for the duration deemed necessary to produce a therapeutic effect. In the design of drug-delivery systems, the knowledge of the amount of drug to be administered, the rate of delivery, the route or site of administration, and the tissue distribution of the drug is essential. For optimum drug delivery, microparticulate systems should be designed on the basis of the desired time course of drug response. The establishment of a correlation between the in vivo absorption rate with an in vitro parameter (e.g., dissolution rate) would provide a basis for the development of microsphere systems designed to optimize the drug delivery rate following administration. Furthermore, it would also serve as a quality control tool for the optimization of formulation parameters during drug development. The focus of this chapter is to evaluate the various methods available that allow an estimation of the fraction of the dose absorbed following extravascular administration for the assessment of in vivo absorption kinetics. An attempt will be made to describe some of the mathematical bases for these techniques, but the emphasis of this chapter is on the evaluation of various techniques to determine the fraction of the dose absorbed. The correlation between the fraction of the dose absorbed in vivo and, for instance, the fraction of the dose released in vitro, allows establishment of some meaningful in vitro–in vivo correlation techniques to evaluate the pharmacokinetics of microsphere systems.

II. BACKGROUND

The availability of sophisticated pharmaceutical microparticulate systems provide a wide choice of drug administration and absorption sites, rates of administration, and temporal profiles of drug absorption. Following administration of the microparticulate system, the drug is initially released into the local environment, and this is followed by absorption of the drug into the systemic circulation, which is detected by the measurement of the blood or plasma concentration of the drug. The time course of the concentration of a drug at the site of action is governed by a variety of processes that can be classified as either processes of absorption or processes of disposition. The amount of drug that reaches the active site is partly determined by the extent of absorption and partly by the rate of absorption of the drug. The extent of absorption, or bioavailability, refers to the fraction of the total dose administered that reaches the systemic circulation following extravascular administration, and is an index of the systemic exposure of the drug. Disposition, which is inclusive of distribution, metabolism, and excretion processes, also plays a role in determining the fraction of drug reaching the active site. However, disposition processes are not amenable to modifications for optimization of drug delivery to the active site. Hence, the only means for controlling the drug delivery to its site of action is by modification of the absorptive processes. The route of administration and the choice of the formulation usually provide the means for controlling absorptive processes for drug-delivery systems.

Also, interestingly, variations in the pharmacological activity of a drug result from variations in the absorptive processes. The role of absorption in the overall activity of the drug and the procedures to control absorption require methods to quantify the absorptive processes. Direct quantification of absorption is not feasible in vivo, and in vitro and in situ preparations are restricted to experimental determinations. Following the administration of a microparticulate system in vivo, evidence of absorption is obtained by the appearance of the drug in the systemic circulation. Mathematical analysis of the measured blood or plasma concentrations is required to characterize the absorptive processes. However, the measured concentrations are a result of both the absorption and disposition processes. Hence, *absorption* can be defined in terms of input measured in terms of either the release of the drug from the microparticulate system, or as a combined process of release and systemic absorption. The release process is usually the rate-limiting step in this process. In fact, microparticulate systems are most effective when the drug release from the particles is the rate-limiting step in making the drug available. Hence, the rate of "absorption" estimated from the microparticulate systems represents the in vivo rate of release from this system, rather than a true rate of absorption of the drug from an immediate-release dosage form. For the purpose of this chapter, the *rate of absorption* referred to henceforth, will be synonymous with the rate of release of the drug from the microparticulate system.

Novel drug delivery systems invariably modify the release rate of the drug from the microparticulate system to allow a better pharmacokinetic profile of the drug. The regulatory assessment of these microparticulate systems requires demonstration of modified release, where possible by both in vitro and *in vivo* methods. It is recommended that one should establish suitable in vitro dissolution tests to define the product specifications. Furthermore, the suitability of the in vitro test should be determined by its relation to the actual in vivo absorption characteristics of the drug. The estimated in vivo input rate is then used to establish a correlation with the in vitro release rate of the drug from the microsphere systems. The latter release rate can subsequently be used to predict the drug concentration–time profile in vivo following modifications in the microparticulate system. Thus, in addition to the safety and efficacy issues, the regulatory approval of a novel microparticulate system requires the demonstration of some biopharmaceutical and pharmacokinetic properties of that microparticulate system. These include the reproducibility of release from the new drug delivery system both in vivo and in vitro, a defined bioavailability profile that excludes the possibility of dose dumping, and the demonstration of good absorption relative to an appropriate standard dosage form.

III. ASSESSMENT OF DRUG INPUT RATE

A measurement of the extent of absorption from a pharmaceutical microparticulate system provides useful, but incomplete, information of the absorption process. Additional information on the rate of absorption (i.e., rate of delivery to the systemic circulation) is needed to obtain a better understanding of the drug input. Some approaches do not evaluate the complete time course of the absorption, and single parameters such as maximal plasma concentration (C_{max}) or time to reach the maximum plasma concentration (t_{max}) are used to assess the rate of absorption. More information is obtained from the statistical moments of the absorption process. This approach allows the calculation of the mean input (absorption or release) times. These mean input times are associated with the errors in the estimations of the mean residence times, and do not provide a full understanding of the absorption process, since they are not associated with the extent of absorption. An understanding of the mechanism of the absorption process requires the evaluation of the time dependence of the absorption rate as a mathematical function or in a numerical form. Furthermore, evaluation of different dose levels would reveal the dose-dependence of the absorption kinetics. Thus, more informative than the single-parameter approaches are the methods that characterize the full time course of the input process. These approaches can be divided into curve-fitting and "k_a methods," mass balance methods, and formal deconvolution methods. All these methods require that the drug disposition should be time- and concentration-independent (i.e., follow linear pharmacokinetics).

A. Simple Measures of Absorption Rate: $C_{\max}$ and $t_{\max}$

Following administration of a microparticulate system, the drug concentration in the blood or plasma increases with time and reaches a maximum. The magnitude of the maximum concentration $C_{\max}$, and the time at which the maximum concentration occurs $t_{\max}$, provide an indication of the absorption rate. As a general rule, the quicker the $t_{\max}$ and the higher the $C_{\max}$, the faster the absorption rate of the drug. However, these parameters do not provide information exclusively about the absorption process because both parameters are dependent on the disposition processes of the drug, and the interpretations about the rate of absorption do not generalize to all situations.

1. Determination of Absorption Rates by the Statistical Moments Approach

The statistical moments theory considers the absorption of a drug from a microparticulate system as a stochastic process. Following the administration of a microparticulate system, it is not possible to predict at what time an individual molecule will be absorbed from the absorption site. However, when many molecules are presented to the absorption site, they behave collectively in a predictable manner. This predictability of the population, with the unpredictability of the individual molecule, is the essential feature of the stochastic process. This can be mathematically expressed using the probability density function. If $p(t)$ represents the probability density function for absorption, then $p(t)dt$ represents the probability that a molecule introduced at time zero is absorbed in the time interval $(t, t + dt)$. If n is the total number of molecules absorbed, then by the frequency interpretation of probability, $np(t)dt$ is the number of molecules absorbed in the time interval $(t, t + dt)$. Usual pharmacokinetic notation represents $n = FD$, where D is the administered dose, and F is the fraction of the dose absorbed. In terms of the rate of absorption, $R_{abs}(t)$, the amount absorbed in the time interval $(t, t + dt)$ is $R_{abs}(t)dt$. Thus,

$$p(t) = \frac{R_{abs}(t)}{FD} \tag{1}$$

Thus, the rate of absorption corrected for the fraction of the dose absorbed represents the probability density function for the absorption of a single molecule. We can define the mean residence time (MRT) of a molecule at the absorption site as

$$\text{MRT}_{\text{abs}} = \left(\frac{1}{FD}\right) \int_0^\infty t R_{\text{abs}}(t) dt \tag{2}$$

The variance of the residence time (VRT) of a molecule at the absorption site may be defined as

$$\text{VRT}_{\text{abs}} = \left(\frac{1}{FD}\right) \int_0^\infty (t - MRT_{\text{abs}})^2 R_{\text{abs}}(t) dt \tag{3}$$

Experimentally, MRT and VRT can be calculated from the observed plasma concentration–time profile of the drug. Assuming linear kinetics, if $C\partial$ is the plasma concentration obtained following unit impulse input into the general circulation (intravenous bolus), and C is the plasma concentration obtained following administration of the microparticulate system, then

$$\text{MRT}_c = \text{MRT}_{c\partial} + \text{MRT}_{\text{abs}} \tag{4}$$

$$\text{VRT}_c = \text{VRT}_{c\partial} + \text{VRT}_{\text{abs}} \tag{5}$$

where

$$\text{MRT}_c = \frac{\int_0^\infty t C(t) dt}{\int_0^\infty C(t) dt} \tag{6}$$

and

$$\text{VRT}_c = \frac{\int_0^\infty (t - \text{MRT}_c)^2 C(t) dt}{\int_0^\infty C(t) dt} \tag{7}$$

Corresponding equations for $\text{MRT}_{c\partial}$ and $\text{VRT}_{c\partial}$ are obtained by replacing C by $C\partial$ in the foregoing equations. These equations show that the additive property of statistical moments allows one to characterize the absorption process as the mean residence time of absorption from

$$\text{MRT}_{\text{abs}} = \text{MRT}_c - \text{MRT}_{c\partial} \tag{8}$$

VRT_{abs} is similarly obtained from

$$\text{VRT}_{\text{abs}} = \text{VRT}_c - \text{VRT}_{c\partial} \tag{9}$$

The MRT_{abs} and VRT_{abs} provide an immediate indication of the absorption rate. A rapid absorption is associated with a smaller value of MRT_{abs}. Furthermore, a small value of VRT_{abs} indicates that most of the absorbed drug is absorbed over a short time period. If first-order absorption is assumed, then the statistical moment parameters allow the estimation of the first-order absorption rate constant as

$$MRT_{abs} = \frac{1}{k_a} \tag{10}$$

and

$$VRT_{abs} = \frac{1}{k^2_a} \tag{11}$$

The relation that $VRT_{abs} = MRT^2_{abs}$ further provides some indication of whether first-order kinetics were followed. A significant departure from this relation suggests that the absorption kinetics may not be first order. Statistical moments also allow the testing of a novel drug-delivery system with a reference dosage form without the need for an intravenous pharmacokinetic study.

The advantages of statistical moment parameters over the simple measures such as t_{max} and C_{max} stems from the independence of the mean and variance of residence times from the disposition of the drug. The primary disadvantage in the use of statistical moments for evaluating absorption processes results from the estimation of the area under the curve (AUC) beyond the last measured data point. Large errors are introduced in the AUC estimate on extrapolation, and integration of the functions to estimate AUC results in even larger errors.

2. Assessment of Absorption Rate by Curve-Fitting Methods

Curve-fitting methods assume an analytical function to describe both the input and disposition processes. The absorption rate is constructed by identifying, on the basis of certain assumptions, the parameters in the fitted function, which refer to the absorption process.

For example, use of the k_a method in a one-compartment model with first-order absorption from the site of administration, gives the plasma concentration, $C(t)$, as

$$C(t) = \left[\frac{(k_a FD)}{(k_a - k)V)} \right] (e^{-kt} - e^{-k_a t}) \tag{12}$$

where k_a is the rate of absorption, k is the rate of elimination, F is the fraction of the dose absorbed, D is the dose administered, and V is the volume of distribution of the drug. Fitting experimental data on the assumption that it follows the foregoing k_a method results in an expression of the form

$$C(t) = B_1 e^{-b_1 t} + B_2 e^{-b_2 t} \tag{13}$$

where B_1, B_2, b_1, and b_2 are constants. One assumes that if $b_1 > b_2$, then absorption is rapid (i.e., $k_a > k$), yielding $k_a = b_1$. Conversely, with the assumption that $k_a < k$, one arrives at a conclusion that $k_a = b_2$. An obvious pitfall with the k_a method is the identification of k_a with either b_1 or b_2. The identification of k_a also arises with drugs that follow multicompartment kinetics. The possibility of misidentification of k_a with any exponent increases with increasing number of compartments required to describe the plasma concentration–time data. A further problem is that of "vanishing" exponentials [33]. Mathematically pharmacokinetic models require the same number of exponential terms in describing the drug concentration–time data following intravenous and extravascular administration. However, it is often likely that one or two of the exponential terms are relatively unimportant and appear to vanish during the fitting of these data. These "vanishing" exponentials lead to ambiguity over which pharmacokinetic model would be assigned to a given data set.

Other curve-fitting methods determine the in vivo input rate from an extrapolation of the analytical functions used to describe the in vitro dissolution data, based either on theoretical physicochemical properties (e.g., cube root law) or the application of empirical gamma or Weibull functions. The modeling approach based upon the convolution theory and the use of Weibull functions to describe drug input rates have been recently adapted to describe complex absorption kinetics [5]. From the assumptions of convolution theory, the plasma concentration is obtained as

$$C(t) = D(t) * I(t) \tag{14}$$

where $C(t)$ is the plasma concentration, $D(t)$ is the time function that describes the drug concentrations following the administration of a unit impulse intravenous dose, and $I(t)$ refers to the input rate, and $*$ refers to the convolution operation. The cumulative amount of drug, $A(t)$, entering the body can be calculated by the integration of the input rate, $I(t)$, so that the foregoing equation may be written as

$$C(t) = \frac{D(t) * dA(t)}{dt} \tag{15}$$

The kinetic processes of absorption and disposition are described in terms of concentrations as

$$D(t) = F \, A e^{-kt} \tag{16}$$

where F is the extent of absorption of the drug, and A and k are the coefficient and exponent of the elimination function. The input rate is estimated using a single or double Weibull function (depending on the complexity of absorption) as

$$I(t) = [l - e^{-(t/td1)\gamma_1}] \tag{17}$$

or

$$I(t) = [1 - e^{-(t/td1)\gamma_1}] + [1 - e^{-(t/td2)\gamma_2}] \tag{18}$$

where td is the time necessary to transfer 63% of the administered drug from the input site to the systemic circulation. The exponent γ is a unitless number ranging from zero (when the system simulates a zero-order input) to 1 (when it simulates a first-order input) or values greater than 1 (when the curve takes on a sigmoidal shape). The estimate for γ increases with increasing sigmoidicity of the curve.

Both the k_a method and the curve-fitting methods are very restrictive in their assumptions and provide unrealistic approaches to determine the input rates following administration of various microparticulate systems. Some of the problems associated with the k_a method are avoided when intravenous data (as reference) are available. However, if this is the case, the mass balance or deconvolution methods (described in the following) appear to be much more preferable in the assessment of the absorption rate.

B. Mass Balance Methods for the Assessment of Input Rate

The problems associated with curve-fitting have prompted the development of mass balance methods for the assessment of input rate. These approaches use compartmental models and determine the amount of drug remaining to be absorbed at any time from the sum of the amounts in each compartment and that which has been eliminated. A compartment depicts the body as a single, kinetically homogeneous unit. Mass balance approaches have the advantage of not requiring any prescribed form of the input function (i.e., zero- or first-order absorption). These methods stem from the observation that following administration of the drug, the total amount of drug eliminated is equal to the total amount of drug absorbed. The associated equations for estimating the input rate by mass balance approaches are based on an instantaneous mass balance. The reader is referred to an excellent review article on the mathematical basis for these mass balance approaches [18].

The amount of drug that has been absorbed up to time t, must either be present in the body, or must be eliminated, as expressed by Eq. (19).

$$A_{abs}(t) = A(t) + A_{el}(t) \tag{19}$$

where $A_{abs}(t)$ is the amount of drug absorbed up to time t, $A(t)$ is the amount of drug in the body at time t, and $A_{el}(t)$ is the total amount of drug eliminated up to time t. The rate of absorption (i.e., rate of entry of the drug into systemic circulation) $R_{abs}(t)$ is related to the total amount absorbed by:

$$A_{abs}(t) = \int_0^t R_{abs}(T)dT \qquad (20)$$

that is,

$$R_{abs}(t) = \frac{dA_{abs}}{dt} \qquad (21)$$

Assuming first-order elimination, the rate of elimination of the drug is given by

$$\frac{dA_{el}}{dt} = Cl\ C(t) \qquad (22)$$

where Cl is the systemic clearance of the drug following an intravenous bolus dose, and $C(t)$ is the plasma concentration of the drug at time t. Assuming that Cl is constant (i.e., linear pharmacokinetics)

$$A_{el}(t) = Cl\int_0^t C(T)dT \qquad (23)$$

Therefore,

$$A_{abs}(t) = A(t) + Cl\ \int_0^t C(T)dT \qquad (24)$$

The estimation of the amount of the drug in the body at time t, $A(t)$, is determined by the measurement of plasma concentrations of the drug following extravascular administration. The estimate of $A(t)$ requires the assumptions of one-compartment or two-compartment models for the disposition of the drug. The Wagner–Nelson method [34] assumes a one-compartment disposition of the drug, whereas the Loo-Riegelman method [19] is applicable to a drug following two-compartment disposition.

1. The Wagner–Nelson Method

If we assume one-compartment disposition, the amount of the drug in the body at time t, $A(t)$ is given by

$$A(t) = V\ C(t) \qquad (25)$$

where V is the volume of distribution of the drug.

Thus, the mass balance equation for the total amount absorbed up to time t, is given by

$$A_{abs}(t) = VC(t) + Cl\int_0^t C(T)dT \qquad (26)$$

where $\int_0^t C(T)\mathrm{d}T$ is the area under the plasma concentration–time curve up to time t (AUC_{0-t}). Thus the foregoing expression may also be written as

$$A_{\mathrm{abs}}(t) = V\,C(t) + Cl \cdot \mathrm{AUC}_{0-t} \tag{27}$$

The total amount of drug ultimately absorbed (at time infinity), noting that the drug concentration in the body at time infinity is zero, is given by

$$A_\infty = Cl\int_0^\infty C(T)\mathrm{d}T \tag{28}$$

where $\int_0^\infty C(T)\mathrm{d}T$ at time infinity is the area under the plasma concentration–time curve ($\mathrm{AUC}_{0-\infty}$). The fraction of drug absorbed up to time t can be estimated from the ratio

$$\frac{A_{\mathrm{abs}}(t)}{A_\infty} = \frac{V\,C(t) + Cl\,(\mathrm{AUC}_{0-t})}{Cl \cdot \mathrm{AUC}_{0-\infty}} \tag{29}$$

The Wagner–Nelson method makes no assumptions for the absorption process. However, if a semilogarithmic plot of percentage drug remaining to be absorbed [i.e., $100(1 - [A_{\mathrm{abs}}(t)/A_\infty])$] versus time yields approximately a straight line, it suggests an apparent first-order absorption process and the apparent first-order rate of absorption, k_{a}, can be estimated from the slope of the line. Similarly, a straight line from the plot of percentage remaining to be absorbed versus time on rectilinear coordinates suggests zero-order absorption of the drug.

To analyze data from microsphere systems used in sustained- or prolonged-release microparticulate systems, one may not be able to sample the plasma samples until the entire dose has been released from the delivery system. For example, subcutaneous injections of microsphere microparticulate systems designed to release hormones (or therapeutic proteins) over a prolonged period, are not amenable to sampling until the entire dose has been released at the site of injection owing to physical limitations of sampling over such long durations. Furthermore, during formulation development, the experiments are conducted in various preclinical animal models, which develop antibodies to these human-derived hormones or proteins that interfere with the measurement of exact plasma concentrations from such formulations. In such cases, it is important to recognize that the amount of drug absorbed at time infinity cannot be estimated from Eq. (28). However, one should recognize that the product of the systemic clearance and the area under the plasma concentration–time curve at time infinity is equal to the total amount of dose administered. Hence, one can substitute the total amount of the dose administered in the denominator of Eq. (29), and the amount absorbed at time t is given by

$$\frac{A_{\mathrm{abs}}(t)}{A_\infty} = \frac{[V\,C(t) + (Cl \cdot \mathrm{AUC}_{0-t})]}{\mathrm{dose}} \tag{30}$$

The fraction of the drug absorbed at time t may also be estimated from the urinary excretion data [18]. The plasma concentration at time t, $C(t)$ may be estimated from the urinary excretion data as

$$C(t) = \frac{[dA_u/dt]}{k_e V} \tag{31}$$

where k_e is the apparent first-order excretion rate constant, and dA_u/dt is the rate of excretion of the drug. Following substitution and rearrangement of the resulting equation, the amount of the drug in the body at time t, is given by

$$A_{abs}(t) = \frac{[(dA_u/dt)_T]}{k_e + (k/k_e)(A_u)_T} \tag{32}$$

where k *is* the first-order elimination rate constant from the body, and $A_{u(T)}$ is the amount recovered unchanged in the urine following administration of the microparticulate system. At time infinity, the amount of the drug excreted in the urine is given as

$$A_{abs\infty} = \left(\frac{k}{k_e}\right) A_{u\infty} \tag{33}$$

Multiplying both sides by $V\,k_e$ (i.e., renal clearance Cl_r) in Eq. (32) and (33), yields

$$Cl_r\,A_{abs}(t) = V\left[\frac{dA_u}{dt_T}\right] + Cl\,(A_u)_T \tag{34}$$

and

$$Cl_r\,A_{abs\infty} = Cl\,A_{u\infty} \tag{35}$$

The ratio of Eqs. (34) and (35) yields the fraction of the drug absorbed at any time t as

$$\frac{A_{abs}(t)}{A_{abs\infty}} = \frac{\{V[dA_u/dt)_T] + Cl\,(A_u)_T\}}{(Cl\,A_{u\infty})} \tag{36}$$

In theory, the fraction of the dose absorbed following administration can be solely calculated from the urinary excretion data. However, urine must be collected over an adequate period to allow the estimation of the pharmacokinetic parameters. This may not be practical for some of the microsphere systems intended for releasing the drug over long time periods.

The most serious limitation of the Wagner–Nelson method is that it is applicable only to drugs displaying one-compartment disposition. In all other cases, it is an approximation. For drugs displaying multicompartment disposition, the

Wagner–Nelson method results in an underestimation of the time at which absorption ceases, and an overestimation of the absorption rate [18]. One approach to overcoming this problem is to use the Loo–Riegelman method [19] described in the following. However, the application of the latter method requires concentration–time data, following both intravenous and the intended route of administration of the drug, which may or may not be available at all times. For this reason, the Wagner–Nelson method serves as a valuable approximation method for the determination of the input rate following administration of pharmaceutical microparticulate systems.

Cohen *et al.* [9] presented a modified version of the Wagner–Nelson method for the assessment of in vivo release rate from liposomal formulations intended for sustained release over a month, following subcutaneous administration. These investigators determined the input rate into the surrounding tissues (s.c. space) and in the plasma following drug absorption. From the mass balance equation, the concentration of the drug in the plasma can be obtained by

$$\frac{dA_p}{dt} = (F k_a V_s C_s) - (k_{el} V_d C_p) \tag{37}$$

where F is the extent of absorption of the drug from a solution of the drug administered subcutaneously, k_a and k_{el} are the absorption and elimination rate constants, V_s and V_d are the volume of the subcutaneous space and the volume of distribution, respectively, and C_s and C_p are the concentrations of the drug in the subcutaneous space and plasma, respectively. Rearranging and integrating Eq. (37) yields the following relation between the plasma and subcutaneous drug concentration:

$$F V_s k_a \left[\int_0^t C_s(T) dT \right] = [V_d C_p(t)] + [k_{el} V_d (\int_0^t C_p(T) dT)] \tag{38}$$

The amount of drug in the subcutaneous space can be estimated from

$$\frac{V_s dC_s}{dt} = R_0 - (k_a V_s C_s) \tag{39}$$

where R_0 is the instantaneous release rate of the drug from the microparticulate system into the subcutaneous space. The release rate into the subcutaneous space can be obtained by rearrangement and integration of Eq. (39) to yield

$$\int_0^t R_0(T) dT = (V_s C_s) + [k_a V_s (\int_0^t C_s(T) dT)] \tag{40}$$

Dividing Eq. (40) by the total dose of the drug administered, the fraction of the dose released from the microparticulate system can be estimated as

$$\frac{A_{abs}(t)}{A_{abs\infty}} = [V_d/F \cdot \text{Dose}] \cdot \{C_p \cdot (1 + k_{el}/k_a) + k_{el} \int_0^t C_p(T) dT + (1/k_a)(\Delta C_p/\Delta t)\} \tag{41}$$

This method is again restricted to compounds with one-compartment disposition and assumes that the estimates for the rates of absorption and elimination remain constant between studies. Furthermore, depending on the error in the estimation of the absorption rate constant, there will be an error introduced in the fraction of the dose absorbed, usually leading to an overestimation of this fraction. Nevertheless, this method allows a reasonable first approximation of the in vivo release rate of the drug from prolonged-release microsphere microparticulate systems.

2. The Loo–Riegelman Method

The Loo–Riegelman method [19] is applicable to a drug following two- or multicompartment disposition. It requires the plasma concentration–time data following intravenous administration to assess the fraction of the dose absorbed at time t. The mathematical basis for a drug with two-compartment disposition is based on the mass balance equation

$$A_{abs}(t) = A_1(t) + A_2(t) + A_{el}(t) \tag{42}$$

where $A_1(t)$ and $A_2(t)$ are the amounts of the drug in the central and peripheral compartments, and $A_{el}(t)$ is the total amount eliminated from the body by all pathways. Assuming elimination from the central compartment only, the amount in the central compartment is given by

$$A_1(t) = V_1 C_1(t) \tag{43}$$

where V_1 and $C_1(t)$ are the volume of the central compartment, and the measured blood or plasma concentration, respectively. The differential equation for the rate of change of the amount of drug in the peripheral compartment is given by

$$\frac{dA_2}{dt} = k_{1.2}A_1 - k_{2.1}A_2 \tag{44}$$

where $k_{1.2}$ and $k_{2.1}$ are the first-order intercompartmental transfer rate constants. Loo and Riegelman [19] used a linear approximation for $C_1(t)$ over each data interval, and the amount of the drug absorbed at time t was estimated as

$$A_{abs}(t) = V_1C_1(t) + V_2 C_2(t) + Cl \int_0^t C(T)dT \tag{45}$$

where

$$C_2(t) = C_2(t-1)e^{-k_{2.1}\Delta t} + [k^{1.2}/k^{2.1}] \{C_1(t-1) (1 - e^{-k_{2.1}\Delta t})\}$$
$$+ [k_{1.2}/(k_{2.1})^2] \{[C_1(t) - C_1(t-1)]/\Delta t\} (e^{-k_{2.1}\Delta t} - k_{2.1}\Delta t - 1) \tag{46}$$

Loo and Riegelman assumed that the sampling period was relatively short so that the term $k_{2.1}\Delta t$ was less than or equal to 0.5; thus allowing the third term in the expression for $C_2(t)$ to be reduced using a Taylor expansion (i.e., $e^{-x} = 1 - x + x^2/2$) to

$$k_{1.2} \cdot \frac{[(C_1(t) - C_1(t-1))/\Delta t] \cdot (\Delta t)^2}{2} \tag{47}$$

Boxenbaum and Kaplan [4] have shown that this approximation is a potential source of error and should be avoided. Although the Loo–Riegelman approach is limited because of the requirement of intravenous concentration–time data, it is a very useful approach for the evaluation of absorption kinetics. Furthermore, the method can be used for drugs that distribute in any number of compartments, provided elimination occurs from the central compartment. Wagner [31] has shown that the Loo–Riegelman method can also be applied when the elimination occurs from the peripheral compartment. Expressions analogous to that presented for the estimation of $C_2(t)$ must be written for each peripheral compartment and the amount of drug absorbed in the body at time t can be estimated from the mass balance equation. A modification of the Loo–Riegelman approach, without the need for intravenous reference data, is presented by Gerardin et al. [17]. However, this approach requires the unrealistic condition that distribution occurs much more slowly than the input [22].

Cutler [11,12] has also shown that both the Wagner–Nelson and Loo–Riegelman methods are applicable to compounds showing a variable clearance or nonlinear disposition. Although the mass balance approaches provide a reasonable approximation of the input rate, they are still restricted by assumptions and are "model-dependent." These limitations may be overcome by "model-independent" methods, termed *deconvolution*, which are described in the following.

C. Assessment of Absorption Rate by Deconvolution Techniques

The process of determination of the drug input rate from a measured response (e.g., plasma concentration–time profile) is referred to as *deconvolution*. The prediction of a response from drug input is known as *convolution*. It is reported that the deconvolution process is very sensitive to noise in the output data, and the error is amplified in the estimation of the drug input rate. On the other hand, the variation in the drug input rate is dampened when estimating the output. The methods used for deconvolution of output data for microsphere systems is similar to the available methods for other traditional drug delivery systems. Several methods have been proposed in the literature for estimating the drug input rate using deconvolution from pharmaceutical dosage forms [11,12,22,26,27].

Deconvolution techniques are used to analyze the pharmacokinetics of drug input following administration of the microparticulate system. These methods arise from considering the body as a linear system relative to drug disposition. It is a model-independent approach and uses two fundamental properties of linear response systems. First is the property of superposition (i.e., the sum or superposition of any two inputs $[f_1(t) + f_2(t)]$ results in a response that is the superposition $c_1(t) + c_2(t)$ of the individual responses for their respective input rates. The second property of linear response systems used in analysis using deconvolution techniques is the convolution integral property. This convolution integral property defines the relation between the input function and the observed response following the input, and is given by Eq. (48).

$$G(t) = \int R(\mathrm{T}) \, G\partial(t - T) \, dT \qquad (48)$$

where $R(t)$ is the input rate of the drug into the body, and $G(t)$ is the resulting response from the input. $G\partial(t)$ is the response following a unit dose impulse input (such as an intravenous bolus dose). Excellent reviews on the mathematical basis of deconvolution are available in the literature [11,12,26,27]. The major assumptions underlying the convolution integral property are that the response is linearly related to the input, and the system is time-invariant [i.e., the unit impulse response $G\partial(t)$, is independent of the time of administration]. The limitations of the deconvolution approaches include its applicability to non-interacting inputs only. Hence, before applying deconvolution techniques to assess the input rate, the system should be evaluated for the principle of superposition and interaction. A superposition test requires a minimum of two different intravenous bolus or infusion inputs. A test for interaction requires two experiments: one involving only a direct intravenous input and the other involving an indirect and direct input combined. The latter experiment requires the drug in either the direct or indirect input to be radiolabeled to evaluate possible interaction(s) between the two inputs.

The convolution integral defined in the foregoing can be evaluated either analytically or numerically, when $R(t)$ and $G\partial(t)$ are known, to predict the response (such as drug concentration) following drug input. This operation is termed *convolution*. When $G(t)$ and $G\partial(t)$ are known, Eq. (48) serves as the basis for the estimation of the input rate. This procedure of estimating $R(t)$ from the convolution integral is termed *deconvolution*. Many methods have been proposed for numerical deconvolution. These include the finite difference methods, the least-squares methods, and Fourier analysis. The objective of this section is to define some of the mathematical basis for these approaches, and some of the approaches that allow the determination of the input rate from an applications standpoint.

1. Rationale for the Development of Methods for Numerical Deconvolution

Vaughan and Dennis [23] discuss the need for numerical evaluation of data to determine the input function. They have shown that even if $G(t)$ and $G\partial(t)$ are known analytically in the convolution integral:

$$G(t) = \int R(T) \, G\partial(t-T) \mathrm{d}T \tag{49}$$

the integral function defined by Eq. (49) cannot be solved, except in certain cases, because the Laplace transform of the equation

$$g(s) = r(s) \, g\partial(s) \tag{50}$$

with solution

$$r(s) = \frac{g(s)}{g\partial(s)} \tag{51}$$

cannot be transformed back into time space by the convolution theorem, because $1/g\partial(s)$ is not a Laplace transform [14]. In general, numerical evaluation of $g(s)$ and $g\partial(s)$, with subsequent numerical calculation of $r(s)$ and inversion, is unsuccessful because of instability [1,10]. Although Coulam *et al.* [10] have shown that Fourier transform methods are reasonably accurate, these methods are cumbersome and the use of simple numerical methods would be advantageous in the deconvolution of blood concentration–time data to assess the input rate [23].

2. Finite Difference Methods for Numerical Deconvolution

The application of the finite difference method for deconvolution was first introduced as the area–area method by Rescigno and Segre [21]. The deconvolution method requires no assumptions concerning the number of compartments or the kinetics of absorption. However, this method needs data following intravenous administration, in addition to that following the intended route of administration. Furthermore, the concentrations must be measured at the same times following both routes of administration. However, the concentrations do not need to be measured at equally spaced intervals.

The mathematical basis for the area–area method is somewhat ambiguous. It approximates the absorption rate by a constant over a time interval. Essentially, this method approximates the absorption rate as an equal pulse length "staircase" function. A *staircase input* is defined as a finite set of rectangular pulses of a duration $p_j - p_{j-1}$, and intensity I_j, beginning at time $t = p_{j-1}$ and ending at time $t = p_j$ ($j = 1, 2, 3...$), where $I_j \geq 0$, $t \geq 0$, and $p_0 = 0$. The staircase function, wherein the area of each step is equal to the area of the function itself, results in the multiplication of the exact output by the pulse lengths. On deconvolution, this product of the pulse lengths and exact output is interpreted as the actual area of the output function itself [23]. Simulations have shown that the area–area method results in large and unpredictable errors in the estimation of the staircase input function [23].

Benet and Chiang [2] have suggested that the area–area method is only appropriate when the characteristic response is a single exponential function. Vaughan and Dennis [23] have shown that the response to a staircase input function can be derived for both, equal and unequal pulse lengths, by the application of the Laplace transformation methods. If $Y(p_j)$ is the concentration (exact output) at time point p_j, then the response can be obtained from

$$Y(p_j) = \sum_{i=1}^{j} I_i \left\{ \int_0^{p_j - p_{i-1}} G\partial(T) \, dT - \int_0^{p_j - p_i} G\partial(T) dT \right\} \tag{52}$$

where $j = 1, 2, 3..., p_o = 0$, and $Y(0) = 0$, and $\int G\partial(T) dT$ is the AUC following intravenous bolus administration over the defined time interval.

If all the staircase pulse lengths are equal (i.e., $p_j - p_{j-1} = a$ for all j,) then the substitution of $p_j = j_a$ and $p_i = i_a$ into Eq. (52) yields

$$Y(j_a) = \sum_{i=1}^{j} I_i \int_{(j-i)a}^{(j-i+1)a} G\partial(T) dT \tag{53}$$

where $j > 0$, $Y(0) = 0$, and $\int G\partial(T) dT$ is the AUC following intravenous bolus administration over the time interval.

Vaughan and Dennis [23] have shown that once the exact output, $Y(p_j)$ at time point p_j, or $Y(j_a)$ at time point j_a, and the response to a unit impulse dose are known, a particular staircase input function can be obtained from either Eq. (52) or Eq. (53) for unequal or equal pulse lengths, respectively. For example, if the blood concentrations following the intended route of administration are collected at equally spaced time intervals ($\Delta t = a$), and the AUC following intravenous bolus administration is for each time interval, $(n-1)a$ to na, is designated as A_n, where $n = 1, 2, 3...$, then the intensity of the staircase input can be estimated by rearranging the Eq. (53) as:

$$I_1 = \frac{Y(a)}{A_1} \tag{54a}$$

$$I_2 = \frac{[Y(2a) - I_1 A_2]}{A_1} \tag{54b}$$

$$I_3 = \frac{[Y(3a) - I_1 A_3 - I_2 A_2]}{A_1} \tag{54c}$$

$$I_n = \frac{[Y(na) - I_1 A_n - I_2 A_{n-1} \cdots - I_{n-1} A_2]}{A_1} \tag{54d}$$

A similar set of equations can be derived for unequal pulse lengths to estimate the intensity of the staircase input.

Proost [20] proposed a general numerical deconvolution equation for drug concentration–time data with unequal lengths, applying model-independent methods. It was demonstrated that it is not necessary to know the response to a unit impulse input as an analytical function (i.e., curve-fitting) and the drug concentration–time data from the unit impulse dose could be used directly. This method is an approximation of the method proposed by Vaughan and Dennis [23], with the exception that the integrals for $G\partial(t)$ are obtained by the application of either linear or log-trapezoidal rule. Furthermore, the concentration at time zero following an intravenous dose is estimated by logarithmic extrapolation. Proost [20] showed that the staircase approximation of the input rate, I_n, obtained by the trapezoidal approximation of the unit impulse data or an analytical solution for the integral of $G\partial(t)$ yielded similar input rates. The cumulative fraction of the dose absorbed was estimated by the numerical integration of I_n as:

$$F_n = \sum_{i=1}^{j} [(I_{i-1 \text{ to } i})(T_i - T_{i-1})] \tag{55}$$

where T_i is the ith sampling time point, and I_{i-1} to $_i$ is analogous to the staircase input function of Vaughan and Dennis [23] in Eq. (53), with the exception that the integral of $G\partial(t)$ is replaced by the trapezoidal approximation of the integral. The method of Proost [20] can be applied without any knowledge of the pharmacokinetic model and does not need any curve-fitting techniques for the intravenous data. It only assumes that the pharmacokinetics of the drug exhibit linear kinetics.

A numerical deconvolution method has been derived for the determination of the in vivo input rates, based on the linear interpolation of the observed drug concentrations and deconvolution of the resulting trapezoidal function [24]. The derived in vivo input functions are discontinuous, and a general expression for the cumulative drug input is also derived. The expression for cumulative drug input is a generalization of the Loo–Riegelman equation, and it yields results similar to the point–area method using the staircase approximation for input functions.

Wagner [32] has shown that the fraction of the drug remaining to be absorbed at the site of administration can be obtained from a similar expression. In terms of the sampling interval (Δt), the fraction unabsorbed (FR) can be obtained from

$$FR_{n\Delta t} = [H_{(n+1)\Delta t}/H_{n\Delta t}] - \sum_{i=2-n+1}^{j=1-n} \{[F_{i\Delta t}/F_{\Delta t}][FR]_{(j-1)\Delta t}\} \tag{56}$$

where $n\Delta t$ is the time after n sampling intervals equal to Δt, H is a function describing the drug concentration–time curve following the intended route of administration, and F is a function describing the drug concentration–time curve following intravenous bolus administration. $F_{n\Delta t}$ may be expressed as the drug concentration at time $n\Delta t$, or as the AUC between $n\Delta t$ and $(n-1)\Delta t$. $H_{n\Delta t}$ can be expressed only in terms of drug concentration. When both, H and F are expressed in terms of drug concentrations, the method is referred to as the point–point method [18]. Benet and Chiang [2] recommended the use of the point–area method, in which F should be expressed as the area under the drug concentration–time curve following intravenous administration, and H should be expressed in terms of drug concentration in Eq. (56). Vaughan and Dennis [23] also designate their method of deconvolution as the point–area method, as specific output data points, $Y(p_j)$, and the integral of $G\partial(t)$, are used to derive the staircase input function.

Chiou [7] proposed a finite difference method for numerical deconvolution using an instantaneous midpoint input principle. This assumes that all the drug absorbed during a given interval, regardless of the complexity of the absorption kinetics, is absorbed instantaneously at the midpoint of the interval. This assumption is based on the finding that the plasma concentration for a one-compartment model system following an intravenous infusion can be approximated by assuming that the entire dose was administered as a bolus at the midpoint of the infusion period [8]. This approximation is valid when the infusion period is much shorter than the half-life of the drug. The deconvolution method proposed by Chiou [7] requires no assumptions about the pharmacokinetic models or the site of drug elimination. The fraction of the dose absorbed at time t (F) can be calculated from

$$F_1 = \frac{C_{t1}}{C_{iv(0.5t1)}} \tag{57}$$

where C_{t1} is the plasma drug concentration at time $t1$ after dosing, and $C_{iv(0.5t1)}$ is the plasma drug concentration at time $0.5t1$, following an intravenous bolus dose. The amount of drug absorbed between times $t1$ and $t2$ can be estimated by a similar comparison, provided the drug concentration is corrected for the contribution from the drug absorbed before $t1$ (C_{p1}). To estimate C_{p1}, it is assumed that all drug before time $t1$ is instantaneously absorbed at time $0.5t1$, and the contribution is estimated as

$$C_{p1} = F_1 C_{iv(t2-0.5t1)} \tag{58}$$

where $C_{iv(t2-0.5t1)}$ is the theoretical plasma concentration at time $(t2-0.5t1)$, when the same extravascular dose is administered as an intravenous bolus dose. This principle is used to estimate the fraction absorbed during other sampling intervals as follows:

$$\frac{F_2 = [C_{t2} - C_{p1}]}{C_{iv0.5(t2-t1)}} \tag{59a}$$

$$\frac{F_3 = [C_{t3} - (C_{p1} + C_{p2})]}{C_{\text{i.v.}0.5(t3-t2)}} \qquad (59\text{b})$$

$$\frac{F_n = [C_{tn} - \Sigma_{i=1}^{n-1} C_{pi}]}{C_{\text{iv}0.5(tn-tn-1)}} \qquad (59\text{c})$$

Vaughan [25] has shown that the method described by Chiou [7] is an approximation of the point–area method [24]. Furthermore, the midpoint input method is similar to the approximation of the area–area method [21], which uses rectangular functions to approximate the integrals of $G\partial(t)$ centered about the midpoint of integration. Vaughan [25] cautions against the use of approximations of the point–area method for the estimation of staircase input, because it leads to large errors in the cumulative drug input functions.

The finite difference methods for numerical deconvolution provide a simple method to estimate the drug input rates without assumptions concerning the pharmacokinetic models or absorption kinetics. Their main drawback is the instability in presence of data noise, which may require smoothing procedures [16]. Cutler [13] has pointed out that the smoothing procedures may yield errors in the estimates of the input functions. Because the finite difference methods are computationally simple, they can be used without the use of complex algorithms and computers, needed for the more complex, but stable, deconvolution methods, such as the least-squares [11,12].

3. Least-Squares Method for Numerical Deconvolution

The least-squares method provides the determination of the best estimate of the true input rate. The criteria for determining the best estimate for the input function relies on the accuracy of the prediction of the response $G(t)$. Thus, the proximity of the predicted $G(t)$ values with the observed $G(t)$ values determine the accuracy of the estimated input rate. The least-squares deconvolution method requires assumptions and specifications of the mathematical form of the input function. Once the function describing the input function is specified, it is introduced into the convolution integral and the observed response $G(t)$ is compared with the predicted response $G'(t)$.

When little or no information is available about the nature of the input function, Cutler [13] has proposed the use of polynomial functions to describe the input function. This is because the polynomials are fairly flexible and represent a wide range of functions with reasonable accuracy. Cutler [11] has shown that an exponential input function and the cube root dissolution law represent adequate approximations of the input rate, with simulated data. He proposed [12] the use of orthogonal functions to derive the polynomial functions for approximations of the input function. This approach was proposed based on the difficulty in estimating the input function using a polynomial approximation by Gamel *et al.* [16], which was attributed to the ill-conditioned set of polynomial equations. The least-squares approach is more stable to data noise, but is computationally more complex and requires the use of fairly extensive and specialized programs to determine the drug input functions.

Veng-Pedersen [26–28] has also proposed a least-squares deconvolution method for estimating the drug input rate using both a polynomial and a polyexponential approximation of the input rate. This approach has led to the development of a method that uses an adaptive least-squares cubic spline function for the approximation of the input rate. Given simulated data, this method is reported to have significant advantages over the method proposed by Cutler [12]. These include the superiority of fitting the bolus input response by a polyexponential expression that is a smoothing function and does not oscillate between the data points as a polynomial expression does. Furthermore, the Veng-Pedersen [26] approach provided a simple and explicit mathematical function for the rate and extent of drug input rate, which alleviates the complexity of the back-transformation and summation of terms in the Cutler approach [12]. Because the input function is calculated directly by linear regression in the Veng-Pedersen method, it is purported to be computationally simpler, and does not require the extensive and specialized methods for deconvolution.

The least-squares method for deconvolution are considered model-independent, even though they require the determination of the unit impulse response parameters by fitting the data to some function (e.g., polyexponential). The unit impulse response parameters require their determination by nonlinear regression. However, these estimated parameters do not require any uniqueness, and their actual values have no influence on the accuracy of the determination of the input function [29]. The input function is approximated by a polynomial expression that uses these estimated parameters collectively, rather than individually. Hence, contrary to the model-dependent approaches, such as curve-fitting and k_a methods, the least-squares method is not affected by the individual errors in the estimation of the parameters. Furthermore, because the input function is determined by linear, rather than nonlinear, regression, it is computationally simpler (for instance, no initial parameter estimates are required), and the problems of multiple minima are eliminated with this approach [29].

Despite the advantage of the stability of the least-squares deconvolution methods in presence of data noise, these methods require complex algorithms to determine the input functions. A simpler algorithm for easier implementation of this deconvolution technique has been provided [30]. It is based on a polyexponential approximation of the absorption response and the response from intravenous bolus or infusion administration. The absorption response is represented by

$$c(t) = \sum_{i=1}^{m} b_i e^{-\beta_i t +} \tag{60}$$

where $t_+ = (t - t_{lag})_+$ and $c(0) = 0$; t_{lag} is the estimated lag time before appearance of the drug into systemic circulation following extravascular administration. The characteristic response from a unit impulse dose $c\partial(t)$ is obtained from the polyexponential approximation of the intravenous bolus response:

$$c_{iv}(t) = \sum_{i=1}^{n} a_i e^{-\alpha_i t} \tag{61}$$

where $\alpha_i > 0$. Thus, $c\partial(t)$ can be obtained as

$$c\partial(t) = \frac{c_{iv}(t)}{D_{iv}} \tag{62}$$

where D_{iv} is the intravenous bolus dose.

The cumulative amount of drug absorbed, expressed as a percentage of the dose (PCT) has been shown to be

$$PCT(t) = [100/D] \int_0^t f(t) dt \tag{63}$$

where $f(t)$ is the rate of direct input. This expression is equivalent to

$$PCT(t) = u_0 + \sum_{i=1}^{m+n-1} u_i e^{-v_i t_+} \tag{64}$$

The absorption rate is

$$f(t) = \left[\frac{D}{100} \right] \sum_{i=1}^{m+n-1} u_i(-v_i) e^{-v_i t_+} \tag{65}$$

where $v_i = \beta_i$ for $i = 1, 2, 3, \ldots, m$; and $v_i = -\gamma_{i-m}$ for $i = (m+1), (m+2), \ldots, (m+n-1)$; u_i is defined as

$$u_i = K_4 b_i \left\{ K_1 - \left(\frac{K_3}{\beta_i} \right) - \sum_{j=1}^{n-1} \left[\frac{g_j}{\gamma_j(\gamma_j + \beta_i)} \right] \right\} \tag{66}$$

for $i = 1, 2, 3, \ldots, m$; and

$$u_i = K_4 \left(\frac{g_{i-m}}{\gamma_{i-m}} \right) \sum_{j=1}^{m} \left(\frac{b_j}{\gamma_{i-m} + \beta_j} \right) \tag{67}$$

for $i = (m+1), (m+2), \ldots, (m+n-1)$; u_0 is obtained as

$$u_0 = -\sum_{i=1}^{m+n-1} u_i \tag{68}$$

The parameters $\{g_i, \gamma_i\}1^{n-1}$ are obtained from the parameters $\{a_i, \alpha_i\}1^n$. The γ parameters are $(n-1)$ roots of the $(n-1)^{th}$ polynomial and can be obtained by conventional numerical methods [30]. The remaining parameters in the expression for PCT [see Eqs. (63) and (64)] are defined by:

$$K_1 = \frac{1}{\sum_{i=1}^{n} a_i} \tag{69}$$

$$K_2 = K_1^2 \sum_{i=1}^{n} a_i \alpha_i \tag{70}$$

$$K_3 = K_2 - \sum_{i=1}^{n-1} \frac{g_i}{\gamma_i} \tag{71}$$

$$K_4 = \frac{100 D_{iv}}{D} \tag{72}$$

where D is the dose administered extravascularly.

This approach to deconvolution is purported to be computationally easier and Veng-Pedersen [30] has provided a computer subroutine for easy implementation of the algorithm. Although the use of least-squares deconvolution approach has been simplified by the use of more "user friendly" algorithms, they still require the use of fairly extensive and specialized computer programs to estimate the input functions. These requirements, along with the a priori assumptions for the approximation of the input function, limits the routine use of this approach for deconvolution of data from extravascular administration of microparticulate systems.

4. Miscellaneous Methods for Deconvolution of Extravascular Data

Berman [3] proposed a deconvolution scheme for the calculation of input function given a system and its response. It is based on the conversion of the convolution integral equation to an initial value problem. This technique is particularly simple and advantageous when dealing with compartmental systems. Foster et al. [15] have illustrated the use of this deconvolution approach, using compartmental analysis, with commonly available numerical integration software. The scheme can be used to determine an unknown input over time, from the measured response and the impulse response of the system. These investigators report that this method requires the solution of two systems of differential equations; those specifying the compartmental systems for the response function and the impulse response of the system. The input function can be calculated using simulations of these compartmental models. The accuracy of the simulations is determined from the goodness-of-fit of the function to the observed response data from the system.

A novel approach for deconvolution, based on the principle of maximum entropy, has been introduced by Charter and Gull [6]. Maximum entropy methods have been used successfully to handle noisy and incomplete data, thus allowing its application to the reconstruction of an input rate (continuous function) from a small number of data points. The resulting input functions are physiologically realistic and smooth. Furthermore, the blood samples are not required to be taken at equally spaced intervals, and no preliminary smoothing or interpolation of data is required. The maximum entropy approach is conceptually complex, and it is proposed that the maximum entropy may offer most reliable deconvolution for drugs exhibiting nonlinear kinetics [22].

D. Significance of Assessment of Input Rates

The mass balance and deconvolution techniques allow the estimation of in vivo input of drug into the biological system. By using the method of preference, one calculates the input rates from the microparticulate system. The input rates allow the estimation of the fraction of the dose absorbed as a function of time. Cumulative amount of drug absorbed (in vivo) versus time plots are constructed and are compared with in vitro dissolution rate plots to establish meaningful in vitro–in vivo correlation. Such correlation precludes the need for the conduct of expensive and time-consuming pharmacokinetic experiments during formulation development or optimization and provides a tool for quality control purposes.

IV. CONCLUSION

This chapter has provided some of the mathematical basis and applications for the various approaches available to a user for the estimation of the in vivo input rate of the drug from a given microparticulate system. The choice of the proper method depends on the availability of the data from intravenous and extravascular routes of administration, as well as the quality of the data. For analysis of in vivo absorption kinetics from microsphere systems, the investigator should conduct proper pharmacokinetic experiments before selecting the most suitable method of data analysis. An implicit assumption in the conduct of these pharmacokinetic experiments is the availability of sensitive and specific analytical methods for the drug. The pharmacokinetic evaluation usually (but not necessarily) entails the determination of the pharmacokinetics of the drug following intravenous or extravascular (intended) administration. The intravenous reference data should be obtained, as and when possible, to best estimate the unit impulse response. Model-independent and model-dependent pharmacokinetic parameter estimates should be obtained from in vivo experiments to obtain a proper handle on the extent of absorption (bioavailability), systemic clearance, and volume of distribution of the drug.

The establishment of meaningful in vitro–in vivo correlation warrants the need for a method to estimate the in vivo drug input rate that would be independent of the disposition processes of the drug, does not require any assumptions about the form of the input rate, is accurate and not affected by noise in the data, and is easy to implement. Additional features would include the applicability of the method to drugs displaying nonlinear kinetics. Because of the complexity of the assessment of the input rate, current investigators prefer the use of various deconvolution techniques to determine the in vivo input rate. Of the many approaches discussed in the text of this chapter, the finite difference methods for numerical deconvolution provide techniques that are independent of the disposition processes, use relatively sparse data, and are computationally simpler, thereby precluding the need for complex algorithms and fairly extensive computer programs to ascertain the input rate. Their main disadvantage is the instability in presence of data noise, which can be minimized by the use of least-squares deconvolution methods, which need complex algorithms and extensive computer programs.

From the standpoint of examination of data, the point–area method, with the staircase approximation [23] provides a useful method for deconvolution. It is fairly independent of the disposition processes, requires no assumptions about the absorption kinetics, and is easy to implement. The Proost [20] modification of the staircase input function offers an additional advantage of the trapezoidal estimation of the integral for the unit impulse response, thus precluding the need for an analytical solution for the integral. This eliminates a curve-fitting step in the data analysis. Another useful first approximation for the assessment of the input rate is the point–area method of Benet and Chiang [2], because of its ease of implementation. A prerequisite for the use of deconvolution techniques is the availability of intravenous reference data.

When the pharmacokinetic parameters following intravenous administration are available, the Loo–Riegelman method provides reasonably accurate estimates of the fraction of the dose absorbed in vivo following extravascular administration. Although the method is model-dependent on the number of compartments, it is applicable to multicompartmental systems with central or peripheral elimination. Furthermore, this method is much simpler to use, the fraction absorbed can be estimated by a handheld calculator, and it can be easily implemented in the form of spreadsheets for routine analysis. When the drug follows one-compartment kinetics and no intravenous reference data are available, the Wagner–Nelson method for the estimation of the fraction of the dose absorbed is remarkably accurate and simple to use. Once again, this mass balance approach provides the in vivo input rate by using handheld calculators or simple spreadsheets, allowing rapid manipulation of the data to obtain the in vitro–in vivo correlation of the amount of drug released from these microsphere systems.

REFERENCES

1. R. E. Bellman, B. E. Kalaba, and J. Lockett, *Numerical Inversion of the Laplace Transform,* Elsevier, New York, 1966.
2. L. Z. Benet and C.-W. N. Chiang, The use of deconvolution methods in pharmacokinetics, *Abstracts of Papers Presented at the 13th National Meeting of the APhA Academy of Pharmaceutical Sciences,* Vol. 2, 1972, 169.
3. M. Berman, A deconvolution scheme, *Math. Biosci.* 40:319–323 (1978).

4. H. Boxenbaum and S. Kaplan, Potential source of error in absorption rate calculations, *J. Pharmacokinet. Biopharm. 3*:257–264 (1975).

5. F. Bressolle, R. Gomeni, R. Alric, M. J. Royer-Morrot, and J. Necciari, A double Weibull input function describes the complex absorption of sustained-release oral sodium valproate, *J. Pharm. Sci. 83*:1461–1464 (1994).

6. M. K. Charter and S. F. Gull, Maximum entropy and its application to the calculation of drug absorption rates, *J. Pharmacokinet. Biopharm. 15*:645–655 (1987).

7. W. L. Chiou, New compartment- and model-independent method for rapid calculation of drug absorption rates, *J. Pharm. Sci. 69*:57–62 (1980).

8. W. L. Chiou, G. W. Peng, and R. L. Nation, Rapid estimation of volume of distribution after a short intravenous infusion and its application to dosing adjustments, *J. Clin. Pharmacol. 18*:266–271 (1978).

9. S. Cohen, H. Bernstein, C. Hewes, M. Chow, and R. Langer, The pharmacokinetics of, and humoral responses to, antigen delivered by microencapsulated liposomes, *Proc. Natl. Acad. Sci. USA 88*:10440–10444 (1991).

10. C. M. Coulam, H. R. Warner, H. W. Marshall, and J. B. Bassingthwaighte, A steady-state transfer function analysis of portions of the circulatory system using indicator dilution techniques, *Comput. Biomed. Res. 1*:124–138 (1967).

11. D. J. Cutler, Numerical deconvolution by least squares: use of prescribed input functions, *J. Pharmacokinet. Biopharm. 6*:227–242 (1978).

12. D. J. Cutler, Numerical deconvolution by least squares: Use of polynomials to represent the input function, *J. Pharmacokinet. Biopharm. 6*:243–263 (1978).

13. D. Cutler, Assessment of rate and extent of drug absorption. *Pharmacol. Ther. 14*:123–160 (1981).

14. G. Doetsch, *Guide to the Applications of Laplace Transforms*, Van Nostrand, New York, 1961, Chap. 7.

15. D. M. Foster, D. G. Covell, and M. Berman, Applications of a general method for deconvolution using compartmental analysis, *Comput. Biol. Med 18*:253–266 (1988).

16. J. Gamel, W. Rousseau, C. Katholi, and E. Mesel, Pitfalls in digital computation of the impulse response of vascular beds from indicator-dilution curves, *Circ. Res. 32*:516–523 (1973).

17. A. Gerardin, D. Wantiez, and A. Jaouen, An incremental method for the study of absorption of drugs whose kinetics are described by a two-compartment model: Estimation of the microscopic rate constants, *J. Pharmacokinet. Biopharm. 11*:401–424 (1983).

18. M. Gibaldi and D. Perrier, *Pharmacokinetics*. (M. Gibaldi and D. Perrier, eds.), 1982, pp. 145–198.

19. J. Loo and S. Riegelman, New method for calculating the intrinsic absorption rate of drugs, *J. Pharm. Sci. 57*:918–928 (1968).

20. J. H. Proost, Application of numerical deconvolution technique in the assessment of bioavailability, *J. Pharm. Sci. 74*:1135–1136 (1985).

21. A. Rescigno and G. Segre, *Drug and Tracer Kinetics*, Blaisdell, Waltham MA, 1966.

22. G. T. Tucker and P. R. Jackson, Pharmacokinetic evaluation of novel drug delivery systems: Assessment of rate, *Novel Drug Delivery and Its Therapeutic Appli-*

cation (L. F. Prescott and W. S. Nimmo, eds.), John Wiley & Sons, London, 1989, pp. 113–120.

23. D. P. Vaughan and M. Dennis, Mathematical basis of point–area deconvolution method for determining in vivo input functions, *J. Pharm. Sci. 67*:663–665 (1978).

24. D. P. Vaughan and M. Dennis, Mathematical basis and generalization of the Loo–Riegelman method for the determination of in vivo drug absorption, *J. Pharmacokinet. Biopharm. 8*:83–98 (1980).

25. D. P. Vaughan, Approximation in point–area deconvolution algorithm as mathematical basis of empirical instantaneous midpoint-input deconvolution method, *J. Pharm. Sci. 70*:831–832 (1981).

26. P. Veng-Pedersen, Model-independent method of analyzing input in linear pharmacokinetic systems having polyexponential impulse response I: Theoretical analysis, *J. Pharm. Sci. 69*:298–305 (1980).

27. P. Veng-Pedersen, Model-independent method of analyzing input in linear pharmacokinetic systems having polyexponential impulse response II: Numerical evaluation, *J. Pharm. Sci. 69*:305–312 (1980).

28. P. Veng-Pedersen, Novel deconvolution method for linear pharmacokinetic systems with polyexponential impulse response, *J. Pharm. Sci. 69*:312–318 (1980).

29. P. Veng-Pedersen, Novel approach to bioavailability testing: Statistical method for comparing drug input calculated by a least-squares deconvolution technique, *J. Pharm. Sci. 69*:318–324 (1980).

30. P. Veng-Pedersen, An algorithm and computer program for deconvolution in linear pharmacokinetics, *J. Pharmacokinet. Biopharm. 8*:463–481 (1980).

31. J. G. Wagner, Application of the Loo–Riegelman absorption method, *J. Pharmacokinet. Biopharm. 3*:51–57 (1975).

32. J. G. Wagner, Do you need a pharmacokinetic model, and, if so, which one? *J. Pharmacokinet. Biopharm. 3*:457 (1975).

33. J. G. Wagner, Linear pharmacokinetic models and vanishing exponential terms: Implications in pharmacokinetics, *J. Pharmacokinet. Biopharm. 4*:395–425 (1976).

34. J. G. Wagner and E. Nelson, Percent absorbed time plots derived from blood level and/or urinary excretion data, *J. Pharm. Sci. 52*:610–611 (1963).

12

Adjuvant-Active Polymeric Microparticulate Vaccine-Delivery Systems

Douglas Kline

Schering-Plough Research Institute
Kenilworth, New Jersey

Justin Hanes and Robert Langer

Massachusetts Institute of Technology
Cambridge, Massachusetts

I. INTRODUCTION

When Edward Jenner began injecting an extract of cowpox lesions into patients to prevent smallpox infection in the late 18th century [1], little could he have known how his crude inoculation would revolutionize the science of disease prevention and control. Since those humble beginnings, the science of vaccination has both spurred and adapted biotechnological advances to produce vaccines that are efficacious and safe.

Most recently, developments in the fields of protein sequencing and genetic engineering have engendered the subunit vaccine approach, in which the whole-killed or attenuated agents often present in vaccine preparations are replaced with a peptide or protein subunit known to elicit an effective immune response toward the parent organism. Because subunit vaccines consist of well-characterized molecules, often produced by recombinant DNA technology, and do not contain the disease-causing agent, their safety profiles are superior to conventional whole-organism vaccines in which the absence of viable infectious agents must constantly be validated. This is of particular relevance as vaccines for more serious illnesses, such as hepatitis B and human immunodeficiency virus infection (HIV), are developed. Unfortunately, the improvement in safety afforded by subunit vaccines often comes at the expense of efficacy. Subunit vaccines are frequently poorly immunogenic, necessitating several booster injections to achieve the desired antibody response and a more frequent revaccination schedule [2].

In the administration of any vaccine program (subunit or conventional), the number and frequency of injections required for protection can become a crucial factor in its success. More booster injections translate into more patient visits, and patient compliance becomes a limiting factor. Usually, this was considered in terms of vaccination programs in developing countries where, even if the problems of inadequate storage and improper vaccination practices [3] could be overcome to ensure potent vaccine doses were correctly administered, the population can be nomadic or difficult to reach by health authorities. However, recent experience has shown that patient compliance can be problematic even with conventional vaccines in an industrialized country, such as the United States. Witness the recent outbreaks of measles among large numbers of unvaccinated children younger than 5 years of age and in unsuccessfully vaccinated children of school and college age [4]. In response to these recent epidemics, the Advisory Committee on Immunization Practices (ACIP) has set a goal for the year 2000 of 90% of children fully vaccinated (four doses of diphtheria, tetanus, and pertussis; three doses of oral polio; three doses of *Haemophilus influenzae*; and one dose of measles, mumps, and rubella) by age 2; however, studies show that the target still far exceeds the practice [5]. Opportunities to immunize children during contact with health care workers (such as at hospital emergency wards) was cited as a means of improving the vaccine coverage, but in one study, almost 40% of adults accompanying children to a pediatric emergency department provided inaccurate information about their measles immunization history [6].

Finally, the increase in patient visits to attain complete coverage will increase the cost of vaccination, whether it be through a private health care provider, a local health clinic, or through subsidized mass vaccination programs on a state or national level.

Clearly, from the standpoint of disease prevention and economics, there is a demonstrated need to reduce the number of injections required for the newly developed subunit vaccines as well as the existing traditional vaccines. More importantly, as subunit vaccines for diseases more serious than measles are developed, and the consequences of incomplete protection include death, their efficacy must be improved to as close to 100% as possible. The primary strategy for achieving these goals is improvement of the immunogenicity of the infectious agent or subunit (the antigen) through the use of immunological adjuvants.

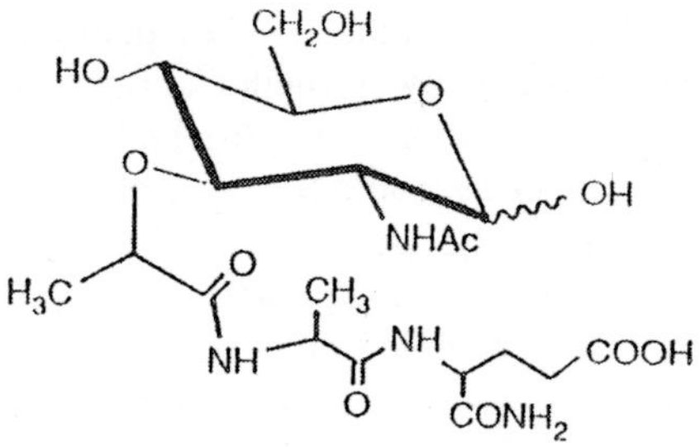

Fig. 1 Structure of muramyl dipeptide (MDP).

II. ADJUVANTS

An *adjuvant* is a compound administered with the antigen, or one that provides a mode of presentation of antigen that will enhance the immune response toward that antigen. The mechanisms of adjuvant action are complex, and as more is discovered about their effects on the immune system, the less likely it appears that a few theories will be able to explain the actions of the entire spectrum of adjuvant–antigen combinations. For simplicity, adjuvant action may be broken down into two general categories: (a) direct stimulatory effect on the immune system and (b) the method of antigen presentation to the immune system.

An example of the former category is muramyl dipeptide (MDP), a component of a peptidoglycan found in the cell wall of several *Mycobacterium* strains used as adjuvants (Fig. 1). Muramyl dipeptide is the minimum monomeric structure that retains the adjuvanticity of the parent peptidoglycan. A water-soluble molecule, MDP interacts with a variety of cells of the immune system, including B cells, T cells, and macrophages. On the cellular level, MDP has a broad spectrum of activity, including mitogenicity and polyclonal B-cell activation. The activity and specificity of MDP can be changed by modifications in its structure; for example, replacement of L-alanine with D-alanine causes immunosuppression. This relation between structure and function is viewed as evidence that interactions between immune cell receptors and MDP are an integral part of its action [7].

Many adjuvants, however, do not appear to have direct interactions with immunocompetent cells. Rather, these adjuvants are hypothesized to act by affecting the manner in which the immune system interacts with antigen. Aluminum-containing compounds, such as aluminum hydroxide, are the adjuvants most widely used in human vaccines and the only ones currently approved by the FDA. Antigen is adsorbed onto an aluminum hydroxide suspension and is thus injected in the context of an antigen–aluminum complex. Aluminum adsorbates have been hypothesized to work primarily by a "depot" effect by which adsorbed antigen is kept at the injection site [8] or within the peripheral lymph nodes [9] for extended periods, and becomes available for more efficient processing by immunocompetent cells. Although the persistence of antigen does seem to play a role in the adjuvanticity of the aluminum salts, the depot theory is only one aspect of what appears to be a complex process, involving interactions of antigen, adjuvant, and the immune system [7]. For example, the antigen's conformation in association with the aluminum compound is expected to have a dramatic effect on its interaction with antigen-presenting cells (APC). Similarly, the capacity of these cells to take up and process antigen is directly influenced by the aluminum compounds themselves. Alum adjuvants increase the circulation of lymphocytes through draining lymphoid tissue [10], induce the production of plasma cell-containing granulomas at the injection site [7], and increase the uptake of antigen by macrophages and the resultant antigen-induced T-cell proliferation [11]. These effects, in conjunction with the depot effect, all result in more efficient processing of antigen by the immune system.

However, alum adjuvants suffer from several drawbacks that limit their commercial use and may render them unsuitable for use in subunit vaccines. Compared with other adjuvants, alum is relatively weak [12] and may not be able to sufficiently enhance the immune response toward a poorly immunogenic antigen. This problem is especially relevant to subunit vaccines. Physicochemical and morphological differences in alum vaccines owing to batch-to-batch variation [13] and aging [14] can cause variability in the immune response that must be overcome by repeated injections, in part defeating the purpose of including the adjuvant in the first place.

A family of more potent adjuvants that also use a depot effect are Freund's adjuvants, which are water-in-oil emulsions. Incomplete Freund's adjuvant (IFA) consists of aqueous antigen emulsified in a low-viscosity, low specific gravity mineral oil [7] with an added emulsifier. Incomplete Freund's adjuvant is hypothesized to act in a manner similar to aluminum compounds, but its superior adjuvanticity may be due to several factors. Because antigen may be released from within a water-in-oil emulsion more slowly than from alum [15], the depot effect of IFA is apparently more prolonged than that of alum. Additionally, the nonmetabolizable mineral oil component of IFA is more inflammatory than alum and, thus, may be more potent in its stimulation of immune cells. The importance of this aspect of IFA is demonstrated when other metabolizable oils, such as peanut oil, are substituted for mineral oil; both the adjuvanticity and the intensity of local reactions were reduced [16–18]. Incomplete Freund's adjuvant was used in an influenza vaccine in Great Britain that was withdrawn from the market after a low incidence of cyst or abcess formation at the injection site was observed [17]. Although the United States armed forces also used an IFA-containing influenza vaccine, no water-in-oil emulsion vaccine has ever been licensed for use in the general population in the United States [7].

Complete Freund's adjuvant (CFA), which is IFA containing killed mycobacteria, demonstrates how the depot and immunostimulatory effects of a water-in-oil emulsion can be combined to maximize adjuvanticity. Complete Freund's adjuvant was developed after Freund used mineral oil to increase the immune response to mycobacteria, the bacterium responsible for tuberculosis [19]. Subsequent studies found that killed mycobacteria–mineral oil combination could increase the immune response toward other antigens when injected in a water-in-oil emulsion and led to the development of both CFA and IFA. Analysis of the immunostimulating properties of the bacterium demonstrated that they are localized within the cell wall. Finally, the adjuvanticity of the whole bacterium was isolated to the subunit MDP (discussed earlier), which is the minimum adjuvant-active structure of the cell wall [20]. In an CFA vaccine, the mineral oil–mycobacteria phase attracts and stimulates a variety of immune cells, including macrophages, dendritic cells, and lymphocytes [18], and the depot effect of the emulsion ensures that antigen is continuously present. (This combination made CFA the strongest first-generation adjuvant in most applications.) Unfortunately, the immunostimulatory effect of mineral oil emulsions is largely due to a heightened local inflammatory reaction, which for CFA is severe enough to cause granuloma formation, ulceration, and pain at the injection site, as well as systemic effects, such as fever [21]. Consequently, CFA is unsuitable for human or veterinary vaccines, and even its use in experimental animals is increasingly being restricted.

Experience with aluminum-based and Freund's adjuvants demonstrates several concepts of vaccination. Sustained delivery of antigen improves the immune response when compared with a bolus injection of an aqueous preparation of antigen. In addition, although concomitant stimulation of the immune system by the vehicle (an aluminum adsorbate or water-in-oil emulsion) further strengthens the immune response, the magnitude of the adjuvant effect practically achievable is limited by local inflammatory reactions. Given the importance of the depot effect to the function of many adjuvants, it is not surprising that one of the first applications of the technology of controlled delivery of macromolecules from polymer systems was in the development of single-step vaccination systems.

III. POLYMERIC SUSTAINED-DELIVERY SYSTEMS AND IMMUNIZATION

By releasing small amounts of macromolecules over sustained periods, ranging from days to years, polymeric controlled-release systems are able to greatly improve the depot effect crucial to the working of adjuvants, such as alum and Freund's. An advantage of these systems is that the dose and release rate can be varied by changing various parameters of each system. This allows the antigen to be reproducibly delivered according to the kinetics for optimum immune response, whereas antigen release from emulsions and alum adsorbates is expected to be of much shorter duration, more variable, and less easily controlled.

The first polymeric controlled-delivery vaccination systems were developed with the concept of using a polymer purely as a matrix to achieve a desired release profile, with no direct effect of the polymer on the immune system desired (immunoinert). Pellets were fabricated from poly(ethylene–vinyl acetate), a nonbioerodible polymer that was inert and biocompatible in previous controlled-release applications, loaded with several different protein antigens, and implanted subcutaneously in both mice [22] and rabbits [23]. In each case, an adjuvant effect over two injections of antigen in saline was seen, occasionally comparable with two injections of antigen in CFA.

Subsequent efforts with immunoinert polymers sought to improve the interaction of antigen and immune system by two methods: (a) the size scale of the delivery vehicle was reduced from millimeters to microns, and (b) a bioerodible polymer (such as polyesters of lactic and glycolic acid or polyanhydrides) was employed as the matrix. These microparticular systems display many advantages over the larger implantable geometries. From a practical standpoint, particles smaller than 100 μm in diameter can be injected through a standard 22-gauge needle, whereas larger systems must be surgically implanted [2]. Because the polymer matrix is designed to erode under physiological conditions, by polymer hydrolysis and solubilization of the degradation products, device removal, usually by an invasive surgical procedure, is unnecessary. In addition, smaller microparticles, those in the size range of 1–10 μm, are readily phagocytosed by macrophages [24–26]. Thus, antigen may remain protected by the delivery system until its targeted delivery to an antigen-presenting cell, avoiding the necessity for release or desorption from the system and possible degradation. The reduced particle size and choice of a bioerodible polymer also give greater flexibility in the release kinetics possible for a given antigen dose. By such methods as varying the copolymer composition and molecular weight [27] and the fabrication conditions [28], bioerodible polymer microparticles can be designed to release antigen continuously, or in a discrete pulses, over 1 week to 1 year. Finally, the most commonly used bioerodible lactic–glycolic acid polyesters have been approved by the FDA for clinical use in various applications, such as absorbable surgical sutures [29,30] and sustained release of leuprolide acetate (Lupron Depot) [31].

The characteristics of the aforementioned bioerodible microparticle-delivery systems make them promising candidates for use as efficacious vaccines, as well as excellent tools for probing the effects of antigen-release patterns and vaccine formulation parameters on the immune response [2]. For example, Eldridge and co-workers [32] used a poly(lactic–glycolic acid) (PLGA) microsphere delivery system for staphylococcal enterotoxin B toxoid, which consisted of small (1- to 10-μm) and larger (20- to 50-μm) particles administered together. The mixture of the two size ranges was able to elicit a strong secondary immune response (Fig. 2), far surpassing those resulting from the administration of either size range alone or of antigen and alum. These results support the rationale of using smaller particles, which are taken up by macrophages, to generate the primary antibody response and primed memory B cells. The larger microspheres, which cannot be phagocytosed, release their antigen more slowly and provide the long-term persistent levels of antigen that stimulate a strong, sustained secondary antibody response [2]. A pulsatile-release pattern, which mimics antigen levels obtained with multiple injections, was attainable with a single injection in a study with tetanus toxoid [33]. After one injection of a mixture of PLGA microspheres of different sizes, monomer ratios, antigen loading, and method of preparation (coacervation vs. spray-drying), a strong initial antigen dose, with two subsequent booster doses at 1 and 3 months, was delivered. For certain strongly immunogenic antigens, this system, by reproducing the antigen levels achieved clinically with multiple injections, may provide a viable one-step vaccination system.

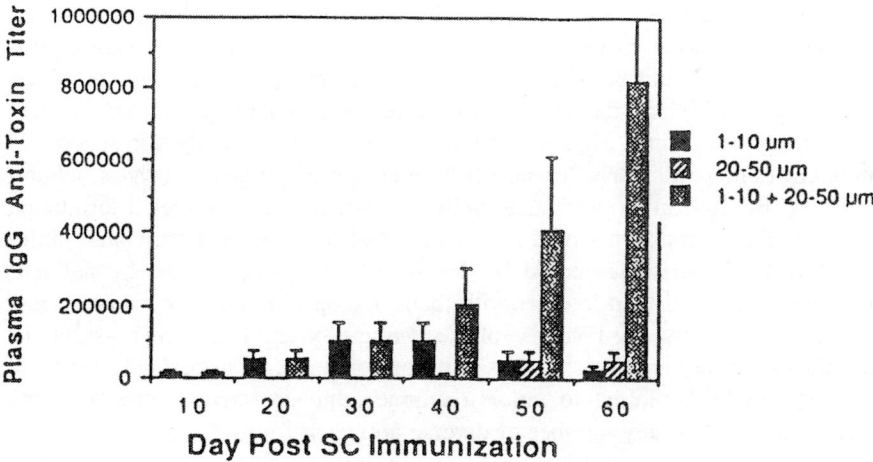

Fig. 2 The IgG antitoxin titer to staphylococcal enterotoxin B (SEB) toxoid in the plasma of mice induced by intraperitoneal immunization with SEB toxoid in 50:50 d, 1-PLGA microspheres 1–10 μm in diameter, 20–50 μm in diameter, or in a mixture of 1–10 and 20–50 μm in diameter. (From Ref. 32.)

Polymers other than PLGA have also been used in microparticulate vaccine systems, such as polyanhydrides [34] and the microencapsulated liposomes (MEL) [35]. In the latter system, model antigen-loaded liposomes were incorporated into ionically cross-linked alginate polymer microspheres. By protecting the liposomes from rapid degradation and clearance from the injection site, the polymer microcapsules acted as a depot for antigen and resulted in a three- to fourfold improvement in the antibody levels over those obtained with unencapsulated liposomes. These and other applications show that bioerodible microparticulate vaccine systems have a great potential for success in reducing the number of booster injections required for a variety of antigens.

IV. ADJUVANT-ACTIVE POLYMERIC MICROPARTICULATE-DELIVERY SYSTEMS

A. Introduction

In all of the microparticulate antigen-delivery systems just discussed, the polymer acts merely as a matrix (depot) to allow sustained antigen release or protection and is postulated to have little or no direct effect on the immune system. If Freund's or alum are used as the paradigm of adjuvant action, the strongest adjuvants combine the antigen depot effect with a direct immunostimulatory or immunoenhancing effect. By comparison, if a polymeric microparticulate vaccine system could be designed so that the delivery system is itself immunoenhancing and releases antigen according to the optimal profile, the resulting immune response should be increased beyond that obtained with an immunoinert release system. If so, the extra immune response obtained may be the key to eliciting sufficiently high and persistent levels of antibody for successful immunization, particularly for poorly immunogenic antigens, such as subunit vaccines. Because microparticulate-delivery systems can be targeted to interact with different components of the immune system by varying their size, adjuvant-active microparticles could be designed to be phagocytosed by antigen-presenting cells and then interact with them in a specific manner, while releasing antigen. Because the response of the immune system is different—subtly or grossly—for every antigen, the versatility of polymeric microparticulate-release systems can be exploited to tailor a unique adjuvant-active microparticulate vaccine for each of any number of diverse antigens.

Research into this emerging field has concentrated on a variety of strategies for producing adjuvant-active polymeric microparticles. A variety of adjuvant-active microparticulate vaccine systems are discussed in the following. Although the ultimate goal of this research is to produce a viable vaccine, the results also provide further insight into the complex mechanisms of adjuvanticity.

B. Poly(methyl methacrylates) and Derivatives

From their clinical use in surgery and dentistry, nonbioerodible polyacrylates were among the first polymers suggested for use as microparticular adjuvants [36]. In a series of studies, Kreuter and co-workers used a poly(methyl methacrylate) system with an average particle size of 50–300 nm to investigate the adjuvant effect with influenza virus and to correlate it with various polymer physicochemical properties. Initial efforts showed that polymerization of methyl methacrylate monomer by gamma-irradiation in the presence of influenza virions resulted in a greater adjuvant effect than adsorption of the virions onto previously fabricated microparticles [37]. Coating of the virion by monomer and polymer during the polymerization process is proposed as an explanation for this phenomenon, as no virion features can be seen on the particle surface by scanning electron microscopy [38].

The source of the adjuvanticity of this polymer system is unknown. The monomer, methyl methacrylate (MMA) (Fig. 3), causes a severe inflammatory response, with cell destruction and abscess formation, when injected at a 1% concentration, and a less intense response at 0.1% [12]. In contrast to CFA, a strong adjuvant that is hypothesized to act by a similarly intense inflammatory effect, methyl methacrylate actually suppresses the immune response when added to conventional and poly(methyl methacrylate) influenza vaccines [12].

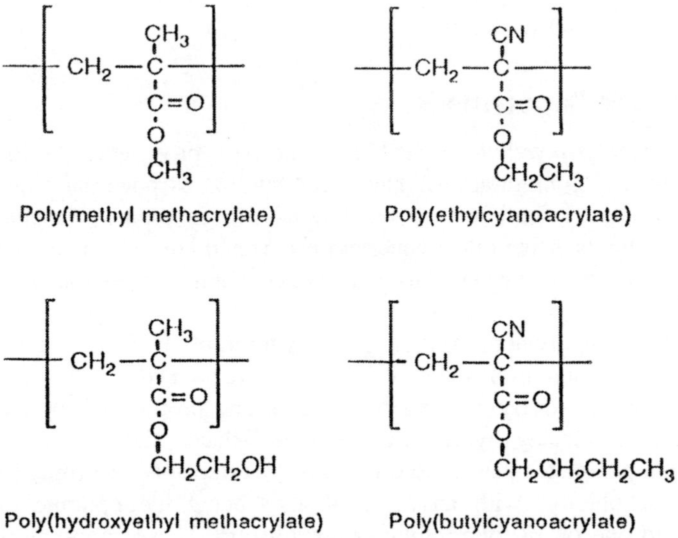

Poly(methyl methacrylate)   Poly(ethylcyanoacrylate)

Poly(hydroxyethyl methacrylate)   Poly(butylcyanoacrylate)

Fig. 3 Structures of poly(methyl methacrylate), poly(hydroxyethyl methacrylate), poly(ethyl cyanoacrylate), and poly(butyl cyanoacrylate).

Although repeated washing and final lyophilization were used to remove residual monomer after polymerization, it is possible that a small amount of residual monomer may cause a subacute inflammatory response that acts to attract or activate immune cells. The basis of the interactions of the poly(methyl methacrylate) particles with these cells is also unclear. Particle size has a profound effect on adjuvanticity, with particles larger than 800 nm showing little or no effect [12]; this may be indicative of direct cell–particle interactions. Studies of the fate of poly(methyl methacrylate) nanoparticles after intravenous administration showed that they were taken up by the liver, spleen, and bone marrow [39]. However, after intramuscular [39] or subcutaneous [40] administration, the particles remained at the injection site.

Further studies were performed by adsorbing antigen onto already prepared microparticles and correlating polymer and microparticle properties with the observed adjuvanticity. As with the copolymerized virion preparations, smaller particles were more adjuvant-active [41]. However, the surface area increases with decreasing particle size, an effect that is critical for systems in which antigen is adsorbed. To investigate the effect of polymer hydrophobicity, copolymer particles of MMA with varying ratios of hydroxyethyl methacrylate (a more hydrophilic monomer) were prepared, as were particles of poly(ethyl cyanoacrylate) or poly(butyl cyanoacrylate) (see Fig. 3). In general, adjuvanticity increased with hydrophobicity as measured by contact angle or hydrophilic monomer content. Because cell adhesion and uptake by macrophages is improved for more hydrophobic systems, this conclusion may be considered further evidence that phagocytosis by immune cells plays a part in the poly(methyl methacrylate) vaccine system [42].

C. Nonionic Block Copolymers

Block copolymers of poly(oxyethylene) (POE) and poly(oxypropylene) (POP), in a variety of structural configurations, have been studied as potential adjuvants. Although not strictly particles per se, they have been used as vaccine adjuvants in microparticulate-like oil-in-water emulsions and are recognized for their usefulness in probing the mechanisms and physicochemical properties that underlie adjuvanticity.

Originally, triblock copolymers consisting of hydrophilic POE segments, flanking a central hydrophobic POP segment (Fig. 4), were studied as adjuvants in aqueous suspensions [43]. By varying the lengths and proportions of the segments, polymers of varying hydrophile–lipophile balance (HLB) can be made. The HLB is a parameter that is used to classify surfactants according to their relative hydrophobicity, with an HLB of zero being hydrophobic (a spreading agent) and one of 18 hydrophilic (a solubilizer). These copolymers were tested as aqueous suspensions in mice for their ability to enhance primary and secondary antibody responses and delayed-type hypersensitivity (DTH) to

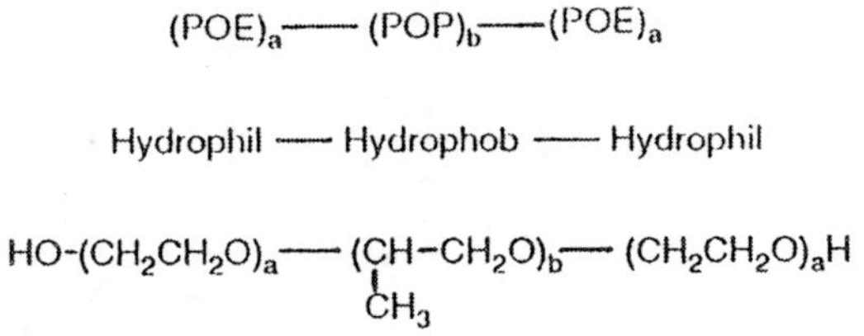

$(POE)_a \text{------} (POP)_b \text{------} (POE)_a$

Hydrophil ------ Hydrophob ------ Hydrophil

$HO\text{-}(CH_2CH_2O)_a \text{------} (\underset{\underset{CH_3}{|}}{CH}\text{-}CH_2O)_b \text{------} (CH_2CH_2O)_aH$

Fig. 4 General structure of polyoxyethylene (POE)–polyoxypropylene (POP) triblock copolymers. (From Ref. 43.)

sheep red blood cells and dinitrophenyl–bovine serum albumin (DNP-BSA). Two structurally similar copolymers of different molecular weight (Fig. 5) were adjuvant-active toward different immune responses; the larger of the two, L121 (HLB = 0.5), enhanced the primary and secondary antibody responses toward both antigens, and the smaller copolymer, L101 (HLB = 1.0), showed adjuvant activity in the DTH reaction. Their similarities in structure, but differences in action, were taken as evidence that active sites on the molecule were not responsible for stimulation of the immune system. Instead, an explanation, based on the HLB, was proposed to account for the fact that copolymers of higher HLB were either not adjuvant-active or suppressed the immune response [43].

Further studies used block copolymers in an oil-in-water emulsion system [44,45]. Unlike in the water-in-oil emulsion adjuvants such as Freund's, in which antigen is concentrated within a dispered aqueous phase, the copolymers were used to stabilize protein antigen within an oil phase and concentrate it on the droplet surface. Both L121 and L101 stabilized BSA in the oil phase and improved the antibody response. L121 was a more potent adjuvant, but L101 induced more granuloma formation, which indicates that, in this situation, adjuvanticity cannot be completely explained by the strength of the inflammatory

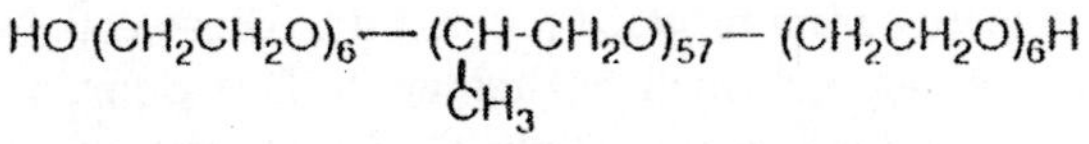

$HO\,(CH_2CH_2O)_6 \text{------} (\underset{\underset{CH_3}{|}}{CH}\text{-}CH_2O)_{57} \text{------} (CH_2CH_2O)_6H$

Triblock L101

$HO\,(CH_2CH_2O)_7 \text{------} (\underset{\underset{CH_3}{|}}{CH}\text{-}CH_2O)_{70} \text{------} (CH_2CH_2O)_7H$

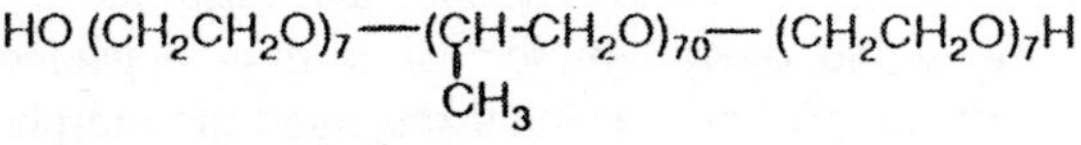

Triblock L121

Fig. 5 Structures of triblock copolymers L121 and L101. (From Ref. 45.)

·········· POP   ━━━━ POE

Fig. 6   Structure of octablock copolymers: POE, poly(oxyethylene); POP, poly(oxy-propylene). (From Ref. 45.)

response. A hypothesis was presented that the oil droplet acts as a depot as well as a surface on which protein, copolymer, and adsorbed host molecules (such as component C) can activate inflammatory mechanisms and be presented to the immune system in a favorable configuration [44].

To further investigate this hypothesis, triblock copolymers, with POP chains flanking the POE chain (so-called reverse triblocks), octablock copolymers (Fig. 6) and triblock copolymers were tested in the oil-in-water system with a more complete characterization of the biological responses. Correlations between chemotaxis, complement activation, antigen retention at the injection site, and adjuvanticity were observed. The more adjuvant-active systems also induced inflammatory responses characterized by local neutrophil infiltration and lymph node germinal center hyperplasia (a site of B-cell differentiation), as opposed to granuloma formation. From a structural standpoint, the reverse triblock copolymers showed a diminished adjuvanticity. In addition, a low HLB was a necessary, but not sufficient, condition for adjuvanticity [45].

The surfaces formed by the nonionic block copolymers were then studied by dispersing them in saline to observe their morphology, or coating them onto supports for physicochemical characterization and adsorptive behavior measurements. In comparison with the other copolymers, the most adjuvant-active copolymers were observed to form fibers with large surface area (instead of spheres) in saline, displayed more hydrophilic surfaces, as determined by contact angle measurements, and adhered serum proteins less tightly. Those copolymers that induced granuloma formation formed crystalline surface domains that were toxic to macrophages. Copolymers with slight structural differences were used to demonstrate the effects of copolymer structure on surface properties. When the POE chains in a triblock copolymer were lengthened, the hydrophilic surface was destabilized because it was too mobile, and adsorbed proteins were shed from it. Changing the overall molecular size of a triblock copolymer, but keeping the relative proportions constant, changed the morphological appearance from a fiber to a sphere because the shortened center POP chain could not fold enough to allow the hydrophilic POE chains to properly orient themselves to form a completely hydrophilic surface. This surface morphological change manifests itself in lower protein binding and lack of complement activation. Finally, antibody-binding studies were used to show that antigen adsorbed onto the more hydrophilic surfaces was more accessible and, thus, its conformation was affected by the hydrophobicity of the surface [46].

These observations were taken as evidence that the orientation of nonionic copolymer segments on the oil drop in the oil-in-water vaccine determines the type of surface formed and, thereby, influences both the orientation of antigen and adsorbed proteins and their interaction with the immune system. Therefore, the nonionic block copolymers provide evidence that the surface structure of an adsorption microparticulate vaccine system is another characteristic that may be investigated to optimize the desired immune response.

D. Tyrosine-Based Adjuvants

1. L-*Tyrosine and Its Simple Derivatives*

L-Tyrosine was first found to be adjuvant-active when used as an adsorbate for grass pollens [47]. Because of their low, but finite, water solubility (5 mg/L) [48], L-tyrosine particulates provide a relatively hydrophobic surface for antigen adsorption, yet can easily be suspended in an aqueous environment. After subcutaneous injection in guinea pigs, significantly higher antibody levels were achieved with antigen–tyrosine adsorbates than with antigen in saline [47,49]. Studies using intraperitoneal injections of radiolabeled antigen demonstrated that adsorbed antigen persisted at the injection site, and its appearance in the bloodstream was delayed relative to antigen in saline [48].

L-Tyrosine embodies many features that make it desirable as an adjuvant. Because it is a naturally occurring constituent of human and animal systems, the toxicity of compounds based on L-tyrosine is expected to be low. Furthermore, it is a relatively simple molecule, yet has functional groups that allow a variety of chemical modifications. Indeed, the finding that L-tyrosine is adjuvant-active when used as a particulate adsorbate has been the genesis for a large volume of adjuvant research.

Because the general trend is that adjuvanticity increases with increasing hydrophobicity, simple *N*-acyl, *O*-acyl, amide, and ester derivatives of L-tyrosine were tested in guinea pigs with the same grass pollen antigens [50]. The findings reproduced the correlation of hydrophobicity with adjuvanticity as the chain length of the derivative group was increased. Wheeler also found [50] that, as the hydrophobicity increased, so did the inflammatory response, as measured by local lesions. In fact, the most adjuvant-active derivatives caused ulceration and were inappropriate for use in animal systems [50], evidence that the adjuvanticity of L-tyrosine and its derivatives involves both the depot effect and immune stimulation by a local inflammatory response.

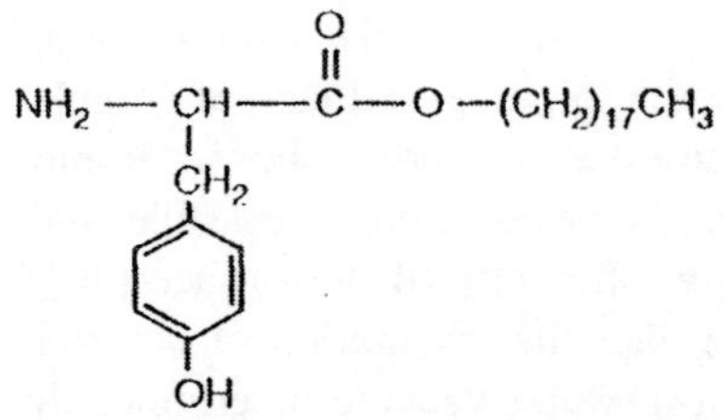

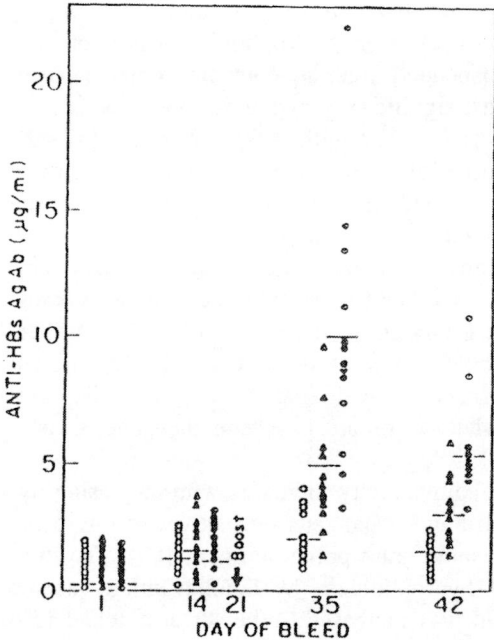

O
‖
NH₂ — CH — C — O — (CH₂)₁₇CH₃
        |
        CH₂
        |
        (ring)
        |
        OH

Fig. 7 Structure of stearyl L-tyrosine.

Interestingly, other researchers have shown good adjuvanticity with fewer problems with biocompatibility using the stearyl ester of L-tyrosine (Fig. 7). Studies using recombinant hepatitis B surface antigen (HBsAg) in mice showed that stearyl L-tyrosine adsorbates produced an adjuvant effect stronger than alum in the secondary response (Fig. 8) [51]. Furthermore, antibodies of the

Fig. 8 Levels of antibodies against hepatitis-B surface antigen (HBsAg) in groups of 15 BALB/c mice immunized with 50 ng of HBsAg in saline (open circles); in 50 μg stearyl tyrosine in a 1-mg/ml suspension (solid circles); and in 100 μg alum (solid triangles). (From Ref. 51.)

IgG2a and IgG2b isotype were preferentially produced; these isotypes are associated with higher affinity and complement fixation [51], with the IgG2a isotype most efficient at antibody-dependent cell-mediated cytotoxicity [52]. Unlike the alum vaccine, the stearyl tyrosine adjuvant did not elicit anti-HBsAg IgE antibodies—a class of antibodies associated with immediate-type hypersensitivity [52]. By using other viral and bacterial antigens in combination with other ester analogues, Penney et al. [52] found that the adjuvant effect of stearyl tyrosine derivatives was quite antigen-specific. Stearyl tyrosine was comparable with alum for bacterial toxoids, but better than alum for viral antigens. A nonaromatic derivative, stearyl glycyl glycine, was more potent than stearyl tyrosine in some viral vaccines. This is noteworthy in that the adjuvant effect of tyrosine has often been explained by the presence of the aromatic moiety, which is known to be involved in T-cell recognition [53] and other immune responses [52]. The stearyl tyrosine derivative has low toxicity and pyrogenicity and does not cause granuloma formation [51], and in none of the foregoing studies do the authors cite adverse local inflammatory reactions.

2. Dityrosine Monomers and Polymers

Because of the demonstrated adjuvanticity of L-tyrosine and its simple derivatives, more complex derivatives were studied for application in controlled release of antigen. Kohn and Langer [54] synthesized a bioerodible dityrosine-based polyiminocarbonate for use in an implantable adjuvant-active vaccine delivery system. Such a system was envisioned to deliver antigen and to enhance the immune response over an extended period, after which the device would degrade and be adsorbed by the body. As part of the characterization of their systems, the adjuvanticity of the monomer, *N*-benzyloxycarbonyl tyrosyl tyrosine hexyl ester (CTTH; Fig. 9), was assessed in particulate form [55]. When used as an adsorbate for BSA in mice, CTTH had an adjuvanticity comparable with that of CFA. There was no histopathological evidence of an acute or chronic inflammatory response toward CTTH particles, which were completely absent from the injection site after 56 weeks [55]. A controlled-release device made of poly(CTTH-iminocarbonate) (Fig. 10), loaded with BSA, was

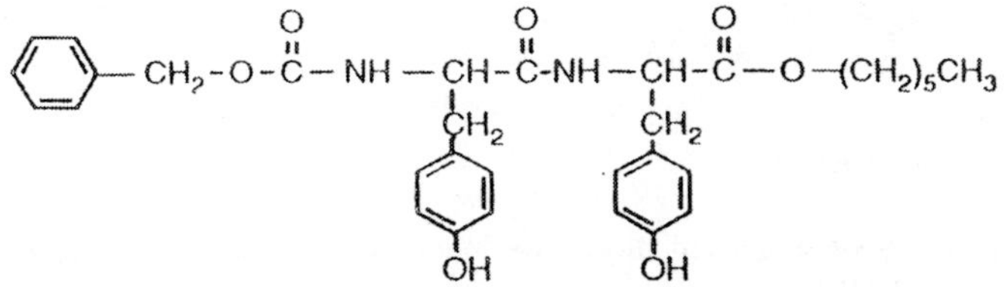

Fig. 9 Structure of *N*-benzyloxycarbonyl tyrosyl tyrosine hexyl ester (CTTH).

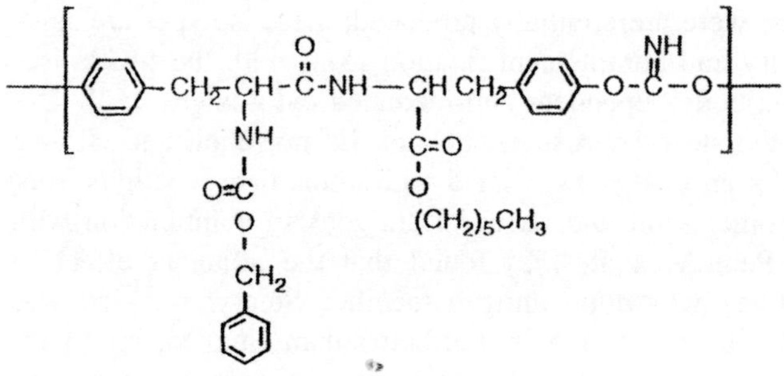

Fig. 10 Structure of poly(CTTH-iminocarbonate).

implanted subcutaneously in mice, and the resulting antibody levels were compared with those from BSA–saline injections and BSA-loaded poly(bisphenyl-*A*-iminocarbonate) [poly(BPA)] implants. Poly(bisphenyl-*A*-iminocarbonate) is a polymer with the same polymerizing bond as poly(CTTH) and with similar in vitro BSA-release profiles [56]. The antibody response from the poly(CTTH)

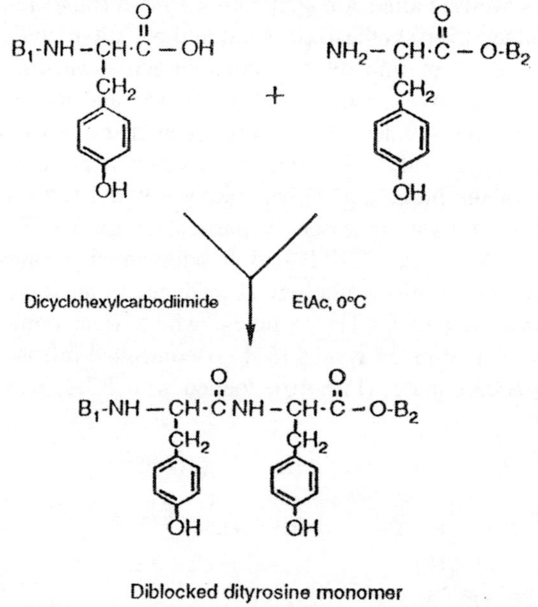

Diblocked dityrosine monomer

Fig. 11 Synthesis pathway of diblocked dityrosine monomers. B_1 can be *n*-propyl, *n*-butyl, *n*-hexyl, or *n*-octyl; B_2 is benzyloxycarbonyl or *t*-butyloxycarbonyl. (From Ref. 54.)

device was superior to that of the injections or the poly(BPA) device. The intrinsic adjuvanticity of the CTTH monomer was promoted as a possible explanation for the superior adjuvanticity of the poly(CTTH) device [55].

To further investigate the relations between structure and adjuvanticity, a systematic investigation of the adjuvanticity of a series of dityrosine monomers and polymers, when used as microparticulate adsorbates for BSA, was undertaken [57]. The synthesis route of Kohn et al. [54] was used (Figs. 11 and 12). Briefly, monomers with either benzyloxycarbonyl or *t*-butylcarbonyl NH_2-terminal blocking groups and with *n*-alkyl ester chains, ranging from propyl to octyl, were prepared and purified by two aqueous and one organic recrystallizations. The tyrosine hydroxyls of the monomers were cyanalated by reaction with cyanogen bromide and then polymerized using a strong base catalyst. The resulting polyiminocarbonates were precipitated into methanol and dried. Particulates were formed by grinding the monomers or polymers in a mortar and pestle and sieving to smaller than 250 μm.

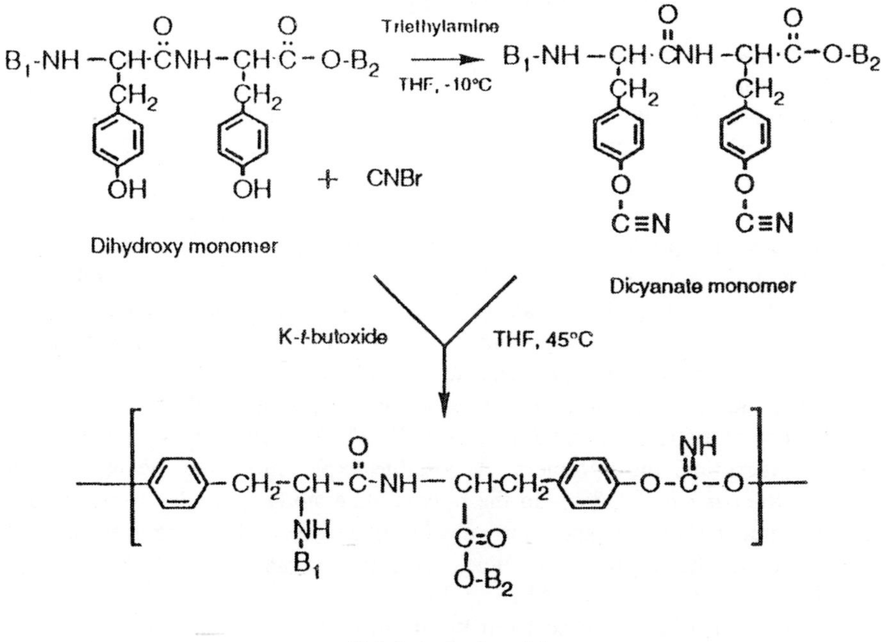

Fig. 12 Synthesis of pathway of poly(iminocarbonate) from dihydroxy monomer. B_1 can be *n*-propyl, *n*-butyl, *n*-hexyl, or *n*-octyl; B_2 is benzyloxycarbonyl or *t*-butyloxycarbonyl. (From Ref. 54.)

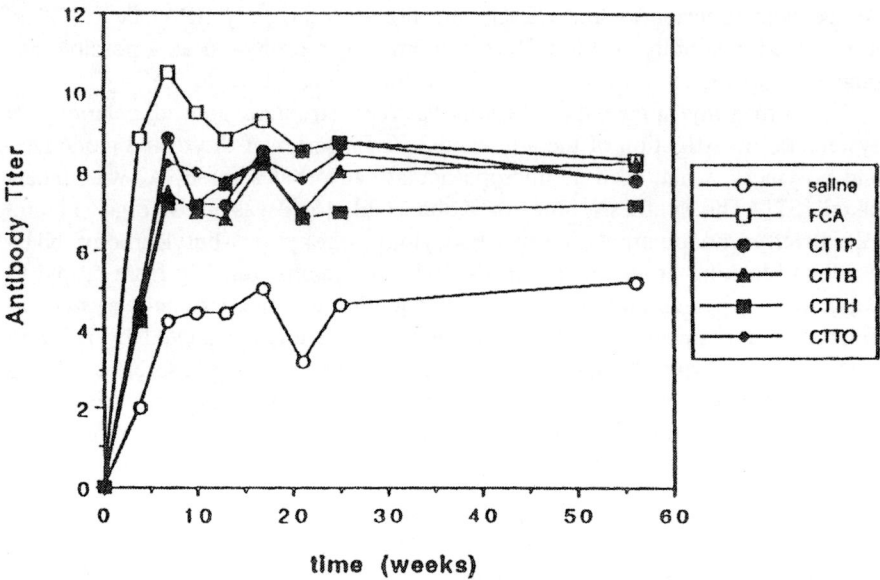

Fig. 13 Anti-BSA levels for monomers with benzyloxycarbonyl (C) NH$_2$-terminal blocking group: P, propyl; B, butyl; H, hexyl; O, octyl. Mice were injected at time zero and at 5 weeks with 50 µg of BSA in saline, CFA, or 40 mg of particles. Data points represent the average of between three and six mice.

Monomer and polymer particulates were adsorbed with BSA and injected subcutaneously into mice, with BSA in saline and CFA as controls, and given as a booster injection after 5 weeks. The antibody levels are found in Figs. 13–16. In direct contrast with previous studies using both L-tyrosine derivatives [50] and polyacrylates [42], the derivatives with the shorter chain lengths (i.e., those that are less hydrophobic) showed the highest antibody response (frequently one statistically comparable with CFA). This trend is seen with both the monomers and the polymers. A possible explanation may lie in the size range of the particles studied. In the polyacrylate study, particles were smaller than 100 nm in diameter and, thus, easily phagocytosed by antigen-presenting cells, whereas the dityrosine particles were in the range of 50–250 µm and would not be expected to be taken up by immunocompetent cells. This means that antigen must be desorbed from the dityrosine particulate before it can be processed by the immune system. Therefore, the enhancing effect of hydrophobicity on particle phagocytosis is not applicable to this system. Instead, the interactions of BSA and the particle surface become paramount. Adsorption of proteins onto hydrophobic surfaces is highly entropically driven, often irrevers-

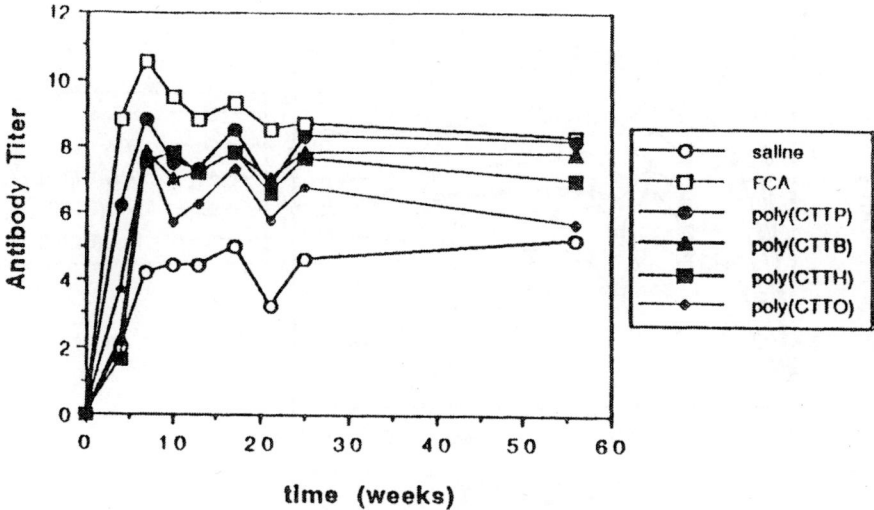

Fig. 14 Anti-BSA levels for polymers with benzyloxycarbonyl (C) NH$_2$-terminal blocking group: P, propyl; B, butyl; H, hexyl; O, octyl. Mice were injected at time zero and at 5 weeks with 50 μg of BSA in saline, CFA, or 40 mg of particles. Data points represent the average of between three and six mice.

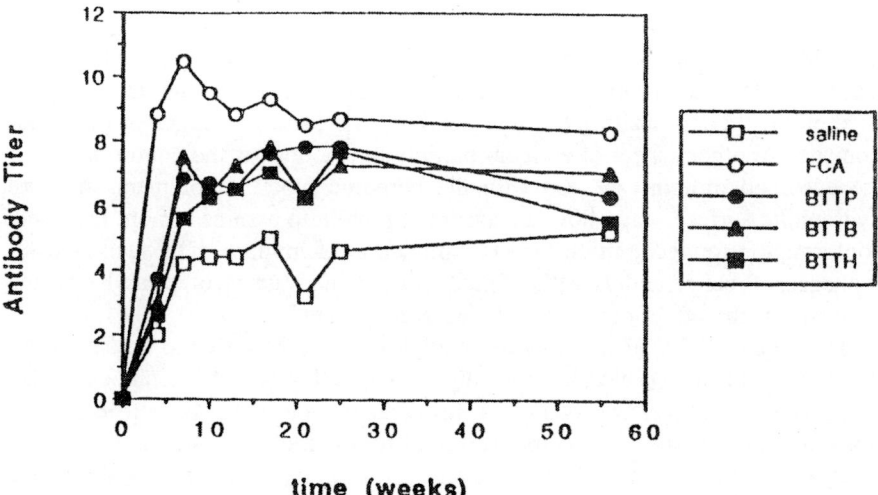

Fig. 15 Anti-BSA levels for monomers with *t*-butyloxycarbonyl (B) NH$_2$-terminal blocking group. Mice were injected at time zero and at 5 weeks with 50 μg of BSA in saline, CFA, or 40 mg of particles. Data points represent the average of between three and six mice.

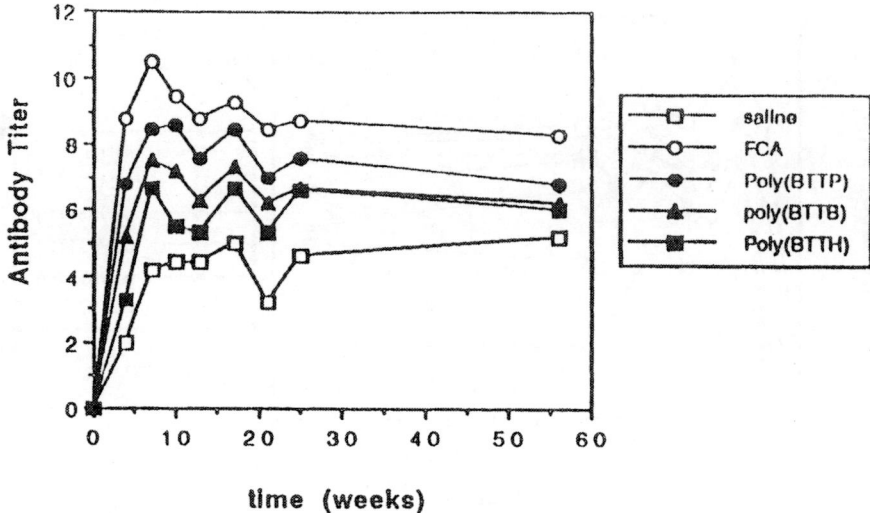

Fig. 16 Anti-BSA levels for polymers with *t*-butyloxycarbonyl (B) NH$_2$-terminal blocking group. Mice were injected at time zero and at 5 weeks with 50 μg of BSA in saline, CFA, or 40 mg of particles. Data points represent the average of between three and six mice.

ible [58,59], and involves interactions with the hydrophobic regions of the protein [60]. Here, a less hydrophobic surface might be expected to enhance BSA desorption, thereby leading to more continuous release of antigen at the injection site. Another aspect of antigen–particle interactions is the conformation of the adsorbed protein, as seen with the nonionic block copolymers. A more hydrophilic surface may allow the adsorbed protein to assume a more favorable conformation for recognition by and stimulation of immune cells such as macrophages, T cells, and B cells. Finally, as in the studies of stearyl tyrosine derivatives, the effect can be highly antigen-specific.

The adjuvanticity of the polymer particles does not appear to be caused by the release of intact monomer during polymer degradation. Iminocarbonate bonds can degrade by two pathways: (a) an acid-catalyzed pathway, leading to transformation of the iminocarbonate bond to a carbonate bond; and (b) base-catalyzed iminocarbonate bond cleavage [56]. In vitro studies of dityrosine polymer particles and slabs at physiological pH show molecular weight decrease and the appearance of a carbonate shoulder on the infrared (IR) spectrum, indicating that both pathways are operating. After a certain time period, the molecular weight reaches a plateau, corresponding to low oligomers, beyond which no further decrease is seen owing to the stability of the carbonate bond at physiological pH. There is little or no weight loss, which would be an indication of solubilization of monomer or monomer degradation products. In fact, 1 year after particles were injected subcutaneously in mice, the polymers persisted at the site of injection, whereas the monomers had disappeared completely [57].

Further development of dityrosine polyiminocarbonate systems, whether devices or particulates, requires improvement of the bioerodibility of the polyiminocarbonate. One possible strategy is to influence the local pH to provide a more basic environment for polymer degradation and favor the base-catalyzed chain cleavage route. Another strategy could be to obtain polymer microparticles of a size that would be phagocytosed by antigen-presenting cells for which the arsenal of degradation mechanisms would be expected to be more varied than would simply pH-driven hydrolysis. These microparticles could be made as adsorbates, or could be used to incorporate and release antigen in the same manner as the poly(lactic–glycolic acids). Toward that end, the synthesis reaction has been modified to increase the polymer molecular weight [61–64] so that conventional microparticulate manufacturing techniques can be applied. Naturally, the immune response toward a microparticular dityrosine system may be quite different from that seen with larger particles, and remains to be seen.

3. Other Tyrosine Polymers

The incorporation of L-tyrosine into biodegradable polymers for use as adjuvant systems is being pursued by a variety of different synthetic strategies. Poly(anhydride-*co*-imide) copolymers containing tyrosine derivatives have been synthesized [65] and have been used to encapsulate and release BSA, a model antigen [34]. Copolymers of trimellitylimido-L-tyrosine (TT) and sebasic acid (SA) or 1,3-bis(*p*-carboxyphenoxy) propane (CPP) in various ratios were synthesized. The microparticulate system was designed to release a soluble tyrosine derivative, by polymer degradation, concomitantly with antigen for the first several days. The in vivo and in vitro adjuvant behavior of this system is currently being investigated and is expected to provide further direction in optimizing the polymer degradation and release characteristics, through the choice and ratios of the copolymers used [34]. In addition, polyphosphates of a variety of dityrosine and serine–tyrosine monomers have been synthesized and used as adsorbates for *Schistosoma japonicum* antigen [66]. The 10-μm–particle adsorbates showed strong adjuvant activity, comparable with CFA and antigen. Depending on the choice of blocking groups, degradation rates at physiological pH can vary from less than a week to much longer.

Because of the flexibility of polymerizations of L-tyrosine-based compounds, there are almost limitless possibilities for novel L-tyrosine-containing polymers for use in microparticulate vaccine systems. As research into the L-tyrosine-mediated adjuvanticity expands the understanding of this phenomenon, polymer synthesis is expected to become more clearly directed by the desired degradation and release rates and physicochemical properties for a given application.

E. Incorporated Immunostimulants

A necessary aspect of any adjuvant-active microparticulate vaccination system is the delivery of antigen to the immune system over a sustained time period. In most of the work described in the foregoing, the adjuvant activity is built into the release matrix, usually a polymer. However, a less elegant, but viable, concept is to use the antigen-delivery system itself to deliver soluble adjuvant simultaneously with antigen, by incorporating them both within the release matrix. The continuous supply of antigen is available for processing and recognition by, and stimulation of, the immune system, whereas the sustained levels of adjuvant in the vicinity of the antigen can directly stimulate such immune cells as macrophages, T cells, and B cells. In fact, controlled-release technology is tailor-made for the delivery of soluble adjuvants, such as MDP, which is rapidly cleared from the body after a bolus injection [67]. The feasibility of controlled delivery of an immunological adjuvant from a microparticulate system has been demonstrated with the release of a muramyl dipeptide derivative from PLGA microspheres for macrophage antitumor activation [68]. In this work, the microspheres were designed to be phagocytosed, thereby localizing the adjuvant within the target cell. However, there is no reason why the components cannot be released extracellularly with a somewhat lower efficiency.

Preliminary work on systems of this type has been reported in the literature. In the development of an antifertility vaccine [18,69], the formulation consists of PLGA microspheres loaded with the antigen (human chorionic gonadotropin) and nor-MDP as an adjuvant. A mixture of fast, moderate, and slow-release microspheres was prepared from polymers of different degradation rates and was able to stimulate antibody production for more than 1 year. Studies with injections of antigen and adjuvant showed that adjuvant was required during the initial immune response to elicit high antibody levels, but that those levels could be sustained for up to a year thereafter, with the delivery of antigen alone [18]. This finding further simplifies formulation in that the sustained delivery of adjuvant needs to be guaranteed only for the first few weeks after injection. This could be achieved by separate populations of antigen- and adjuvant-releasing microspheres designed for long- and short-term release, respectively. Additionally, a vaccine for human immunodeficiency virus type 1 (HIV-1) consisting of a subunit antigen, MN rgp 120, was encapsulated in PLGA microspheres with QS21, a purified component of a saponin adjuvant [70]. A single injection of the microsphere formulation is able to elicit higher neutralizing antibody levels than three injections of the solubilized components at much higher doses [71].

The advantages of this adjuvant strategy are its relative simplicity and the potential to use "off-the-shelf" polymers that are already in clinical use, such as PLGA. However, a key requirement is the availability of a soluble adjuvant for which the immunological effects are appropriate for the antigen being delivered. Also, the simultaneous delivery of two compounds by a release pattern that is efficacious for both involves a complex interplay of formulation parameters, such as dose and loading, polymer degradation behavior, and the chemical properties of the antigen and adjuvant. If the desired delivery patterns and doses for antigen and adjuvant are very different, optimization of their simultaneous delivery may not be practical.

V. CONSIDERATIONS IN THE DEVELOPMENT OF ADJUVANT-ACTIVE MICROPARTICULATE VACCINE SYSTEMS

Because the primary purpose of immunization is to prevent disease, rather than treat an existing one, a vaccine must be designed such that the risks of any adverse effects involved in receiving it are greatly outweighed by the risks involved in actually contracting the disease. And because a polymeric microparticulate vaccine system is more complex than a standard vaccine preparation, care must be taken to design a system that safely and efficaciously imparts protection against disease to the recipient. Furthermore, a method of manufacture for the vaccine must be developed to result in a reproducible, stable, sterile product. Some considerations in the design, formulation, and manufacture of safe adjuvant-active polymeric microparticulate vaccine systems are discussed in the following sections.

A. Design Considerations

During the evaluation of candidate polymers, biocompatibility becomes an important issue. If a nonadjuvant polymer, such as PLGA, is to be used as a release matrix, the polymer is usually chosen to have minimal interaction with the body. Adjuvant-active polymers, though, have been designed to interact in some way with the immune system, antigen, or both, and the biocompatibility of these systems is no longer just a question of being bioinert. In nonerodible polymer systems, the persistence of microparticulates at the injection site or throughout the lymphatic system must be determined, and its effect on the surrounding tissues must be evaluated. All aspects of the continuing presence of an immune-stimulating polymer surface in a living system must be thoroughly studied, especially potential links to autoimmune complications. One need only look to the ongoing controversy over the systemic health effect of silicone in breast implants and vascular devices to see the heightened awareness of this area of materials evaluation [72,73]. When the polymer is bioerodible, then its degradation products must be identified, quantified, and characterized, and their toxicity determined. If the adjuvant-active polymer is composed of an adjuvant-active monomer, the activity of the degraded forms must be studied, as small changes in a molecule (MDP, for example) can profoundly affect its adjuvant and inflammatory actions [74]. Finally, because a polymeric microsphere system may contain a large dose of antigen for release over long periods, either device failure and dose dumping must be shown to have a vanishingly low probability, or the safety of administration of the complete dose as a bolus must be proved.

B. Formulation Considerations

The formulation of a conventional vaccine generally involves choosing generally regarded as safe (GRAS) excipients and preparing them in a form that ensures the potency of the vaccine throughout its shelf life. These requirements often present quite a challenge in and of themselves; in the formulation of polymeric microparticulate vaccine system, the level of complexity raises additional safety and stability considerations.

If an existing polymer is used, a compendial grade should be used when possible. If none is available, a reputable manufacturing source must be found, and critical product parameters, such as polydispersity and contaminants levels, must be identified and their effects on the vaccine quantified. On the other hand, if a novel adjuvant-active polymer is to be used, the vaccine developer has the additional responsibility of determining a method for preparing it according to a well-controlled reaction and purification scheme that results in reproducible material over many batches.

The final form of the vaccine will dictate which stability considerations must be addressed during the formulation phase. An aqueous suspension is feasible if the antigen is adsorbed onto nonerodible polymer microparticles. Here, the stability of the adsorbed antigen must be ensured. Desorption of the antigen from the microparticulate surface is to be avoided, as this may alter the processing of the antigen after injection. An aqueous environment increases the likelihood of chemical degradations, such as oxidation or deamidation [75,76]. The adsorbed antigen, particularly proteins, should be stabilized toward conformational changes that may irreversibly disrupt three-dimensional epitopes or lead to insoluble aggregate formation or altered desorption behavior. Ideally, the foregoing stability considerations would also result in a product that is stable at ambient temperature or higher, allowing it to be easily distributed and stored in geographical areas where immunization needs are great, but refrigeration is difficult.

Dry powder or lyophilized formulations are appropriate when bioerodible microspheres are used or the aqueous stability of adsorbed antigen is not sufficient. Generally, a molecule, once in a dried environment, tends to be relatively stable. However, the processing steps involved in getting to that low-water–content state are often themselves deleterious to antigen or polymer stability. Antigen-loaded polymer microspheres are usually prepared in the presence of organic solvents and lend themselves to forming dry powders. However, during the incorporation process, antigen in contact with organic phases may undergo denaturation or aggregation. Frequently, stabilizers used during this process will be incorporated in the final device, and their effect on product performance must be determined [71]. Lyophilization requires that an aqueous suspension of the vaccine be frozen, and the concentration of components during this process can cause pH shifts, ionic strength effects, and aggregation [77]. The proper choice of cryoprotectant and buffer species will minimize these effects. The formulation of either dry powders or a lyophilized solid must also ensure that the vaccine preparation is efficacious on reconstitution and that no harmful changes in antigen structure during the drying process have been introduced.

C. Manufacturing Considerations

Among the many difficulties involved in manufacturing microparticulate pharmaceuticals for parenteral administration, sterility assurance of the final product may be the most potentially troublesome. Unless the system consists of extremely small nanoparticles, one will not be able to aseptically filter the formulated vaccine through a 0.22-μm-sterilizing filter, as is commonly done with most parenterals. Thus, the sterility will have to be imparted either earlier in the manufacturing process, before the production of the microparticulates, or at the end by terminal sterilization. Formation of the microparticulates under aseptic conditions adds many levels of complexity, as more product intermediates must be sterilized, and their sterility after manufacturing and handling manipulations must be validated. From a unit operations perspective, terminal sterilization would be the preferred method of product sterilization, but it carries its own liabilities. If a suspension formulation is used, moist-heat sterilization is viable, but the cycle must be carefully developed so that the desired sterility assurance level (10^{-6}) is guaranteed, without affecting the stability of the polymer and the antigen. For a heat-labile protein or peptide antigen, this may not be possible. The other option for sterilization is gamma irradiation, a process that forms free radicals that can lead to antigen degradation and aggregation; polymer reactions, such as molecular weight decrease and cross-linking

[78]; and antigen–polymer reactions. Changes in polymer properties can lead to alteration of the release characteristics of a microparticulate [79], and this effect of radiation has been observed with poly(lactide–glycolide) microspheres [80]. For a detailed treatment of the manufacturing issues associated with microparticulate vaccines, the reader is referred to a description of the development of a polylactide microsphere vaccine system [71].

VI. CONCLUSIONS

The development of adjuvant-active microparticulate vaccine-delivery systems seeks to combine the versatility in release profiles obtained with microparticulate-release systems with the additional adjuvant effect sometimes necessary for an effective immune response. These safe, single-injection vaccines would find widespread application in the place of existing conventional vaccines, the efficacy of which is sometimes compromised by patient compliance or reproducibility, and in new subunit vaccines against illnesses such as hepatitis B and HIV for which the antigen is only weakly immunogenic. Besides their potential for clinical use, these vaccine systems are also being used to further probe the immune response to vaccination and to find the optimum antigen- and adjuvant-release profiles.

Antigen can either be adsorbed onto the particulate surface or incorporated within the microsphere to be released over time. The adjuvanticity can be designed into the polymer itself by using a monomeric unit known to be adjuvant-active, such as derivatives of L-tyrosine. Another strategy is to release soluble adjuvant simultaneously with antigen.

Controlled vaccine delivery from bioinert PLGA microspheres leads to levels of immunity far higher and longer-lasting than those obtained by immunization with the conventional adjuvants, such as alum and CFA. This is most likely due to the ability of controlled-release systems to provide a much longer-lasting antigen depot. The next generation of vaccine vehicles may combine an optimized depot effect with adjuvant-active polymers to stimulate the immune system and further improve the immune response.

As with any emerging technology, many formidable challenges in the design, formulation, and manufacture of adjuvant-active polymeric microparticulate vaccine systems for parenteral use remain to be addressed as promising candidates proceed further along the developmental process. The end result, though, will be a new generation of vaccine systems that may not only improve vaccination with existing antigens, but may very well make the difference between success and failure with poorly immunogenic subunit antigens.

REFERENCES

1. J. W. Kimball, *Introduction to Immunology*, Macmillan Publishing, New York, 1986, p. 564.
2. J. Hanes, M. Chiba, and R. Langer, Polymer microspheres for vaccine delivery, *Vaccine Design: The Subunit and Adjuvant Approach* (M. F. Powell and M. J. Newman, eds.), Plenum Press, New York, 1995, pp. 389–412.
3. S. C. Arya, Human immunization in developing countries: Practical and theoretical problems and prospects, *Vaccine 12*:1423–1435 (1994).
4. T. N. V. A. Committee, The measles epidemic: The problems, barriers, and recommendations, *JAMA 266*:1547–1552 (1991).
5. E. R. Zell, V. Dietz, J. Stevenson, S. Cochi, and R. Bruce, Low vaccination levels of US preschool and school-age children: Retrospective assessments of vaccination coverage, 1991–1992, *JAMA 271*:833–839 (1994).
6. K. P. Goldstein, F. J. Kviz, and R. S. Daum, Accuracy of immunization histories provided by adults accompanying preschool children to a pediatric emergency department, *JAMA 270*:2190–2194 (1993).
7. R. Edelman, Vaccine adjuvants, *Rev. Infect. Dis. 2*:370–383 (1980).
8. A. T. Glenny, G. A. H. Buttle, and M. F. Stevens, Rate of disappearance of diphtheria toxoid injected into rabbits and guinea-pigs: Toxoid precipitated with alum, *J. Pathol. Bacteriol. 34*:267–275 (1931).
9. R. Bomford, Aluminum salts: Perspectives in their use as adjuvants, *Immunological Adjuvants and Vaccines* (G. Gregoriadis, A. C. Allison, and G. Poste, eds.), Plenum Press, New York, 1988, pp. 35–41.
10. P. Frost and E. M. Lance, On the mechanism of action of adjuvants, *Immunology 35*:63–68 (1978).
11. J. W. Mannhalter, H. O. Neychev, G. J. Zlabinger, R. Ahmad, and M. M. Eibl, Modulation of the human immune response by the non-toxic and non-pyrogenic adjuvant aluminum hydroxide: Effect on antigen uptake and antigen presentation, *Clin. Exp. Immunol. 61*:143–151 (1985).
12. J. Kreuter and I. Haenzel, Mode of action of immunological adjuvants: Some physicochemical factors influencing the effectivity of polyacrylic adjuvants, *Infect. Immun. 19*:667–675 (1978).
13. S. Shirodkar, R. Hutchinson, D. Perry, J. White, and S. Hem, Aluminum compounds used as adjuvants in vaccines, *Pharm. Res. 7*:1282–1288 (1990).
14. S. L. Nail, J. L. White, and S. L. Hem, Structure of aluminum hydroxide gel. II. Aging mechanism, *J. Pharm. Sci. 65*:1192–1195 (1976).
15. W. J. Herbert, The mode of action of mineral-oil emulsion adjuvants on antibody production in mice, *Immunology 14*:301–318 (1968).
16. H. Fukumi, Effectiveness and untoward reactions of oil adjuvant influenza vaccines, *Symp. Ser. Immunobiol. Stand. 6*:(1967).
17. C. H. Stuart-Harris, Adjuvant influenza vaccines, *Bull. WHO 41*:617–621 (1969).
18. V. C. Stevens, Vaccine delivery systems: Potential methods for use in antifertility vaccines, *Am. J. Reprod. Immunol. 29*:176–188 (1993).
19. J. Freund, J. Casals, and E. P. Hosmer, Sensitization and antibody formation after

injection of tubercle bacilli and paraffin oil, *Proc. Soc. Exp. Biol. Med. 37*:509–513 (1937).

20. F. Ellouz, A. Adam, R. Ciorbaru, and E. Lederer, Activity of bacterial peptido-glycan derivatives, *Biochem. Biophys. Res. Commun. 59*:1317 (1974).

21. H. Ogonuki, S. Hashizume, and H. Abe, *International Symposium on Adjuvants of Immunity 6*:125–128 (1966).

22. I. Preis and R. Langer, A single-step immunization by sustained antigen release, *J. Immunol. Methods 28*:193–197 (1979).

23. S. M. Niemi, J. G. Fox, L. R. Brown, and R. Langer, Evaluation of ethylene–vinyl acetate copolymer as a non-inflammatory alternative to Freund's complete adjuvant in rabbits, *Lab. Anim. Sci. 35*:609–612 (1985).

24. Y. Tabata and Y. Ikada, Effect of the size and surface charge of polymer micro-spheres on their phagocytosis by macrophage, *Biomaterials 9*:356–362 (1988).

25. Y. Tabata and Y. Ikada, Macrophage phagocytosis of biodegradable microspheres composed of L-lactic/glycolic acid homo- and copolymers, *J. Biomed. Mater. Res. 22*:837–858 (1988).

26. J. H. Eldridge, J. A. Meulbroek, J. K. Staas, T. R. Tice, and R. M. Gilley, Vaccine-containing biodegradable microspheres specifically enter the gut-associ-ated lymphoid tissue following oral administration and induce a disseminated mu-cosal immune response, *Adv. Exp. Med. Biol. 251*:192–202 (1989).

27. S. Cohen, M. J. Alonso, and R. Langer, Novel approaches to controlled release antigen delivery, *Intl. J. of Technology Assessment and Health Care 10*:121–130 (1994).

28. S. Cohen, T. Yoshioka, M. Lucarelli, L. H. Hwang, and R. Langer, Controlled delivery systems for proteins based on poly(lactic/glycolic acid) microspheres, *Pharm. Res. 8*:713–720 (1991).

29. A. K. Schneider, Polylactide sutures, U.S. Patent 3,636,956 (1972).

30. D. K. Gilding and A. M. Reed, Biodegradable polymers for use in surgery: Poly-glycolic/poly(lactic acid) homo- and copolymers: 1. *Polymer 20*:1459–1464 (1979).

31. H. Okada, Y. Inoue, T. Heya, H. Ueno, Y. Ogawa, and H. Toguchi, Pharmaco-kinetics of once-a-month injectable microspheres of leuprolide acetate, *Pharm. Res. 8*:787 (1991).

32. J. H. Eldridge, J. K. Staas, J. A. Meulbroek, T. R. Tice, and R. M. Gilley, Biodegradable and biocompatible poly(DL-lactide-*co*-glycolide) microspheres as an adjuvant for staphylococcal enterotoxin B toxoid which enhances the level of toxin-neutralizing antibodies, *Infect. Immun. 59*:2978–2986 (1991).

33. B. Gander, C. Thomasin, H. P. Merkle, Y. Men, and G. Corradin, Pulsed tetanus toxoid release from PLGA-microspheres and its relevance for immunogenicity in mice, Proceedings International Symposium on Controlled Release and Bioactive Materials, Controlled Release Society, Washington DC, 1993, pp. 65–66.

34. J. Hanes, M. Chiba, and R. Langer, Degradation and erosion of tyrosine-con-taining polyanhydrides: Implications for vaccine delivery, Proceedings Interna-tional Symposium on Controlled Release and Bioactive Materials, Controlled Re-lease Society, Nice, France, 1994, pp. 44–45.

35. S. Cohen, H. Bernstein, C. Hewes, M. Chow, and R. Langer, The pharmacoki-

netics of, and humoral responses to, antigen delivered by microencapsulated liposomes, *Proc. Natl. Acad. Sci. USA 88*:10440–10444 (1991).

36. J. Kreuter, R. Mauler, H. Gruschkau, and P. Speiser, The use of new polymethylmethacrylate adjuvants for split influenza vaccines, *Exp. Cell Biol. 44*:12–19 (1976).

37. J. Kreuter and P. Speiser, New adjuvants on a polymethylmethacrylate base, *Infect. Immun. 13*:204–210 (1976).

38. J. Kreuter and P. Speiser, In vitro studies of poly(methyl methacrylate) adjuvants, *J. Pharm. Sci. 65*:1624–1627 (1976).

39. J. Kreuter, U. Tauber, and V. Illi, Distribution and elimination of poly(methyl-2-^{14}C-methacrylate) nanoparticle radioactivity after injection in rats and mice, *J. Pharm. Sci. 68*:1443–1447 (1979).

40. J. Kreuter, M. Nefzger, E. Liehl, R. Czok, and R. Voges, Distribution and elimination of poly(methyl methacrylate) nanoparticles after subcutaneous administration to rats, *J. Pharm. Sci. 72*:1146–1149 (1983).

41. J. Kreuter, U. Berg, E. Liehl, M. Soliva, and P. P. Speiser, Influence of the particle size on the adjuvant effect of particulate polymeric adjuvants, *Vaccine 4*:125 (1986).

42. J. Kreuter, E. Liehl, U. Berg, M. Soliva, and P. P. Speiser, Influence of hydrophobicity on the adjuvant effect of particulate polymeric adjuvants, *Vaccine 6*:253–256 (1988).

43. H. Snippe, M. J. DeReuver, F. Strickland, J. M. N. Willers, and R. L. Hunter, Adjuvant effect of non-ionic block copolymer surfactants in humoral and cellular immunity, *Int. Arch. Allergy Appl. Immunol. 65*:390–398 (1981).

44. R. Hunter, F. Strickland, and F. Kezdy, The adjuvant activity of nonionic block polymer surfactants. I. The role of hydrophile–lipophile balance, *J. Immunol. 127*:1244–1250 (1981).

45. R. L. Hunter and B. Bennett, The adjuvant activity of noionic block polymer surfactants. II. Antibody formation and inflammation related to the structure of triblock and octablock copolymers, *J. Immunol. 133*:3167–3175 (1984).

46. R. L. Hunter and B. Bennett, The adjuvant activity of non-ionic block polymer surfactants. III. Characterization of selected biologically active surfaces, *Scand. J. Immunol. 23*:287–300 (1986).

47. A. C. M. L. Miller and E. C. Tees, A metabolisable adjuvant: Clinical trial of grass pollen tyrosine adsorbate, *Clin. Allergy 4*:49–55 (1974).

48. A. W. Wheeler, D. M. Moran, B. E. Robins, and A. Driscoll, L-Tyrosine as an immunological adjuvant, *Int. Arch. Allergy Appl. Immunol. 69*:113–119 (1982).

49. A. C. M. L. Miller, A. P. Hart, and E. C. Tees, *D. pteronyssinus*–tyrosine adsorbate: Biological and clinical properties, *Acta Allergol. 31*:35–43 (1976).

50. A. W. Wheeler, N. Whittall, V. Spackman, and D. M. Moran, Adjuvant properties of hydrophobic derivatives prepared from L-tyrosine, *Int. Arch. Allergy Appl. Immunol. 75*:294–299 (1984).

51. A. Nixon-George, T. Moran, G. Dionne, C. Penney, D. LaFleur, and C. Bona, The adjuvant effect of stearyl tyrosine on a recombinant subunit hepatitis B surface antigen, *J. Immunol. 144*:4798–4802 (1990).

52. C. L. Penney, D. Ethier, G. Dionne, A. Nixon-George, H. Zaghouani, F. Michon, H. Jennings, and C. A. Bona, Further studies on the adjuvanticity of stearyl tyrosine and ester analogues, *Vaccine 11*:1129–1134 (1993).

53. J. L. Maryanski, A. S. Verdini, P. C. Weber, F. R. Salemme, and G. Corradin, Competitor analogs for defined T cell antigens: Peptides incorporating a putative binding motif and polyproline or polyglycine spacers, *Cell 60*:63 (1990).

54. J. Kohn and R. Langer, Polymerization reactions involving the side chains of alpha-L-amino acids, *J. Am. Chem. Soc. 109*:817–820 (1987).

55. J. Kohn, S. M. Niemi, E. C. Albert, J. C. Murphy, R. Langer, and J. G. Fox, Single-step immunization using a controlled release, biodegradable polymer with sustained adjuvant activity, *J. Immunol. Methods 95*:31–38 (1986).

56. J. Kohn and R. Langer, Poly(iminocarbonates) as potential biomaterials, *Biomaterials 7*:176–182 (1986).

57. D. F. Kline, Tyrosine-based, adjuvant-active polyiminocarbonates and controlled release of antigen, PhD Thesis, Massachusetts Institute of Technology, 1993.

58. D. R. Absolom, W. Zingg, and A. W. Neumann, Protein adsorption to polymer particles: Role of surface properties, *J. Biomed. Mater. Res. 21*:161–171 (1987).

59. W. Norde and J. Lyklema, The adsorption of human plasma albumin and bovine pancreas ribonuclease at negatively charged polystyrene surfaces. I. Adsorption isotherms. Effects of charge, ionic strength, and temperature, *J. Colloid Interfacial Sci. 66*:257–265 (1978).

60. F. MacRitchie, The adsorption of proteins at the solid/liquid interface, *J. Colloid Interface Sci. 38*:484–488 (1972).

61. J. Kohn, The synthesis and characterization of pseudopoly(amino acids): New polymers for medical applications, *Polym. Preprints 31*:178–179 (1990).

62. C. Li and J. Kohn, Synthesis of poly(iminocarbonates): Degradable polymers with potential applications as disposable plastics and as biomaterials, *Macromolecules 22*:2029–2036 (1989).

63. S. Pulapura, C. Li, and J. Kohn, Structure–property relationships for the design of polyiminocarbonates, *Biomaterials 11*:666–678 (1990).

64. S. Pulapura and J. Kohn, Biomaterials based on "pseudo"-poly(amino acids): A study of tyrosine derived polyiminocarbonates, *Polym. Preprints 31*:233–234 (1990).

65. A. Staubli, E. Ron, and R. Langer, Hydrolytically degradable amino acid containing polymers, *J. Am. Chem. Soc. 112*:4419–4424 (1990).

66. H.-Q. Mao, R.-X. Zhuo, and C.-L. Fan, Synthesis and biological properties of polymer immunoadjuvants, *Polym. J. 25*:499–505 (1993).

67. M. Parant, F. Parant, L. Chedid, A. Yapo, J. F. Petit, and E. Ledever, Fate of the synthetic immunoadjuvant, muramyl dipeptide (^{14}C-labeled) in the mouse, *Int. J. Immunopharmacol. 1*:35–47 (1979).

68. Y. Tabata and Y. Ikada, Activation of macrophage in vitro to acquire antitumor activity by a muramyl dipeptide derivative encapsulated in microspheres composed of lactide copolymer, *J. Controlled Release 6*:189–204 (1987).

69. V. C. Stevens, J. E. Powell, A. E. Lee, P. T. P. Kaumaya, D. H. Lewis, M. Rickey, and T. J. Atkins, Development of a delivery system for a birth control

vaccine using biodegradable microspheres, Proceedings International Symposium on Controlled Release and Bioactive Materials, Controlled Release Society, Orlando, 1992, pp. 112–113.

70. C. Kensil, S. Soitysik, U. Patel, and D. Marciani, Structure/function relationship in adjuvants from *Quillaja saponaria* Molina, *Vaccines 92: Modern Approaches to New Vaccines Including Prevention of AIDS* (F. Brown, R. M. Chanou, H. Ginsberg, and R. Lerner, eds.), Cold Spring Harbor Laboratory Press, Cold Spring Harbor, NY, 1992, pp. 35–40.

71. J. Cleland, Design and production of single immunization vaccines using polylactide microsphere systems, *Vaccine Design: The Subunit and Adjuvant Approach* (M. F. Powell and M. Newman, eds.), Plenum Press, 1995, pp. 439–462.

72. R. M. Goldblum, R. P. Pelly, A. A. O'Donell, D. Pyron, and J. P. Heggers, Antibodies to silicone elastomers and reactions to ventriculoperitoneal shunts, *Lancet 340*:610–613 (1992).

73. N. Touchette, Silicone implants and autoimmune disease: Studies fail to gel, *J. NIH Res. 4*:89–92 (1992).

74. F. Audibert, L. Chedid, P. Lefrancier, and J. Choay, Distinctive adjuvanticity of synthetic analogs of mycobacterial water soluble components, *Cell. Immunol. 21*:243 (1976).

75. Y. J. Wang and M. A. Hanson, Parenteral formulations of proteins and peptides. Stability and stabilizers, *J. Parenter. Sci. Technol. 42*:S2 (1988).

76. M. C. Manning, K. Patel, and R. T. Borshardt, Stability of protein pharmaceuticals, *Pharm. Res. 6*:903–917 (1989).

77. M. J. Pikal, Freeze-drying of proteins. II. Formulation selection, *Biopharmaceutics 3*:26 (1990).

78. I. Horacek and L. Kudlacek, Influence of molecular weight on the resistance of polylactide fibers by radiation sterilization, *J. Appl. Polym. Sci. 50*:1–5 (1993).

79. R. Langer and N. Peppas, Chemical and physical structure of polymers as carriers for controlled release of bioactive agents: A review, *J. Macromol. Sci. 23*:61–125 (1983).

80. S. R. Hartas, J. H. Collett, and C. Booth, The influence of gamma-irradiation on the release of melatonin from poly(lactide–*co*-glycolide) microspheres, Proceedings International Symposium on Controlled Release and Bioactive Materials, 1992, pp. 321–322.

13

Oral Vaccination by Microspheres

Hitesh R. Bhagat and Paresh S. Dalal

OraVax, Inc.
Cambridge, Massachusetts

Ranjani Nellore

University of Maryland
Baltimore, Maryland

I. INTRODUCTION

Immunization is the most important preventive action for protection against disease, disability, and death resulting from an infection. It is the act of artificially inducing an active or passive immune response, desired to ward off or even eradicate an infectious pathogen in the system. Passive or short-term immunity is conferred on an individual by exogenously formed antibodies that prevent or ameliorate the advent of infection. Two situations in which passive immunization occurs are the transplacental transmission of antibodies to the fetus and the injection of immunoglobulins (Ig) for a specific preventive purpose. The protection bestowed by passive immunity usually lasts for a short term and, therefore, may require repeated immunization each time the protection from the infection wanes. Active immunization involves the induction of antibodies to develop defensive capability against the infection, and it is accomplished by exposing an individual to the immunogens. This concept serves as a cornerstone of immunoprophylactic or active vaccination [1]. The three major approaches to active immunization employ the use of live-attenuated, inactivated, killed, or detoxified infectious agents; the extracts of infectious agent; and the use of live vectors capable of producing specific antigens to stimulate the immune response against the infectious agent. Live-attenuated vaccines produce an immunological response most like that occurring from a natural infection and generally confer a lifelong protection with a single dose [1]. By contrast, other forms of immunogens do not induce permanent immunity with one dose, making repeated vaccination and boosters necessary to develop and maintain sufficient levels of antibody.

II. BACKGROUND

A. Types of Immunity

Traditionally, vaccine development has been concerned mainly with the induction of systemic immunity by parenteral immunization. Because most infectious agents were detected parenterally following the course of infection, it appeared to be the most direct mode of warding off the infection. Consequently, a substantial number of vaccines developed employ the parenteral route of administration. Until the 1960s it was perceived that only systemic mechanisms existed for the production of immunity following an antigenic stimulus, and the detection of antibodies in external secretions was the result of an overflow of the conferred immunity [2]. However, the success of polio vaccine by oral administration and the poor induction of mucosal immunity by parenteral administration of cholera vaccine suggested the presence of an immunological system characteristic of certain external secretions. Considerable evidence has now been gathered that indicates the existence of an independent mucosal immune system [3] that can be anatomically and functionally divided into at least two distinct interconnected compartments. The IgA derived from mucosal effector sites represents more than 75% of all antibody isotypes produced in humans [4]. These mucosal inductive sites include the gut-associated and the nasal-associated lymphoreticular tissue, strategically placed in the gastrointestinal tract and the nasopharyngeal tonsil area, respectively. Antigenic stimulation of these tissues cause dissemination of the T-helper cells and the IgA precursor B cells to the effector tissue and to the secretory glands for subsequent antigen-specific antibody response. The distribution of the effector tissue in the bronchus-associated lymphoreticular region may represent another site for immune response to antigenic stimuli in the respiratory tract [4].

B. Potential of Mucosal Immunization Therapy

Because most infectious agents enter the body through mucosal surfaces, immunization of these surfaces will represent a potent mechanism of warding off the pathogen at the site of entry. The gastrointestinal, nasopharyngeal, pulmonary, and genitourinary surfaces are bathed in mucus that contains immunoglobulins, almost exclusively of secretory IgA derived from the plasma cells underlying the mucosal membrane [3]. Manifestation of this IgA response has been achieved by a direct antigenic stimulus of the mucosally associated lymphoid tissue. However, the development of delivery systems targeted to these immunologically active sites remains a practical challenge.

C. Routes of Mucosal Immunization

The theoretical basis for oral immunization is ascribed to the existence of the compartmentalized common mucosal immune system. Following this concept, antigenic stimulus in the gut-associated lymphoid tissue or bronchus-associated lymphoid tissue will stimulate IgA precursor cells that migrate to distant surfaces where they express secretory IgA specific for that antigen. Oral immunization is both practical and relatively safer and, therefore, is a relatively preferred route compared with parenteral administration. Oral immunization thus should effectively stimulate antibody production in all associated secretory sites [5–7]. The feasibility of such oral immunization has been examined in a limited number of studies. Moderate immunogenic response achieved by this route has prompted the concurrent administration of adjuvants, such as cholera toxin, to amplify the immune response.

D. Mucosal Immunization Delivery Systems

The current strategy for the induction of a mucosal immune response involves the use of particulate-delivery systems. Formulation of antigens into particulate carrier systems offers the potential of optimizing delivery to immunoresponsive sites and also protection of the antigen against proteolytic degradation in the gastrointestinal tract.

1. Replicating Antigen-Delivery Systems

The replicating antigen-delivery systems proliferate in the host tissues following immunization, resulting in a more prevalent and a effective presentation of the antigens [8] and, therefore, are more likely to be effective at stimulating an immune response following oral delivery. Replicating systems may also be able to induce a cytotoxic T-lymphocyte response that can result in longer-lasting immunity. Typically, a replicating-delivery system involves the genetic alteration of a live virus to function as a vector. Multiple foreign genes can replace the nonessential regions of the viral genome, resulting in an immune response against multiple pathogens. This approach enables the recombinant virus to function as a vaccine for two or more infectious agents. The consequences of the host recombinant virus, however, need to be better defined before replicating systems can gain broad acceptability as a preferred antigen delivery system.

2. *Nonreplicating Antigen-Delivery Systems*

Recent advances in DNA recombinant technology has led to some novel approaches that are focused on the production of particulate antigen-presentation systems [8] (e.g., Ty-VLP; [Ty- virus-like particles], HBsAg [hepatitis B surface antigen] and others). The Ty-VLPs are produced by transposing the fusion proteins of the yeast on *Ty*-encoded particle-forming protein and the viral protein of interest. Each particle is approximately 50 nm in diameter and contains several hundred copies of the coupled antigen, expressed in a polyvalent form on the particulate surface. For HBsAg, the antigenic proteins self-assemble to form 22-nm particles that serve as the immunogen. Another interesting approach of antigen presentation involves construction of solid matrix antibody–antigen (SMAA) complexes by attachment of monoclonal antibodies (MAbs) to a suitable solid matrix. The desired antigens are complexed onto the monoclonal antibodies to produce SMAA particulates that are capable of presenting multiple antigens to the immune system and, thereby, evoking an enhanced multivalent immune response. The enhancement of the immune response is believed to be due to the particulate character of the system and the presence of the antibody–antigen complex, which is a potent stimulator of the humoral and cell mediated immunity [9].

A newer approach to the presentation of peptide antigens is by linking multiple radially branching peptide epitopes, 9–15 residues in length, to a core matrix of a trifunctional amino acid, such as lysine. These multiple antigen–peptide systems may also be sufficiently flexible to permit the inclusion of multiple antigenic peptides, such as the T- and B-cell epitopes to produce an enhanced antibody response to the proteins from which these peptides were derived. The use of biodegradable polymeric particulate carrier systems has been pursued fairly vigorously during the past decade. The choice of the polymeric material has been primarily serum albumin, polylactide and polyglycolides, and poly(acrylic acid) systems [9]. Liposomal systems have also been investigated for the delivery of the antigenic load [10]; however, the instability of these systems in the gastrointestinal tract has been the constraining factor in its successful implementation. Lectin-mediated delivery has also been the subject of several studies. In general, lectins resist proteolytic hydrolysis in the gastrointestinal tract, they readily bind to the endothelial cells, and are rapidly endocytosed. Their absorption is not limited to preexisting secretory or systemic immunity. However, the prevalence of toxic symptoms undermines the use of the lectins as a readily suitable particulate carrier. Furthermore, the consequences of conjugation of the lectins to antigens on the immune responses need to be further evaluated before its successful implementation as a suitable antigen delivery system.

III. ORAL IMMUNIZATION USING MICROSPHERES

Potentials and Pitfalls of Microsphere Delivery Systems

Potentials	Pitfalls
Protection of antigen during transit	Poor uptake of microparticles
Controlled release of antigen	Potential loss of antigenicity during encapsulation and storage
Potential for pulsed release	Variability of uptake among individuals
Incorporation of multiple antigens and immunostimulants possible	Expensive and complicated manufacturing process
Polymer can enhance immunogenicity of synthetic subunit vaccines	Toxicity of residual organic solvents

A. Biodegradable Polymers

When attempting to design a microparticulate antigen delivery system for targeting mucosal sites, it is important to choose a polymer that is biodegradable, biocompatible, and safe for use in humans. Implantable polymers have frequently been used for various applications where localized delivery of drug is desired, but the nonbiodegradable systems need to be surgically removed from the body once complete release has occurred. A biodegradable polymer is ideal for immunization purposes, for it can release the antigen at the desired rate and does not necessitate an additional surgical step for retrieval of the depleted system. Although various classes of biodegradable polymers (e.g., polyesters, polyamides, polyorthoesters and polyanhydrides) are available, homo- and co-polymers of lactic and glycolic acid that fall in the category of biodegradable polyesters are the most popular candidates for vaccine delivery [11–13]. These polymers have demonstrated excellent tissue compatibility, are nontoxic, and have been used for several years as resorbable sutures. In addition, these are currently the only biodegradable polymers approved for human use [14].

Polylactide, polyglycolide, and their copolymers are easily prepared by cyclic dimerization of the lactic and glycolic acid monomers, followed by ring-opening polymerization of the dimers in presence of a catalyst. The monomers can either be D-, L-, or DL-isomers, thus offering possibilities for synthesizing polymers of various characteristics by changing variables such as monomer stereochemistry, comonomer ratio, polymer chain linearity, and molecular weight. Release of entrapped molecules occurs primarily by bulk erosion of the polymer, although a combination of surface erosion, diffusion of the active agent through the polymer itself, or its release through the pores may also play a part. Random, nonenzymatic, hydrolytic scissioning of the ester groups to form the water-soluble monomers, lactic and glycolic acid, which are endogenous metabolites, is perceived to be the primary mechanism of biodegradation.

Degradation kinetics and release times differ for the various polymers in this class, with poly(L-lactide) exhibiting the slowest degradation rate, followed by poly(DL-lactide), poly(DL-lactide-*co*-glycolide), and lastly polyglycolide with the fastest degradation rate [15]. Detailed discussions on the different polymers of the lactide–glycolide class and their properties have been published [15].

B. Microspheres

Biodegradable polymers can be used in many different forms, depending on the application, with microspheres or microcapsules being the most popular form. These are free-flowing spherical microparticles, ranging in size from a few micrometers to about 200 μm. The term *microcapsule* usually refers to a reservoir-type system in which the active molecules are enclosed in the cavity surrounded by the polymeric membrane, whereas the term *microsphere* usually implies a monolithic system in which the active agent is uniformly distributed through the polymeric matrix. Microcapsules are expected to provide diffusion-controlled zero-order release, whereas microspheres are expected to provide erosion-controlled first-order release of incorporated drug [16], although in practice, both mechanisms coexist in various proportions. Particles that are smaller than 1 μm are usually termed *nanoparticles*, but in general, the polymers as well as the processing techniques used for nanoparticles differ considerably from the ones used for microparticles. The properties, characterization, and application of biodegradable nanoparticles have been reviewed [17].

C. Microencapsulation

Several techniques are available for microencapsulation, and the choice of a method depends on the physical and chemical properties of the polymer and antigen to be encapsulated, and the function and desired size of the microspheres. A high ratio of antigen to polymer is preferred to minimize the amount of mass that needs to be administered, without compromising the release kinetics. In addition, the microencapsulation technique must afford a pharmaceutically acceptable product relative to residual solvents and processing aids, batch-to-batch reproducibility, ease of scale-up, and high encapsulation efficiency and yields. For commercialization, cost-effectiveness is also an important requirement, especially for product isolation and drying and for solvent disposal.

The microencapsulation techniques in use can be broadly classified as

1. Solvent extraction and solvent evaporation
2. Phase separation
3. Spray drying

1. Solvent Extraction and Solvent Evaporation

The extraction–evaporation microencapsulation methods have been widely used because they can be easily set up in a laboratory and do not require any specialized equipment. A good review of procedures and modifications of solvent extraction and evaporation methods used for microsphere manufacture has been presented [18]. In both these processes, the polymer is first dissolved in a suitable volatile solvent, usually methylene chloride for solvent evaporation, or acetonitrile for solvent extraction. The active agent can be incorporated into the polymer solution either as an aqueous solution, to form a primary emulsion, or as a solid matrix, which forms a dispersion. In such a system, droplet formation is a dynamic process in which droplets constantly form, collide, and coalesce or redivide. The size distribution of the droplets at steady-state depends on various parameters of the experimental setup, including shear, viscosity of the two immiscible phases, interfacial tension between the liquids, and the presence of stabilizer(s). In the solvent extraction process, the solvent for polymer is dissolved away when the emulsion (or dispersion) is added to a suspension medium that is a nonsolvent for the polymer (e.g., heptane). This leads to the formation of solid microspheres in a short period, the microspheres can be recovered either by filtration or centrifugation. Solvent extraction has been used for encapsulation of various peptides and proteins [19]. On the other hand, in the solvent evaporation process, droplet solidification occurs by evaporation of the volatile solvent at the continuous-phase–air-phase interface. Most commonly, the primary water-in-oil (W/O) emulsion is formed by the aqueous solution of the antigen in the polymer solution, which is later emulsified into a large volume of aqueous phase (typically, aqueous solution of a suitable emulsifier, such as polyvinyl alcohol) to form an W/O/W emulsion. In general,

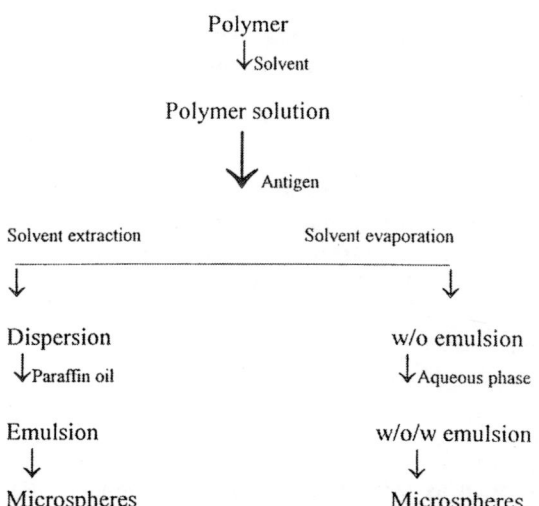

Fig. 1 Steps in manufacturing microspheres using solvent extraction and solvent evaporation processes.

longer-processing times are required to obtain solid microspheres by solvent evaporation. The mixing rate and evaporation time need to be carefully controlled for reproducibility. Figure 1 shows a schematic diagram of the various steps involved in microsphere manufacture using either the solvent extraction or solvent evaporation process. Considerable effort has been directed toward the assessment of various-processing conditions and techniques for manufacture of microspheres using these methods [20,21]. Solvent evaporation has been used to successfully encapsulate proteins, including bovine serum albumin, ovalbumin, tetanus toxoid, staphylococcal enterotoxin B toxoid, and peptides, such as leuprolide acetate [22–25]. Entrapped volatile solvent in the microspheres can be significantly reduced by subjecting the microspheres to vacuum drying; however, trace amounts of organic solvents are often difficult to remove. To minimize the safety and regulatory concerns when dealing with organic solvents, liquid carbon dioxide under supercritical conditions has recently been used as a nonsolvent for the polymer.

2. Phase Separation

The earliest efforts at microencapsulation used coacervation–phase separation techniques to produce pressure-sensitive dye microcapsules for applications such as carbonless carbon paper. Gelatin and ethyl cellulose are the polymers most commonly used to make microcapsules using this technique [26]. Robinson et al. [27] used a nonsolvent-induced coacervation method to encapsulate gentamicin sulfate in poly(lactide) microspheres. In this method, the polymer was dissolved in a suitable solvent (methylene chloride), to which the active drug was added to form a suspension. Hexane was used as a nonsolvent to cause separation of the polymer around the active drug particles. Once phase separation occurred completely, the microspheres were allowed to harden and then separated from the medium.

3. Spray-Drying

The spray-drying technique has not been exploited to the fullest potential for microencapsulation because of the need for specialized equipment; however, this is the easiest method to scale-up. Spray-drying converts a liquid into a powder in a single step. It involves the following four stages:

1. Atomization of feed into a spray
2. Spray–air contact
3. Drying of the spray
4. Separation of dried product

Several atomization systems, such as rotary, pressure, and pneumatic, are available. The feed that is pumped through the atomizer, using a suitable pumping system, is either an emulsion or dispersion of the active agent in polymer solution. The feed is sprayed into the drying chamber, where air is blown in either cocurrent or countercurrent pattern. Because of the availability of a large surface area for evaporation, both heat and mass transfer occur rapidly, leading to formation of dry microspheres in a very short time. The product is separated from the airstream by a cyclone separator [28].

The method of microencapsulation significantly affects the characteristics of the microspheres. In general, porous, spherical particles, with a broad size distribution, that provide rapid release of incorporated active agent are obtained by the solvent extraction technique, whereas less porous microspheres are obtained by solvent evaporation and phase separation techniques. All of these techniques provide high yields on a laboratory scale, but are difficult to scale up. Spray-drying provides porous powders that are approximately spherical with a narrow size distribution. The technique is also easily scalable; however, the microsphere yields are significantly lower on the laboratory scale.

IV. M CELL BIOLOGY

Because mucosal surfaces serve as the portals of entry for a variety of bacterial, viral, and parasitic organisms, it seems logical to target these surfaces for development of immunity. But most current vaccination protocols call for parenteral immunizations, which do not effectively induce mucosal antibodies. The mucosal system differs from the systemic immune system in its main immunoglobulin, its method of stimulation, and also its function. In spite of the distinct advantages offered by oral immunization, only limited success has been achieved in the local immunization strategies that have been pursued. To design a successful antigenic delivery system for oral immunization, one needs to first understand the basics of the common mucosal immune system, with special emphasis on the gut-associated lymphoid tissues (GALT). The largest mass of lymphoid tissue found along the gastrointestinal tract (GIT) are the Peyer's patches (PP) which are the major sites of antigen uptake. The PP (Fig. 2) typically consist of a dome region over the follicle, where the B-cell and T-cell zones are located. Macrophages, plasma cells, and B lymphocytes are also found in the dome region. A unique epithelium, made up of cuboidal epithelial cells and microfold cells (M cells), separates the dome region from the gut. Some degree of antigen uptake occurs by intracellular digestion through the cuboidal epithelial cells, which have fewer goblet cells, thus limiting mucous secretion to facilitate the process. However, the M cells are the major sites of antigen uptake and differ both morphologically and functionally from the rest of the epithelial cells. M cells have short microvilli, small cytoplasmic vesicles, and fewer lysosomes. The cytoplasm of the M cell engulfs one or more intrusive particles, such as viruses, bacteria, and other macromolecules that impinge on the M cell. Antigen uptake by the M cells does not lead to digestion, but

INDUCTIVE SITES (GALT)

Antigens (Vaccine)

M Cell

Dome

T-Cell Zone

B-Cell Zone

MLN → TD → Blood

Homing

sIgA+ B cells
CD4+ T cells
(IL-5, IL-6)

EFFECTOR SITES

Lamina Propria
Gastrointestinal tract
Upper respiratory tract
Genitourinary tract

Glandular Tissue

Mammary
Salivary
Lacrymal
Sweat (?)

Plasma Cell

Epithelial Cells

SC

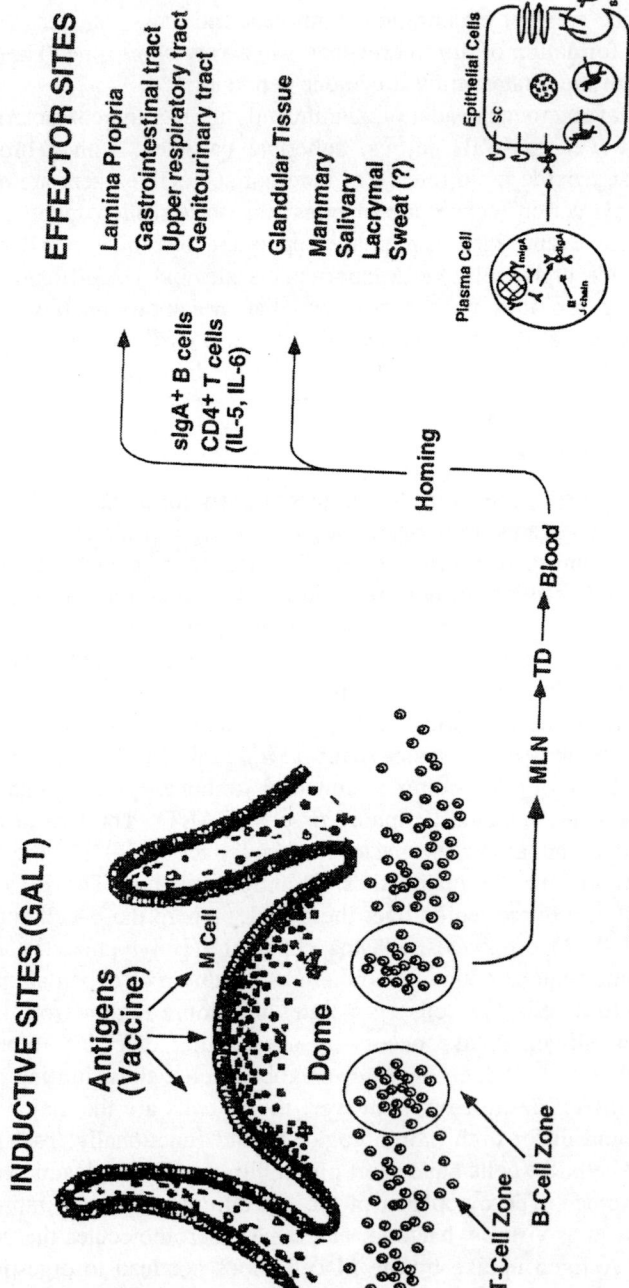

Fig. 2 Inductive sites in gut associated lymphoid tissue.

actually results in transport of the intact particle from the gastrointestinal lumen to the underlying lymphoid tissue to present the particles to the cells of the mucosal immune system. The follicles contain germinal centers where a large proportion of B cells that are committed to IgA synthesis reside. After antigen stimulation in PP, the antigen-induced B and T cells leave the PP through the efferent lymphatics and reach the systemic circulation through the thoracic duct. Circulating B cells then enter distant mucosal tissues, where they are preferentially retained. In the mucosal effector sites, B cells clonally expand and mature into IgA plasma cells. Thus, antigen-sampling occurs at the M cells of GALT, with the IgA-committed lymphoblasts migrating to mucosal effector sites where they secrete IgA locally and provide immunity. This sequence of events constitutes the common mucosal immune system that is responsible for conferring immunity at mucosal sites.

Because the PP have been proved to be the primary sites of particle uptake, considerable effort has been directed toward specifically targeting antigens to these sites to facilitate antigenic stimulation, as well as to provide a means that will result in enhanced uptake by the M cells. In general, uptake of antigens in particulate form is superior to that in soluble form. Microspheres, because of their inherent particulate characteristics, serve as suitable vehicles for transport of antigens across the intestinal mucosa by way of M-cell uptake. However, the total particulate uptake remains quite insignificant and, therefore, quantitation of the extent of uptake of orally administered particles and evaluation of the factors affecting it have been of paramount interest.

V. QUANTITATION OF MICROSPHERE UPTAKE

Chess et al. [53] suggested in 1950 that ingested, finely divided particles are taken up from the GIT and are responsible for production of chronic enteritis and systemic lesions. Although the phenomenon of particle uptake from the GIT was studied from time to time, it was not until the 1980s that definitive studies were conducted. Although the extent of uptake is still an area of disagreement, it is generally accepted that intact particle uptake does occur for particles smaller than 10 μm in diameter. It is also generally understood that the uptake is rather limited. This realization has led to several efforts directed toward quantitation of the extent of particle uptake and studies in improving particle characteristics to increase uptake. Because of poor bioavailability of orally delivered microspheres, there is a significant increase in modeling in vivo microsphere uptake using laboratory animals. Approaches to quantitate the uptake have been developed at the cellular and morphological [31,32] as well as whole-tissue levels [33,34]. Morphological studies have commonly been performed using the rabbit ligated-loop model. Typically, Peyer's patch sections are tied carefully in an anesthetized rabbit to maintain continuation of blood flow, but minimize movement of contents in the intestine into which a suspension of microspheres is introduced. The animal is kept alive for the duration of study (usually, not more than 2 h), after which the Peyer's patches are removed and processed for microscopic observation and counting of particles. Generally, fluorescent microspheres are used to aid in visualization. Howard et al. [32] developed a unique approach for quantitation of particle uptake in which mesenteric vessels of ligated rabbit loop were monitored for fluorescent polystyrene microspheres. These morphological methods of determining uptake, although quantitatively definitive and anatomically specific, are tedious and require considerable skill. Additional pitfalls of the method are the artificiality of the ligated-loop approach and difficulty in providing longer-term observation of particle uptake. There is also a potential for recounting a microsphere split into consecutive tissue sections. To overcome some of these disadvantages, several researchers have taken a whole-tissue approach. Here, the animals are fed the microparticles orally, and the tissue in which particle uptake is to be quantitated is retrieved and then dissolved or handled otherwise to extract the microspheres, or their components. Jani et al. [34] extracted freeze-dried ground tissue and quantitated uptake through analysis for polystyrene from the polystyrene microspheres. Ebel [33] dissolved the tissue in a mixture of ethanol and potassium hydroxide and quantitated fluorescent microspheres by flow cytometry. Bhagat et al. [35] modified the approach to make it suitable for quantitation of polyester microspheres. They evaluated uptake in four different animal groups (Wistar rats, Syrian hamsters, "young" Balb/c mice, 6 to 8-weeks old; and "old" Balb/c mice; 14 to 16 weeks old) and quantitated microsphere uptake using a fluorescence assay. Fluorescent microspheres were found in Peyer's patches (Fig. 3), but not in spleen, mesenteric lymph or, intestine without Peyer's patches. The arithmetic mean microsphere uptake was highest in rats, followed by hamsters, old mice, and young mice. The microsphere uptake, as a percentage of dose administered, varied from 7.4 to 2.8×10^{-5} and the percentage uptake appeared to be in agreement with those of Ebel [33] and Howard [32].

Particle-related factors that may affect the extent of uptake include the particle size, surface charge, hydrophobicity, and attachment of ligands. Extraneous factors affecting uptake include the formulation vehicle and its volume [36,37], as well as fasted versus fed state of the animal [33]. In general, smaller-sized particles have been reported to have a higher extent of uptake. Jani et al. [38] quantitated uptake of fluorescent polystyrene microspheres of 0.05, 0.1, 0.3, 0.5, 1.0, and 3.0 μm. Although 34% of the 0.05-μm particles were taken up, only about 5% of the 1-μm particles were taken up. The distribution of the particles was also size-dependent. The effect of surface charge has been investigated using liposomes [39], and it was observed that negatively charged liposomes were preferentially taken up by Peyer's patches. The effect of hydropho-

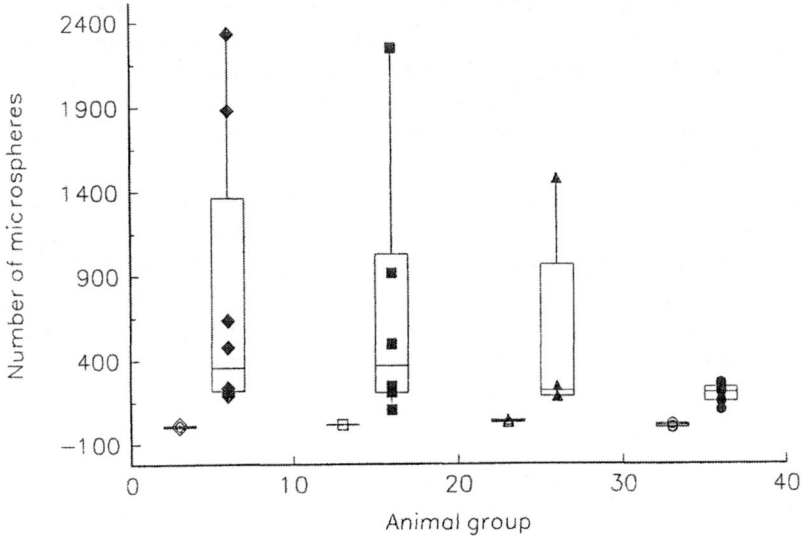

Fig. 3 Uptake of orally administered fluorescent polystyrene microspheres in Peyer's patches of rats (◆); rat controls (◇); hamsters (■); hamster controls (□); large mouse (▲); large mouse control (△); mouse (●); and mouse control (○).

bicity on uptake was reported by Eldrich et al. [40] from which it was concluded that hydrophobic polystyrene microspheres were preferentially taken up, poly(lactic acid) (PLA) and poly(lactide-*co*-glycolide) (PLG) microspheres were taken up to a lesser extent, and hydrophilic cellulose matrices were not taken up at all. A study in rabbits has confirmed that polystyrene (PS) microspheres are preferentially taken up over PLG microspheres [41]. Several ligands, with specific affinity to GALT, have been investigated. Generally, the ligand is attached to the surface of the microspheres, either by adsorption or, by covalent linkage. Pappo et al. [42] developed a monoclonal antibody directed against rabbit Peyer's patches. Surface adsorption of this antibody to PS microspheres resulted in a two- to threefold increase in uptake of the particles.

VI. ORAL VACCINATION BY MICROSPHERES

A. Influenza

Haemophilus influenzae assails the mucosal surfaces of the respiratory tract causing considerable distress and sometimes precipitating chronic bronchitis. Clancy et al. [43] tested killed preparation of *H. influenzae* in a clinical study on patients prone to recurrent acute bronchitis. The study demonstrated that oral administration of killed *H. influenzae* was partially effective in reducing the incidences of acute bronchitis when compared with placebo treatment over a 6-month period. Farag-Mahmod et al. [44] studied the immunogenic response following oral and duodenal administration of formalin-inactivated influenza virus in mice. The study concluded that protective immunity could result if the antigen is protected from degradation in the gastrointestinal environment. In a set of experiments [44,45], comparing the oral immunization of free versus microencapsulated virus, administered by the systemic and oral routes, in mice; the results were more dramatic, with the primary experimental population exhibiting greater protection against the challenge infection. The study indicated an enhanced induction of protective immunity by administration of microencapsulated antigen when compared with the free-antigen suspension. Other studies that used protenoid-encapsulated influenza virus antigen [47] have also demonstrated a similar enhancement of immune response following oral administration when compared with the unencapsulated preparations of the antigen.

B.   Staphylococcal Enterotoxin B

Eldridge et al. [40,48] used biodegradable microspheres constructed of the copolymer poly (DL-lactide-*co*-glycolide), containing formalized staphylococcal enterotoxin B to examine the potential for raising an immune response to microencapsulated antigen in mice. Microspheres in a size range of $1-10$ μm were demonstrated as an effective delivery vehicle for the antigen, and its immunopotentiating action was concluded to be due to the protection of the labile antigen by the wall material during gastrointestinal transit and the efficient uptake of the microspheres by the Peyer's patches. In contrast, the soluble antigen was relatively ineffective as an immunogen following oral administration. The data also suggested induction of both systemic and mucosal antibody responses following the oral administration of the microencapsulated antigen [49]. Similar results were also observed when poly(DL-lactide-*co*-glycolide) (PLG) microspheres were used to deliver staphylococcal enterotoxin B toxoid in an experiment involving the immunization of rhesus monkeys.

C.   Enterotoxigenic *Escherichia coli*

The potential of PLG microspheres in immunization against enterotoxigenic *E. coli* was investigated [19]. Colonization factor antigen (CFA/II) was obtained from *E. coli* and encapsulated into biodegradable PLG by a solvent extraction process. The microspheres were characterized in vitro for size distribution, antigen loading, antigen release, moisture, and residual solvent content. The microspheres, evaluated in vitro using a naive rabbit model, were safe and immunogenic when administered both intramuscularly as well as intraduodenally. The CFA/II-loaded microspheres were investigated in the first clinical trial of oral PLG microsphere vaccine. Both serology data and challenge with *E. coli* in human subjects indicated a potential for immunization with this approach [50,51].

VII. MICROSPHERES TO PROVIDE RELEASE OF ANTIGEN IN THE GASTROINTESTINAL TRACT FOR IMMUNIZATION

Another alternative approach by which microspheres could be useful for oral immunization is through providing release of antigen in the GIT. In this instance, microsphere size is not limited by the upper limit of particle size (5–10 μm) for phagocytosis, and the microspheres can be nonbiodegradable. Furthermore, the microspheres may be enteric-coated to prevent degradation in the stomach. On entry into duodenum, the enteric coating can dissolve to release the microspheres and initiate antigen delivery. Although conventional oral dosage forms, such as tablets and capsules, may also be used for release of antigens in the GIT, an advantage of the microsphere delivery system is prolonged residence time in the GIT. The residence time may be further increased by attachment of adhesions or bioadhesives to the surface of the microspheres [52].

VIII. DELIVERY SYSTEMS FOR MICROSPHERES

For commercial use, microspheres should be placed in a pharmaceutically acceptable oral delivery system. Potential oral delivery systems include tablets, capsules, and dry powder for reconstitution into a suspension. The former two are the most widely used oral delivery systems, but could pose significant technological challenges for delivery of microspheres. Tablets are prepared by compression of free-flowing granules, and the compression forces may deform and fuse the microspheres, rendering them into larger aggregates on disintegration in aqueous media. Also, compression forces and heat generated during compression may affect the stability of encapsulated antigen.

A capsule formulation, although providing more favorable manufacturing conditions, could lead to chemical instability. Commercial capsules are made of gelatin or starch and may have up to 12% moisture to maintain flexibility. The moisture maintains a high relative humidity inside the capsule, promoting chances for hydrolytic cleavage of the biodegradable polymer. Hydrolysis of the polymer may increase antigen release rate and make the polymeric matrix acidic, resulting in potential stability problem for the antigen. Placing the microspheres in a moisture-resistant container for reconstitution with a suitable vehicle before administration presents a suitable alternative. This approach is commercially used with hydrolytically unstable antibiotics (such as erythromycin and cefprozil) and has been extended to clinical investigation of oral microsphere delivery [50,51].

REFERENCES

1. S. A. Spector, Immunoprophylaxis and immunotherapy, *Infectious Diseases and Medical Microbiology* (A. I. Braude, C. E. Davis, and J. Fierer, eds.), W. B. Saunders, Philadelphia, 1986, p. 679.
2. J. R. McGhee and J. Mestecky, In defense of mucosal surfaces, *Infect. Dis. Clin. North Am. 4*:315 (1990).
3. T. B. Tomasi, Jr., E. M. Tan, A. Solomon, and R. A Prendergast, Characteristics of immune system common to certain external secretions, *J. Exp. Med. 121*:101 (1965).
4. J. R. McGhee, J. Mestecky, M. T. Dertzbaugh, J. H. Eldridge, M. Hirasawa, and H. Kiyono, The mucosal immune system: From fundamental concepts to vaccine development, *Vaccine 10*:75 (1992).
5. J. R. McGhee and H. Kiyono, New perspectives in vaccine development: Mucosal immunity to infections, *Infect. Agents Dis. 2*:55 (1993).
6. C. F. Cuff, D. C. Hooper, D. Kramer, D. H. Rubin, and J. J. Cebra, Functional and phenotypic analyses of the mucosal immune response in mice: Approaches to studying the immunogenicity of antigens applied by the enteric route, *Vaccine Res. 1*:175 (1992).
7. R. H. Waldman, J. Stone, K. Bergman, R. Khakoo, V. Lazzell, A. Jacknowitz, E. R. Waldman, and S. Howard, Secretory antibody following oral influenza immunization, *Am. J. Med. Sci. 292*:367 (1986).
8. D. T. O'Hagan, Oral delivery of vaccines, *Clin. Pharmacokinet. 22*:1 (1992).
9. D. T. O'Hagan, Novel nonreplicating antigen delivery systems, *Curr. Opin. Infect. Dis. 3*:393 (1990).
10. C. A. Gilligan and A. L. Wan Po, Oral vaccines, *Int. J. Pharm. 75*:1 (1991).
11. S. J. Holland and B. J. Tighe, Polymers for biodegradable medical devices. 1. The potential of polyesters as controlled macromolecular release systems, *J. Controlled Release 4*:155 (1986).
12. M. R. Brophy and P. B. Deasy, Biodegradable polyester polymers as drug carriers, *Encyclopedia of Pharmaceutical Technology*, 2 (J. Swarbrick and J. C. Boylan, eds.), Marcel Dekker, New York, 1988, p. 1.
13. A. G. Thombre and J. R. Cardinal, Biopolymers for controlled drug delivery, *Encyclopedia of Pharmaceutical Technology*, 2 (J. Swarbrick and J. C. Boylan, eds.), Marcel Dekker, New York, 1988, p. 61.
14. T. R. Tice and S. E. Tabibi, Parenteral drug delivery: Injectables, *Treatise on Controlled Drug Delivery* (A. Kyodenieus, ed.), Marcel Dekker, New York, 1992, p. 315.
15. R. W. Baker, ed., Biodegradable systems, *Controlled Release of Biologically Active Agents* Wiley Interscience, New York, 1987, pp. 84–94.
16. M. T. Aguado and P. H. Lambert, Controlled-release vaccines—biodegradable

polylactide/polyglycolide (PL/PG) microspheres as antigen delivery vehicles, *Immunobiology 184*:113 (1992).

17. P. Couvreur and C. Vauthier, Polyalkylcyanoacrylate nanoparticles as drug carrier: present state and perspectives, *J. Controlled Release 17*:187 (1991).
18. R. Arshady, Preparation of biodegradable microspheres and microcapsules: 2. Polylactides and related polyesters, *J. Controlled Release 17*:1 (1991).
19. R. Reid, E. C. Boedeker, C. E. McQueen, D. Davis, L. Y. Tseng, J. Kodak, K. Sau, C. L. Wilhelmsen, R. Nellore, P. Dalal, and H. R. Bhagat, Preclinical evaluation of microencapsulated CFA/II oral vaccine against enterotoxigenic *E. coli*, *Vaccine 11*:159 (1993).
20. T. Sato, M. Kanke, H. G. Schroeder, and P. DeLuca, Porous biodegradable microspheres for controlled drug delivery. I. Assessment of processing conditions and solvent removal techniques, *Pharm. Res. 5*:21 (1988).
21. R. Bodmeier and J. W. McGinity, Solvent selection in the preparation of poly(D,L-lactide) microspheres prepared by the solvent evaporation method, *Int. J. Pharm. 43*:179 (1988).
22. S. Cohen, T. Yoshioka, M. Lucarelli, L. H. Hwang, and R. Langer, Controlled delivery systems for proteins based on poly(lactic/glycolic acid) microspheres, *Pharm. Res. 8*:713 (1991).
23. M. J. Alonso, R. K. Gupta, C. Min, G. R. Siber, and R. Langer, Biodegradable microspheres as controlled-release tetanus toxoid delivery systems, *Vaccine 12*:299 (1994).
24. J. H. Eldridge, J. K. Staas, J. A. Meulbroek, T. R. Tice, and R. M. Gilley, Biodegradable and biocompatible poly(DL-lactide-*co*-glycolide) microspheres as an adjuvant for staphylococcal enterotoxin B toxoid which enhances the level of toxin-neutralizing antibodies, *Infect. Immun. 59*:2978 (1991).
25. Y. Ogawa, M. Yamamoto, H. Okada, T. Yashiki, and T. Shimamoto, A new technique to efficiently entrap leuprolide acetate into microcapsules of polylactic acid or copoly(lactic/glycolic) acid, *Chem. Pharm. Bull 36*:1103 (1988).
26. R. Arshady, Microspheres and microcapsules, a survey of manufacturing techniques, Part II: Coacervation, *Polym. Eng. Sci. 30*:905 (1990).
27. S. S. Sampath, K. Garvin, and D. H. Robinson, Preparation and characterization of biodegradable poly(L-lactic acid) gentamicin delivery systems, *Int. J. Pharm. 78*:165 (1992).
28. R. Bodmeier and H. Chen, Preparation of biodegradable poly(±)lactide microspheres using a spray drying technique, *J. Pharm. Pharmacol. 40*:745 (1988).
29. M. E. Le Ferve and D. D. Joel, Peyer's patch epithelium, an imperfect barrier, *Intestinal Toxicology* (C. M. Schiller, ed.), Raven Press, New York, 1984, p. 45.
30. D. T. O'Hagan, Microparticles as oral vehicles, *Novel Delivery Systems for Oral Vaccines* (D. O'Hagan, ed.), CRC Press, Boca Raton FL, 1994, p. 175.
31. J. Pappo and T. H. Ermak, *Clin. Exp. Immunol. 76*:144 (1989).
32. K. A. Howard, N. W. Thomsa, S. S. Davis, and D. T. O'Hagan, *Proc. Int. Symp. Controlled Release Bioact. Mater. 20*:296–297 (1993).
33. J. P. Ebel, *Pharm. Res. 7*:848 (1990).
34. P. Jani, G. W. Halbert, J. Langridge, and A. T. Florence, *J. Pharm. Pharmacol. 42*:821 (1990).

35. H. R. Bhagat, W. Williams, D. Metelitsa, and T. P. Monath, Investigation of microsphere uptake in animals, *Proc. Int. Symp. Controlled Release Bioact. Mater. 20*:579–580 (1994).

36. H. O. Alpar, W. N. Field, R. Hyde and D. A. Lewis, The transport of microspheres from the gastrointestinal tract to inflammatory air pouches in the rat, *J. Pharm. Pharmacol. 41*:194 (1989).

37. D. A. Lewis, J. Eyles, W. N. Field, and H. O. Alpar, Observations on the effect of the volume of water and tonicity in microsphere uptake in rat gut, *J. Pharm. Pharmacol. 44 (Suppl.)*:1986 (1992).

38. P. U. Jani, G. W. Halbert, J. Langridge, and A. T. Florence, Nanoparticle uptake by the rat gastrointestinal mucosal: Quantitation and particle size dependence, *J. Pharm. Pharmacol. 42*:821 (1990).

39. H. Tomizawa, Y. Aramaki, Y. Fujii, T. Hara, N. Suzuki, K. Yachi, H. Kikuchi, and S. Tsuchiya, Uptake of phosphatidylserine liposomes by rat Peyer's patches following intraluminal administration, *Pharm. Res. 10*:549 (1993).

40. J. H. Eldridge, C. J. Hammond, J. A. Meulbroek, J. K. Staas, R. M. Gilley, and T. R. Tice, Controlled vaccine release in the gut-associated lymphoid tissues. I. Orally administered biodegradable microspheres target the Peyer's patches, *J. Controlled Release 11*:205, (1990).

41. M. A. Jepson, N. L. Simmons, D. T. O'Hagen, and B. H. Hirst, Comparison of poly(lactide-*co*-glycolide) and polystyrene microspheres targeting to intestinal M cells, *J. Drug Target 1*:245 (1993).

42. J. Pappo, T. H. Ermak, and H. J. Steger, Monoclonal antibody-directed targeting of fluorescent polystyrene microspheres to Peyer's patch M cells, *Immunology 73*:277 (1991).

43. R. L. Clancy, A. W. Cripps, and V. Gebski, Protection against recurrent acute bronchitis after oral immunization with killed *Haemophilus influenzae, Med. J. Aust. 152*:413 (1990).

44. F. I. Farag-Mahmod, P. R. Wyde, J. P. Rosborough, and H. R. Six, Immunogenicity and efficacy of orally administered inactivated influenza virus vaccine in mice, *Vaccine 6*:262 (1988).

45. Z. Moldoveanu, J. K Stass, R. M. Gilley, R. Ray, R. W. Compans, J. H. Eldridge, T. R. Tice, and J. Mestecky, Immune response to influenza virus in orally and systemically immunized mice, *Curr. Top. Microbiol. Immunol. 146*:91 (1989).

46. Z. Moldoveanu, M. Novak, W. Huang, R. Gilley, J. Stass, D. Schafer, R. Compans, and J. Mestecky, Oral immunization with influenza virus in biodegradable microspheres, *J. Infect. Dis. 167*:84 (1993).

47. N. Santiago, S. Milstein, T. Rivera, E. Garcia, T. Zaidi, H. Hong, and D. Butcher, Oral immunization of rats with protenoid microspheres encapsulating influenza virus antigen, *Pharm. Res. 10*:1243 (1993).

48. J. H. Eldridge, R. M. Gilley, J. K. Stass, Z. Moldoveanu, J. A. Meulbroek, and T. R. Tice, Biodegradable microspheres: Vaccine delivery system for oral immunization, *Curr. Top. Microbiol. Immunol. 146*:59 (1989).

49. J. H. Eldridge, J. K. Stass, J. A. Meulbroek, J. R. McGhee, T. R. Tice, and R.

M. Gilley, Biodegradable microspheres as a vaccine delivery system, *Mol. Immunol. 28*:287 (1991).

50. E. Boedeker, R. Reid, H. Bhagat, C. Tucket, G. Losonsky, J. Nataro, R. Edelman, and M. Levine, Safety, immunogenicity and efficacy in human volunteers of biodegradable, biocompatible microspheres containing colonization factor antigen/II (CFA/II) as an enteral vaccine against enterotoxigenic *E. coli* (ETEC), *Vaccine* (in press).

51. C. O. Tacket, R. H. Reid, E. C. Boedeker, G. Losonsky, J. P. Nataro, H. Bhagat, and R. Edelman, Enteral immunization and challenge of volunteers given enterotoxigenic *E. coli* CFA/II encapsulated in biodegradable microspheres, *Vaccine 12*:1270 (1994).

52. D. E. Chickering, J. S. Jacob, and E. Mathiowitz, The use of radio-opaque barium sulfate to study GI transit of bioadhesive microspheres, *Proc. Int. Symp. Controlled Release Bioact. Mater. 20*:244 (1993).

53. S. Chess, D. Chess, G. Olander, W. Benner, and W. H. Cole, Production of chronic enteritis and other systemic lesions by ingestion of finely divided foreign materials, *Surgery 27*:221 (1950).

14

Biodegradable Gelatin Microspheres for Drug Delivery to Macrophages

Yasuhiko Tabata and Yoshito Ikada

Kyoto University
Kyoto, Japan

I. INTRODUCTION

When a foreign material enters or is placed in the body, the living system initiates host defense mechanisms that involve what may be called inflammatory or immune responses. Thus, the presence of foreign materials, irrespective of their property, may cause local damage, leading to acute inflammatory changes; an evoked immune reaction; or after an acute inflammatory reaction, one that is associated with an infiltration of host cells, predominantly of the mononuclear phagocyte linage. The mononuclear phagocyte system consists of circulating monocytes in the blood and macrophages in the tissue. Both types of cells are derived from myeloid progenitor cells. The progenitor cells differentiate into monocytes in the blood after leaving the bone marrow. After circulation for about 8 h in the bloodstream, the monocytes enlarge and then migrate into the tissues as they differentiate to become macrophages. Macrophages serve different functions in different tissues, and are named to reflect their tissue location (e.g., Kupffer cells in the liver, alveolar macrophages in the lung, microglial cells in the brain, and lymphoid macrophages in the spleen) [1].

Macrophages ($M\phi$) were initially thought to function simply as phagocytic cells. Recently, however, it has become clear that their phagocytic function is only the beginning of their functional role in the immune response. Following phagocytosis, $M\phi$ serve a vital function, both as antigen-presenting cells and as secretory cells. This function is indispensable to initiate the immune response, and the cells, together with lymphocytes, are essential for the development of both cellular and humoral immunocompetence [2–5]. Although phagocytosis initially activates Mϕ, their activity can be further enhanced by various factors. This enhancement of Mϕ activity is defined as *Mϕ activation*, and the activated Mϕ exhibit their morphological, biochemical, and functional changes, expressing the enhanced resistance against virus-infected cells, tumor cells, and intracellular bacteria, compared with the baseline values of resident and stimulated Mϕ [6–8]. Taken together, Mϕ regulate their biological functions in response to microenvironmental stimuli (e.g., action of soluble factors and contact with foreign materials or other cells) and play a central role in controlling the immune response through their functions, such as phagocytosis, antigen presentation, and secretion of factors. Thus, if there are some drugs that will act on Mϕ, their use will enable one to manipulate these functional activities, leading to regulation of the degree of humoral and cellular immune reaction. However, if the drugs are injected into the body, in their soluble form one cannot always expect their high efficacy in Mϕ activation because of their poor ability to reach the Mϕ. Therefore, to achieve this goal, we must develop carrier systems for delivering the drugs to Mϕ.

The role of Mϕ in the host defense against neoplasms has attracted increasing attention. A large amount of evidence has been accumulated that Mϕ activation is potentially of great importance in the host defense against primary and metastatic tumors [9,10]. Macrophages can be activated by a variety of Mϕ-activating agents (immunomodulatory agents) to exert antitumor functions [11–17]. These agents comprise the major substances currently identified as biological response modifiers (BRM) [18]. However, several problems are associated with trials for use of the immunomodulatory agents to activate Mϕ. Among them are the rapid catabolism and clearance of the agents and their serious side effects when given in the high-dosage regimens required for significant therapeutic efficacy. It is likely that this in vivo instability of diffusive agents is mainly responsible for the disappointing results of recent clinical trials to augment host resistance to tumors by immunotherapy. One possible way to overcome the problems associated with Mϕ-activating agents is to target them directly to the Mϕ, and particle carriers are the most suitable for this purpose. It is conceivable that Mϕ targeting of the agents is achieved by taking advantage of the inherent avidity of Mϕ to ingest foreign materials. Macrophages are quantitatively superior in their endocytic capacity to other cell types present in the body. You have only to incorporate the labile Mϕ-activating agents into a biodegradable microsphere that is susceptible to Mϕ phagocytosis. The foreign microspheres will be phogocytosed by Mϕ. Moreover, if they are prepared from a material that is well recognized by Mϕ, more ready and selective phagocytosis will be possible. The ingested microspheres will be degraded in Mϕ, followed by the slow release of the entrapped drug in the cells and, consequently, Mϕ will be effectively activated to acquire antitumor activity, compared with Mϕ-activating agents in the soluble form (Fig. 1). This administration, not only prevents the premature degradation of the activating agents, but also reduces the dosage necessary to activate the Mϕ. Most types of microspheres tend to be rapidly cleared from body fluids by Mϕ belonging to the mononuclear phagocyte system (MPS). This is an inevitable and a natural consequence. Thus, if delivery of drugs to other sites by use of microsphere carriers is required, the normal rapid MPS-mediated clearance must be prevented. On the other hand, if a drug becomes effective only when internalized into phagocytic cells, the drug–microsphere composite must be of a size that is susceptible to phagocytosis, with a surface structure that is readily recognized by the cells as a target for ingestion. Our research is based on a concept that will actively use this Mϕ-mediated clearance mechanism, which is equipped inherently in the body, to achieve the therapeutic goal.

Gelatin is a biodegradable polymer used extensively for industrial, pharmaceutical, and medical purposes, and its in vivo safety has been proved through its long clinical application as a plasma expander [19]. Another unique advantage of gelatin for drug targeting to Mϕ is that it works as a very strong opsonin [20]. Thus, as an efficient carrier for delivering drugs to Mϕ, we have tried to prepare microspheres from this opsonic gelatin with a size range susceptible to Mϕ phagocytosis. Microspheres, composed of gelatin as well as other biodegradable polymers, work very effectively as carriers for several immunomodu-

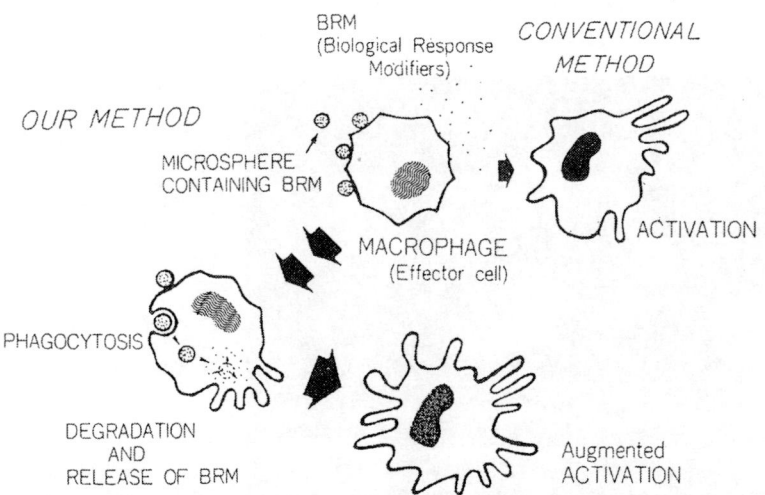

Fig. 1 Our research concept for Mϕ activation with biodegradable gelatin microspheres containing BRM.

latory agents in activating the antitumor function of Mϕ [21]. This chapter focuses on gelatin microspheres and reviews their efficacy as a drug carrier in activating the antitumor function of Mϕ, together with our knowledge about their properties and interactions with Mϕ. In addition, chemical conjugation of the immunomodulatory agents with gelatin is mentioned as another carrier for their delivery to Mϕ, and this feature of Mϕ activation of the drug–gelatin conjugate is compared with that of gelatin microspheres. We also discuss the potential of gelatin microspheres as an immunological adjuvant capable of delivering antigens to Mϕ.

II. PREPARATION AND CHARACTERIZATION

A. Preparation

The Mϕ-activating agent used here is a recombinant human interferon alfa A/ D (IFN or A/D-IFN, 1.5×10^8 IU/mg protein) which is effective on mouse cells as well as human cells. The gelatin used in this study, is an alkaline type, although any type of gelatin can be applied for this procedure. Gelatin microspheres incorporating IFN were prepared through cross-linking of gelatin with glutaraldehyde in a water–oil (w/o) emulsion, in which an aqueous gelatin solution containing IFN is dispersed in the organic solution [22]. Cross-linking of gelatin microspheres is carried out by the addition of glutaraldehyde to the organic phase after generating the w/o emulsion, to avoid their aggregation during microsphere preparation. According to this method, the shape of microspheres is invariably spherical and the size can be widely changed by sonication of the emulsification (Fig. 2). This sonication process reduces the average di-

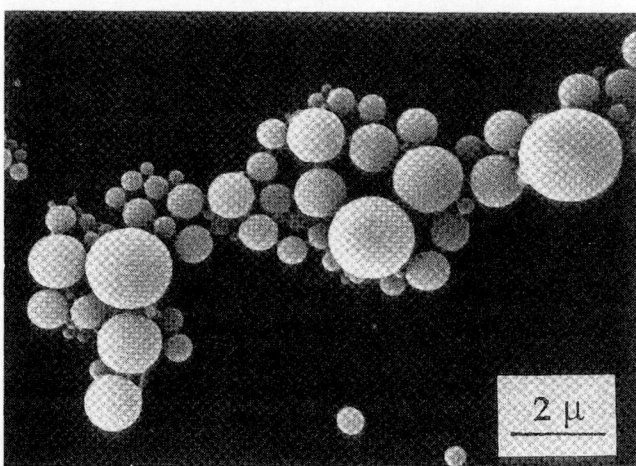

Fig. 2 A scanning electron micrograph of glutaraldehyde–cross-linked gelatin microspheres containing IFN.

ameter of microspheres to about 1.5 μm, which is a size susceptible to Mϕ phagocytosis [23]. In addition, the extent of gelatin cross-linking can be controlled by changing the amount of gelatin and glutaraldehyde added, resulting in a satisfactory regulation of microsphere degradation.

Delivery of immunomodulatory agents to Mϕ has also been attempted through chemical modification of the agents with gelatin itself to create a soluble carrier that is different from the water-insoluble microsphere approach. The water-soluble agent–gelatin conjugate is prepared according to a water-soluble carbodiimide method [24]. The agents used for this purpose are muramyl dipeptide (MDP) [24,25], IFN [26], and interleukin-1 (IL-1) [27].

B. Microsphere Degradation and Interferon Release

Degradation of gelatin microspheres was examined at 37°C in phosphate-buffered saline solution (PBS; pH 7.4), or in one that contained collagenase (Figs. 3 and 4). No degradation of microspheres was observed in PBS within 2 days, irrespective of the extent of cross-linking. However, the addition of collagenase in PBS enabled their degradation, at a rate that decreased as the amount of glutaraldehyde and gelatin added in microsphere preparation increased. Release experiments using gelatin microspheres containing ^{125}I-labeled IFN demon-

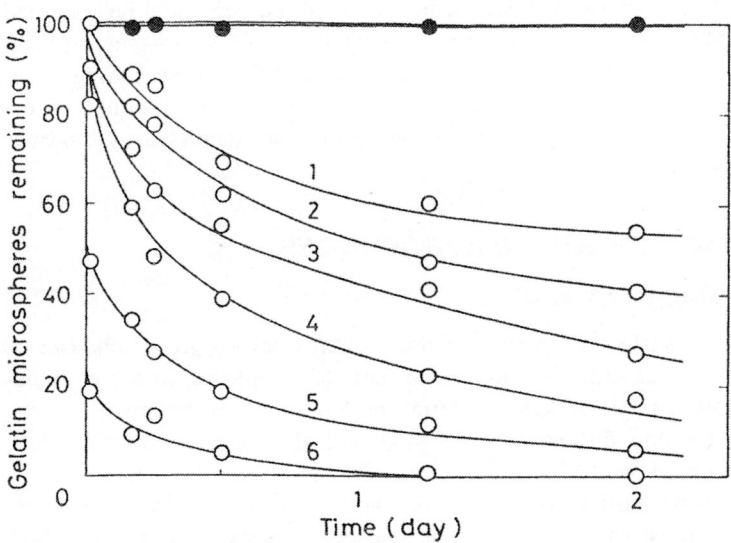

Fig. 3 In vitro degradation profiles of IFN-loaded gelatin microspheres cross-linked with glutaraldehyde in concentrations of (1) 1.33, (2) 0.71, (3) 0.28, (4) 0.14, (5) 0.05, and (6) 0.03 mg/mg gelatin. (open mark, in PBS containing collagenase; solid mark, in PBS). (From Ref. 22.)

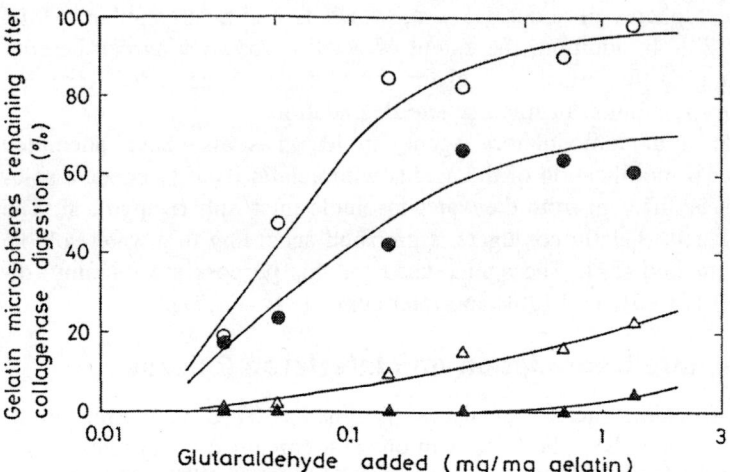

Fig. 4 Effect of gelatin and glutaraldehyde concentrations, at different degrees of cross-linking, on collaganase digestion of gelatin microspheres for 30 min. The concentration of gelatin used in microsphere preparation is (○) 20, (●) 10, (△) 5, and (▲) 2 wt%. (From Ref. 22.)

strated that IFN was released slowly with time, and the rate could be regulated by changing the extent of gelatin cross-linking in the PBS–collagenase solution, whereas no release was observed in PBS alone [22]. These findings indicate that gelatin microspheres are fairly stable, and IFN incorporated in the microspheres does not leak out readily unless they are degraded by enzymes such as collagenase.

III. INTERACTION WITH MACROPHAGES

A. Macrophage Uptake

The size and the surface properties of the particles have a great influence on Mϕ phagocytosis [28–32]. However, the particles employed in these studies are mostly commercial latexes, bacteria, pollen, and carbon microspheres. Thus, it was extremely difficult to study systematically the effects of their physicochemical properties on Mϕ phagocytosis. To investigate the detailed behavior of Mϕ toward particle phagocytosis, well-characterized polymer microspheres are the most suitable not only because their size is easily controlled, but also because microspheres with different surface natures are readily prepared from the same starting material. We have prepared a variety of polymer microspheres with different sizes, surface charges, or hydrophobicities to inves-

tigate the effect of microspheres' properties and the presence of bioactive proteins on their Mϕ phagocytosis [33]. A series of phagocytosis experiments
demonstrated that precoating or surface immobilization with gelatin was the
most effective method to enhance the extent of microsphere phagocytosis
among all other types of microspheres. The opsonic effect of gelatin for
Mϕ phagocytosis was stronger than that of immunoglobulin or fibronectin,
which are well known as opsonic serum proteins [20,33]. Fibronectin and other
adhesive proteins present in serum might be bound to the surface of gelatin
molecules covering the microspheres, for gelatin has a high affinity for such
proteins. As a result, it is possible that microspheres, precoated or surface-
modified with gelatin, undergo increased opsonization in the presence of
serum.

It is highly conceivable that microspheres prepared from gelatin with such a
strong opsonic ability are well phagocytosed by Mϕ. Thus, we prepared gelatin
microspheres through chemical cross-linking to assess their phagocytosis by
Mϕ. Every gelatin microsphere prepared was phagocytosed by Mϕ within the
incubation time, and the amount of microspheres ingested was proportional to
the amount of microspheres added to the Mϕ culture, until saturation of phago-

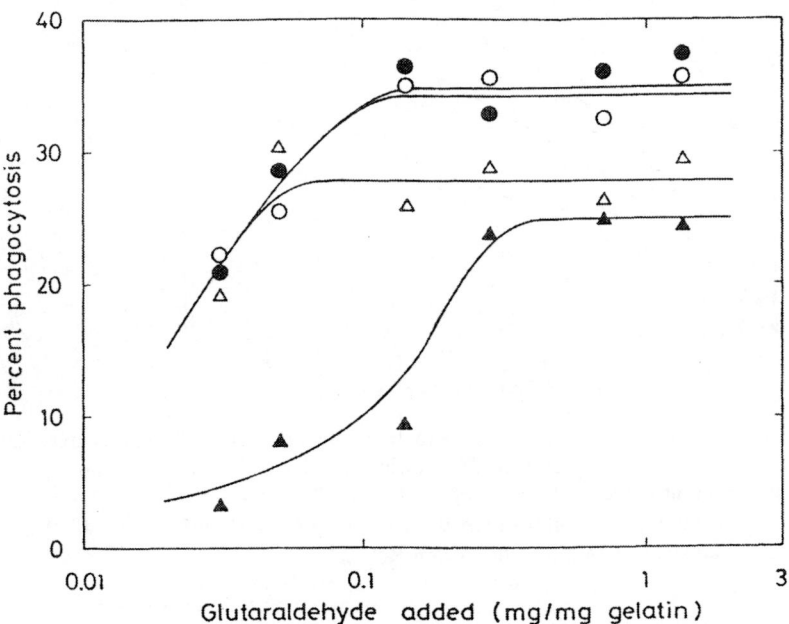

Fig. 5 Macrophage phagocytosis of gelatin microspheres with different degrees of
cross-linking (see Fig. 4 for symbols). (From Ref. 22.)

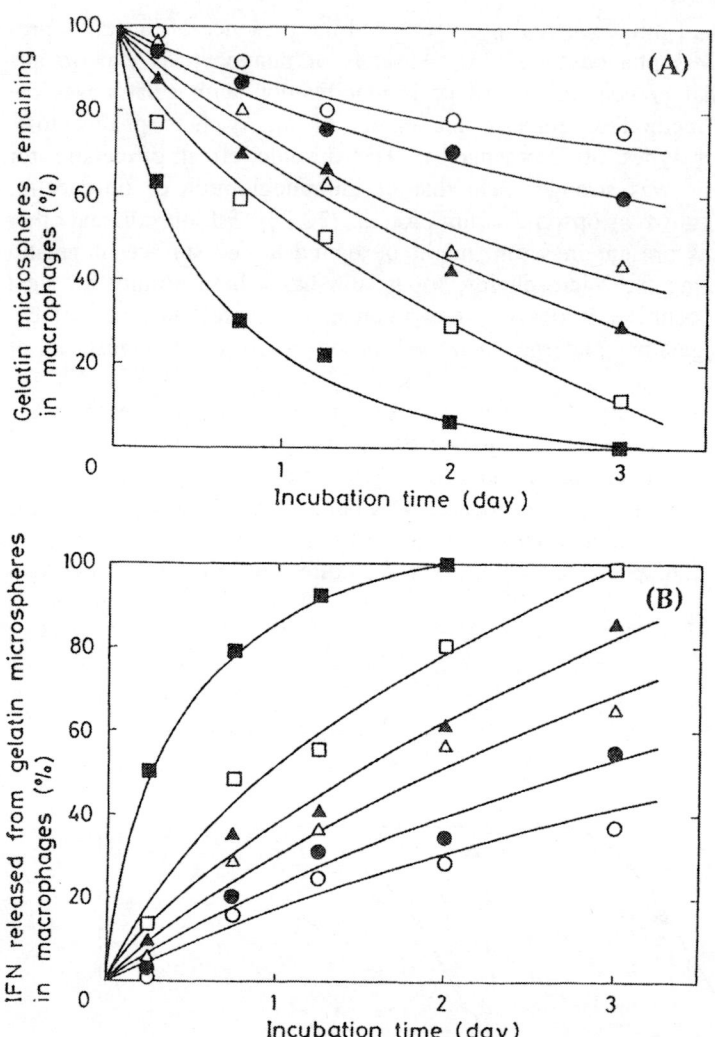

Fig. 6 (A) Degradation profiles of gelatin microspheres with different degrees of cross-linkage in mouse peritoneal Mφ. (B) Release profiles of IFN from gelatin microspheres with different degrees of cross-linkage in Mφ. The concentration of glutaraldehyde used in the microsphere preparation is (○) 1.33, (●) 0.71, (△) 0.28, (▲) 0.14, (□) 0.05, and (■) 0.03 mg/mg gelatin. (From Ref. 22.)

cytosis was observed at high microsphere doses. Figure 5 shows the effect of the degree of gelatin cross-linkage on Mφ phagocytosis. The extent of phagocytosis decreased with the decreasing concentration of gelatin and glutaraldehyde used in their preparation. Every microsphere prepared was well ingested by Mφ, although the proportions were lower as the concentrations of gelatin and glutaraldehyde decreased. When using low concentrations, the amount of cross-linkage in microspheres is small. It is likely that the sizes of microspheres enlarge because they swell in the culture medium, resulting in a reduction of phagocytosis [22].

Macrophages also ingests water-soluble substances by a so-called pinocytotic process [28]. Pinocytosis experiments demonstrated that gelatin, in soluble form, was taken up by Mφ to a greater degree than opsonic serum proteins, in vivo as well as in vitro [24,25]. Gelatin injected intraperitoneally was pinocytosed by Mφ in the peritoneal cavity, and the Mφ pinocytosis increased with an increase in the gelatin's molecular weight [34]. It is possible that the long retention of large gelatin molecules in the peritoneal cavity results in the high amount of Mφ pinocytosis, because a similar susceptibility of gelatin to Mφ pinocytosis in vitro was observed, irrespective of the molecular weight. These findings strongly indicate the potential for using gelatin itself as a water-soluble carrier to target drugs to Mφ.

B. Microsphere Degradation and Interferon Release in Macrophages

Figure 6 shows the degradation of gelatin microspheres containing IFN and the subsequent release of IFN from the microspheres within the macrophages. The phagocytosed microspheres were gradually degraded in the Mφ, leading to the slow release of IFN in the cells. Both the degradation and IFN release changed, depending on the extent of cross-linkage in the microspheres, which can be controlled by the concentration of glutaraldehyde added in their preparation [22]. Neither microsphere degradation nor IFN release was observed in PBS without collagenase. These findings suggest that enzymatic hydrolysis is responsible for degradation of gelatin. Gelatin microspheres are degraded in the cells by lysosomal enzymes, such as collagenase, after Mφ ingestion, leading to intracellular slow release of IFN.

IV. MACROPHAGE ACTIVATION OF ANTITUMOR FUNCTION

A. In Vitro Macrophage Activation

Mouse peritoneal Mφ were pretreated in vitro with various doses of free IFN or with microspheres containing IFN, with or without 10 ng/ml lipopolysaccharide

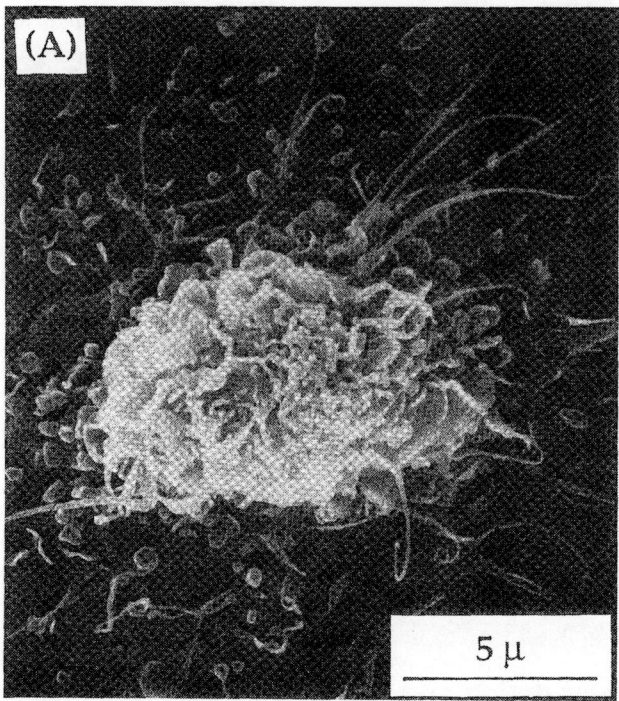

Fig. 7 SEM micrographs of a mouse peritoneal macrophage pretreated with gelatin microspheres containing IFN. (A) After 24-h incubation, the microspheres had been completely phagocytosed by a macrophage, leading to its activation. (B) After 24-h cocultivation with Meth A tumor cells, the activated macrophage is binding to and delivering an attack on the Meth A tumor cell.

(LPS), and the inhibitory activity of Mφ on the growth of mouse Meth A fibrosarcoma cells (Meth A) was assessed. Figure 7 shows scanning electron micrographs of macrophages activated through phagocytosis of gelatin microspheres containing IFN and of the activated macrophage after cocultivation with Meth A cells. It is apparent that the activated macrophage is binding to and delivering an attack on the tumor cell.

Figure 8 shows the dose responses for Mφ activation. The Mφ pretreated with free IFN alone scarcely inhibited the growth of tumor cells, whereas the inhibitory activity on tumor growth was drastically augmented by addition of LPS (10 ng/ml), although LPS at this dose alone or with low doses of IFN could not induce any substantial Mφ activation. On the other hand, a significant augmentation of the antitumor activity of Mφ pretreated with gelatin mi-

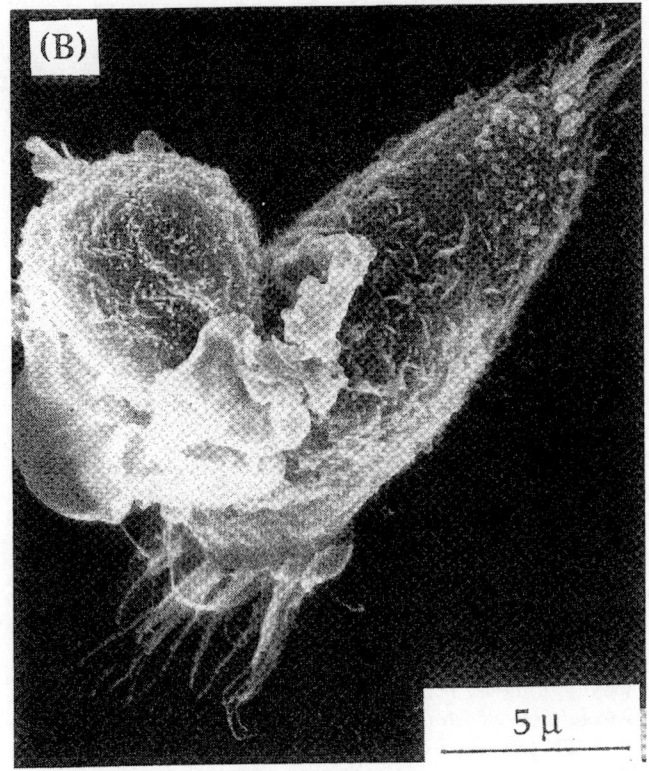

crospheres containing IFN was observed, even without LPS addition. More-over, the amount of IFN required in the microspheres for activity induction was about several hundred times less than that of free IFN given to Mϕ with LPS. The addition of LPS shifted the direction of the dose–response curve for the IFN-loaded microspheres toward the lower dose by about one order of magnitude. Pretreatment of Mϕ with IFN-free empty microspheres brought about a feeble activation of Mϕ, and addition of IFN-free microspheres had no influence on the activation induced by free IFN plus LPS, or on the weak activation by free IFN. This indicates that the phagocytosis of the microsphere matrix itself neither contributed to nor interfered with the Mϕ activation. In addition, the high efficiency of the microspheres containing IFN for Mϕ activa-tion, in comparison with free IFN and LPS, was also shown by the short incu-

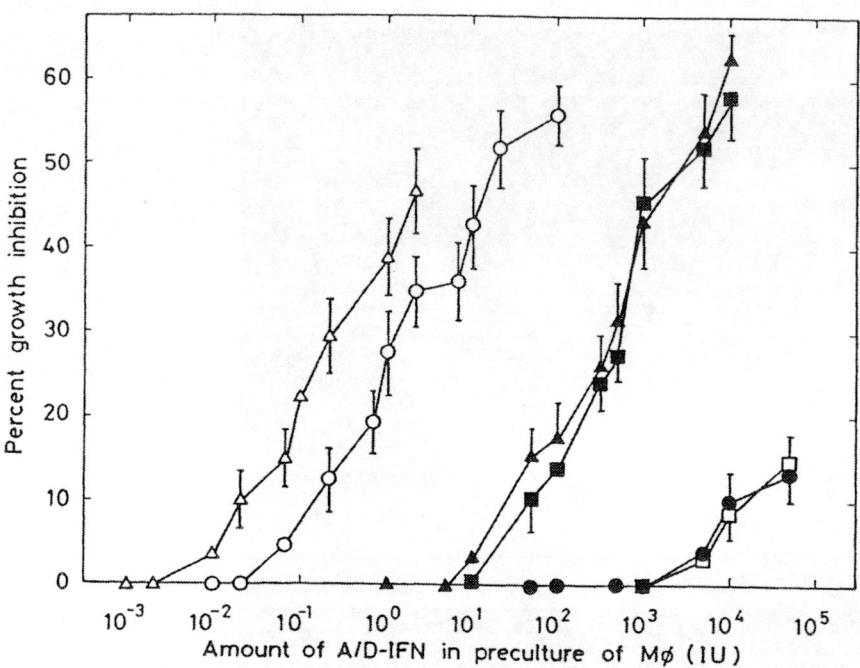

Fig. 8 The inhibitory activity of mouse peritoneal Mφ on in vitro growth of Meth A tumor cells. Macrophages were pretreated for 8 h at 37°C with (●) IFN; (▲) IFN plus 10 ng/ml LPS; (□) IFN plus IFN-free microspheres; (■) IFN plus 10 ng/ml LPS plus IFN-free microspheres; (○) microspheres containing IFN; and (△) microspheres containing IFN plus 10 ng/ml LPS. The percentage growth inhibition of Mφ pretreated with 10 μg/ml LPS was 51% in this experiment. (From Ref. 35.)

bation time for Mφ activation and by the long retention of the activated state of Mφ [35]. These results clearly demonstrate that gelatin microspheres containing IFN enable Mφ to exhibit the in vitro inhibitory activity on tumor cell growth far more efficiently than does free IFN. Chemical conjugation of IFN with gelatin was also effective in enhancing the antitumor function of Mφ in vitro. Similarly to gelatin microspheres containing IFN, the water-soluble IFN–gelatin conjugates effectively induced Mφ activation, compared with free IFN, irrespective of the IFN dose and the time required for induction of activity [26].

We investigated the effect of anti-IFN antibody on Mφ activation. Macrophages were treated with either free IFN or IFN-loaded microspheres, each of which had been pretreated with an excess amount of the antibody. The antibody prevented the Mφ activation induced by free IFN plus LPS, and the slight activation by free IFN alone was also abrogated. The antibody, however, did

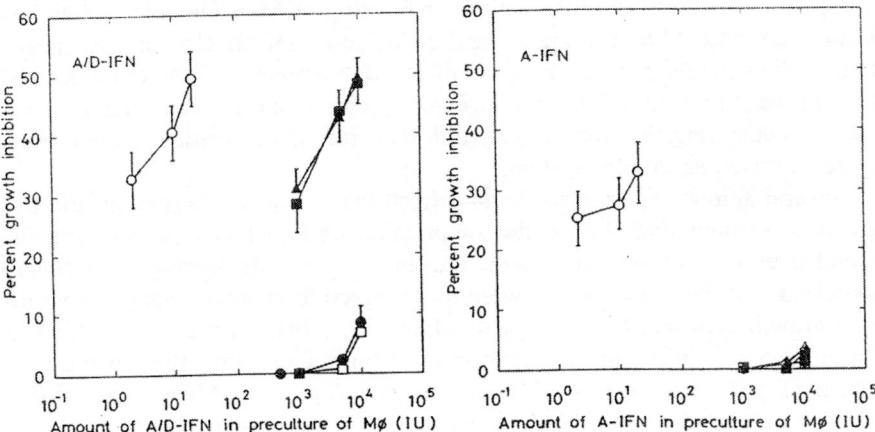

Fig. 9 Effectiveness of human-specific IFN-A for mouse peritoneal Mφ when incorporated in gelatin microspheres. Macrophages were pretreated for 8 h at 37°C with (●) IFN or IFN-A; (▲) IFN or IFN-A plus 10 ng/ml LPS; (□) IFN or IFN-A plus IFN-free microspheres; (■) IFN or IFN-A plus 10 ng/ml LPS plus IFN-free microspheres; and (○) microspheres containing IFN or IFN-A. (From Ref. 35.)

not interfere with the Mφ activation by the IFN-incorporated microspheres [35]. This indicates that Mφ activation by microspheres does not depend on the binding of the IFN, on the microsphere surface, to the IFN-receptor of Mφ. Thus, Mφ activation by gelatin microspheres containing IFN is caused by the encapsulated IFN following microsphere ingestion by Mφ.

Macrophage activation by the microspheres containing IFN does not require the aid of LPS, in contrast to free IFN, as is shown in Fig. 8. In addition, another IFN-A, which is specifically effective for human cells, but not for mouse cells, was effective in activating mouse Mφ when included in the microspheres, although free IFN-A was incapable of doing so even in the presence of LPS (Fig. 9). This implies that the mechanism of Mφ activation by microspheres containing IFN is different from that of free IFN. The species specificity of IFN may be attributed to the specificity of binding between IFN and the IFN receptor, and it can be abrogated when the IFN is internalized into Mφ through phagocytosis. This phenomenon is identical with that for liposome-encapsulated IFN-γ reported by Fidler et al. [36].

B. In Vivo Macrophage Activation

Gelatin microspheres containing IFN or free IFN were injected into the peritoneal cavity of normal mice to estimate whether or not they are effective in intraperitoneally activating the antitumor function of Mφ. The microsphere injection activated Mφ in the peritoneal cavity more effectively than IFN injection in the soluble form, irrespective of the IFN amount and the time required for Mφ activation [37]. Thus, it was clearly confirmed that the gelatin microspheres containing IFN were effective in inducing Mφ activation in the in vivo system as well as in vitro system.

Several animal experiments demonstrated the in vivo efficacy of gelatin microspheres containing IFN in the suppression of tumor cell growth. Intratumoral injection of the microspheres into mouse footpads bearing solid tumors indicated that the microspheres were more effective in suppressing in vivo tumor growth than free IFN (Fig. 10). Moreover, a distinct preventive effect of the intravenous injection of gelatin microspheres containing IFN on the incidence of pulmonary metastasis was observed (Fig. 11). Many macrophages exist in or around neoplastic tissues. They are of prime importance in the host resistance against the tumor cells [38,39]. It is conceivable that Mφ are effectively activated by injected IFN-loaded microspheres to suppress tumor cell growth, and the activated state of Mφ is maintained for a long period compared

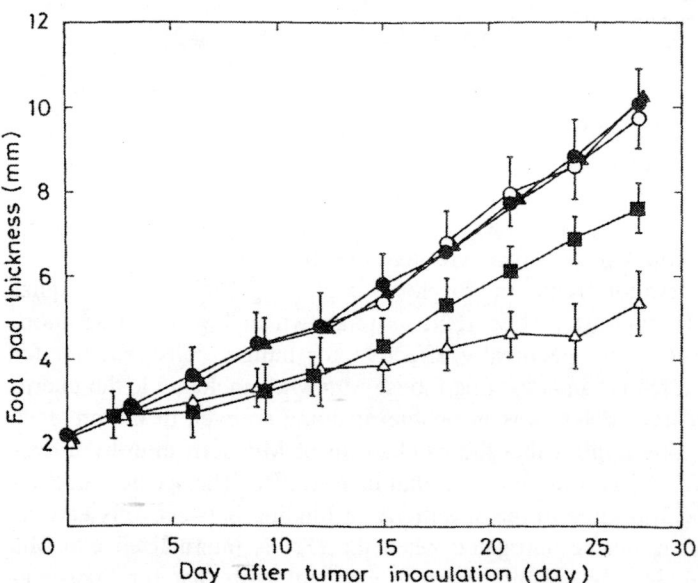

Fig. 10 Suppressive effect of intratumoral injection of IFN or gelatin microspheres containing IFN on the growth of solid Meth A tumors: (○) saline; (■) 5×10^4 IU of IFN; (△) 177 μg of microspheres containing 260 IU of IFN; (▲) 166 μg of IFN-free microspheres; and (●) 177 μg of IFN-free microspheres plus 260 IU of IFN. Each of the groups consists of seven mice. (From Ref. 37.)

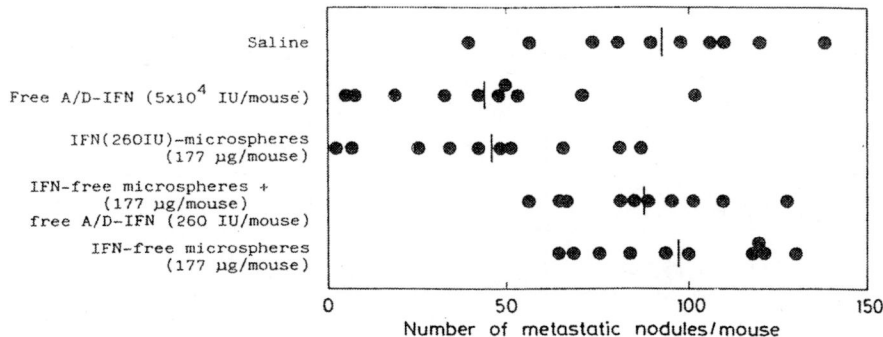

Fig. 11 Preventive effect of intravenous injection of IFN or gelatin microspheres containing IFN on the pulmonary metastasis of B16 melanoma cells. (From Ref. 37.)

with that of Mφ given free IFN. It may be concluded that this efficient antitumor activation of Mφ leads to suppression of the in vivo tumor cell growth.

On the other hand, in a therapeutic experiment using mice with ascitic tumor, the gelatin microspheres containing IFN were not always effective in suppressing the tumor growth in the mouse peritoneal cavity. An antitumor effect of the microspheres, as strong as that shown for the foregoing tumor models, was not observed, although Mφ collected from the peritoneal cavity of tumor-bearing mice were activated strongly to acquire antitumor activity through intraperitoneal injection of gelatin microspheres containing IFN alone. The number of Mφ infiltrated into the peritoneal cavity was not increased by the microsphere injection. It is possible that incorporation into the microspheres prevents IFN from exerting inherent systemic effects, such as direct cytostatic effects against tumor cells and augmentation of Mφ recruitment and natural killer (NK) activity. This finding indicates that the ability to target gelatin microspheres to Mφ permits high Mφ localization of IFN, unfortunately leading to blockage of its systemic antitumor effects. As expected, the microsphere injection, in combination with free IFN, in a low dose at which no suppressive effect was observed, enabled us to improve the survival rate of tumor-bearing mice [37].

Thus, a water-soluble form of administration was explored to target IFN to Mφ through its direct conjugation with gelatin, because gelatin is more readily pinocytosed by Mφ than other opsonic proteins [24,25]. The IFN–gelatin conjugate effectively augmented both Mφ recruitment and NK activity, in contrast to gelatin microspheres containing IFN [26]. Figure 12 shows that the conjugate-type dosage form is superior to that of the microsphere type. The injection of conjugate alone suppressed the ascitic tumor growth in the peritoneal cavity

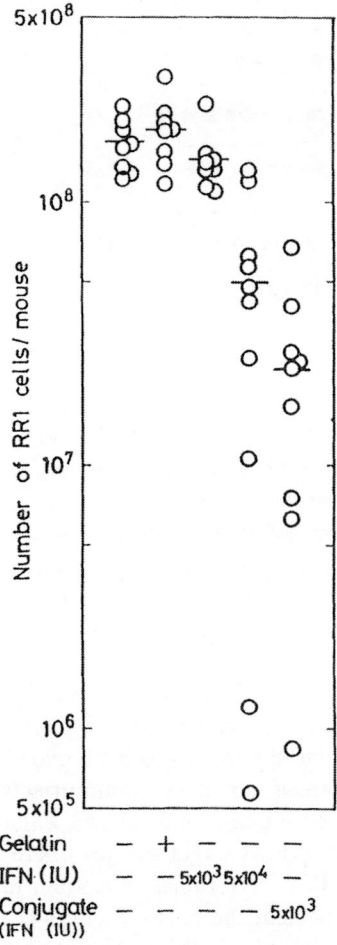

Gelatin   —   +   —   —   —
IFN (IU)   —   — 5x10³ 5x10⁴ —
Conjugate   —   —   —   — 5x10³
(IFN (IU))

Fig. 12 In vivo growth of Meth A tumor cells in the peritoneal cavity of mice receiving intraperitoneal injection of IFN–gelatin conjugate or other agents. The agents were intraperitoneally injected daily from day 3 to day 6. Each of the groups consists of nine mice. (From Ref. 26.)

of mice. It was confirmed that the water-soluble IFN–gelatin conjugate is an effective dosage form for delivering IFN to Mϕ without losing its systemic effects. In addition, this water-soluble dosage form will resolve the difficulty in reducing the size of microspheric dosage forms that restricts their therapeutic use.

V. ANTIBODY PRODUCTION

For an antigen to function as an efficient vaccine, it must stimulate a persistent, long-term immune response. One way to maintain high antibody responses for long periods after vaccination is to create an antigen depot at the injection site. The persistence of the antigen in such a depot and its slow rate of release ensure continuous stimulation of the immune system. This effect is achieved by incorporating the antigen into an appropriate vehicle. Conventionally, mineral salts and water-in-oil emulsions have been extensively used as the antigen vehicle, which is generally called an immunological adjuvant. Currently, aluminum salts are the only adjuvants approved for human use, but they are not effective for many vaccines, and their use for booster immunization has been questioned [40]. Thus, various polymer microspheres have been assessed as vehicular candidates for depot induction and release of antigens, in place of conventional adjuvants, to demonstrate their enhanced adjuvanticity [41–47]. Polymer microspheres can be manufactured in a more reproducible yield that shows more reproducible adjuvant effects than those currently used (e.g., aluminum hydroxide).

Not only the property of microspheres to control the release of antigen, but also their size and surface properties have a great influence on the microsphere-induced adjuvanticity [41,42,48]. These phenomena can be explained in terms of antigen recognition by antigen-presenting cells such as Mϕ and dendritic cells. The Mϕ recognition must be a key step to initiate the immune response against the invaded antigens (e.g., antibody production). Thus, if an antigen is incorporated into biodegradable microspheres susceptible to Mϕ recognition, more effective Mϕ phagocytosis of the antigen will be achieved, followed by the intracellular release of antigen with the microspheres' degradation, leading to enhanced production of the antigen-specific antibody. The microspheres composed of opsonic gelatin are the most suitable carrier for this purpose.

Table 1 shows the level of IgG antibody produced in the mouse serum 10 days after the primary immunization with gelatin microspheres containing the model proteinaceous antigen human gamma globulin (HGG), or other agents. Subcutaneous injection of gelatin microspheres containing HGG was effective in enhancing the production of HGG-specific IgG antibody, and the level of antibody produced was significantly higher than that when HGG was incorporated in the conventional Freund's incomplete adjuvant (FIA). Injection of

Table 1 Production of IgG Antibody by HGG Injection in Different Forms

Injection with	IgG produced[a] (titer $\times 10^5$)
PBS	<0.001
Empty gelatin microspheres	<0.001
FIA[b]	<0.001
Free HGG (100 μg/mouse)	2.41 ± 1.35[c]
Gelatin microspheres containing HGG (100 μg/mouse)	21.6 ± 8.37
Free HGG (100 μg/mouse) + empty gelatin microspheres	1.42 ± 0.07
FIA containing HGG (100 μg/mouse)	6.22 ± 2.46

[a] 10 days after injection.
[b] Freund's incomplete adjuvant.
[c] SEM.

HGG-free, empty microspheres neither induced antibody production, nor brought about significant change in the level of antibody production elicited by HGG injection. This indicates that antigen incorporation into gelatin microspheres is essential for augmenting the antibody response to the antigen. In addition, a single injection of gelatin microspheres maintained a high serum level of antibody over 35 days, compared with that of FIA containing HGG.

The ability of gelatin microspheres to enhance antibody production was greatly influenced by their extent of cross-linking, which can be varied by changing the concentration of glutaraldehyde used in their preparation. When microspheres were swollen in aqueous solution, the swollen microspheres were smaller as the glutaraldehyde concentration was increased, and their water content decreased with an increase of the glutaraldehyde concentration, in good accordance with cross-linking theories. Figure 13 shows the effect of the microsphere size on the production of antibody. The ratio of the antibody produced by microsphere injection at day 10 to that produced by FIA injection is plotted against the diameter of microspheres swollen in aqueous solution. Apparently, production of IgG antibody was greatly dependent on the size of microspheres. This phenomenon can be explained in terms of interactions between gelatin microspheres and Mϕ. Figure 14 shows the result of Mϕ phagocytosis of gelatin microspheres containing HGG, together with the initial rate of HGG released from them as a function of the size of water-swollen microspheres. The microspheres were ingested, to a larger extent, as their size became smaller (see Sec. III.A). Conversely, the release rate of HGG from microspheres decreased with an increase in their size. Gelatin microspheres cross-

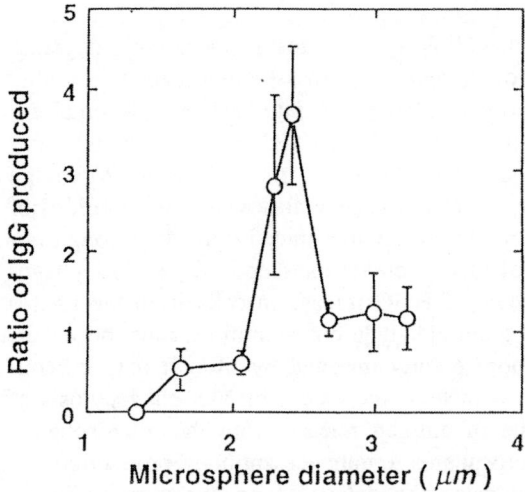

Fig. 13 The effect of microsphere size on the production of IgG antibody induced by gelatin microspheres containing HGG: 100 μg of HGG was injected to each mouse.

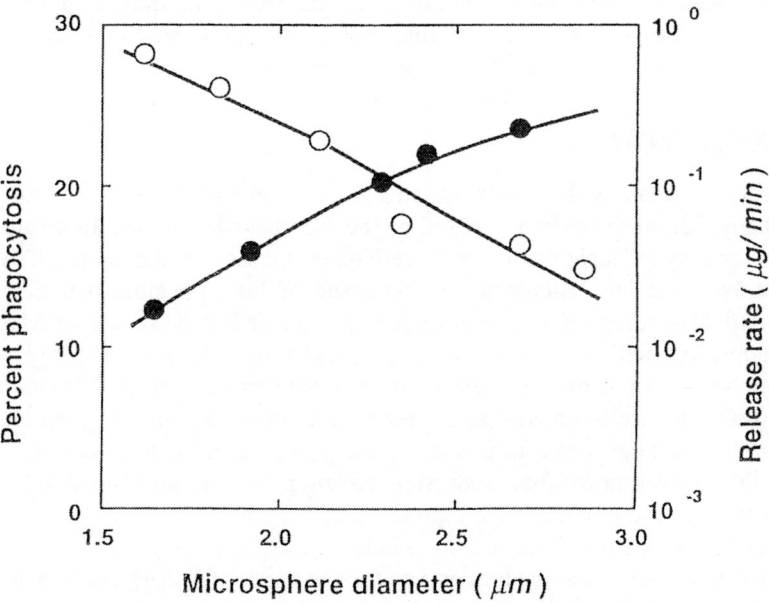

Fig. 14 The effect of microsphere size on Mϕ phagocytosis of gelatin microspheres containing HGG (○) and the rate of HGG released from the microspheres (●).

linked with lower concentrations of glutaraldehyde have higher water contents because of their low amount of cross-linking. As a result, more rapid degradation of the microspheres with a lower degree of cross-linking leads to a higher rate of HGG release. As gelatin microspheres are degraded by lysosomal enzymes (e.g., collagenase) after ingestion by Mϕ, HGG will be released intracellularly following the enzymatic degradation of microspheres. When the HGG-loaded gelatin microspheres were prepared with lower concentrations of glutaraldehyde, HGG could not be effectively introduced into Mϕ through microsphere phagocytosis because of their decreased susceptibility to Mϕ phagocytosis. On the other hand, the rate of HGG release in cells from the microspheres cross-linked with higher glutaraldehyde concentrations must be lower, although the microspheres are more readily ingested by Mϕ. It may be concluded that the balance of the two factors, the extent of Mϕ phagocytosis of gelatin microspheres and the rate of antigen release from the microspheres, affects the activity of gelatin microspheres to enhance antibody production.

Gelatin microspheres also enhanced the delayed type hypersensitivity response to the antigen to a larger extent than did FIA. Incorporation of the antigen into gelatin microspheres can deliver a large amount of the antigen to Mϕ and effectively induce their stimulation, leading, for example, to increased secretion of soluble factors. It is possible that targeting of antigen–carrier complexes to Mϕ results in an enhancement of a specific cellular immune response. These findings indicate that the gelatin microsphere is a promising adjuvant to enhance both humoral and cellular responses to antigen.

VI. CONCLUSION

This chapter has dealt with carrier systems for delivery of drugs to macrophages. The design of the carrier is based on the fact that Mϕ are quantitatively superior in endocytic capacity to other cell types present in the body. If a drug is incorporated into microspheres composed of biodegradable materials susceptible to Mϕ recognition, the drug will be inevitably delivered to Mϕ and, consequently, will be able to selectively exert its action on Mϕ only. Gelatin functions as a strong opsonin compared with fibronectin or immunoglobulin, which are well-known opsonic proteins in the body. In this chapter, we adduced two examples to estimate the efficacy of gelatin microspheres as a carrier for drug targeting to Mϕ. Both Mϕ activation of antitumor function by IFN and antibody production against a proteinaceous antigen were enhanced by the incorporation of each substance into gelatin microspheres, providing sound evidence that the gelatin microsphere is an efficient carrier for drug delivery to Mϕ. In addition, gelatin itself is also efficient enough to target drugs to Mϕ, even if not formulated into microspheric forms. The water-soluble drug–gelatin conjugate was effective in activating the antitumor function of Mϕ, to a similar or sometimes higher extent than water-insoluble gelatin microspheres containing drugs. It will be necessary to choose these two types of gelatin carriers to suit one's objective.

Macrophages play an important role in the disposal of damaged or aged cells, hematopoiesis, blood coagulation, atherogenesis, wound healing, and metal ion transport, in addition to development of immune responses and host resistance against virus-infected cells, tumors, or microorganisms. Experimental evidence has been accumulated that the macrophage is one of the essential cells controlling homeostatic balance, and its functions can be regulated by biological response modifiers (BRM). A significant effort is now underway to develop effective BRM that will regulate functional activities of Mϕ. However, the inability to target Mϕ prevents BRM from inducing effective activation of Mϕ and causes serious side effects attributed to the high-dosage regimens of the drugs. Thus, it is necessary to develop drug-delivery systems capable of Mϕ targeting before Mϕ activation using BRM can be applied successfully for immunotherapy. It is likely that such a carrier system can reduce the BRM dosage required for Mϕ activation. Research on drug-delivery systems for BRM to manipulate functional activities of Mϕ has just started.

REFERENCES

1. I. Carr, *The Macrophage: A Review of Ultrastructure and Function*, Academic Press, London, 1973.
2. C. W. Pierce, Macrophages: Modulators of immunity, *Am. J. Pathol. 98*:10 (1980).
3. E. R. Unanue, Cooperation between mononuclear phagocytes and lymphocytes in immunity, *N. Engl. J. Med. 303*:977 (1980).
4. A. S. Rosenthal, Regulation of the immune response—role of the macrophage, *N. Engl. J. Med. 303*:1153 (1980).
5. A. Lasser, Progress in pathology. The mononuclear phagocytic systems, a review, *Hum. Pathol. 14*:108 (1983).
6. R. J. North, The concept of the activated macrophage, *J. Immunol. 121*:806 (1978).
7. M. L. Karnovsky and J. K. Lazdins, Biochemical criteria for activated macrophages, *J. Immunol. 121*:809 (1978).
8. Z. A. Cohn, The activation of mononuclear phagocytes: Fact, infancy, and future, *J. Immunol. 121*:813 (1978).
9. I. J. Fidler and G. Poste, Macrophage-mediated destruction of malignant tumor cells and new strategies for the therapy of metastatic disease, *Springer Semin. Immunopathol. 5*:161 (1982).
10. I. J. Fidler, Macrophages and metastasis—a biological approach to cancer therapy: Presidential address, *Cancer Res. 45*:4741 (1985).
11. R. J. David, Macrophage activation by lymphocyte mediators, *Fed. Proc. 34*:1730 (1975).

12. W. F. Dow and P. M. Henson, Macrophage stimulation by bacterial lipopolysaccharides. I. Cytolytic effect on tumor target cells, *J. Exp. Med. 148*:544 (1978).

13. F. S. Elouz, A. Adam, R. Ciorbaru, and E. Ledever, Minimal structural requirements for adjuvant activity of bacterial peptideglycan derivatives, *Biochem. Biophys. Res. Commun. 59*:1317 (1974).

14. S. Kotani, Y. Watanabe, and T. Kinoshita, Immunoadjuvant activities of synthetic *N*-acetyl-muramyl peptides or -amino acids, *Biken J. 19*:9 (1976).

15. I. J. Fidler, J. H. Darnell, and M. B. Budman, Tumoricidal properties of mouse macrophages activated with mediators from rat lymphocytes stimulated with concanavalin A, *Cancer Res. 36*:3608 (1976).

16. S. Sone and I. J. Fidler, Tumor cytotoxicity of rat alveolar macrophages activated in vitro by endotoxin, *J. Reticuloendothel. Soc. 27*:269 (1980).

17. E. Ledever, Synthetic immunostimulants derived from the bacterial cell wall, *J. Med. Chem. 23*:819 (1980).

18. R. K. Oldham, Biological response modifiers, *JNCI 70*:789 (1983).

19. D. Zekorn, Modified gelatin as plasma substitutes, *Bibl. Haematol. 33*:131 (1969).

20. Y. Ikada and Y. Tabata, Phagocytosis of bioactive microspheres, *J. Bioact. Compt. Polym. 1*:32 (1986).

21. Y. Tabata and Y. Ikada, Drug delivery systems for antitumor activation of macrophages, *Crit. Rev. Ther. Drug Carrier Syst. 7*:121 (1991).

22. Y. Tabata and Y. Ikada, Synthesis of gelatin microspheres containing interferon, *Pharm. Res. 6*:422 (1989).

23. Y. Tabata and Y. Ikada, Effect of the size and surface charge of polymer microspheres on their phagocytosis by macrophage, *Biomaterials 9*:356 (1988).

24. Y. Tabata and Y. Ikada, Macrophage activation for antitumor function by muramyl dipeptide–protein conjugate, *J. Pharm. Pharmacol. 42*:13 (1990).

25. Y. Tabata and Y. Ikada, Targeting of muramyl dipeptide to macrophages by gelatin conjugation to enhance their in vivo antitumor activity, *J. Controlled Release 27*:79 (1993).

26. Y. Tabata, K. Uno, T. Yamaoka, Y. Ikada, and S. Muramatsu, Effects of recombinant α-interferon–gelatin conjugate on in vivo murine tumor cell growth, *Cancer Res. 51*:5532 (1991).

27. Y. Tabata, K. Uno, Y. Ikada, T. Kishida, S. Muramatsu, Potentiation of in vivo antitumor effects of recombinant-1α by gelatin conjugation, *Jpn. J. Cancer Res. 84*:681 (1993).

28. R. J. North, Endocytosis, *Semin. Hematol. 7*:161 (1970).

29. Z. A. Cohn, *Mononuclear Phagocytes* (R. van Furth, ed.), Blackwell Scientific, Oxford, 1970, p. 121.

30. F. M. Griffin, Jr., J. A. Griffin, J. E. Leider, and S. C. Silverstein, Studies on the mechanism of phagocytosis. I. Requirements for circumferential attachment of particle-bound ligands to specific receptors on the macrophage plasma membrane, *J. Exp. Med. 142*:1263 (1975).

31. T. P. Stossel, Phagocytosis: Recognition and ingestion, *Semin. Hematol. 12*:83 (1975).

32. C. J. van Oss, *Phagocytic Engulfment and Cell Adhesiveness*, Marcel Dekker, New York, 1975.

33. Y. Tabata and Y. Ikada, Phagocytosis of polymer microspheres by macrophages, *Adv. Polym. Sci. 94*:107 (1990).
34. Y. Tabata and Y. Ikada, Enhanced macrophage activation for antitumor function by muramyl dipeptide chemically modified with gelatin of different molecular weights, *Proc. Int. Symp. Controlled Release Bioact. Mater. 22*:552 (1995).
35. Y. Tabata, K. Uno, Y. Ikada, and S. Muramatsu, Potentiation of antitumor activity of macrophages by recombinant interferon alpha A/D contained in gelatin microspheres, *Jpn. J. Cancer Res. 79*:636 (1988).
36. I. J. Fidler, W. E. Fogler, E. S. Kleinerman, and I. Sakai, Abrogation of species specificity for activation of tumoricidal properties in macrophages by recombinant mouse or human interferon-encapsulated in liposomes, *J. Immunol. 135*:4289 (1985).
37. Y. Tabata, K. Uno, S. Muramatsu, and Y. Ikada, In vivo effects of recombinant interferon alpha A/D incorporated in gelatin microspheres on murine tumor cell growth, *Jpn. J. Cancer Res. 80*:387 (1989).
38. A. Mantovani, N. Polentrutti, G. Peri, Z. B. Shavit, A. Vecchi, G. Bolis, and C. Manggioni, Cytotoxicity on tumor cells of peripheral blood monocytes and tumor-associated macrophages in patients with ascites ovarian tumors, *JNCI 64*:1307 (1980).
39. I. J. Fidler, S. Sone, W. E. Folger, and Z. L. Barnes, Eradication of spontaneous metastasis and activation of alveolar macrophages by intravenous injection of liposomes containing muramyl dipeptide. *Proc. Natl. Acad. Sci. USA 78*:1680 (1981).
40. H. Bergstrand, I. Anderson, I. Nystrom, R. Pauwels, and H. Bazin, The nonspecific enhancement of allergy, *Alllergy 38*:246 (1983).
41. J. Kreuter, U. Berg, E. Liehl, M. Soliva, and P. P. Speiser, Influence of the particle size on the adjuvant effect of particulate polymeric adjuvants, *Vaccine 4*:125 (1986).
42. J. Kreuter, U. Berg, E. Liehl, M. Soliva, and P. P. Speiser, Influence of hydrophobicity on the adjuvant effect of particulate polymeric adjuvants, *Vaccine 6*:253 (1986).
43. M. E. D. Martin, J. B. Dewar, and J. F. E. Newman, Polymerized serum albumin beads possessing slow release properties for use in vaccine, *Vaccine 6*:33 (1986).
44. J. H. Eldridge, J. K. Staas, J. A. Meulbroek, T. R. Tice, and R. M. Gilley, Biodegrable and biocompatible poly (D,L-lactide-*co*-glycolide) microspheres as an adjuvant for staphylococcal enterotoxin B toxoid which enhanced the level of toxin-neutralizing antibody, *Infect. Immun. 59*:2978 (1991).
45. D. T. O'Hagen, D. Rahman, J. P. Acgee, H. Jeffery, M. C. Davis, P. Williams, and S. S. Davis, Biodegradable microparticles as controlled release antigen delivery systems, *Immunology 73*:239 (1991).
46. M. T. Aguado and P. H. Lambert, Controlled-release vaccines—biodegradable polylactide/polyglycolide(PL/PG) microspheres as antigen vehicles, *Immunobiology 184*:113 (1992).
47. R. V. Nellore, P. G. Pande, D. Young, and H. R. Bhagat, Evaluation of biodegradable microspheres as vaccine adjuvant for hepatitis B surface antigen, *J. Parent. Sci. Technol. 46*:176 (1992).
48. Y. Tabata, R. Nakaoka, and Y. Ikada, Potentiality of gelatin microspheres as an immunological adjuvant, *Vaccine 13*:653 (1995).

15

Bioadhesive Liposomes for Topical Treatment of Wounds

Rimona Margalit

Tel Aviv University
Tel Aviv, Israel

I. INTRODUCTION

Since their "birth" three decades ago, liposomes have been in the front line of microparticulate carriers investigated for and developed as drug-delivery systems. Furthermore, the majority of efforts and resources have been, and are, directed toward the systemic administration of liposomes for the treatment of tumors and infectious diseases [1–5]. It is beyond dispute that the severity of the problems when such diseases are treated by free drugs merits all past, current, and future efforts for development of delivery systems that would overcome them and lead to substantial improvement in clinical outcomes. However, the strong focus on therapies that require systemic liposome administration has cast a shadow on other situations that might also benefit from the replacement of free drug by drug in a delivery system.

This chapter is dedicated to such "off the mainstream" cases, in particular to the topical treatments of wounds and burns. (The term wound will be used throughout this chapter as including both wounds and burns.) Accordingly, the first part of this chapter will outline the therapeutic targets, the needs for and rationale behind the use of delivery systems. The second part will focus on liposomes as the delivery system of choice, examining their advantages and drawbacks for the proposed tasks and summarize the available in vivo data. The third part will discuss ways and means to overcome (or at the least sufficiently reduce) those drawbacks, introducing the concept of bioadhesive liposomes. Critical pertinent issues that concern the bioadhesive liposomes will be the subject of the fourth part. Issues such as the rationales behind the selection of specific bioadhesive ligands and the development of more than one liposomal species; practical aspects of such modified liposomes as pharmaceutical products; research and development strategies for such systems, together with selected examples of experimental data. Approaches to functional testing and discussion of future prospects will be offered at the concluding part of this chapter.

II. TOPICALLY ADMINISTERED DRUGS IN THE TREATMENT OF WOUNDS AND BURNS: DEFICIENCIES OF TREATMENTS WITH FREE DRUGS AND THEIR POTENTIAL RESOLUTION BY THE USE OF DRUG-DELIVERY SYSTEMS

Several principles of wound management, when applied together, have been recognized as critical to successful healing [6–13]. Principles such as (a) effective wound cleansing, including debridement, if needed; (b) proper pre- and postoperative procedures (for surgical wounds); (c) minimal interference with the natural self-healing processes (postcleansing); (d) maintenance of a moist wound environment, which is most conducive to healing; (e) the use of occlusive dressings that are changed at relatively low frequencies (such as once every few days, rather than once or more a day); and (f) good nutrition and general care of the patient. In addition, medical intervention for wound healing can also have drug-associated components, the two major drug classes being growth factors and antibiotics.

A. Wound Healing: Deficiencies of Treatment with Free Growth Factors

Growth factor treatment of wounds, mainly by topical administration, is a therapy still in development. After maturation into established clinical modalities, growth factors will be prescribed for two distinct categories of patient populations. One category would include patients who have trauma or surgical wounds in which the self-healing processes have not been significantly jeopardized [14,15]. For these patients the growth factors will be given to accelerate the processes, the benefits of faster healing being self-evident. The other category includes patients who suffer from nonhealing, often chronic, wounds, with different types of underlying causes, such as: diabetes, venous insufficiencies, immunocompromised state (temporary or permanent), and others [16–20]. For these patients the growth factor therapy is anticipated to reverse the situation from nonhealing into healing wounds.

Several growth factors, most prominent among them being epidermal growth factor (EGF), platelet-derived growth factor (PDGF), fibroblast growth factor

(FGF; a and b) and transforming growth factor (TGF; α and β) are under investigation for these tasks. More than a decade of studies, conducted in vitro and in vivo and including experimental treatments, have secured their potential in wound healing [21–29]. Yet, the road from the laboratory to the clinic is still strewn with obstacles responsible for the current status of clinical trials that have been described by terms running from unsatisfactory to failure. Major among these obstacles is that lack of effective delivery systems, an impediment not only in therapy but also in research and development, especially into the biology of wound repair and the roles of growth factors in it. Tracing the fate of a growth factor administered topically to a wound, as will be done in the following, illustrates the need for an effective carrier, provides the definition of effective delivery, and sets the essential requirements a proposed carrier would have to meet.

For the sake of the present discussion, a simplified, schematic representation of a wound is illustrated in Fig. 1. Concentrating, first, on the bottom section

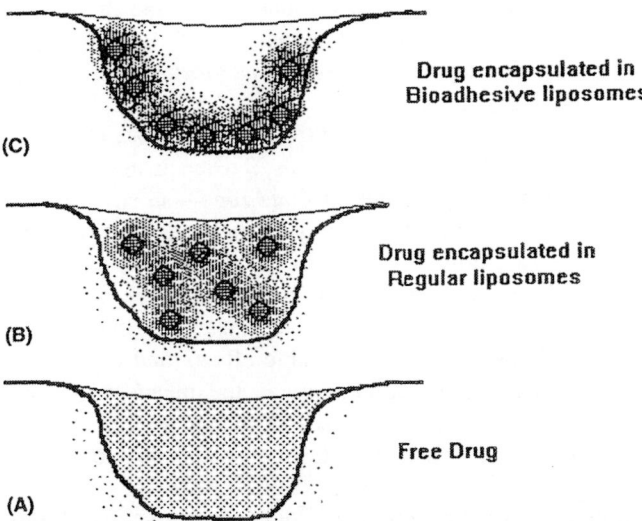

Fig. 1 A schematic representation of a wound and drug-delivery systems within it. (A) The bell-shaped (inverted) curve illustrates the border between the wound—which is represented by the space within "the bell"—and the surrounding, healthy, tissues. The "ladles" on both sides of the bell represent the uppermost layers of the healthy skin surrounding the wound. The dots represent soluble free drug inside the wound and in some of the surrounding tissues. (B) Same as panel A, but also containing drug-encapsulating regular liposomes that have been administered topically to the wound and are in the process of drug release. (C) Same as panel A, but also containing drug-encapsulating bioadhesive liposomes that are adhering to their matrix or cellular binding sites and are in the process of drug release.

(see Fig. 1A), the wound is represented by the interior of the bell-shaped curve, the curve itself represents the border between the injured and the surrounding healthy tissues. The two top ladle-shaped extensions represent uppermost layers of the healthy skin surrounding the wound. The wound can be envisioned as composed of three zones: (a) The drug administration zone which is, naturally, the topmost section of the wound. (b) The target zone, occupying a thin layer inside the bell-shaped line, which consists of healthy cells that have been recruited from the surrounding tissues. These cells adhere to the matrix, are heavily involved in the healing processes, and have embedded in their membranes receptors for the growth factors (i.e., the actual sites of action for this particular class of "drugs"). (c) The intermediate zone, that will be discussed in detail later on, which consists of the space between the two other zones and occupies, before effective healing becomes active, most of the wound.

Current dosage forms of topically administered growth factors, especially those that are designated for the eventually established treatment, are mostly in the form of solutions (poured onto the wound or soaked into the dressings); solutions made more viscous, such as cellulose gels; and growth factors soaked into collagen sponges [30–33]. The natures of these dosage forms are such that immediately after their introduction through the administration zone, or shortly thereafter, the entire dose is free within the wound; therefore, it is exposed to all its elements, as illustrated by the dots in Fig. 1A. To reach their targets, the growth factors have to travel across the intermediate zone—an environment that is enzymatically hostile to polypeptides, catalyzing their degradation [34]. Coupled with clearance processes, only a fraction of the initial dose can reach the targets in active form. With growth factors, the margin between doses that are low enough to produce beneficial effects alone, and doses that can already cause reduction in the level of response or undesirable effects and toxicity, is quite narrow [25,34]. This makes dose escalation, as the means to deliver enough drug to the target to compensate for losses of degradation and clearance, a limited option, as even the transient high doses could do more harm than benefit.

Even if dose escalation were a valid option, it would not suffice to address the timing factor. Growth factors are agents that act at a specific stage of the cell cycle [34–36]. Although investigators might vary in the detailed assignment of specific factors to stages, the assignments offered by Gartner and associates [35] serve as an example for the tie between growth factor and stage of cell cycle: PDGF is classified as a competence factor exerting a positive effect on the move from the G_0 to the G_{1a} stage; EGF and insulin are classified as entry factors, facilitating the advance form the G_{1a} to the G_{1b} stage; insulin-like growth factor-1 (IGF-1) is classified as a progression factor, acting toward the end of the G_{1b} stage. A continuous supply of active growth factor, of a day or more, is required to make a clinical difference [23,25,31]. The consequences of this situation is that, to supply sufficient growth factor to the wound for a sufficient duration, growth factor therapy currently requires treatment regimens of multi- and frequent dosing. For example, treatment regimens of the clinical trials discussed earlier (which, were not satisfactory) consisted of twice or more daily administrations, for prolonged periods that have reached 100 days [18,30,37,38]. Needless to say, the principles of good wound management cannot be implemented with the current growth factor treatments. This leads to a situation in which an agent indicated for improvement in healing has to be administered in a treatment regimen that counteract its therapeutic efficacy.

B. Wound Bacterial Infections: Deficiencies of Treatment with Free Antibiotics

Infections, mainly bacterial, are the major complication in wounds and cannot be ignored. A distinction that is significant for diagnosis, prognosis, and treatment goals, exists between infected and colonized wounds. The term *infection* is reserved for wound bacterial burdens that exceed 10^5 organisms per gram tissue or 10^5-10^6 organisms per square centimeter of wound [39–42]. Lower bioburdens are termed *bacterial colonization*. Obviously, the cutoff between these two categories might be shift to lower burdens for patients with compromised host defense mechanisms.

In infected wounds, host defense mechanisms alone cannot control the wound bacterial burdens. If left untreated such wounds will not heal and, unless appropriate intervention is implemented, the bacteria can invade surrounding healthy tissue and eventually reach the circulation, becoming whole-body sepsis [15,39–44]. In contrast, if the wound is merely colonized, host defense mechanisms can control the situation and keep it from deteriorating into an infection with its resultant damage, and the wound will heal [15,39–44].

Intervention for the treatment of infected wounds by topically administered antiseptics has been ruled out and is being phased out owing to their now well-recognized lack of efficacy that, moreover, is coupled with severe detrimental effects [15,45–46]. Antibiotics are the preferred treatment, whether for an infected wound or as a prophylactic measure (in surgery) and are frequently given systemically [15,45–46]. However, internal routes are often inadequate to supply the wound site with enough drug to meet the therapeutic goals, be they preventive measures or efficacious bioburden reduction to the colonization level [15,45–47]. As notably stated by clinicians [see, for example, Ref. 15] topical antibiotics might do better, either as stand-alone therapies or (as the case requires) in conjunction with systemic antibiotics.

Unlike the therapy-in-development status of growth factors, topical treatment of wounds with antibiotics is an established therapy, albeit not frequently prescribed with products on the market, most of which are in the forms of creams and ointments [48,49]. Several major drawbacks of these dosage forms can be held responsible for the marked preference clinicians have for oral and IV administration of antibiotics (for wounds) despite the principle recognition that topical treatment could be a viable and favored option.

As with the current dosage forms of growth factors, the total antibiotic dose becomes free in the wound immediately or shortly after its administration. The free drug is then prone to inactivation or degradation and to fast clearance. The lack of effective targeting leads to indiscriminate distribution over the whole wound. Thus, only a fraction of the administered dose reaches its sites of action (i.e., the bacterial colonies), a fraction that is too small to be effective, as evidenced by the well-documented treatment regimens (when topical antibiotics are used) of two to six daily applications [48,49]. Evidently, and similar to the situation with growth factors, such high frequency of dosing does not allow one to practice the principles of good wound management. In addition, most creams and ointments are not biodegradable, which can cause further trauma to the wound when removal of the remnants of the previous dose is required before reapplication.

C. Delivery Systems for Topical Treatment with Growth Factors and Antibiotics

To provide a significant improvement in the clinical outcome, an antibiotic or growth factor delivery system would have to successfully tackle the problems that arise from the exposure of the free drug to the wound and, at the same time, address the issues of treatment regimens. Needless to say, to offer effective solutions, rather than shifting the weight of the problem from one category to another, the drug carrier should nor brings major obstacles of its own.

Successful resolution of the aforediscussed problems that afflict topical treatment of wounds with free drugs requires the carrier to (a) provide the drug with protection from the hostile environment; (b) slow down the clearance; (c) target itself, still carrying most (if not all) of its original drug load to the sites of drug action and deliver the drug there, over a time span that fits the demands of the therapy. If the carrier would not only target itself to the proper zone(s), but would also be capable of retention there for several days, during which it would act as a sustained-release depot, it would address the timing issue, especially (but not only) for to the growth factors, and also fit with, rather than against, the principles of good wound management, thereby allowing their implementation. If the carrier would also be biodegradable and biocompatible, nontoxic, and nonimmunogenic, it would diminish the risks of introducing new carrier-related drawbacks.

III. REGULAR LIPOSOMES AS DRUG-DELIVERY SYSTEMS FOR TOPICAL ADMINISTRATION TO WOUNDS AND BURNS

A. Liposomes: Fit Between Properties and Task Performance Requirements

Among microparticulate carriers, liposomes are proposed here as prime candidates for that carrier. Liposomes are biodegradable and biocompatible and are sufficiently nontoxic and nonimmunogenic for systemic and nonsystemic administrations [see, e.g., Refs. 1–5, 50–54; and references therein]. Liposomes have, in principle, the ability to protect their encapsulated matter from the external environment and to act as sustained-release depots [1–5,50–60].

As will be shown in the following by two specific examples, evaluation of the protective ability of liposomes, with their encapsulated matter, can be done by testing the ability of the encapsulated matter to retain both its structural integrity and its indicated biological (therapeutic) activity.

By using EGF for itself and as a general prototype for growth factors involved in wound healing, EGF-encapsulating unilamellar and multilamellar liposomes were incubated in a mixture of nonspecific proteases used as a model of the hostile enzymatic environment in a wound. The following three systems were subsequently subjected to gel-exclusion chromatography and were studied separately for the kinetics of EGF efflux from the liposomes: (a) EGF encapsulated within the enzyme-exposed liposomes; (b) EGF encapsulated within liposomes that were not exposed to the enzymes; and (c) free EGF, also not exposed to the enzymes. The chromatograms of all three systems were identical, as were the diffusion parameters of all three [55,60, Okon et al., in preparation]. In testing of the biological activity of EGF (i.e., stimulation of cell proliferation) in the classic assay developed by Carpenter and Zendegui [62] in cultures of Balb/MK cells, liposome-encapsulated EGF not only maintained its activity, but surpassed that of free EGF. For example, at the end of 4 days, a single dose of 0.08 ng of free EGF gave an eightfold increase in cell proliferation, whereas under the same conditions and for the same dose liposome-encapsulated EGF gave a 30-fold increase [Okon et al., in preparation]. Taken together, these data are a clear indication that liposomes can protect encapsulated peptides and polypeptides from a hostile enzymatic environment and retain the biological activity of the encapsulated matter.

With ampicillin itself and as a model for liposome-encapsulated antibiotics, high-performance liquid chromatography (HPLC) chromatograms of liposome-encapsulated ampicillin were identical with those of free (intact) drug [63; Schumacher and Margalit, in preparation]. Liposome-encapsulated ampicillin, was also as active as free ampicillin in the classic antibacterial assay for this drug, using the recommended *United States Pharmacopeia* (USP) test organisms [63,64; Schumacher and Margalit, in preparation]. As with EGF, such data are clear indication that encapsulation within liposomes can retain the structural integrity and the antibacterial activity of liposome-encapsulated antibiotics.

Evaluation of the ability of liposomes to act as sustained-release depots is essentially a diffusion study that requires the ability to separate this process from drug release that is due to liposomal damage. Specific examples and favored experimental designs will be presented later in this communication. For the topical applications discussed herein, as with many other potential clinical applications of liposomes, targeting remains a major unresolved issue. Despite the lack of targeting, regular liposomes have already been tested in vivo as topical delivery systems of growth factors and of antibiotics in wounds. Those studies are summarized in the following. What can be, and has been, done to address the issue of targeting in wound healing is the subject of a later section.

B. In Vivo Studies of Regular Liposomes in Wound Healing and Wound Infections

Brown et al. [65] have tested liposomes as a growth factor-delivery system for the topical treatment of incisional wounds in mice. They monitored the progress in healing by following the changes in tensile strength. An interesting dual-carrier system was devised in which insulin, serving as an intraliposomal carrier, was first complexed with EGF, then the insulin–EGF complex was encapsulated within the liposome. The liposomal formulation provided a transient increase in tensile strength, compared with empty liposomes or saline. [No comparisons with free drug, or with empty liposomes suspended in free drug, were reported for this experimental set.] Combined with exceptionally high intraliposomal retention of the EGF reported in this study, it could be that the internal carrier induced a sustained-release that was too slow for effective therapy. Evaluation of the merits of this dual-carrier concept for EGF, and for similar-sized polypeptides, and the extent to which it can be optimized to go beyond the transient effect, awaits further studies.

In a study of healthy nonwounded mice, Grayson et al. [66] tested gentamicin-encapsulating multivesicular liposomes (DepoFoam) as a prophylactic anti-infective treatment for surgical wounds. The liposomal formulation was injected subcutaneously to provide local depots and was challenged 48 h later by a bacterial inoculum, also injected subcutaneously, to the same location. Evaluation of bioburden reduction 48 h later showed that the liposomal formulation was significantly superior to empty liposomes or saline.

Price et al. reported a series of studies [67–69] in which they explored the potential of a liposome-like lipid-based carrier, denoted solvent dilution microcarrier (SDMC), in the treatment of infected wounds. In the first study [67] two treatment regimens were applied to rats bearing infected wounds. In the regimen (a) used for the free drug (tobramycin or silver sulfadiazine) there were twice-daily applications over a total treatment period of 72 h. In the regimen used for the drug–carrier formulation (b), the system was formed in situ by the separate addition of free drug and the carrier-forming material onto a sponge inserted into the wound, and the sponge was wetted twice daily for a total period of 72 h. The same level of reduction in bioburden was achieved in both regimens, for equal total doses (cumulative in regimen a, and initial in regimen b) of the free and of the carrier-mediated, drug. The authors propose that these results, together with the differences in the treatment regimens, show that there might be an advantage to the liposomal formulations. In the second study [68], albeit with rats bearing noninfected burns, tobramycin, when given in a preformed carrier formulation, was retained unchanged at the wound site for between 24 and 72 h after administration, and most of the lipid marker remained at the wound area, mostly on the dressing. In the third study [69] conducted with rats bearing bacteria-inoculated incisional wounds, the effects of a single dose of tobramicin on wound bioburden at the end of 1, 2, and 3 days, were compared for free drug and for drug in the preformed carrier formulation. The authors did not provided any physicochemical details on the level of tobramycin encapsulation, or on the kinetics of drug release. Nor is the issue of whether unencapsulated drug was removed before treatment addressed. This makes it difficult to make any comparisons between this carrier system and other antibiotic-encapsulating regular liposomes. The in vivo data are encouraging, showing the carrier formulation is twofold better than free drug (and both are better than no treatment at all) in reducing the bioburden under conditions designed to model surgical prophylaxis. Similar future studies in which bioburdens in the absence of treatment will be clearly in the infection range will tell whether the potential of this drug–carrier formulation is limited to prophylaxis or also extends to the treatment of infected wounds.

Although all these studies can be taken as encouragement for exploring liposomes as drug-delivery systems for topical treatment of wounds and burns, they cannot be taken as a completed evaluation of liposomes for such tasks. There are limitations in the models used: in one the ability of liposomal systems to prevent or control infection was studied in animals that received the bacteria, but had no wounds, whereas in another the ability of the liposomal systems to deliver their "goods" was tested in animals that had wounds, but no bacterial infection. That liposomal systems were at least as satisfactory as free drug was clearly shown in the study in which they were applied to animals bearing infected wounds. However, the twice-daily interference with the wound, even when the liposomal systems were applied, indicates that such systems would not fit with the principles of good wound management. Also, none of the cited systems were simple, regular liposomes that encapsulated the drug alone; rather, Grayson et al. [66] used multivesicular liposomes; Price et al. [67–69] used lipid-based microcarriers; whereas Brown et al. [65] used classic multila-mellar vesicles (MLV), but with the dual-carrier approach.

IV. THE CONCEPT OF BIOADHESIVE LIPOSOMES AND FIRST-GENERATION SYSTEMS

A. Local Targeting: Definition, Requirements, and the Limitations of Regular Liposomes

Target definition was deemed essential before beginning any discussion on the deficiencies of regular liposomes relative to drug targeting in topical wound and burn therapies and of potential means to remedy the situation. Especially since, as will be seen from the following discussion, carrier-mediated drug targeting is concerned with two distinct types of targets.

The first type is the ultimate therapeutic target (i.e., the site of drug action). This would be the relevant receptor when a growth factor is concerned, as would the bacterial colonies when antibiotics are concerned. The second target is the binding site of the drug carrier. Such targets should obviously be in close proximity to the site of drug action (i.e., to the first target), but need not, and often should not, coincide with it. For growth factors, targeting the carrier to the receptor would actually block sites of drug action, and targeting the carrier to the interior of the cell (if possible at all) would miss the drug target alto-gether. Thus, a feasible target for the carrier would be extracellular matrices (ECM) that are close to the cells carrying the receptors, or receptors on that same (or close-by) cells that respond to ligands other than the specific growth factor delivered. [The term *receptor* is taken here in its broadest form. It refers to membrane-embedded entities for which receptor–ligand interactions start and end in ligand–receptor binding, as well as those for in which ligand–receptor binding is merely the first step in a cascade of extra- and intracellular events.] For antibiotics, the carrier should be targeted to the location of the colony. For extracellular infection this would be the extracellular matrix, where the bacteria themselves adhere. For intracellular infections, receptors on the cell membranes that do not interfere with the drug action are a reasonable option. For the latter, an ability of the cells to internalize the liposomes would be an attractive benefit; however, such internalization is not an absolute requirement. Holding the drug-encapsulating liposomes as a local depot at the cell surface can suffice to pro-vide a significant improvement over treatment with free drug. To remove confusion in referral to each type of target and to avoid repetition of lengthy definitions, the following terminology will be adopted: the first type of targets will be simply addressed as *drug sites*, and the second type of targets will be simply addressed as *carrier sites*.

The following combination of factors are required, to achieve both carrier and drug targeting that would lead to successful wound therapy with topically administered liposomes: (a) arrival of the carrier at its sites while still containing most (at the least) of its original drug load in active form; (b) high-affinity binding of the carrier to its sites; (c) retention of the carrier at those sites for the duration of the treatment, despite tissue and fluid dynamics; and (d) sustained drug release. The time spans of the latter two (i.e., site retention and drug release) should not only coincide, but should be within the range befitting the optimal time intervals dictated by principles of wound management (see Section II). Currently, a span of 1–4 days for both seems reasonable. Shorter spans might be too similar to treatment with free drug to provide improvements in the clinical outcomes. Significantly longer spans of drug release might deliver too little drug at a time to effectively induce healing or combat continuously proliferating bacteria.

The critical need for targeting can be appreciated by envisioning scenarios that might ensue for treatment with regular (i.e., nontargeted) liposomes, taking into account that such liposomes would probably be dispersed over the whole wound, as illustrated in Fig. 1B. For instances for which the diffusion of the encapsulated drugs is fast enough to result in liposomal depletion within a time span that is significantly less than a day, the outcome of liposomal treatment might be no worse than with free drug. However, if these liposomes act as sustained-release depots, then the results of the treatment might actually be worse than with free drug. Thus, replacing treatment with free drug by treatment with drug encapsulated in regular nontargeted liposomes, might defeat the purpose of using a drug-delivery system at all, even if it is capable of meeting all other requirements, such as the mutual drug–environment protection and others, discussed in previous sections.

The concept of bioadhesive liposomes, which is the subject of the next section, has evolved as a consequence of the needs to target both the drug and the carrier.

B. The Concept of Bioadhesive Liposomes

After having defined the carrier sites, the obvious starting point for the design of means that would endow the liposomes with sufficient affinity to those sites, is the wide repertoire of bioadhesive macromolecules, naturally occurring and synthetic, that are capable of high-affinity binding to those sites. The conceptual approach was to modify the liposome, by covalently anchoring such a bioadhesive macromolecule (ligand) to its surface. By being made an integral part of the liposome, it was anticipated that the ligand would confer on the complete delivery system its ability to bind with high affinity to the designated carrier sites, as illustrated in Fig. 1C. For such surface-modified liposomes, denoted *bioadhesive liposomes*, and schematically illustrated in Fig. 2, to have (theoretically at least) all the components and properties required for that ideal growth factor or antibiotic delivery system, the following two conditions have to be simultaneously fulfilled: (a) the transition from the free to the liposome-bound state should not lead to any significant loss in the ligand's bioadhesivity, and (b) the surface modification should not significantly compromise any of the other advantages of liposomes that are critical for task performance.

The availability of many macromolecules with the capability of adhesion to components within the extracellular matrix or to receptors embedded in the membranes of the relevant cells necessitated the formulation of selection guidelines. The adopted guidelines were as follows: (a) The choice of bioadhesive ligands was restricted to naturally occurring macromolecules that would not significantly compromise the nontoxic, nonimmunogenic, biocompatible, and biodegradable nature of liposomes. (b) Among the naturally occurring ligands, preference was given to those that are already approved for human use (at the least topical) or are on the road to approval. (c) Among the ligands that fit guidelines (a) and (b) preference was given to ligands that would result in

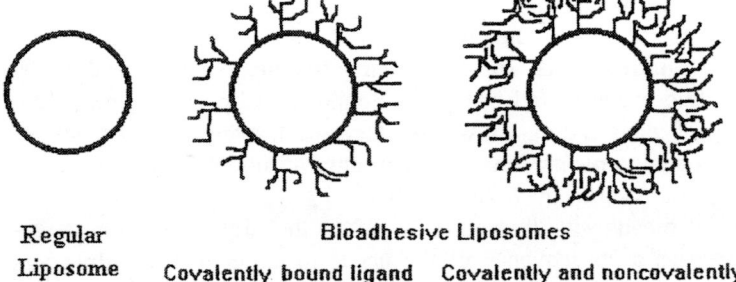

Regular
Liposome

Bioadhesive Liposomes

Covalently bound ligand   Covalently and noncovalently
                            bound ligand

Fig. 2 Schemes of regular and bioadhesive liposomes. The two species of bioadhesive liposomes illustrate two classes of bioadhesive liposomes: (left) Liposomes with a relatively modest ligand surface density, for which it can be assumed that the ligand molecules are bound covalently to the liposomal surface, and (right) liposomes with sufficiently high ligand surface density, for which it can be assumed that there are ligand molecules covalently bound at the liposomal surface that serve, themselves, as anchors for noncovalent binding of additional ligand molecules. This manifestation of self-adhesion is expected for species of bioadhesive ligands that are known, in their free form, to be capable of self-interaction.

liposomal products that would be applicable (each) to wide patient populations.

To implement these guidelines, the design and study of four species of bio-adhesive liposomes was initiated, selecting the following macromolecules for the first-generation series: collagen, gelatin, hyaluronic acid, and EGF. Collagen, gelatin, and hyaluronic acid are known for their potential to bind to components of the extracellular matrix [70–78], fibroblasts have receptors for collagen [70,80–82], and there are, obviously, EGF receptors within wounds and burns. Lest there be some confusion, it is stressed that EGF and other growth factors involved in wound healing, can serve in two different roles. Encapsulated within the liposomes, a given growth factor is a therapeutic agent for wound healing. For this role, bioadhesive ligands other than that specific growth factor would be suitable. Anchored to the liposomal surface, that same growth factor is a bioadhesive ligand engaged in the local targeting, for delivery of drugs other than that particular growth factor.

Starting with several, rather than a single, potential system, is a reasonable policy whenever the process of translation a concept into reality is initiated. An added incentive to the investment of efforts and resources is the benefits that would come from having more than one carrier system. The wound is a continuously changing environment, and it is quite feasible that even for the delivery of the same drug, there would be periods when one bioadhesive ligand would be better that another, and periods when the reverse would hold. For example, the presence and levels of collagenase and hyaluronidase in a wound vary with time (with progression of the healing) as do the presence and levels of collagen and hyaluronic acid. As more becomes known on the composition and cycles of wound extracellular matrix relative to its components and on wound enzymology, it should be possible to fit and rotate the species of bioadhesive liposome used according to the situation in the wound, to maximize the beneficial effects of these delivery systems.

C. Liposome Surface Modification

There are, essentially, two major approaches for the preparation of surface-modified liposomes: One is to first form a lipid–ligand conjugate, then use it as one of the components of the liposome formulation. The other is to first form the liposomes, then perform the modification by cross-linking the ligands, to the liposomal surface. In either approach the lipid or the liposome requires functional groups for the cross-linking. The second approach has been adopted for the bioadhesive liposomes discussed here, using phosphatidylethanolamine (PE) at concentrations of 2–30 mol% in the formulation to supply functional residues for cross-linking [57–59,61].

Stability concerns for the encapsulated drug, the liposome, and the bioadhesive ligand, together with the realization that to progress to the clinic and become established treatment, these liposomes have to be manufactured as pharmaceutical products, were taken into account for the modification procedures. Efforts were made to select and, if need be, revise, simplified procedures that would require the minimal possible steps and would adhere as close as possible to physiological pH and mild temperatures. Given these considerations, glutaraldehyde was selected as the cross-linker of choice for the proteins and polypeptides among the bioadhesive ligands, in a one-step process, at neutral pH, and temperatures (depending on the ligand) in the range of 4–37°C [57,59,82]. For the nonproteinaceous matter, hyaluronic acid, a water-soluble carbodiimide was the preferred cross-linker, and it required a two-step process: preactivation of the hyaluronic acid with the carbodiimide under mildly acidic conditions (pH range of 3–4) followed by binding of the activated hyaluronic acid to the liposomes under mildly basic conditions (pH range of 8–8.5) [59]. A potential drawback of modifying preformed liposomes is the loss of encapsulated drug that can occur during the postmodification steps taken to remove byproducts and excess reagent(s). This can be rectified by making use of pregathered data on drug release kinetics for the particular drug and liposome in question. For drug–liposome systems in which there is significantly slow diffusion of the encapsulated matter that drawback might be practically nonexistent. For drug–liposome systems suffering from fast and extensive loss of encapsulated matter, inclusion of free drug in all separation and wash steps at a dose level corresponding to that initially used in the liposome-swelling solutions can prevent, or at the least significantly minimize, depletion.

V. BIOADHESIVE LIPOSOMES: ESSENTIAL REQUIREMENTS FOR TASK PERFORMANCE AND THEIR MOLECULAR AND IN VITRO TESTING

A. Essential Requirements

To act as designated and to achieve optimal response to the drug under the best treatment regimens, the bioadhesive liposomes have to meet several essential requirements that can be divided into two categories: the ability to act as sustained release drug depots and bioadhesivity.

As already discussed in general terms relative to sustained drug release, the rate of drug release has to be set between two limits: It cannot be too fast; fast depletion, such as within few hours or less from administration is ruled out, as it might be too similar to treatment with free drug, defeating the purpose of using a delivery system. It also cannot be too slow, or it might be worse than treatment with free drug, delivering too little drug between dosages to make any significant effect with growth factors or to effectively combat the continuously proliferating bacteria. Thus, the optimal time span for sustained-release would be one such that most (if not all) of the drug load released would coinciding with the dressing change regimen. A balance should be struck between the amount of drug that can be packed into each dose, the amount of drug and time that would be best for treatment, and the longest time interval between changes of dressings that the former provides for and the clinician allows. Taking into consideration that the latter time interval should be longer than a day, and that realistically clinicians might not opt for longer than several days, half-lives of drug release in the range of 1–4 days seem the most reasonable.

When considering bioadhesivity, designing liposomes to be bioadhesive and naming them as such, does not guarantee that they will act as such. This ability has to be proved in the field in a three-part test. First, the bioadhesive liposomes have to be capable of a high-affinity binding and adherence to the designated carrier sites that would be significantly better than that of regular liposomes. Next, to make a real difference from free drug, the carrier has to remain at its site for the duration during which it acts as the sustained-release depot. Two types of dynamic processes that take place in vivo threaten to prematurely detach the bioadhesive liposome from its site. One type is tissue related: Cells in a healing wound migrate, die, and proliferate, the extracellular matrix undergoes changes in its composition and in the extent to which it is occupies or is interspersed with cells. All of these events might detach cell- and matrix-bound liposomes. The other type is fluid dynamics, the flow of the biological fluid over or through the region where the liposomes adhere might also lead to their detachment, well beyond the mere displacement of binding equilibria. Thus, besides testing for high-affinity binding, the bioadhesive liposomes also have to be tested for their ability to be retained at their sites for the specified duration (in the 1–4 day span suggested earlier) despite those tissue and fluid dynamics.

Experimental approaches for quantitative determinations of these essential requirements and for the identification and implementation of parameters that can be used to control and adjust those properties, together with selected data examples, are the subject of the next section.

B. Kinetics of Drug Release

Drug release from intact liposomes into the medium within which the liposomes are placed (or suspended), is essentially a process of diffusion in a heterogeneous system, the studies of which can be started at the test tube level. To realistically model the in vivo expected situation, the experimental conditions in such experiments should be set to (a) prevent any significant liposomal dilution (if at all) and (b) to create and maintain, throughout an experiment, unidirectional drug efflux. The first condition would serve to mimic the in vivo situation in topical administration of liposomes to wounds where, as opposed to IV administration, liposomal dilution is expected to be minimal. The second condition would serve to mimic the cumulative effect of several processes that in vivo would reduce the concentration of free drug at the vicinity of the liposomes to levels that would preclude any significant back-diffusion of the drug into the liposomes. These processes include binding of the drug coming out of the liposome to its sites of action (the major event), binding to other nontarget entities, dilution, clearance, metabolism, or other drug-inactivating processes.

A simple way to provide for these conditions and to have control over the liposome, and separately the initial drug doses, is through the use of an appropriate dialysis setup. Briefly [the interested reader is referred elsewhere for extended technical details; 55–60], a desired sample of a liposomal preparation is introduced into a dialysis sac, selecting a membrane with a pore size that will withhold the liposomes inside, but not be a barrier to the drug, and the sac is immersed into a reservoir of an aqueous drug-free medium of choice. Sac/reservoir volume ratios are selected to provide sufficient dilution of the drug diffusing out of the sac to maintain unidirectional flux conditions. This can be done either by selection of a large enough reservoir to begin with, or by transferring the sac at designated periods to reservoirs of fresh, drug-free, medium. The duration of an experiment, as well as the method of sampling at time points along the way (from the sac, from the single large reservoir, and at each of the many smaller reservoirs) as well as verification of other details are up to the investigator. Regardless of the specific details, the investigator ends up with the cumulative release as the dependent parameter with time—obviously, the independent one—that, for the sake of the following discussion, will be denoted f and t; f expressed in fractions or percentile normalized to the total drug initially present in the system.

The process of drug release has been defined in the foregoing as a diffusion in a heterogeneous system, and before discussion of data processing, this needs clarification. First, it should be taken into account that when introduced into the sac, the drug in the liposomal preparation occupies two types of pools. One type, which consists of a single pool, is the drug that was unencapsulated at that time and is simply free and soluble in the aqueous medium within the sac. The other type of drug pool, of which there could be one or more, accounts for the various forms of the liposome-associated drug. Before addressing the liposome-associated drug pools, attention is drawn here to the unencapsulated drug. Although many experimental procedures have been traditionally used to separate a liposomal preparation from its share of unencapsulated drug, it should be recognized that the removal of drug from the external aqueous medium actually triggers drug efflux out of the liposomes, because it sets an electrochemical gradient. The extent to which drug will escape from the liposomes within the time frame of interest, replenishing the unencapsulated pool in supposedly "clean" systems, depends on several factors, such as the diffusion properties of the drug, the specific liposomal species, the relative volume of the external aqueous phase, and the time that will elapse from the "cleaning" step until the liposomes are taken to another use. For small molecular weight drugs, especially if not highly lipophilic, the period of grace during which there is no detectable unencapsulated drug might be quite short. Rather than fight the situation, at least for the sake of the present type of studies, the experimental setup just described, and the data-processing, detailed in the following, allow one to account for the share of the unencapsulated drug and separate its quantitative contribution from that of real interest; namely, the liposome-associated drug.

Focusing solely on the liposome-associated drug, one has to contend with the possibility that it could occupy several different pools, at least to begin with. For such situations (i.e., multiintraliposomal original drug pools), the heterogeneous nature of drug diffusion, in the different media that the drug will cross until it is released from the liposome, is quite clear. That the multiphase situation is also encountered in the most simple cases, can be seen from the following description. The least complex case would be for a lipophilic drug residing in the lipid bilayer of a unilamellar liposome. Its release from the liposome would amount to diffusion in a nonmixing two-phase heterogeneous system, organic (i.e., the lipid bilayer) and aqueous (i.e., the external medium). Next in line, in terms of relative simplicity, would be an aqueous-soluble drug that resides in the aqueous phase inside a unilamellar liposome. Its release from the liposome would amount to diffusion in a nonmixing three-phase heterogeneous system, the drug diffusing from one aqueous phase, across a lipid bilayer, and into the another (the external).

Among the theoretical frameworks available for analysis of diffusion data, two properties make the theoretical approach developed by Eyring particularly suitable for drug diffusion from liposomes. It allows one to deal simultaneously with drug release from homogeneous (unencapsulated drug) and heterogeneous (liposome-associated and liposome-encapsulated drug) systems. It yields parameters that allow the direct determination of the fraction of encapsulated drug and of the half-life of drug release. In the long run, such parameters are especially useful for defining optimization criteria and for designing dose ranges and treatment regimens with liposomal systems that are destined to serve as therapeutic entities.

Several mechanisms can be proposed for the case in which, initially, the drug is located within the liposomes in several different pools. Two categories are discussed in the following sections.

1. Category 1: A Mix of Parallel and Sequential Processes

The basic assumptions for mechanisms within this category are (a) the diffusion of the drug from each pool is completely independent of the other original pools, and (b) the diffusion of the drug from each original pool progresses through a series of intermediate pools. Accordingly, the latter assumption can be expressed as a series of sequential first-order processes, the first is the diffusion of the drug from its original pool into the first intermediate one, the last of which is the diffusion of the drug from the aqueous medium inside the sac, to the external bulk medium in which the sac is immersed. The total diffusion, which takes both assumptions into account, can be expressed as a series of parallel processes, each representing an original pool, and within each process the series of sequential steps described in the foregoing. This mix of parallel and sequential processes is demonstrated below by three original pools, denoted DP_{11}, DP_{i1}, and DP_{n1}:

$$DP_{11} \xrightarrow{k_{11}} DP_{12} \xrightarrow{k_{12}} DP_{13},\ldots,DP_{1m} \xrightarrow{k_f} D \tag{1a}$$

$$DP_{i1} \xrightarrow{k_{i1}} DP_{i2} \xrightarrow{k_{i2}} DP_{i3},\ldots,DP_{ih} \xrightarrow{k_m} D \tag{1b}$$

$$DP_{n1} \xrightarrow{k_{n1}} DP_{n2} \xrightarrow{k_{n2}} DP_{n3},\ldots,DP_{nD} \xrightarrow{k_f} D \tag{1c}$$

where the "first reactant" is (as already defined) the original drug pool denoted, for example, DP_{i1}, and the end product is the drug in the external aqueous medium denoted D. DP_{i2} to DP_{im} are the intermediate drug pools associated with this particular original pool and k_{i1} to k_f are the respective first-order rate constants. The last step is, obviously, the diffusion of the drug that is free in the aqueous medium within the sac (in which the liposomes are suspended), into the external aqueous medium in which the sac is immersed. For this mechanism, at time 0, all intermediate pools are empty, except the last one, which corresponds to the unencapsulated drug in the system at time 0.

For the most simple case, there would be only one original drug pool; hence, a single sequential process. The most simple case would also dictate a single, limiting step, and the diffusion of the liposome-associated drug would be given by:

$$DP_{i1} \xrightarrow{k_{i1}} D \tag{2}$$

2. Category 2: Sequential Processes

The basic assumption for mechanisms in this category is dependency and connection among all original liposome-associated drug pools, so that excepting the innermost drug pool, each of the other pools serves both as an original pool and as an intermediate pool for the original ones preceding it. This mechanism can also be described as a series of sequential first-order steps, similar to those provided in one of the lines of Eq. (1) (i.e., 1a, or 1b, or 1c), but with the following significant differences: (a) There would be only one, sequential, process; and (b) all "intermediate" pools would already be occupied at time 0. For the most simple case, there would be only one rate-limiting step, and the process could also be described by Eq. (2). Thus, for the most simple cases, both categories lead to the same quantitative expression.

3. The Unencapsulated Drug

The discussion in Sections V.B.1 and V.B.2 was restricted to the liposome-associated drug. There remains the need to also account for the unencapsulated drug present in the sac at time 0, especially for those cases where the diffusion of the unencapsulated drug out of the sac is slow enough to contribute to the overall rate of drug accumulation in the external medium.

One approach to deal with the unencapsulated share, can be defined as the "two-systems in one" [55]. The situation prevailing in the sac at time 0 is taken as the outcome of having introduced two different and independent systems into the same space. One, the free drug soluble in an aqueous medium; the other, the liposomes (with their associated drug). The overall accumulation of drug in the external bulk medium for the liposomal case described by Eq. (2), would then be the contribution of two parallel first-order processes, with a common product (free drug in the external bulk medium) and different reactants, one accounting for the encapsulated, the other for the unencapsulated, drug. These processes are described in Eq. (3).

$$\text{DP}_\text{e} \xrightarrow{k_\text{e}} \text{D} \quad \text{DP}_\text{u} \xrightarrow{k_\text{u}} \text{D} \tag{3}$$

where DP_e and DP_u represent the encapsulated and (original) unencapsulated drug pools and k_e and k_u the respective rate constants. An advantage of this approach is that it does not demand restrictions on the relative magnitudes of these two rate constants.

Solving for the case in which the measured dependent variable is the product (i.e., the cumulative released drug), will result in the following expression (see Appendix for details):

$$f = f_\text{u}(1 - e^{-k_\text{u}t}) + f_\text{e}(1 - e^{-k_\text{e}t}) \tag{4}$$

where f is the cumulative release, normalized to the total drug in the system at time 0, and f_u and f_e are the fractions of the total drug at time 0 occupying the unencapsulated and encapsulated pools. Obviously, the latter two should sum up to unity.

Another way to approach the issue of the unencapsulated drug, can be defined as "single system, sequential process" and is essentially a return to Eq. (1), for the case of a single original liposome-associated drug pool, with two rate-limiting steps, the first and the last in the sequence:

$$\text{DP}_\text{e} \xrightarrow{k_\text{e}} \text{DP}_\text{u} \xrightarrow{k_\text{u}} \text{D} \tag{5}$$

where both pools are occupied at time 0, by encapsulated and unencapsulated drug, respectively. Obviously, this approach is relevant only if the two rate constants are sufficiently close enough in magnitude that both contribute to the overall rate of drug accumulation in the bulk external medium. Solving, as before, for the case where the measured dependent variable is the product (see the Appendix for details of the derivations) will yield the following expression:

$$f = f_e \frac{k_e(1 - e^{-k_u t})}{k_e - k_u} + f_u(1 - e^{(k_e - k_u)t}) \tag{6}$$

It is strongly argued that Eqs. (4) and (6) are sufficiently different from each other that a single set of experimental data (f vs. t) cannot fit both. Thus, data analysis should make it possible to distinguish between these two mechanisms.

4. Selected Experimental Data

The kinetics of drug diffusion have been studied in the experimental design just described, for two species of EGF (mEGF and urogasterone), and for a host of small molecular weight molecules, among them antibiotics, other anti-infective agents, peptides, hormones, and chemotherapeutic agents, in regular and bioadhesive liposomes that were multi- and unilamellar [55–60,63,82]. To account for the unencapsulated drug, the data fit unambiguously with the approach of two systems in one. Selected examples of the magnitudes obtained for the ratio of k_u to k_e for several drugs in regular and in bioadhesive uni- and multilamellar liposomes, are listed in Table 1. It is stressed that to show the span of ratios obtained, the data in the table were taken from systems that were studied under different experimental conditions, especially relative to their liposomal concentration, a parameter that affects the magnitude of k_e [55]. This is the basis of the differences in the magnitudes of the ratio for the systems (see Table 1) that are seen even for the same liposome species and the same drug. The significance of the data listed in Table 1 is proposed to lie in the span of magnitudes obtained for k_u/k_e, as even those at lower end are already too large to fit with the basic assumption for a sequential mechanism. Consequently, a single expression can be used for data analysis (see Appendix for details):

$$f = \sum_{j=1}^{n} f_j(1 - e^{-k_j t}) \tag{7}$$

where n is now the numbers of independent drug pools within the sac, at time 0, one of which is the unencapsulated drug and all the others are liposome-associated drug.

For most of the systems studied, the data fit $n = 2$, indicating one encapsulated and one unencapsulated drug pool at time 0 [55–60,63,82]. On a few occasions, corresponding only to bioadhesive liposomes that are unilamellar and have collagen, hyaluronic acid, or gelatin as the bioadhesive ligand, the

Table 1 The k_u/k_e Ratios for Regular and Bioadhesive Liposomes

Drug	Liposome type[a]	Bioadhesive ligand	k_u/k_e
EGF	ULV	None	100
	MLV	None	1360
	MLV	Hyaluronic Acid	1360
	MLV	Collagen	330
Cefazolin	MLV	None	60
	MLV	Collagen	130
Ampicillin	MLV	None	50
Fluconazole	MLV	None	340
	MLV	Collagen	140
	ULV	None	100
	ULV	Gelatin	90
Progesterone	ULV	None	60
	ULV	None	290
	MLV	None	140
	MLV	None	1130

[a]ULV, unilamellar vesicle; MLV, multilamellar vesicle.

data fit $n = 3$, corresponding to two independent liposome-associated pools, in addition to the unencapsulated one [58,82]. The existence of two liposome-associated pools can be rationalized with the aid of the schemes for the bioadhesive liposome, illustrated in Fig. 2. It is proposed that the bioadhesive ligands can form a bioadhesive layer surrounding the liposomal surface, which can then be a drug reservoir, constituting that second liposome-associated drug pool [58]. For the bioadhesive ligands anchored at discrete locations on the liposomal surface to evolve into a layer that can serve as a drug reservoir, the following conditions have to be met: The surface density of the ligand and the size of the ligand should be of sufficient magnitudes to allow neighboring molecules to have some overlap and the ligand has to have a capability of self-interaction (self-association) that can then come into effect among the overlapping molecules. It is reasonable that, for similar ranges of ligand/lipid ratios, ligand overlap is more likely in unilamellar, than in multilamellar, liposomes. Among the four species of bioadhesive ligands studied so far (EGF, collagen, gelatin, and hyaluronic acid), the latter three have sizes significantly larger than EGF and can self-interact. These properties fit well with the experimental observations that the bioadhesive layer was observed only with unilamellar liposomes, and with those three ligands, but not with EGF [60].

Selected examples of the rate constants for the release of anti-infectives and

EGF (drugs that are relevant for wound therapy) from bioadhesive liposomes, listed in Table 2 in the form of half-lives, make it clear that the such systems can meet the essential requirement for sustained drug release at the optimal time span defined in Section V.A. The data also show the aforediscussed and previously studied [55] relation between the rate constant for the release of an encapsulated drug and the liposome concentration. This relation can be used to adjust, as best allowed by each specific drug, the rate of drug release to the optimal therapeutic range. For example, for treatment with EGF, reduction of liposome concentrations will serve to accelerate the rate of release to bring it within the $t_{1/2}$ range of 1–4 days. In contrast, for treatment with cefazolin, increase in the liposomal concentrations will serve to slow down the rate of release to bring it within that optimal range.

In closing, attention is drawn to the following points: As long as the liposomes remain intact, because it is liposome-dependent, the efflux of the liposome-associated drug should be independent of the medium into which the liposomes are introduced, be it the dialysis system discussed in the foregoing, in vivo, or otherwise. In contrast, the in vivo dissipation of the unencapsulated drug from the site of administration is expected to be totally different from the corresponding data measured in the dialysis sac, and from what would be found in differently designed in vitro systems. On this background, it might seem as if undue attention has been given in this section to the unencapsulated drug. It is suggested that this is not true. There is a need to know (in vitro) as much as

Table 2 Sustained Release Behavior of EGF and of Anti-infective Drugs from Regular and from Bioadhesive Liposomes

Encapsulated drug	Bioadhesive ligand	Liposome type	Lipid concentration (mM)	$t_{1/2}$ (days)
EGF	None	MLV	3	2.4
	None	ULV	17	2.9
	None	MLV	11	4.9
	Collagen	MLV	11	5.9
	Hyaluronic acid	MLV	50	17
Cefazolin	Collagen	MLV	80	0.6
	Collagen	MLV	120	1.8
	Collagen	MLV	250	3.1
Fluconazole	Gelatin	ULV	20	1.2
	Gelatin	ULV	80	4.2
	Collagen	MLV	26	4.4

possible about the unencapsulated drug, not only because of its almost unavoidable presence, especially in liposomal systems used as delivery systems for small molecular weight drugs, but more so because of the need to minimize the effects of this fraction to distance the treatment with the liposomes as far as possible from that of treatment with free drug. "Knowing the enemy" is the first step toward minimizing its undesirable effects.

Administering a liposomal preparation in which a small share of the drug is unencapsulated is not necessarily undesirable. There could be situations in which an initial burst of drug, followed by sustained-release, might prove beneficial to the therapy.

Finally, although pursuit of the sustained-release properties of liposomal preparations is in the sphere of kinetics, if the systems investigated are the original preparation, before any attempt has been executed to separate the unencapsulated fraction, then the fraction of total drug in the system that occupied the liposomal pool(s) at time 0 [one of the parameters determined through data processing according to Eq. (7)] is also the efficiency of encapsulation. This is suggested to be among the most accurate procedures to determine this efficiency, as it is does not require a physical separation of the liposomes from the external medium containing the unencapsulated drug before quantitative determinations. It should be free from overestimation that can occur if separation is by centrifugation, in the course of which unencapsulated drug can precipitate with the liposomes owing to nonspecific adsorption to the liposomal surface. It is also free from the underestimation that can occur when separation is by gel exclusion chromatography or by dialysis, during which the dilution enforced by such separations can trigger the release of encapsulated drug during the process of separation.

C. Bioadhesivity

Before beginning in vivo studies of bioadhesivity, some model of the in vivo designated binding sites is required. Two types of in vitro systems can, in principle, serve as those models. One type consists of naturally derived extracellular matrix (ECM) layers (commercially available or prepared by the investigator). This type of model would fit studies with bioadhesive liposomes in which the ligand is directed toward ECM components. The second type of model consists of cells in culture that can be grown into confluent monolayers. Depending on the specific line, the monolayer not only serves to model a layer in a wound, but can also provide receptors specific to the bioadhesive ligands of interest, as well as cell–cell adhesion proteins and some ECM.

*1. Binding: Conceptual Approaches and Selected Examples of
   Experimental Data*

The experimental designs and data processing of classic receptor–ligand binding studies in cell cultures used for free (soluble) ligands, such as hormones,

antibodies, or neurotransmitters, can be adopted for the first part of the bioadhesivity evaluation. Data processing can be done according to the following
form of Langmuir's binding isotherm:

$$B = \sum_{j=1}^{m} \frac{B_{\text{max}_j}[L]}{K_{d_j} + [L]} \tag{8}$$

where B is the quantity of bound ligand and $[L]$ is the concentration of ligand
remaining unbound at equilibrium; m is the number of different types of bind-
ing sites the cell monolayer has for a given ligand, K_{d_j} and B_{max_i} are the equilibrium constant for the dissociation of the bound ligand from sites of the jth type
and the number of sites of the jth, type respectively. B and B_{max} are both
normalized to the same parameter used to quantitate the cells or the ECM; for
the present example, liposomes play the part of the ligand in Eq. (8).

Adaptation of this approach to the bioadhesive liposomes requires a reexami-
nation of the terms and the concepts behind the issue of specific and nonspecific
binding. When the ligand is an entity, such as a hormone, an antibody, or a
neurotransmitter, the objectives of studying receptor–ligand interactions usually
go beyond the binding stage. This has led to a sharp distinction between binding to the receptor and binding to other cell membrane entities. The former has
been denoted *specific binding* that can appear as high-affinity, or both high-
and low-affinity, interactions. The latter has been denoted *nonspecific binding*,
and also appears as a low-affinity interaction, which is (usually) significantly
lower than that of the lowest in the specific binding. For such cases, the contribution of the nonspecific binding can be subtracted from the total binding at
the experimental stages. This is possible (and is frequently done) for those
situations in which the ligands can be modified (or synthesized) to carry a
suitable label (such as a radioisotope or a fluorescent marker), and it can be
verified that the labeled and nonlabeled versions are identical in the relevant
properties. An alternative approach is to execute the resolution into the specific
and nonspecific components of the overall binding at the data-processing stage,
for which Eq. (8) can be put into the following form:

$$B = \sum_{j=1}^{m} \frac{B_{\text{max}_j}[L]}{K_{d_j} + [L]} + K_{\text{ns}}[L] \tag{9}$$

where the second term on the right describes the contribution of the nonspecific
binding that is distinguished by the constant K_{ns}, and the first term on the right-
hand side (i.e., the summation) describes the contribution of the specific binding alone; m now giving the number of different types of binding sites the
receptor has for a given ligand.

However, remember that the division into specific and nonspecific binding
is subjective, reflecting the investigator's interest, and is not a thermodynamic

property. For the bioadhesive liposomes, the binding alone is the objective. Consequently, the affinity itself suffices to distinguish between different types of binding sites (if those exist at all), and there is no need to introduce the "specific" and "nonspecific" categories. Data processing, therefore, can be done, applying Eq. (8) or Eq. (9) with the bioadhesive liposomes now being, as defined earlier, *the ligand*. For liposomes separating the contributions of low- versus high-affinity binding is more feasible at the data processing, rather than the experimental, stages, as the latter has substantial drawbacks. Preparation (and verification) of identically labeled and nonlabeled versions of a given species of bioadhesive liposome can easily become a monumental task that might not be worth the effort.

Time- and concentration-dependence studies of the binding of bioadhesive unilamellar liposomes and multilamellar liposomes, with EGF, gelatin, collagen, and hyaluronic acid, to monolayers of a suitable cell line (the A431 line) as well as to layers of ECM showed consistently that the bioadhesive liposomes have significantly higher affinity than control, regular liposomes [57–59,60,82]. Data analysis invariably fit Eq. (8) for a single-type binding site, and the standard free energy released on binding was in the range of -30 to -50 kJ/mol, depending on the specific ligand [57–59,60,82]. The maximal binding capacity determined for the bioadhesive liposomes was significantly higher than the corresponding magnitudes observed for regular liposomes, as illustrated by the selected examples listed in Table 3. In particular, among the three ligands expected to interact with ECM components, the affinity ranking

Table 3 Saturation Binding Levels of Regular and of Bioadhesive Liposomes to Monolayers of A431 Cells and to Layers of Biologically Derived ECM

Liposome species	Bioadhesive ligand	Binding matrix	B_{max}[a] (nmol lipid/10^5 cells)
ULV	None	A431 cells	0–1.3
MLV	None	A431 cells	0–0.7
MLV	None	ECM	0.3[b]
ULV	Collagen	A431 cells	14
MLV	Collagen	A431 cells	48
MLV	Collagen	ECM	4[b]
ULV	Gelatin	A431	16
MLV	Gelatin	A431 cells	14
MLV	Hyaluronic acid	A431 cells	30

[a] B_{max} is the experimental value for the regular liposomes, and the result of data processing according to Eq. (8) in the text, for the bioadhesive liposomes.
[b] For the ECM data the B_{max} unit is nmoles lipid per micrograms ECM protein.

was hyaluronic acid > collagen > gelatin, similar to what is known for these molecules in their free form.

2. Retention in the Face of Tissue and Fluid Dynamics:
Experimental Designs and Examples of Experimental Data

In vitro evaluation of the ability of bioadhesive liposomes to be retained in the face of cell-related dynamics requires a model system that can mimic the events of cell proliferation, cell migration, and cell death. Fortunately, such events are normally provided by adherent cell cultures. It is required to simply incubate bioadhesive liposomes with cell cultures and simultaneously determine, at selected time points within a designated time span, the number of viable cells and the corresponding level of bound liposomes. With the cell line A431, and testing for bioadhesive liposomes in which collagen or hyaluronic acid were the bioadhesive ligands, over a period of 30 h, during which cells were in contact with bioadhesive liposomes, the number of viable cells underwent decreases and increases that altogether spanned 300% [60; Yerushalmi and Margalit, unpublished data]. At the same time, once the time point for liposomal binding to attain saturation has been reached (1–2 h from the onset of incubation) the quantity of bioadhesive liposomes bound per given level of viable cells remained quite constant, up to termination of the study, which was at the end of 30 h.

In vitro evaluation of the ability of bioadhesive liposomes to be retained in the face of fluid dynamics requires, in addition to liposomes and a model of the in vivo designated binding area (i.e., the cell monolayer or the ECM layers), the ability to flow fluid over the region in which the liposomes are bound. An experimental setup designed for this task is illustrated in Fig. 3. The lipo-

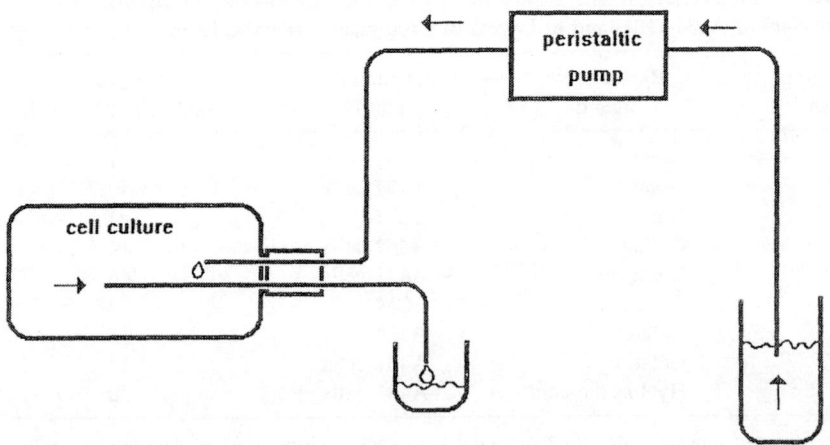

Fig. 3 A scheme of the setup designed for flowing fluid at a controlled rate over a monolayer of cells in a culture flask.

Table 4 Retention of Bioadhesive
Liposomes at Monolayers of A431,
Under Continuous Buffer Flow, as a
Function of Flow Time

Flow period (min)	Liposome retention (μmol lipid)
0	4.5
1	2.70 ± 0.40
5	1.30 ± 0.20
10	0.73 ± 0.25
15	0.63 ± 0.15
30	0.53 ± 0.04
45	0.48 ± 0.21

some "dose" given to the cells will settle into bound and unbound fractions, the distribution of which will depend on the size of the monolayer (i.e., quantity of cells) and on the particular liposomal species and liposomal dose. To mimic the in vivo anticipated situation, the unbound liposomes were not removed before the flow start. Rather, it was anticipated that the flow would first sweep away those liposomes that were unbound, only thereafter affecting the bound fraction. This system was used to test the retention of bioadhesive liposomes in which hyaluronic acid was the bioadhesive ligand, to monolayers of the A431 line [Yerushalmi and Margalit, unpublished data]. Typical data of the quantity of liposomes retained at the cell monolayer after exposure to different periods of fluid flow at the rate of 0.67 ml/min (Table 4), bear out these expectations. Over the first 10 min, the flow induces relatively fast removal of liposomes, with the sharpest drop occurring over the first minute. Once that fraction is swept away, extensive removal of liposomes ceases, and even a fourfold increase in flow time induces only a minor loss of liposomes, which can be attributed to dilution-induced equilibria perturbation. Such data demonstrate that those liposomes that have succeeded in adhering to the cell monolayer are retained there, despite fluid flow which, at the rate used (i.e., 0.67 ml/min) is proposed to be significantly faster that would occur in a real wound, in vivo.

VI. APPROACHES TO FUNCTIONAL TESTING, CONCLUSIONS, AND FUTURE PROSPECTS

As shown by the selected examples of data (together with the cited references) implementation of the experimental designs and data-processing frameworks discussed in Section V, has shown quite clearly that the bioadhesive liposomes can meet the essential requirements (also defined in the foregoing) for task performance. These systems, therefore, have matured into the next step; namely, evaluation of their functionality. Specifically, testing the ability of bioadhesive liposomes encapsulating growth factors or antibiotics, to provide improvements in treatment compared with the free drugs.

Wound healing, stimulated by growth factors, can be evaluated not only in animal models, but also in cell culture-based models, using two basic approaches [83–85]. In one, cell monolayers are grown to a subconfluent state, a growth factor is added to the medium, and its effects on the continued fate of the cultures are thereafter quantitatively evaluated at designated time points. In the other, cell monolayers are grown to confluency, a selected area within the monolayer is wounded, the growth factor is added to the medium, and its effects on recovery within the wounded area are thereafter quantitatively evaluated. Cells lines selected for either approach can have absolute dependence on a given growth factor, or can be grown under conditions that render them sensitive to a selected growth factor. Preliminary studies adopting the first approach were conducted with EGF and the fibroblast cell line NIH3T3 [86]. When grown in a low serum medium (0.13%), cultures of this line become sensitive to EGF. The treatment protocol tested, in line with the dosing regimens toward which the bioadhesive liposomes have been designed, was a single dose, the effects of which (i.e., EGF-stimulated cell proliferation) were evaluated at the end of 4 days from dosing. Treatment with free EGF, over the dose range of 1–10 ng/ml yielded the expected [62] ascending and descending arms of the dose response curve. Stimulation peaked at the dose range of 4–6 ng/ml and reached 180% of untreated controls. The stimulation generated by similar treatments with EGF encapsulated in regular liposomes did not differ significantly from that of free EGF. In contrast, the same treatment with EGF encapsulated in bioadhesive liposomes (hyaluronic acid serving as the bioadhesive ligand) resulted in significantly higher stimulations that, at the dose of 6 ng/ml were 500% higher than the stimulations generated by EGF in its free form or in regular liposomes.

Preliminary studies evaluating the response of infected wounds to treatment with cefazolin-encapsulating bioadhesive liposomes (collagen serving as the bioadhesive ligand) were studied in full-thickness wounds in mice infected with *Staphylococcus aureus* [Margalit et al., in preparation]. The treatment protocol, in keeping with the wound management protocols toward which these liposomes were designed, consisted of a single dose, and the wound bacterial count was determined at the end of 3 days after dosing. The goal was to reduce the bacterial count from the level of infection to the level of colonization. In the untreated controls, wound bacterial counts at the end of 3 days were 2×10^7 to 5×10^7 organism per square centimeter of wound, well within the range of bacterial infection. Topical treatment with a single dose of the antibiotic cefazolin, evaluated 3 days after dosing, produced reductions in wound bacterial count. The magnitude of the reductions and, moreover, achievement of the therapeutic goal were strongly dependent on the formulation used. For example, reductions achieved with 1 mg cefazolin, either free or in regular liposomes (80% encapsulated), were at the most tenfold from untreated controls; therefore, still within the infection domain. In contrast, the wound bacterial counts after treatment with 1 mg of cefazolin in the bioadhesive liposomes (80% encapsulated) were down 100-fold from untreated controls, to the colonization–infection boundary.

In conclusion, it is offered that the molecular and in vitro evidence for the ability of the bioadhesive liposomes to meet essential requirements for task performance, together with the positive results from the preliminary functional studies, attest to the feasibility of this approach. In principle, the future course for these systems is no different from the road taken by developers of drug-delivery systems, irrespective of the therapeutic goals. The road ahead consisting of in vivo studies that, if successful, will mature into clinical studies, together with the development of second- and higher-generation systems. Yet, the nature of the pathological situations addressed in this chapter, together with the changing climate of health care will add additional burdens on the road to established clinical modalities. For a substantial share of the patient population, despite the suffering and anguish, their wounds and burns do not put them in an immediate life-threatening situation. Consequently, it might require more than mere improvements in clinical outcomes to justify making bioadhesive liposomes into critical components of wound therapy. This can be demonstrated by the following considerations: It is anticipated that growth factor therapy, when it will become established, will be a substantially high-cost treatment. This is based on a combination of factors, such as the cost of the growth factor itself, the length of the treatment (weeks of more than one daily dose), hospitalization or clinic visits, and the labor of the care givers. The more benefits a delivery system will be able to provide, by the way of cutting the total dose required, reducing the length of the treatment, the length of hospitalization, or the number of clinical visits and labor costs, the higher will be its chances of advancing from a concept to a high-valued component of wound therapy. Needless to say, added benefits would come from faster healing that will, in turn, reduce the risks of infections and speed the return of the patient to the workforce.

Thus, the combination of improvements in the clinical and economic aspects of wound healing will govern the future prospects of bioadhesive liposomes, as well as other novel treatments in the wound care area.

APPENDIX: DRUG DIFFUSION FROM LIPOSOME SUSPENSIONS, IN A DIALYSIS SET-UP: PARALLEL AND SEQUENTIAL MECHANISMS

The objective of this Appendix is to derive expressions for the dependence of cumulative release of drug from liposomal preparations on time, obtained from dialysis experiments conducted under unidirectional flux conditions. Two types of mechanisms, parallel and sequential, are explored.

A. Parallel Processes

Case 1: Two Drug Pools

Basic Assumptions

1. Two independent systems, one of free, unencapsulated drug, and one of encapsulated drug are present in the sac at time 0.
2. For each system, the diffusion of the drug from the sac into the external bulk medium (i.e., the dialysate) is a single first-order process, with its own unique rate constant.

Given these assumptions, this mechanism can be expressed as two parallel first-order processes with two different reactants, denoted A and B (the encapsulated and unencapsulated drug, respectively) and a common product (the free drug in the dialysate), denoted C:

$$A \xrightarrow{k_1} C \tag{1A}$$

$$B \xrightarrow{k_2} C \tag{2A}$$

Conservation of matter gives

$$[T] = [A_0] + [B_0] \tag{3A}$$

where $[T]$ is the total drug within the sac, at time 0. $[C]$, the parameter of interest at any given time, can also be expressed through conservation of matter:

$$[C] = [A_0] + [B_0] - [A] - [B] \tag{4A}$$

Introducing the results of first-order integration for $[A]$ and $[B]$ and rearranging will give:

$$[C] = [A_0](1 - \exp^{-k_1 t}) + [B_0](1 - \exp^{-k_2 t}) \tag{5A}$$

Which, after division by $[T]$, will furnish the final expression:

$$f = f_1(1 - e^{-k_1 t}) + f_2(1 - e^{-k_2 t}) \tag{6A}$$

where f is the product, normalized to the total drug in the system at time 0, and f_1 and f_2 are the fractional distributions of the drug, at time 0, between the encapsulated and unencapsulated pools. Equation (6A) has three independent variables, as $f_1 + f_2 = 1$.

Case 2: *More Than One Liposome-Associated Drug Pool*

For the case here there is more than one original liposome-associated drug pool, the diffusion from each being an independent first-order process, the overall diffusion can be described by a series of parallel first-order processes with different reactants and a common product. Equation (6A) would then take the form of

$$\sum_{j=1}^{n} f_j (1 - e^{-k_j t}) \tag{7A}$$

where n is the total number of drug pools in the system, one of which is the unencapsulated, and the rest of which are the liposome-associated, and

$$\sum_{j=1}^{n} f_j = 1.$$

B. Sequential Processes

Case 1: *Two Drug Pools*

  Basic Assumptions

1. The diffusion of the drug from the sac is a sequential process, with two rate-limiting steps, in which the encapsulated pool is the initial reactant and the unencapsulated pool is the intermediate product.
2. Even though the unencapsulated pool is an intermediate product, it is already present in the sac at time 0.

This mechanism can be expressed in the following form:

$$A \xrightarrow{k_1} B \xrightarrow{k_2} C \tag{8A}$$

Conservation of matter and the concentration of the final product $[C]$ at any designated time point are given by Eqs. (3A) and (4A), respectively. The dependence of $[A]$ on time is a simple first-order process:

$$[A] = [A_0] e^{-k_1 t}$$

$$\tag{9A}$$

The increase of $[B]$ with time is given by

$$\frac{d[B]}{dt} = k_1 [A] - k_2 [B] \tag{10A}$$

The following transformation is introduced to integrate Eq. (10A):

$$R = [B] e^{k_1 t} \quad \text{or} \quad [B] = R e^{-k_1 t} \tag{11A}$$

Introducing Eqs. (11A) into (10A) will give

$$\frac{d[B]}{dt} = k_1[A] - k_2Re^{-k_1t} \tag{12A}$$

Taking the differential of Eq. (11A) for $d[B]/dt$ will give:

$$\frac{d[B]}{dt} = e^{-k_1t}\frac{dR}{dt} - k_1Re^{-k_1t} \tag{13A}$$

Comparing Eqs. (12a) and (13A) will yield:

$$k_1[A_0]e^{-k_1t} - k_2Re^{-k-1t} = e^{-k_1t}\frac{dR}{dt} - k_1Re^{-k-1t} \tag{14A}$$

Equation (14A) can be divided by e^{-k_1t}, which, after rearranging, will take the form:

$$\frac{dR}{k_1[A_0] + (k_1 - k_2)R} = dt \tag{15A}$$

Integrating Eq. (15A) over R from R_0 to R and over t from 0 to t, and rearranging will yield:

$$\frac{k_1[A_0] + (k_1 - k_2)R}{k_1[A_0] + (k_1 - k_2)R_0} = e^{(k_1 - k_2)t} \tag{16A}$$

If we use (10A) to replace R and R_0 by $[B]$ and $[B_0]$ and solve for $[B]$ this will give:

$$[B] = \frac{k_1}{k_1 - k_2}[A_0](e^{-k-2t} - e^{-k_1t}) + [B_0]e^{(k_1 - k_2)t} \tag{17A}$$

By introducing Eqs. (9A) and (17A) into (4A) and rearrange it will give:

$$[C] = [A_0]\frac{k_1(1 - e^{-k_2t})}{k_1 - k_2} + [B_0](1 - e^{(k_1 - k_2)t}) \tag{18A}$$

Dividing by $[T]$, will furnish the final expression:

$$f = f_1\frac{k_1(1 - e^{-k_2t}) - k_2(1 - e^{-k_1t})}{k_1 - k_2} + f_2(1 - e^{(k_1 - k_2)t}) \tag{19A}$$

Similar to Eq. (6A), Eq. (19A) has three independent variables, as $f_1 + f_2 = 1$.

In closing, it is stressed that, if data analysis yields $k_1 \ll k_2$, then clearly the data does not bear out the basic assumption of the sequential process; namely, that both processes are rate-limiting. In that case the sequential mechanism is ruled out, and Eq. (19A) can be reduced to Eq. (6A), as illustrated in the following:

Dropping k_1 where the term $(k_1 - k_2)$ appears in Eq. (19A) and some rearranging will give:

$$f = f_1(1 - e^{-k_1 t}) + (1 - e^{-k_2 t}) * \left(f_2 - f_1 \frac{k_1}{k_2} \right) \qquad (20A)$$

Substituting $(1 - f_2)$ for f_1 in the term within the last bracket and taking into consideration that for the case where $k_1 \ll k_2$, the ration (k_1/k_2) should also be sufficiently small relative to f_2, and some rearranging will give (for that term alone)

$$\left(f_2 - f_1 \frac{k_1}{k_2} \right) = \left(f_2 - \frac{k_1}{k_2} \right) \cong f_2 \qquad (21A)$$

Substituting back into Eq. (20A) will give

$$f = f_1(1 - e^{-k_1 t}) + f_2(1 - e^{-k_2 t}) \qquad (22A)$$

REFERENCES

1. J. H. Senior, Fate and behavior of liposomes in vivo: A review of controlling factors, *CRC Crit. Rev. Ther. Drug Carrier Syst. 3*:123 (1987).
2. G. Gregoriadis, Overview of liposomes, *J. Antimicrob. Chem. 28*:39 (1991).
3. A. Gabizon and D. Papahadjopoulos, Liposome formulations with prolonged circulation time in blood and enhanced uptake by tumors, *Proc. Natl. Acad. Sci. USA 85*:6949 (1988).
4. G. Gregoriadis and A. T. Florence, Liposomes in drug delivery, *Drugs 45*:15 (1993).
5. A. D. Bangham, Liposomes: The Babraham connection, *Chem. Phys. Lipids 64*:249 (1993).
6. O. M. Alvarez, P. M. Mertz, and W. H. Eaglestein. The effects of occlusive dressings on collagen synthesis and re-epithelization in superficial wounds, *J. Surg. Res. 35*:142 (1983).
7. P. M. Mertz and W. H. Eaglestein, The effect of a semiocclusive dressing on the microbial population in superficial wounds, *Arch. Surg. 119*:287 (1984).
8. W. H. Eaglestein, P. M. Mertz, and V. F. Falanga, Occlusive dressings. *Am. Fam. Physician 35*:211 (1987).
9. S. F. Swain, Bandages and topical agents, *Plast. Reconstr. Surg. 20*:47 (1990).
10. G. C. Xakellis and M. A. Chrischilles, Hydrocolloid versus saline–gauze dressings in treating pressure ulcers: A cost-effective analysis, *Arch. Phys. Med. Rehab. 72*:436 (1992).
11. L. Hulten, Dressings for surgical wounds, *Am. J. Surg. 176*:42S (1994).
12. C. K. Field and M. D. Kerstein, Overview of wound healing in a moist environment, *Am. J. Surg. 176*:2S (1994).

13. C. S. Burton, Venous ulcers, *Am. J. Surg. 176*:37S (1994).
14. D. B. Wijetunge, Management of acute and traumatic wounds: Main aspects of care in adults and children, *Am. J. Surg. 167*:56S (1994).
15. J. W. Alexander and E. P. Dellinger, *Textbook of Surgery: The Biological Basis of Modern Surgical Practice*, 14th ed. (D. C. Sabiston, Jr., ed.), W. B. Saunders, Philadelphia, 1992, pp. 221–232.
16. P. Laing, Diabetic foot ulcers, *Am. J. Surg. 167*:31S (1994).
17. R. L. Poucher, J. D. Leahy, and G. Howells, Active healing of diabetic wounds utilizing growth factor therapy, *Wounds 3*:65 (1991).
18. D. R. Knighton, K. Ciresi, V. D. Fiegel, S. Schumerth, E. Butler, and F. Cerra, Classification and treatment of chronic nonhealing wounds. Successful treatment with autologous platelet-derived wound healing factors, *Surg. Gynecol. Obstet. 170*:56 (1990).
19. V. Falanga and W. H. Eaglestein, Management of venous ulcers, *Am. Fam. Physician 33*:274 (1986).
20. D. Krasner, ed. *Chronic Wound Care*, Health Management Publications, Philadelphia, 1990.
21. G. L. Brown, L. B. Nanney, J. Griffen, A. B. Cramer, J. M. Yancy, L. J. Curtsinger III, L. Holtzin, G. S. Schultz, M. J. Jurkiewicz, and J. B. Lynch, Enhancement of wound healing by topical treatment with epidermal growth factor. *N. Engl. J. Med. 321*:76 (1989).
22. J. M. Davidson, M. Klagsbrun, K. E. Hill, A. Buckley, R. Sullivan, P. S. Brewar, and S. C. Woodward, Accelerated wound repair, cell proliferation, and collagen accumulation are produced by a cartilage-derived growth factor, *J. Cell Biol. 100*:1219 (1985).
23. T. K. Hunt, Basic principles of wound healing, *J. Trauma 30*:S122 (1990).
24. A. N. Kingsnorth and J. Slavin, Peptide growth factors and wound healing, *Br. J. Surg. 78*:1286 (1991).
25. D. R. Knighton and V. D. Fiegel, Regulation of cutaneous wound healing by growth factors and the microenvironment, *Invest. Radiol. 26*:604 (1991).
26. G. Schultz, D. S. Rotatori, and W. Clark, EGF and TGF-alpha in wound healing and repair, *J. Cell. Biochem. 45*:346 (1991).
27. R. A. Yates, L. B. Nanney, R. E. Gates, and L. E. King, Epidermal growth factor and related growth factors, *Int. J. Dermatol. 30*:687 (1991).
28. D. T. Graves and D. L. Cochran, Mesenchymal cell growth factors, *Crit. Rev. Oral Biol. Med. 1*:17 (1990).
29. S. T. Feldman, The effect of epidermal growth factor on corneal wound healing: Practical considerations for therapeutic use, *Refract. Corneal Surg. 7*:232 (1991).
30. D. L. Steed, J. B. Goslen, G. A. Holloway, J. M. Malone, T. J. Bunt, and M. W. Webster, Randomized prospective double-blind trial in healing chronic diabetic foot ulcers, *Diabetes Care 15*:1598 (1992).
31. R. L. Brown, M. P. Breeden, and D. G. Greenhalgh, PDGF and TGF-alpha act synergistically to improve wound healing in the genetically diabetic mouse, *J. Surg. Res. 56*:562 (1994).
32. V. Falanga, W. H. Eaglstein, B. Bucalo, M. H. Katz, B. Harris, and P. Carson,

Topical use of human recombinant epidermal growth factor (h-EGF) in venous ulcers, *J. Dermatol. Surg. Oncol. 18*:604 (1992).

33. K. Okumura, Y. Kiyohara, F. Komada, S. Iwakawa, M. Hirai, and T. Fuwa, Improvement in wound healing by epidermal growth factor (EGF) ointment. I. Effects of nafamostat, gabexate or gelatin on stabilization and efficacy of EGF. *Pharm. Res. 7*:1289 (1990).

34. S. A. Servold, Growth factor impact on wound healing, *Clin. Podiatr. Med. Surg. 8*:937 (1991).

35. M. H. Gartner, J. D. Shearer, M. F. Bereiter, C. D. Mills, and M. D. Caldwell, Wound fluid amino acid concentrations regulate the effect of epidermal growth factor on fibroblast replication, *Surgery 110*:448 (1991).

36. P. A. Falcone and M. D. Caldwell, Wound metabolism, *Clin. Plast. Surg. 17*:443 (1990).

37. D. H. Herndon, P. G. Hayward, R. L. Rutan, and R. E. Barrow, Growth hormones and factors in surgical patients, *Adv. Surg. 25*:65 (1992).

38. J. C. Pastor and M. D. Calonge, Epidermal growth factor and corneal wound healing. A multicenter study, *Cornea 11*:311 (1992).

39. D. P. Lookhill, S. H. Miller, and R. C. Knowles, Bacteriology of chronic leg ulcers, *Arch. Dermatol. 114*:1765 (1978).

40. W. A. Berk, R. D. Welch, and B. F. Bock. Controversial issues in the clinical management of the simple wound, *Ann. Emerg. Med. 21*:95 (1992).

41. D. J. Leaper, Prophylactic and therapeutic role of antibiotics in wound care, *Am. J. Surg. 167*:15S (1994).

42. P. D. Thompson and D. J. Smith, What is infection, *Am. J. Surg. 167*:7S (1994).

43. M. C. Robson, B. D. Stenberg, and J. P. Heggers, Wound healing alterations caused by infection, *Clin. Plas. Surg. 17*:485 (1990).

44. R. L. Nichols, Surgical wound infections, *Am. J. Med. 91*:3B (1991).

45. H. A. Pitt, R. G. Postier, W. A. L. MacGowan, L. W. Frank, A. J. Surmak, J. V. Sitzman, and D. Hayes-Bouchier, Prophylactic antibiotics in vascular surgery, *Ann. Surg. 192*:356 (1980).

46. A. Alinovi, P. Bassissi, and M. Pini. Systemic administration of antibiotics in the management of venous ulcers, *J. Am. Acad. Dermatol. 15*:186 (1986).

47. W. W. Monafo and M. A. West, Current treatment recommendations for topical burn therapy, *Drugs 40*:364 (1990).

48. *Physician's Desk Reference*, 48th ed. Medical Economics Data Production, Montvale NJ, 1994.

49. *Drug Facts and Comparisons*, 1994 ed. Facts and Comparisons, St. Louis MO, 1994.

50. S. Kim, Liposomes as carriers of cancer chemotherapy. Current status and future prospects, *Drugs 46*:618 (1993).

51. S. de Marie, R. Janknegt, and I. A. Bakker-Woudenberg, Clinical use of liposomal and lipid-complexed amphotericin B, *J. Antimicrob. Chemother. 33*:907 (1994).

52. R. T. Mehta, T. J. McQueen, A. Keyhani, and G. Lopez-Berestein, Phagocyte transport as mechanism for enhanced therapeutic activity of liposomal amphotericin B, *Chemotherapy 40*:256 (1994).

53. K. M. Wasan and G. Lopez-Berestein, Modification of amphotericin B's therapeutic index by increasing its association with serum high-density lipoproteins, *Ann. N.Y. Acad. Sci. 730*:93 (1994).

54. S. M. Sugarman and R. Perez-Soler, Liposomes in the treatment of malignancy: A clinical perspective, *Crit. Rev. Oncol. Hematol. 12*:231 (1992).

55. R. Margalit, M. Okon, and R. Alon, Liposomes as delivery systems for peptides and polypeptides: Evaluation of essential requirements through molecular level studies, *Proc. Int. Symp. Controlled Release Bioact. Mater. 17*:83 (1990).

56. R. Margalit, R. Alon, M. Linenberg, I. Rubin, T. J. Roseman, and R. W. Wood, Liposomal drug delivery: Thermodynamic and chemical kinetic considerations, *J. Controlled Release 17*:285 (1991).

57. R. Margalit, M. Okon, N. Yerushalmi, and E. Avidor, Bioadhesive liposomes for topical drug delivery: Molecular and cellular studies, *J. Controlled Release 19*:275 (1992).

58. N. Yerushalmi and R. Margalit, Bioadhesive, collagen-modified liposomes: Molecular and cellular level studies on the kinetics of drug release and on binding to cell monolayers, *Biochim. Biophys. Acta 1189*:13 (1994).

59. N. Yerushalmi, A. Arad, and R. Margalit, Molecular and cellular studies of hyaluronic-acid modified liposomes as bioadhesive carriers of growth factors, for topical delivery in wound healing, *Arch. Biochem. Biophys. 313*:267 (1994).

60. A. Lichtenstein and R. Margalit, Liposome-encapsulated SSD for the topical treatment of infected burns: Thermodynamics of drug encapsulation and kinetics of drug release, *J. Inorg. Biochem. 60*:185 (1995).

61. M. Okon, A study of liposomes as drug delivery systems for systemic, topical and regional administrations, PhD Thesis, Tel Aviv University, 1992.

62. C. Carpenter and J. Zendegui, A biological assay for epidermal growth factor/uorogasterone and related polypeptides, *Anal. Biochem. 53*:279 (1985).

63. I. Schumacher, HPLC and antibacterial assays for liposome-encapsulated antibiotics using ampicillin as the test case, MSc. Thesis, Tel Aviv University, 1993.

64. *United States Pharmacopeia (USP) XXI*, U.S. Pharmacopeial Convention, Mack Printing, 1986, pp. 59–63.

65. G. L. Brown, L. J. Curtsinger, M. White, R. O. Mitchell, J. Pietsch, R. Notdquist, A. Fraundhofer, and G. S. Schultz, Acceleration of tensile strength of incisions treated with EGF and TGF-β, *Ann. Surg. 208*:788 (1988).

66. L. S. Grayson, J. F. Hansbrough, R. L. Zapata-Sirvent, T. Kim, and S. Kim, Pharmacokinetics of DepoFoam gentamicin delivery system and effect on soft tissue infection, *J. Surg. Res. 55*:559 (1993).

67. C. I. Price, J. W. Horton, and C. R. Baxter, Topical liposomal delivery of antibiotics in soft tissue infection, *J. Surg. Res. 49*:174 (1990).

68. C. I. Price, J. W. Horton, and R. B. Charles, Liposome delivery of aminoglycosides in burn wounds, *Surg. Gyncecol. Obstet. 174*:414 (1992).

69. C. I. Price, J. W. Horton, and R. B. Charles, Liposome encapsulation: A method for enhancing the effectiveness of local antibiotics, *Surgery 115*:480 (1994).

70. L. W. Cunningham and D. W. Frederiksen, eds., Structural and Contactile Proteins. Part A: Extracellular Matrix, *Methods Enzymol. 82* (1982).

71. S. Dedhar, E. Ruoslahti, and M. D. Pierschbacher, A cell surface receptor complex for collagen type I recognizes the Arg-Gly-Asp sequence, *J. Cell Biol.* *104*:585 (1987).
72. C. J. Doillon, F. H. Silver, and R. A. Berg, Fibroblast cell growth on a porous collagen sponge containing hyaluronic acid and fibronectin, *Biomaterials 8*:195 (1987).
73. K. Park and J. R. Robinson, Bioadhesive polymers as platforms for oral-controlled drug delivery: Method to study bioadhesion, *Int. J. Pharm. 19*:107 (1984).
74. K. A. Piez and A. H. Reddi, eds., *Extracellular Matrix Biochemistry*, Elsevier Science, New York, 1984.
75. T. C. Laurent, Biochemistry of hyaluronan, *Acta Otolaryngol. (Stockh.) 442*:7 (1987).
76. T. C. Laurent and J. R. E. Fraser, The properties and turnover of hyaluronan, *Ciba Found. Symp. 124*:9 (1986).
77. T. C. Laurent and J. R. E. Fraser, Hyaluronan, *FASEB J. 6*:2379 (1992).
78. B. P. Toole, Hyaluronan and its binding proteins, the hyaladherins, *Curr. Opin. Cell Biol. 2*:839 (1990).
79. B. Goldberg, Binding of soluble type I collagen molecules to the fibroblast plasma membrane, *Cell 16*:265 (1979).
80. B. D. Goldberg, Binding of soluble type I collagen to fibroblasts: Effects of thermal activation of ligand, ligand concentrations, pinocytosis and cytoskeletal modifiers, *J. Cell Biol. 95*:747 (1982).
81. B. D. Goldberg and R. E. Burgeson, Binding of soluble type I collagen to fibroblasts: Specificities for native collagen types, triple helical structure, telopeptides, propeptides and cyanogen bromide-derived peptides, *J. Cell Biol. 95*:752 (1982).
82. E. Avidor, Gelatin-liganded bioadhesive liposomes for topical drug delivery: Molecular and cellular studies, MSc. Thesis, Tel Aviv University, 1991.
83. K. Cohen and B. A. Mast. Models of wound healing. *J. Trauma S149* (1990).
84. C. Ciacci, S. E. Lind, and D. K. Podolsky, Transforming growth factor β regulation of migration in wounded rat intestinal monolayers, *Gastroenterology 105*:93 (1993).
85. J. B. Soltau and B. J. McLaughlin, Effects of growth factors on wound healing in serum-deprived kitten corneal endothelial cell cultures, *Cornea 12*:208 (1993).
86. N. Yerushalmi, Bioadhesive liposomes as drug carriers for topical administration, Ph.D. Thesis, Tel Aviv University, 1995.

16

Ultrasound-Triggered Delivery of Peptides and Proteins from Microspheres

Joseph Kost

Ben-Gurion University of the Negev
Beer-Sheva, Israel

I. INTRODUCTION

Finding a way to mimic the body's pulsatile patterns is high on the agenda of researchers and drug companies. They believe that emulating natural rhythms will enhance the effectiveness of hormones, as well as other proteins and even classic chemical compounds. Diabetes mellitus, for instance, is a chronic disease, with major vascular and degenerative complications. It is treatable inasmuch as the acute metabolic disorders can be kept under control. There is, however, no cure for diabetes, nor are we capable of fully preventing the long-term complications of the disease, which are believed to be the consequence of chronically elevated blood glucose levels. None of the present modes of treatment, including insulin pumps, fully mimics the physiology of insulin secretion. Therefore, even the most compliant patients who check their blood glucose six to seven times daily fail to achieve a completely normal glucose metabolism. The realization that blood sugar levels must be kept in normal limits if diabetic complications are to be avoided has led to the approach of self-regulated insulin delivery systems. These systems are capable of mimicking the natural release pattern in which insulin release is in response to glucose levels in the blood.

Many other hormones, such as the pituitary growth or fertility hormones, are also released in pulses. Yet current methods of therapeutically replacing these hormones in persons who lack them bear little resemblance to the way in which the healthy body performs.

Recent studies in the field of chronopharmacology indicate that the onset of certain diseases exhibit strong circadian temporal dependency. Thus, drug-delivery patterns can be further optimized by regulated delivery, adjusted to the staging of biological rhythms.

In recent years, several research groups have been developing responsive systems that would more closely resemble the normal physiological process in which the amount of drug released can be affected according to physiological needs [1–3]. The responsive polymeric-delivery systems can be classified as open- or closed-loop systems. The open-loop (externally controlled) systems apply external triggers for pulsatile delivery, such as magnetic, ultrasonic, thermal, and electronic; whereas in the self-regulated devices, the release rate is controlled by feedback information, without any external intervention. The self-regulated systems use several approaches as rate-control mechanisms: pH sensitive polymers, enzyme–substrate reactions, pH-sensitive drug solubility, competitive binding, and metal concentration-dependent hydrolysis. In this chapter, the strategies of ultrasound for induced peptide delivery from microspheres will be described.

II. ELEMENTARY ULTRASOUND PHYSICS

Ultrasound is defined as any sound that is of a frequency beyond 16 kHz. The ultrasound wave is longitudinal (i.e., the direction of propagation is the same as the direction of oscillation) [4,5]. Although transverse waves are also propagated in liquids, they need not be normally considered because their attenuation with distance is extremely high. The longitudinal sound waves cause compression and expansion of the medium at half a wavelength's distance, leading to pressure variations in the medium. The wavelength of ultrasound is expressed by the relation;

$$\lambda f = C$$

where:

λ = wavelength
f = frequency
C = speed of propagation

The velocity of sound in liquids is typically close to 1500 m/s, with associated acoustic wavelength of roughly $10 - 0.01$ cm. Clearly, no direct coupling of acoustic field with chemical species on a molecular level can account for its chemical and physical effects. Instead, the effects of ultrasound derive from several different physical mechanisms, depending on the nature of the system [5].

Cavitation is one of several possible mechanisms through which ultrasound can interact with liquid medium. Acoustic cavitation can be considered to involve at least three discrete stages: nucleation, bubble growth and, under proper conditions, implosive collapse. Nucleation of bubbles occurs at weak points in the liquid, such as gas-filled crevices in suspended particulate matter, or from transient microbubbles from prior cavitation events.

Several phenomena associated with nonlinear acoustics, including enhanced transfer of heat and mass, changes in reaction rates, emulsification, depolymerization, and sonically produced biological effects, are directly related to acoustic streaming [5]. Acoustic streaming is the time-independent flow of fluid induced by a sound field. Its origins lie in the conservation of momentum dissipated by the absorption and propagation of sound.

The mass density of the medium ρ and the specific acoustic impedance Z determine the resistance of the medium to sound waves. The mass density also partly determines the speed of propagation C. The higher the mass density, the higher the speed of propagation. The specific acoustic impedance, which is a material parameter, depends on the mass density and the speed of propagation $Z = \rho C$.

As ultrasound energy penetrates into the body tissues, biological effects can be expected to occur only if the energy is absorbed by the tissues. The absorption coefficient a is used as a measure of the absorption in various tissues. For ultrasound consisting of longitudinal waves with perpendicular incidence on homogeneous tissues the following formula applies.

$$I(x) = I_0 e^{-ax}$$

where:

$I(x)$ = intensity at depth x
I_0 = intensity at the surface
a = absorption coefficient

A different value relating to absorption is the *half-value depth* $D_{1/2}$, defined as the distance in the direction of the sound beam at which the intensity in a certain medium decreases by half. For skin the $D_{1/2}$ is 11.1 mm at 1 MHz and 4 mm at 3 MHz. In air $D_{1/2}$ is 2.5 mm at 1 MHz and 0.8 mm at 3 MHz.

To transfer the ultrasound energy to the body, it is necessary to use a contact medium because of the complete reflection of the ultrasound by air. The many types of contact media currently available for ultrasound transmission can be broadly classified as follows: oils, water-oil-emulsions, aqueous gels, and ointments.

III. ULTRASOUND-INDUCED DELIVERY

We have suggested [6–11] the feasibility of ultrasonic-controlled polymeric-delivery systems in which release rates of substances can be repeatedly modulated at will from a position external to the delivery system. Both bioerodible

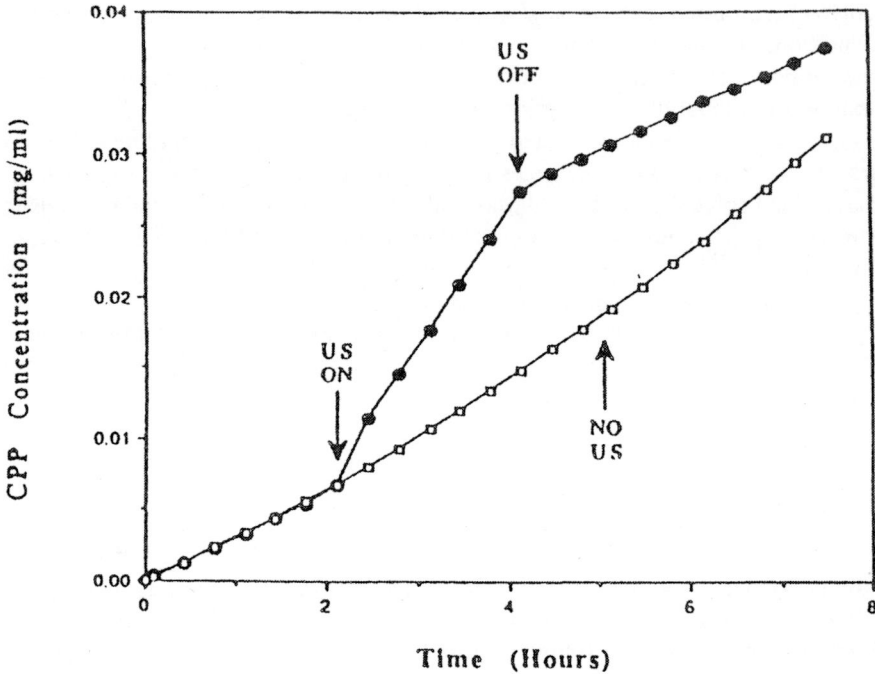

Fig. 1 Effect of ultrasound (1 MHz, 1.7 W/cm²) on the erosion of a polyanhydride device (copolymer of 1,3-bis(*p*-carboxyphenoxy)propane (CPP) and sebacic acid (SA) in the ratio of 20:80 CPP/SA) compared with a nonexposed control. (From Ref. 9.)

and nonerodible polymers were used as drug-carrier matrices. The bioerodible polymers evaluated were polyglycolide, polylactide, poly(bis[*p*-carboxyphenoxy]) alkane anhydrides and their copolymers with sebacic acid. The releasing agents were bovine serum albumin (BSA), insulin, *p*-nitroaniline, *p*-aminohippurate, and norethisterone.

Enhanced polymer erosion (Fig. 1) and drug release were observed when the bioerodible samples were exposed to ultrasound. The systems response to the ultrasonic triggering was rapid (within 2 min) and reversible. The enhanced release was also observed in nonerodible systems exposed to ultrasound for which the release is diffusion-dependent. The nonerodible polymer examined was ethylene-vinyl acetate copolymer (EVAc). Release rates of zinc bovine insulin from ethylene vinyl acetate copolymer matrices were 15 times higher when exposed to ultrasound compared with the unexposed periods. The extent of enhancement can be regulated by the intensity, frequency, or duty cycle of the ultrasound [7].

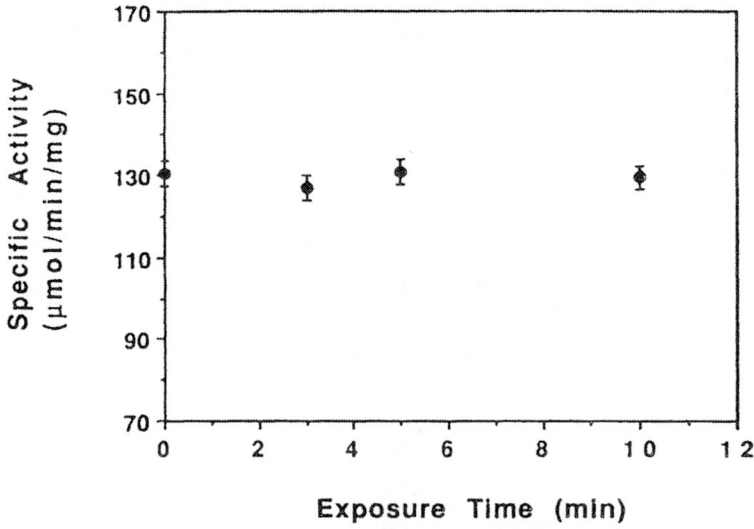

Fig. 2 Specific activity of phospholipase A_2 solution subjected to ultrasound (1 MHz, 2.5 W/cm^2). (From Ref. 12.)

Although the therapeutic ultrasound applied in these experiments is routinely used on humans in physiotherapy, its possible effects on protein integrity and biological activity were also examined [12]. To assess the effect of ultrasonic energy on the integrity of the releasing molecules, insulin samples were evaluated by high-pressure liquid chromatography (HPLC). No significant difference was detected between insulin samples exposed to ultrasound and unexposed samples, suggesting that the ultrasound is not degrading the releasing peptides or proteins [6]. Figure 2 shows that exposure of phospholipase A_2 (PLA$_2$) to ultrasound for 10 mins, at intensities comparable with those used for enhanced release, did not affect its biological activity.

In vivo studies on a model drug have suggested the feasibility of ultrasound-mediated drug-release enhancement [6]. Implants composed of polyanhydride polymers loaded with 10% *para*-aminohippuric acid (PAH) were implanted subcutaneously in the back of catheterized rats. As a urine function marker, PAH is not metabolized and is excreted in the urine, permitting direct in vivo monitoring of the released substance. When exposed to ultrasound, a significant increase in the PAH concentration in urine was detected (Fig. 3). A histopathological examination of the ultrasound-treated area of rats' skins after an exposure of 1 h at 5 W/cm^2, did not reveal any structural differences between the treated and untreated skin.

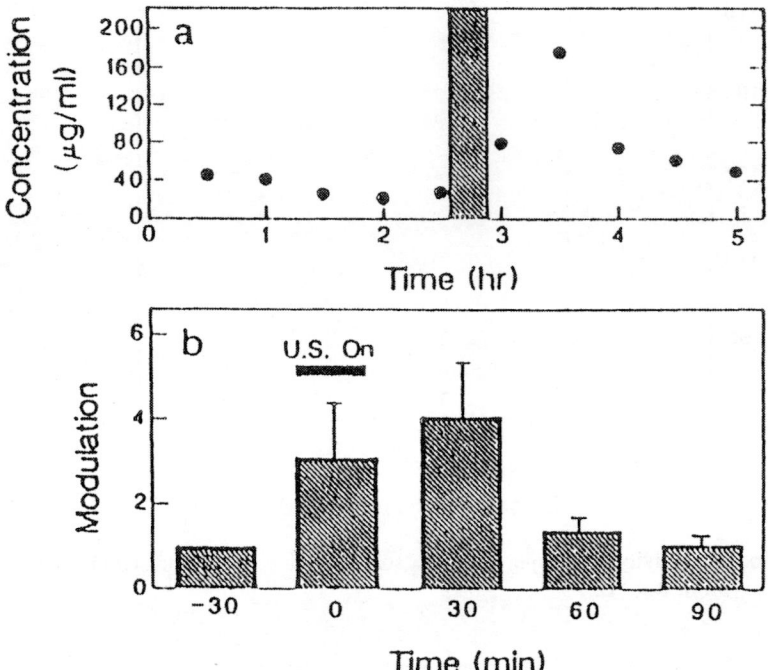

Fig. 3   (a) *p*-Aminohippurate (PAH) concentration in the urine of Sprague–Dawley rats as a function of time before, during, and after an exposure of 20 min to ultrasound. (The shaded area is the ultrasound exposure period.) (b) Modulation vs. time, expressed as a mean and standard deviation of four experimental rats. (Modulation was defined as the ratio of PAH concentration during and after the ultrasound exposure and the mean of the PAH concentration before the exposure.) The implants were copolymers of [bis(*p*-carboxy-phenoxy)propane anhydride] with sebacic acid (80:20) loaded with 10 wt% PAH. (From Ref. 7.)

IV. APPLICATION OF ULTRASOUND AS A TRIGGER TO TERMINATE THE ACTIVITY OF CONTRACEPTIVE-DELIVERY IMPLANTS

The area of contraceptive-controlled drug-delivery systems has been a field of increasing interest. However, one problem central to the field of controlled-release is that it is often extremely difficult to discontinue release on demand once release has commenced. This disadvantage is of special importance for contraceptive microspheres used intramuscularly, because if a medical or personal reason arises that caused one to wish to discontinue treatment, it is not possible to physically remove the injected microspheres. However, because of the high margin of safety with these progestins, it would be medically permissible to accelerate the depletion over a short time period.

The most important feature of the contraceptive-implantable or -injectable controlled-delivery systems is that the minimal effective dose of the contraceptive agent can be administered subcutaneously at a fairly constant level for inhibition of conception over years. Furthermore, the daily dose of medication administered by the implant are considerably lower when compared with use of daily oral pills [13,14]. The objective of this study [11] was to evaluate the possibility to use ultrasound, on demand, as an external trigger to terminate the activity of contraceptive-delivery microspheres in a shorter time. The exposure to ultrasound should enhance the release rates of the contraceptive from the depot and, therefore, deplete the depot faster.

Norethisterone (NET) microspheres 25–45 μm in diameter were prepared at Stolle Research and Development Corporation, Cincinnati, Ohio, by solvent evaporation, using 50:50 (wt%) poly(lactide-*co*-glycolide) [15]. The NET loading was approximately 45% (wt). The microspheres were designed to release drug for approximately 30 days.

For in vivo studies, ten New Zealand white male rabbits 2.3–3.2 kg (5–7 lb) were injected with NET microspheres subcutaneously through an 18-gauge needle at the center of the shaved spot. The injections contained 1–2 mg NET-30 microspheres per kilogram body weight, suspended (13 mg/ml) in a dextran injection vehicle (Stolle R & D). The in vivo experimental results (Fig. 4) demonstrate the feasibility of the ultrasound approach as a possible technique to manipulate the depletion time of implantable contraceptive delivery systems. Serum norethisterone radioimmunoassays were undertaken with norethisterone antiserum and [^{3}H]norethisterone. The NET release in the experimental group that was exposed for a total of 2.5 h from the fourth to the ninth day (30 min each day, 20% duty cycle), were higher and, therefore, the NET-30 microspheres were depleted faster (day 21) relative to controls, which released NET for the initially designed period of 30 days.

The effect of ultrasound energy (duration and intensity) on depletion time, as evaluated in vitro, is displayed in Fig. 5. The experiments were performed in a jacketed cell containing pH 7.4 0.1 M phosphate buffer at 37°C, which was measured and maintained at a constant temperature by circulating water through the jacket. Norethisterone release was detected spectrophotometrically (248 nm) while circulating the phosphate buffer at the rate of 14.0 ml/min with a peristalic pump. A higher ultrasound intensity or duration results in faster release rates and, therefore, faster depletion. Two-hours of exposure to ultrasound at 3 W/cm^2 for 6 consecutive days results in depletion times fourfold shorter than microspheres that were not exposed to ultrasound.

Scanning electron micrographs of microspheres exposed to ultrasound and controls, which were not exposed to ultrasound, are displayed in Fig. 6. As can

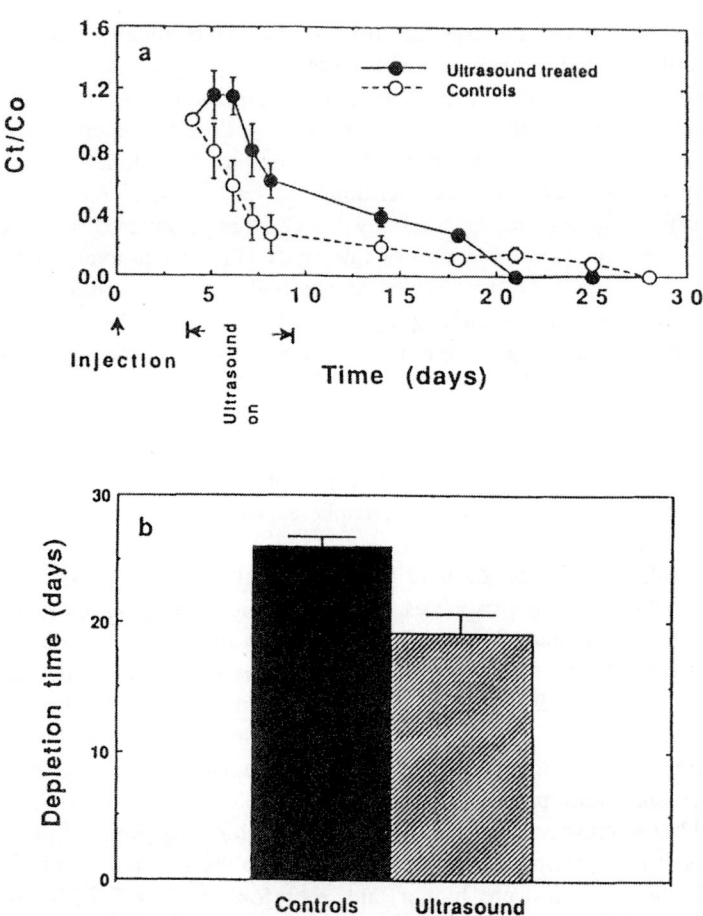

Fig. 4 (a) Norethisterone (NET) concentration (C_t) in serum divided by the concentration before ultrasonic probe treatments (C_0) vs. time, presented as the mean and standard deviation of five rabbits in the experimental and control groups. (b) NET depletion time of implants exposed to therapeutic ultrasound. The rabbits were exposed to ultrasound (1 MHz, 2.5–3 W/cm^2, 20% duty cycle) for 30 min/day for 5 days, from the fourth day after NET microsphere injection, vs. control group that was not exposed to ultrasound. (From Ref. 11.)

be seen, there is a significant effect of ultrasound on the surface morphological appearance of the degrading microspheres. Microspheres, immersed in phosphate buffer and exposed to ultrasound (1 MHz, 1.7 W/cm^2) for 30 min/day for 6 days, showed significant erosion after 1 week, whereas controls (similar

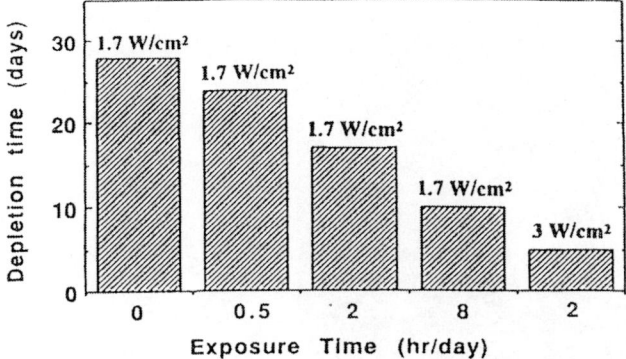

Fig. 5 NET depletion time of 50:50 DL-lactide–glycolide microspheres exposed in vitro to therapeutic ultrasound (1 MHz, 20% duty cycle) at several durations per day for 6 days. (From Ref. 11.)

microspheres immersed in the same buffer for the same duration), but were not exposed to ultrasound, showed no significant erosion.

V. MECHANISM OF ENHANCEMENT

Recently, D'Emanuele et al. [9] demonstrated that significant effects on polyanhydride erosion can be achieved using therapeutically acceptable intensities and frequencies (1 MHz and intensities below 2 W/cm^2). The therapeutic level of ultrasound also significantly decreased the induction period of polymer erosion, the results indicate that ultrasound enhances the rate of degradation within the polymer matrix. The morphological appearance of polymer devices, both exposed and nonexposed to ultrasound, were examined by environmental scanning electron microscopy (ESEM). The ESEM allows samples to be imaged directly in their natural state and in a relatively low-vacuum environment. The authors suggest possible mechanisms, including facilitated penetration of water in the polymer matrix, leading to enhanced anhydride bond hydrolysis, or possibly, a direct rupture of either the anhydride or other sensitive bonds induced by ultrasound, or mechanochemical degradation. The effect may be the result of more than one mechanism, including enhanced polymer degradation and mechanical disintegration of the polymer surface. Acoustic streaming, which accompanies cavitation, may also be involved.

Liu et al. [10] conducted studies on the mechanism of ultrasound-enhanced solid polymer degradation in the presence of a liquid. The erosion rate of the polymers, as well as the release rate of the incorporated drugs when exposed to ultrasound increased with the solution gas content. In polymer solutions, in which the effect of ultrasound has been extensively studied [16], similar behav-

(a)

(b)

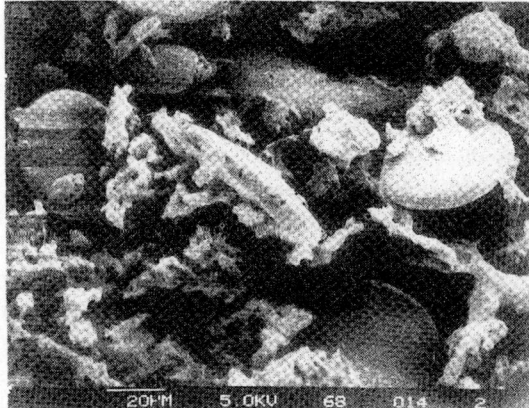

(c)

ior is attributed to an increased number of nucleation centers that increase with solution gas content, causing increased cavitation density, indicating the importance of the cavitation mechanism. The hypothesis that ultrasound affects the release and degradation kinetics as a result of radical formation was also evaluated. No effect of free radicals concentration on polyanhydride degradation was detected. To examine the contribution of ultrasound to each of the separate steps involved in polymer degradation (water diffusion to the labile bond, hydrolysis, diffusion, and dissolution of the degraded products), kinetic experiments were performed in pure dimethyl sulfoxide (DMSO) in which hydrolysis is impossible and polymer degradation is due to mechanical erosion. The enhancing effect of polymer erosion by ultrasound in pure DMSO suggests that the effect of ultrasound cannot be attributed only to hydrolysis. In conclusion, the authors proposed that the ultrasound-enhanced polymer degradation and consequent release of incorporated substances are based on hydrolysis, the mechanical stress, and the cooperation of both factors. Some physicochemical factors, such as solution gas content and solution vapor pressure, do not affect the hydrolysis of the polymer directly, but strongly alter the cavitation density and consequently, the erosion rate of polymer and the release rate of incorporated drug.

Levy et al. [17] also performed in vitro studies to gain insight into the mechanism of ultrasonically enhanced transdermal delivery. The authors consider three factors that might contribute to the ultrasound-enhanced permeability: mixing, cavitation, and temperature. To examine whether ultrasound might affect a boundary layer in the neighborhood of the skin and, thereby, cause higher permeabilities, experiments were performed in vitro under controlled mixing rates. The temperature of the skin exposed to ultrasound was monitored. Cavitation effects were evaluated in in vitro permeability experiments in degased buffer, for which cavitation was minimized. The authors concluded that the small increase in surface skin temperature observed after ultrasound application ($1-2\,^{\circ}C$) is not likely to cause dramatic changes in skin permeability. The ultrasound-enhancing phenomenon was mainly attributed to mixing and cavitation effects.

Mortimer et al. [18] showed that ultrasound exposure led to an increase in the rate of oxygen diffusion through frog skin. The authors found that the oxygen permeability increase is dependent on the ultrasound average intensity, but does not depend on the peak acoustic pressure. Given this finding, they

Fig. 6 Scanning electron micrographs of NET microspheres: (a) as received after processing, before microsphere were immersed into buffer; (b) after a week in phosphate buffer pH 7.4, 37°C; (c) after a week in phosphate buffer combined with exposures for 30 min/day for 6 days to ultrasound at 1.7 W/cm², 20% duty cycle. (From Ref. 1.)

conclude that it is not likely that cavitation is the dominant mode of action, as cavitation mechanisms are a function of acoustic pressure, rather than average intensity. In addition, transient cavitation was not observed through the measurement of OH radicals [19]. Because diffusion increased with increasing average intensity, the most likely mechanism proposed by the authors is acoustic streaming (quartz wind), leading to stirring action in the vicinity of the membrane, affecting the boundary layer (reducing the concentration gradient in the immediate neighborhood of the membrane).

Although it is difficult to be certain whether the data obtained in the in vitro experiments with synthetic polymeric membranes are extendable to the in vivo situation with skin, it is likely that because both involve diffusion through membranes, those factors that ultrasound affects most significantly in vitro with synthetic polymeric membranes may also play a significant role in vivo.

Fogler and Lund [20] proposed that the enhancement of mass transport by ultrasound was due to ultrasonically induced convective transport created by acoustic streaming, in addition to diffusional transport. Acoustic streaming is a secondary flow that produces time-independent vortices when an acoustic wave is passed through the medium [21]. The formation of these vortices or cells inside ducts, tubes, and pores can increase the rate of mass transfer through these enclosures. An analytical solution of the proposed model showed that, with the application of ultrasound, an increase of up to 150% above the normal diffusive transport could be obtained.

Lenart and Auslander [22] found that ultrasound enhances the diffusion of electrolytes through cellophane membranes. They proposed the mechanism to be diminution of the hydration sphere surrounding the electrolytes, thus increasing the electrolyte mobility and diffusion coefficient. The authors also proposed that a local temperature effect caused by the implosion of cavitation bubbles may be a possible mechanism.

Julian and Zentner [23] systematically investigated the effect of ultrasound on solute permeability through polymer films. The increase in permeability was unique to the ultrasonic perturbation and was not attributed to disruption of stagnant aqueous diffusion layers, increased membrane or solution temperature, or irreversible changes in membrane integrity. Recently the authors [24] suggested that the ultrasonically enhanced diffusion is a result of a decrease of the activation energy necessary to overcome the potential energy barriers within the solution–membrane interfaces.

VI. CONCLUSIONS

The results suggest the feasibility of ultrasound to enhance the release of peptides and proteins from bioerodible and nonerodible polymers. These phenomena have appealing therapeutic and commercial possibilities, as many peptides and proteins (hormones, growth factors) may benefit from more subtle delivery patterns under temporal control to obtain optimal therapeutic effects.

As most of the reported results applied ready-made ultrasonic units that were not designed for this specific application, we believe that specifically designed units will enable higher enhancement and permeability mediation that will lead to the preparation of delivery systems linked to miniature power sources that can be externally adjusted for a wide range of clinical applications. Such efforts at developing miniature and relatively inexpensive power sources will also be important for patient use and convenience.

Unfortunately, in spite of the numerous, recently published studies on the effect of ultrasound on release of peptides and other substances from polymers, the mechanism of the enhancing phenomenon is still not well understood nor characterized. The main factors contributing to this phenomenon include mixing, temperature, cavitation, acoustic streaming, and polymer morphological changes. As these are complex phenomena involving several parameters that are hard to separate, carefully designed studies, accompanied by theoretical approaches, are essential to obtain the knowledge needed to design and optimize ultrasonically induced drug-delivery systems.

REFERENCES

1. R. Langer, New methods of drug delivery, *Science 249*:1527–1533 (1990).
2. J. Kost and R. Langer, Responsive polymeric delivery systems, *Adv. Drug Deliv. Rev. 6*:19–50 (1991).
3. J. Heller, Modulated release from drug delivery devices, *Crit. Rev. Ther. Drug Carrier Syst. 10*:253–305 (1993).
4. R. Hoogland, *Ultrasound Therapy*, B. V. Enraf Nonius, Delft, Holland, 1986.
5. K. S. Sislick, *Ultrasound: Its Chemical, Physical and Biological Effects*, VCH Publishers, 1988.
6. J. Kost, K. Leong, and R. Langer, Ultrasonically controlled polymeric drug delivery, *Makromol. Chem. Makrom. Symp. 19*:275–285 (1988).
7. J. Kost, K. Leong, and R. Langer, Ultrasound-enhanced population degradation and release of incorporated substances, *Proc. Natl. Acad. Sci. USA 86*:7663–7666 (1989).
8. J. Kost and R. Langer, Magnetically and ultrasonically modulated drug delivery systems, *Pulsed and Self Regulated Drug Delivery* (J. Kost, ed.), CRC Press, Boca Raton FL, 1990, pp. 3–16.
9. A. D'Emanuele, J. Kost, J. L. Hill, and R. Langer, An investigation of the effect of ultrasound on degradable polyanhydride matrices, *Macromolecules 25*:511–515 (1992).
10. L.-S. Liu, J. Kost, A. D'Emanuele, and R. Langer, Experimental approach to elucidate the mechanism of ultrasound-enhanced polymer erosion and release of incorporated substances, *Macromolecules 25*:123–128 (1992).
11. J. Kost, L.-S. Liu, H. Gabelnick, and R. Langer, Ultrasound as a potential trigger

to terminate the activity of contraceptive delivery implants, *J. Controlled Release* *30*:77–81 (1994).

12. J. Kost, L.-S. Liu, J. Ferreira, and R. Langer, Enhanced protein blotting from Phastgel media to membranes by irradiation of low-intensity ultrasound, *Anal. Biochem. 216*:27–32 (1994).

13. C. C. Chang and F. A. Kincl, Sustain release hormonal preparation. 4. Biological effectiveness of steroid hormones, *Fertil. Steril. 21*:134 (1970).

14. H. R. Asch, B. B. Saxena, N. G. Gupta, P. J. Balmaceda, R. Rivera, B. H. Croxatto, E. M. Coutinho, and R. Landesman, Clinical evaluation of subcutaneous NET implants: Phase I multicenter clinical studies, *Long Acting Contraceptive Delivery Systems* (L. G. Zatuchni, A. Goldsmith, D. J. Shelton, and J. J. Sciarra, eds.), Harper & Row, Philadelphia, 1984, pp. 431–440.

15. L. R. Beck, Y. Z. Pope, C. E. Flowers, D. R. Cowsar, T. R. Tice, and D. H. Lewis, Polyl(DL-lactide-*co*-glycolide)/norethisterone microcapsules: An injectable biodegradable contraceptive, *Biol. Reprod. 28*:186–195 (1983).

16. F. R. Young, *Cavitation*, McGraw-Hill, London, 1989, pp. 150–173.

17. D. Levy, J. Kost, Y. Meshulam, and R. Langer, Effect of ultrasound on transdermal drug delivery to rats and guinea pigs, *J. Clin. Invest. 83*:2074–2078 (1989).

18. A. J. Mortimer, B. J. Trollope, E. J. Villneuve, and O. Z. Roy, Ultrasound-enhanced diffusion through isolated frog skin, *Ultrasonics 26*:348–351 (1988).

19. A. J. Mortimer and J. A. Maclean, A dosimeter for ultrasonic cavitation, *J. Ultrasound Med. 5*:137 (1986).

20. S. Fogler and K. Lund, Acoustically augmented diffusional transport, *J. Acoust. Soc. Am. 53*:59–64 (1973).

21. W. L. Nyborg, *Acoustic Streaming*, Academic Press, New York, 1965, pp. 265–332.

22. I. Lenart and D. Auslander, The effect of ultrasound on diffusion through membranes, *Ultrasonics 18*:216–218 (1980).

23. T. N. Julian and G. M. Zentner, Ultrasonically mediated solute permeation through polymer barriers, *J. Pharm. Pharmacol. 38*:871–877 (1986).

24. T. N. Julian and G. M. Zentner, Mechanism for ultrasonically enhanced transmembrane solute permeation, *J. Controlled Release 12*:77–85 (1990).

17

Polyacrylate Microcapsules for Cell Delivery

Julia E. Babensee and Michael V. Sefton

University of Toronto
Toronto, Ontario, Canada

I. INTRODUCTION

A. Why Microencapsulation of Cells?

Microencapsulation of living cells within a synthetic membrane is proposed as a means of using these cells as a source of therapeutic biomolecules. The cells are protected from the host's immune system by the presence of the polymer wall (Fig. 1). This allows the cells, after their transplantation, to function as a drug-delivery system or a hybrid artificial organ, without the need for immuno-suppressive drugs. By using cells to deliver the molecules, we can take advantage of the physiological control mechanisms by which cells function. Glucose stimulation of insulin secretion by islets of Langerhans is perhaps the best known, but there are many other features of cells that, at least in principle, can be exploited in cell-based delivery systems. Phenotypical gene expression, feed-back loops, paracrine and autocrine growth factor secretion, extracellular matrix or cytokine effects, and the secretion of angiogenic factors by trans-planted cells are some of the ways in which cell-based delivery can be modulated. Such factors are not available to be exploited with conventional drug-delivery systems. On the other hand, cell encapsulation methods are limited by the sensitivity of the cells to the "stress" of the encapsulation process (osmotic, pH, thermal, solvent, and shear), the diffusion barrier imposed by the capsule wall and the new in vivo environment created by the effect of the presence of the capsule (through its biocompatibility) on the surrounding tissue, and the effect of the extracapsular tissue on the intracapsule environment.

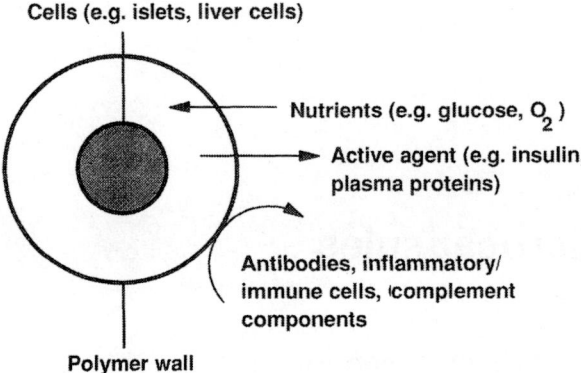

Cells (e.g. islets, liver cells)

Nutrients (e.g. glucose, O_2)

Active agent (e.g. insulin, plasma proteins)

Antibodies, inflammatory/ immune cells, complement components

Polymer wall

Fig. 1   The microcapsule–immunoisolation concept.

Cell transplantation is under consideration for a variety of disorders: insulin-dependent diabetes [1], Parkinson's disease, and other neurological applications [2] (e.g., using BHK cells transfected with the gene for nerve growth factor) [3]. The need for immunosuppressive therapy and the limited availability of donor tissue are the primary limitations of cell transplantion. Immunoisolation is an alternative to immune suppression or immune alteration [4]. Immunoisolation units, typically hollow fibers and frequently with blood contact, have been used with a variety of cells—islets [5,6,8–10], hepatocytes [11], PC12 cells [12–15]—to separate them from the host's immune system. Microcapsules have a high surface/(total) volume ratio and, thus, are expected to result in much smaller implants than is possible with hollow fibers, and with faster response times. By avoiding contact with blood, capsules also circumvent the problems of coagulation and vascular anastomoses.

Microencapsulated cells could be injected or implanted intraperitoneally (IP), for example, although other sites may be feasible or preferred. The issues of retrievability and the number of cells needed and capsule size will influence the choice of implant site and mode of access (e.g., capsules may be contained in an implanted "tea bag" [16]). The implant volume is strongly dependent on capsule size and on how many cells are contained in each capsule. This calculation is particularly important for pancreatic islets, because more than 3×10^5 islets are thought to be necessary [17]. With one islet per capsule, and a low 300,000 islets per transplant, the transplant volume needs to be an unmanageable 1150 ml with 900-μm capsules, a reasonable 30 ml with 400-μm capsules, and only 2.3 ml with 170-μm "capsules." Loss of islet efficacy because of transport limitations or damage, increases the implant volume needed, although packing more than one islet per capsule can compensate. The geometric limitations may be much worse if hollow fibers or flat geometries are considered.

From the pioneering work of Chang [18] numerous biologically active species (e.g., enzymes and whole microbial or plant cells) have been encapsulated or immobilized in polymeric matrices. The more fragile mammalian cells have been encapsulated most frequently in a polylysine-stabilized calcium alginate. Lim and Sun [19] used this process to microencapsulate pancreatic islets to protect transplanted islets from the immune system of rats for as much as 2 years [20]. The feasibility of this approach has been shown with a variety of cells and in various animals [21–24]. The long-term success has occasionally been questioned, however, because of biocompatibility and biostability issues (see following section). A different concern, raised with the use of PC12 cells, is the potential tumorigenicity of such cell lines. The intact capsule wall is a means of limiting cell growth and precluding tumor formation. The strength of hydroxymethyl methacrylate–methyl methacrylate (HEMA-MMA) capsules is an advantage in this context relative to the mechanically weaker alginate capsules.

B. Biocompatibility and Immunoisolation

Biocompatibility of the microcapsule is a critical concern. This is highlighted by the evidence from several groups [25–28] indicating the in vivo failure of alginate–polylysine encapsulated islets, in contrast with the results of Sun and others. Specific concerns include the purity (endotoxin level), source, chemical heterogeneity [29], and intrinsic biocompatibility of alginate–polylysine, the inherent difficulty of producing reproducible cell-containing capsules [30], and the variability in tissue reaction [27]. Others suggest a lack of biostability, with some alginate–polylysine capsules disintegrating after 3 months.

The appropriate host response to capsules (i.e., "biocompatibility" implies the absence, or at least minimization, of tissue reaction and fibrous capsule formation (negligible diffusion resistance), macrophage activation (negligible interleukin-1 [IL-1], cytokine, or cytotoxic agent release [31,32]), and capsule biodegradation. Cytokines released locally may penetrate the polymer wall to affect encapsulated cell secretory behavior and viability, possibly resulting in graft failure, as has been suggested for IL-1 and alginate-encapsulated islets [28]. The presence of cells and added extracellular matrix may also influence the tissue reaction, through the shedding of antigens or the products of dying cells, or through resorbing proteins.

Although the absence of a fibrous capsule may ultimately be desirable, this may not be necessary if the (presumably thin) fibrous capsule is a negligible additional resistance to diffusion (relative to the capsule itself). Alternatively, a vascularized tissue reaction may be preferred, as this would enhance the absorption of the secreted agent. A vascularized membrane–tissue interface has indeed been induced with particular membrane architectures [33]. Also the presence of certain cells (such as rat pancreatic islets) stimulated increased vascularization, perhaps through the secretion of angiogenic factors, such as basic fibroblast growth factor (bFGF) [34].

Immunoisolation is frequently interpreted in terms of IgG (150 kDa) exclusion [5], although a broader definition, involving the entire immune system, may be necessary. Hence, immunoprotection of encapsulated cells has typically been defined by a molecular weight cutoff of less than 150 kDa. For example, alginate–polylysine capsules prepared using polylysine with a molecular weight of 17,000 were found to be impermeable to albumin, hemoglobin, and IgG, indicating an appropriate molecular weight cutoff of less than 67 kDa [35]. Alginate–polylysine capsules, prepared with poly-L-lysine of 21,000–390,000 molecular weight were not permeable to IgG antibodies in vitro [36]. However, implantation of similar capsules in vivo showed positive intracapsular IgG and extracapsular IgG and IgM staining, associated with rejecting xenografts within alginate–polylysine capsules, indicating a possible role of xenoantibodies in graft rejection [37].

The exclusion of antibodies by the polymer membrane may not be essential if the effector functions of antibody binding are prevented from occurring. Specifically, bound aggregates of IgG molecules or individual IgM molecules on the surface of encapsulated cells may not affect the cells unless the larger molecule C1q (410 kDa) also permeates the polymer membrane to bind to the immunoglobulin. This binding would initiate the classic pathway of complement activation which, with the addition of the subsequent complement components (of lower molecular weight), would potentially result in the cell-lysing membrane attack complex. The prevention of complement-mediated lysis of alginate–polylysine microencapsulated cells has been demonstrated [38]. Moreover, Heald and co-workers [39] showed that the alginate–polylysine membrane can protect porcine islets from the antiporcine antibody-dependent complement-mediated lysis of human serum. This experiment was an in vitro simulation of a discordant xenograft transplantation with immunoisolation provided by the alginate–polylysine membrane.

Contact of encapsulated cells, targeted by antibody binding, with cytotoxic T lymphocytes or natural killer (NK) cells would result in their lysis; therefore, these cellular elements, which effect antibody-dependent cell-mediated cytotoxicity, must be excluded by the membrane. Although the physical barrier of the polymer membrane may exclude an entire immune or inflammatory cell, because of the ability of the cells to form cellular processes that may extend over large distances, it has been suggested that membrane-spanning pores larger than 0.1 μm must be avoided [5].

Immunoisolation may also have an advantage in preventing the autoimmune destruction of the transplanted cells, such as islets [40], by separating them from the host (T cells or autoantibodies) by the permselective membrane. Although the detailed mechanism of islet cell damage is unknown, the rapid failure of islet transplants, even in twins [41], may be due to a recurrence of autoimmune disease, rather than the conventional transplant rejection phenomenon [42].

On the other hand, IgG exclusion may be inadequate if antigen-shedding or cytokine effects are crucial. Thus, the source of the cells (i.e., whether the transplant involves autologous tissues or allografts or xenografts) may become an issue. Interleukin-1 (IL-1) inhibition of insulin secretion from islets in vitro [43] is thought to play a role in their autoimmune destruction, so that failure to inhibit IL-1 transport may obviate the antibody exclusion effect of a capsule wall with the "ideal" approximately 100-kDa cutoff. Interestingly, some microcapsules appear to prevent IL-1 inhibition of insulin secretion [44], suggesting that the intracapsule microenvironment is such that the sensitivity of islets to IL-1 is lessened after encapsulation.

C. Intracapsule Cell Behavior

The extracellular matrix (ECM) through binding to cell surface integrins, has a strong influence on cell behavior. Accordingly, adding ECM components to cells before encapsulation or otherwise altering the substrate on or within which the cells are grown provides a means to control the behavior of the transplanted cells. For example, Matrigel-grown hepatocytes have elevated mRNA synthesis when compared with cells grown on plastic or in collagen gel [45]. Cytochrome P-450 activities were also induced in Matrigel but not collagen I [46]. Alteration of the ECM or other features of the cell microenvironment thus provides another means for controlling the performance of the encapsulated cell.

The ECM is a dynamic environment, undergoing constant remodeling (e.g., through the indirect effect of cytokines and growth factors) [47]. The capsule wall will likely affect the extent of remodeling (e.g., by impeding the inward transport of interstitial proteases that otherwise cause ECM turnover); consequently, stabilizing the intracapsule ECM. On the other hand, necrotic cells may release significant amounts of lysosomal enzymes that can degrade biomolecules. The capsule wall will impair their outward diffusion and could cause a build-up of degrading enzymes or of the acidic lysosomal environment.

Minimization of the diffusion limitations associated with microencapsulation is also critical to the success of encapsulation. For example, depletion of a particular nutrient (e.g., oxygen) may cause cells to grow at a lower, diffusion-limited rate or the center of a cell cluster may become necrotic. This has been observed [48] and is similar to what occurs with tumor spheroids [49]. Because the cells in spheroids are at a distance from the surrounding medium, gradients of critical nutrients, growth factors, and metabolites (and also pH) are set up [50]. As a spheroid grows, the number of proliferating cells decreases, the proportion of quiescent cells increases and, eventually, owing to nutrient deprivation and waste product accumulation, necrosis develops at the center of spheroids [50]. The diffusion of larger molecular weight species (such as growth factors, hormones, and cytokines) into spheroidal regions, their interaction with the receptors of cells in the outer few layers of an aggregate [51], and cell contact with the ECM are also important.

The microenvironment of spheroidal aggregates appears to maintain the in vivo differentiated cell characteristics for longer times than in monolayer culture [52,53], presumably, owing to the three-dimensional arrangement and the corresponding diffusion gradients and ECM–cell and high cell–cell contact. With the capsule wall controlling the aggregate size and strongly influencing the diffusion gradients, such differentiated cells with prolonged life spans would presumably be ideally suited for microencapsulated cell applications. Microcapsules are also characterized by a locally high cell density, and this, too, is expected to influence intracapsule cell behavior.

II. ALGINATE–POLYLYSINE MICROCAPSULES AND OTHER METHODS

The use of microencapsulated islets of Langerhans for the treatment of diabetes was first proposed in 1980, by Lim and Sun [19]. This study started the ongoing research of the most popular microencapsulation technique for living cells. It has been further developed by Sun and co-workers (see next paragraphs) and adopted and modified by several other research groups.

The alginate–polylysine microencapsulation technique developed by Sun and co-workers [19] is based on capsule membrane formation by interfacial adsorption of cationic polylysine onto the surface of an anionic alginate cell-containing bead. Microcapsules were prepared by forming droplets of cells suspended in 1.5% (w/v) $CaCl_2$, formed spherical, gelled beads as the alginate complexed Ca^{2+} ions. Electrostatic interactions of the anionic alginate, on contact with an aqueous solution of polylysine, resulted in their complexation to form a homogeneous outer membrane on the surface of the alginate bead. These capsules were then treated with polyethyleneimine to stabilize the membrane for a more durable capsule. The pronounced tissue response to capsules coated with polyethyleneimine [54], and the resultant early failure (2–3 weeks) of allogeneic-microencapsulated islet transplantations in diabetic rats [19], prompted its later replacement with polylysine, such that islet allografts then functioned in a stretozocin-induced diabetic rat for up to 1 year [20]. A uniform coating with alginate was required to cover the positively charged amine groups of the polylysine, which would otherwise support inflammatory cell adherence [55] and a tissue reaction [56,57]. The microcapsules were subsequently treated with a solution of the Ca^{2+}/Mg^{2+} chelator, sodium citrate, to liquify the alginate core and extract residual alginate, leaving a alginate–polylysine–alginate hydrogel membrane of about 4 μm. This microencapsulation technique facilitated the gentle entrappment of sensitive mammalian cells in an aqueous environment without the action of forces, such as shear or pressure, such that they remain functional. This pioneering work demonstarted the conceptual feasibility of islet microencapsulation and therapeutic benefits of transplantation in diabetic animals. Further work has been necessary to refine the process and translate the concept into a practical reality.

Modifications to this alginate–polylysine microencapsulation technique have been aimed at enhancing the cytocompatibility of the encapsulation process and at improving microcapsule properties, such as the intracapsular environment, for cellular viability and function [58–60], mechanical strength [61,62], permeability [61,63,64], size [65], and microcapsule biocompatibility (without encapsulated cells) [55,66–69]. The use of the electrostatic droplet generator to produce small-diameter and stronger alginate-entrapped islets has given renewed impetus to this area [70].

Cytocompatibility of the process (i.e., cell survival during the process of encapsulation) was improved by replacing saline and citrate (used in the original microencapsulation procedure [19]) with Ca^{2+}-free Krebs–Ringer bicarbonate buffer and 1 mM EGTA, respectively, such that insulin secretion by microencapsulated islets was at the same rate as control (unencapsulated) islets, unlike those prepared by the original method [71]. Anchorage-dependent cells, such as hepatocytes, require support for cellular attachment. Because the capsule membrane does not support cellular attachment, this requirement was fulfilled by the coencapsulation of collagen [58,59] or Matrigel [60]. The molecular weight cutoff of the alginate–polylysine membrane was determined [61] to be a function of the molecular weight of polylysine used to prepare the capsules. Microcapsules that were prepared using polylysine with a molecular weight of 17,000 were impermeable to albumin, hemoglobin, and IgG, indicating an appropriate membrane molecular weight cutoff of less than 67 kDa; this was presumed not to allow the passage of immunoglobulins to contact encapsulated cells after their transplantation [61]. Furthermore, the molecular weight cutoff of the alginate–polylysine capsules was lowered by using a two-step polylysine coating, with polylysine of different molecular weights, or by replacing polylysine with poly(L-ornithine) (cutoff lowered to 50 kDa) [63], or by increasing the alginate–polylysine reaction time (from 3 to 6 min) [64]. The strength of microcapsules was enhanced by increasing the alginate–polylysine reaction time and the concentration of the polylysine (from 0.03 to 0.05% [w/v]) [61], or by replacing polylysine with chitosan [62].

The biocompatibility of alginate–polylysine capsules has been questioned, mainly because of the solubility of the individual polyelectrolytes in an aqueous environment. Calcium alginate gels are sensitive to chelating compounds, such as phosphate, citrate, and lactate, as well as antigelling cations, such as Na^+ and Mg^{2+} [72]. Hence, the ionic environment surrounding implanted alginate polylysine capsules formed by Ca^{2+} complexation, may result in biodegradation of the microcapsule; dissolution has even been described during their in vitro culture [73]. However, the complexation of the alginate with polylysine may stabilize the microcapsule membrane [72] such that biodegradation of the complexed membrane would be much less than that expected, based on the individual components. However, the issue of alginate–polylysine microcapsule biostability has been addressed [66], by replacing $CaCl_2$ for gelation with $BaCl_2$, for the preparation of nonbiodegradable barium alginate beads for islet transplantation [66]; barium has a higher affinity for alginate than calcium [72]. The outer surface of the alginate polylysine capsule may also influence the biocompatibility of the capsules through the composition of the alginate coating. Commercial alginate (purified to remove lipopolysaccharides and polyphenols) stimulated human monocytes to release the cytokines IL-1 and tissue necrosis factor (TNF)-α, which are mediators of fibrosis, by stimulating fibroblast proliferation [67]. Mannuronic acid monomers, but not guluronic acid monomers, stimulated cytokine release [67], and capsules with an outer layer of high-mannuronic–content alginate elicited a tissue response after implantation [56,67]. In another attempt to enhance microcapsule biocompatibility, alginate microcapsules were coated with polyethyleneimine and subsequently treated with protamine sulfate and heparin [68]. Although transplantation of protamine–heparin microencapsulated rat islets into the peritoneal cavity of streptozocin-induced diabetic mice maintained normoglucemia for 4 months, a comparison of the tissue response to protamine–heparin microcapsules with that toward alginate–polylysine–alginate microcapsules was not shown [74].

Another attempt to modify the surface of alginate–polylysine capsules for the enhancement of biocompatibility has been described [55,69]. The surface of the microcapsule was modified by the polyelectrolyte complexation of the polyanionic alginate with a polycationic graft copolymer, lysine co-monomethoxy poly(ethylene glycol) (lysine-MPEG) [55]. The MPEG treatment of alginate–polylysine microcapsules reduced fibrinogen and albumin adsorption by 80 and 88%, compared with polylysine, respectively, and in a similar manner when compared with alginate. Furthermore, binding of the inactive complement fragment, iC3b, was reduced on capsules treated with MPEG when compared with polylysine or alginate, indicating a reduced protein adsorption, but not necessarily a lower level of complement activation. The inability of the MPEG-treated surface to support fibroblast attachment was indicated in their rounded morphological appearance and reduced attachment to MPEG-treated capsules, compared with the other two surfaces. These effects of PEG-grafted polymer surfaces are consistent with the observations of others of PEG associated reductions in protein adsorption [75] and mammalian cell adhesion [76].

A more stable surface modification was obtained by the photopolymerization of polyethylene glycol tetraacrylate onto the polylysine capsule surface [69]; an example of interfacial polymerization for microcapsule preparation. Alginate–polylysine capsules that contained islets were coated with eosin Y and suspended in an aqueous solution of polyethylene glycol tetraacrylate (MW, 18,500), triethanolamine and *n*-vinyl pyrolidinone. Photopolymerization was initiated by the exposure of capsules to the visible light of an argon laser, resulting in a capsule coating of between 5 and 10 μm. Cellular overgrowth occured within 4 days on 60–80% of alginate–polylysine capsules recovered from the peritoneal cavity of mice. In contrast PEG-coated microcapsules were free-floating and without attached cells, even up to 20 days postimplantation.

Agarose microcapsules have been prepared in which cells suspended in an agarose solution were extruded into a flowing hydrophobic medium to form spherical droplets that solidified on cooling [77]. The potentially water-soluble and enzyme-sensitive agarose bead was subsequently coated with water-insoluble polyacrylamide by interfacial polymerization [78]. This interfacial polymerization was achieved by light-induced free-radical polymerization of acrylamide and bisacrylamide to form the outer polymer membrane. The expectation was that such a surface modification would result in a microcapsule that was more stable in the physiological aqueous environment and more biocompatible. In vitro biocompatibility studies indicated potential biocompatibility of these capsules; however, there was complement activation in human serum [79], but IL-1 secretion by phorbol-12-myristate-13-acetate-differentiated, U937 macrophage-like cells occurred only in the presence of lipopolysaccharide (LPS) [80]. Macrophage culture in the presence of the materials only, would have more clearly elucidated material-dependent macrophage activation. The agarose microcapsules were reproducible in diameter (96% were 300–500 μm in diameter). Islets encapsulated by this method secreted insulin in response to glucose stimulation [77]; however, owing to fibrosis, explanted islets were unresponsive to glucose increases [79]. The level of cytotoxic free-radical production during the photopolymerization of the high-concentration monomer solution ($\sim 30\%$ v/v), necessary to produce a membrane of sufficient mechanical integrity, was lowered by incorporating polyacrylamide microlatex beads (hydrodynamic diameter 320 nm) into a now 5% acrylamide, 0.25% bisacrylamide aqueous solution [81]. Improved cytocompatibility of this modified microencapsulation process was indicated by an increased prolactin secretion by encapsulated pituitary cells.

Beads of agarose without a outer microcapsule membrane have also been prepared. Agarose droplets are formed containing islets on emulsification in paraffin oil and solidified as they are cooled in an ice bath to form beads [82]. Beads prepared from a 7.5% (w/v) agarose solution, instead of 5% (w/v), were more dense and, thus, less antibody- and complement-permeable. This was interpreted as responsible for the observed increase in the length of xenograft survival, from 32.2 ± 17.8 days to 85.3 ± 23.1 days. Longer survival times of allograft transplantations [83], as compared with xenograft transplantations [84], appeared to indicate that agarose provided immunoisolation to the immune cells that mediate allograft rejection, but were permeable to the xenoantibodies and complement that mediate the rejection of xenografts. This antibody permeability may be due to the absence of an immunoprotective membrane on these agarose beads. The sensitivity of agarose to enzymes, combined with its instability at elevated temperatures and water solubility, raises concerns over the long-term in vivo function of these beads.

Recently, a new method to prepare agarose beads with a microcapsule membrane that is less permeable to antibodies has been described. A mixture of agarose (5% (w/v)) and 10% (w/v) poly(styrene sulfonic acid) (PSSa) was used to prepare microcapsules, which were then treated with Polybrene (PB) to stabilize the surface structure and minimize the leaching of PSSa [85,86]. To enhance the biocompatibility of the microcapsules, the surface was modified by the treatment with chondroitin sulfate [85], or with the potentially more biostable, carboxymethyl cellulose [86].

Cell-containing microcapsules have also been prepared by adding droplets of cell suspension in an aqueous solution of one polyelectrolyte to an aqueous solution of an oppositely charged polyelectrolyte to form the capsule wall by complexation, as is done for alginate–polylysine. However, here, the cell-containing core is liquid, unlike that of alginate–polylysine capsules in which the capsule core is gelled before the second step of membrane formation. From the liquid core of these capsules, unreacted polyelectrolyte is leached out, similar to the removal of unreacted alginate from within the liquefied alginate–polylysine capsule core. Microcapsules have been prepared using combinations of synthetic polyacrylate-based polyelectrolyte polymers [87–89], polyphosphazenes [90], or natural polyelectrolyte combinations of collagen, chitosan, and alginate [91]; carboxymethyl cellulose and chitosan [92]; or Matrigel–carboxymethyl cellulose–chondratin sulfate A and chitosan–polygalacturonate [93]. The first polyelectrolyte is that of the cell-containing core, the second, the corresponding basic polyelectrolyte. The acidic polymer, whether it be synthetic or natural, is used as the interior member of the capsule-forming pair owing to the expected cytocompatibility of acidic polysaccharides, such as alginate, and carboxymethyl cellulose. Cells survived the encapsulation and proliferated within the capsules: hybridoma cells produced antibody [87,92], hepatocytes demonstrated urea and protein synthesis and drug metabolism [93], and hepatocytes of transgenic mice secreted hepatitis B surface antigen (HbsAg), even after capsule transplantation into congenic mice for up to 2 weeks [91]. However, the synthetic polyelectrolyte membranes, although highly permeable, were mechanically weak. Execpt for the collagen–chitosan and alginate [94] combination, which exhibited poor biocompatibility in vivo, the other capsule systems have not been evaluated after implantation. Presumably, similar to alginate–polylysine capsules, there is a concern about the stability of the polyelectrolyte complex in an aqueous environment.

The size of the microcapsule should be at a minimum to minimize diffusion and implant volume constraints. The minimum size is that of an individual islet. Therefore, islet immunoprotection with a conformal coating has been suggested. An interfacial photopolymerization technique to conformally coat islets with poly(ethylene glycol) that is based on the same principles as those, described earlier, to coat the surface of alginate–polylysine capsules has been described [95]. The resultant PEG coating on the surface of the pancreatic islets was approximately 30 μm. Live–dead staining of conformally coated islets with fluorescein diacetate and ethidium bromide, respectively, indicated that 90% of the islets were viable. Furthermore, the conformally coated islets responsed to glucose with stimulated insulin secretion. These results indicated that this process was compatibile with the sensitive nature of the islets, as cell viability and function were maintained. In another method to conformally coat islets, islets suspended in an alginate solution are centrifuged through discontinuous density layers, passing through a $BaCl_2$–Ficoll solution in which entrained alginate surrounding the islet is gelled, and coated islets are finally collected in RPMI 1640 containing 20% (w/v) bovine serum albumin [96]. Islets, conformally coated with 10 μm of alginate, exhibited a biphasic insulin response, identical with control, uncoated islets, after an increase in glucose concentration.

Recently, AN69, a copolymer of poly(acrylonitrile-sodium methylsulfonate) has been used to prepare microcapsules for cell transplantation [97]. AN69, dissolved in DMSO, was coextruded with cell suspension to form capsule droplets, which passed through a layer of ethanol (a moderate precipitant), and into a physiological saline precipitation bath. Insulinoma, RINm5F cells, survived encapsulation, remained viable, and secreted insulin on stimulation with Krebs buffer-containing arginine and theophylline.

III. HYDROXYETHYL METHACRYLATE–METHYL METHACRYLATE-(HEMA–MMA)

A. Why Polyacrylates?

The suitability of a polymer for microencapsulation of cells is determined by its processability (e.g., its viscosity and solubility, especially in solvents that are tolerated by the cells), its permselectivity in the form of the microcapsule wall, and its biocompatibility. The capsule wall must have a high permeability to nutrients and cell-derived products, yet exclude antibodies, complement components, and inflammatory and immune cells. Because the surface chemistry has a large role in defining the capsule biocompatibility, it, too, is significant. An understanding of the influence of various factors, such as polymer chemistry and microencapsulation conditions, on encapsulated cells, is our research objective. We have focused on using polyacrylates for microencapsulation, with the view to deepening our understanding of these issues.

The polyacrylate on which our current research is based, is a thermoplastic hydroxyethyl methacrylate–methyl methacrylate (~75 mol% HEMA) copolymer prepared by solution polymerization after careful monomer purification to reduce the cross-linker content [98]. This copolymer is hydrophilic with an approximate 25–30% (w/w) water uptake [99], consistent with the poly(-HEMA) content, but has mechanical strength, toughness, and elasticity, imparted by the poly(MMA) component. These properties lead to adequate permeability of the polymer capsules to aqueous solutes for cellular sustenance [100] and sufficient mechanical durability to tolerate normal handling and stresses in vivo. The critical requirement of microcapsule biocompatibility was considered likely owing to the common use of the homopolymers poly(MMA) and poly(HEMA) in biomedical applications (e.g., bone cement and intraocular and contact lenses, respectively). The water-insolubility of the HEMA–MMA polymer provides stability in the aqueous physiological environment, but necessitates the use of an organic solvent to prepare the polymer solution. The ultimate success of this material depended on the ability to select a tolerable solvent and to design a gentle encapsulation process.

B. Microencapsulation Processes

We have prepared microcapsules using three different water-insoluble polyacrylates, namely, Eudragit RL [101–103], copolymers of dimethylaminoethyl methacrylate and methyl methacrylate (DMAEMA-MMA) [104], and most often, HEMA–MMA. Common to all these polymers has been the use of coaxial extrusion of cell suspension and polymer solution, shearing of the capsule droplet, and polymer wall formation by interfacial precipitation on nonsolvent contact. However, each polymer has different properties, such as its solubility in solvents, that would be suitable for microencapsulation with cells (diethyl phthalate or polyethylene glycol); the solution viscosity; and the precipitation characteristics in an appropriate nonsolvent. Therefore, their application to microencapsulation required appropriately selected polymer solvents and nonsolvents and droplet-shearing conditions.

Initial microencapsulation studies involved commercially available Eudragit RL, which is a methacrylic acid ester–acrylic acid ester copolymer containing 5% quarternary ammonium groups [101]. Eudragit RL, dissolved in diethyl phthalate, was delivered coaxially around the capsule core, and capsule droplets were sheared off the needle assembly with a coaxial air stream. Capsules were formed as the polymer wall precipitated around the core solution on contact with the nonsolvent free fatty acid–free corn oil–mineral oil mixture. Fibroblasts, encapsulated with this polymer, survived and proliferated if collagen was present in the capsule core as a cell attachment and growth scaffold [101]. Encapsulated islets were viable and insulin secreting to the same extent as control (unencapsulated) islets, but with a delayed response time to increased glucose levels [102]. The limited permeability of Eudragit RL capsules was further illustrated by the failure of CHO cells to proliferate, except within mechanically flawed or leaky capsules [103]. The limited permeability was consistent with the low water uptake of the polymer (in the range 15–20%). Moreover, the unsuitability of Eudragit RL for transplantation of encapsulated cells was suggested by its poor biocompatibility. Implantation of polymer films into rats for 2–3 weeks, resulted in a pronounced tissue reaction. The capsules were also brittle and easily broken. Although Eudragit RL was unsuitable, its use did demonstrate that living cells could survive the rigors of coextrusion and the limited contact with organic solvents during interfacial precipitation.

A second polymer system used for encapsulation was based on dimethylaminoethyl methacrylate (DMAEMA; 16–25 mol%), methyl methacrylate (MMA) and methacrylic acid (MAA: 2.2–4 mol%), with the expectation of promoting cell attachment and growth on the inner capsule surface [104]. This was seen as a simpler way to provide for this requirement of anchorage-dependent cells than the incorporation of collagen within Eudragit RL capsules, as the latter was awkward to coencapsulate. For microencapsulation, DMAEMA–MMA (16 mol% DMAEMA), dissolved in diethyl phthalate and cell suspension were pumped to the tip of the coextrusion assembly. Capsule droplets were sheared off the tip of the assembly by a coaxial filtered air stream and passed through an overlayer of hexadecane, into a mixture of 75:25 (v/v) corn oil–paraffin oil and, finally, into the precipitation bath of corn oil where the enclosing polymer wall was formed. Fibroblasts survived encapsulation, but grew in only about one-third of the capsules, the remainder had cells spreading on the inner capsule surface. Low permeability of the polymer wall was suggested by the low water uptake by the polymer (11% in PBS). Moreover, toxic effects of the incompletely purified polymer wall may have been an additional factor affecting cell growth. Adding methacrylic acid (DMAEMA–MMA–MAA, 78.6:17:4.4 mol%, respectively) increased the polymer water uptake to 27% (w/w), resulting in a polymer that was more permeable to glucose, but was an unsuitable substrate for cellular attachment.

From these experiences, it was apparent that encapsulation of living cells within water-insoluble polyacrylate polymers was feasible, but that improvement of polymer properties were necessary in two key areas; namely, biocompatibility and permeability. The expected properties of HEMA–MMA suggested its suitability for microencapsulation purposes. However, several modifications to the coaxial extrusion, interfacial precipitation process, used for Eudragit RL, were required to accommodate HEMA–MMA. First, because the polymer did not dissolve in diethyl phthalate, the water-soluble polyethylene glycol 200 (PEG-200) was chosen as the solvent, owing to its reasonable tolerance by cells. An advantage of this choice was that a nonsolvent precipitation bath based on physiological phosphate-buffered saline (PBS; containing the Pluronic surfactant, L101) could be used. Second, because the HEMA–MMA–PEG solution was more viscous than Eudragit RL–diethyl phthalate ($\sim$ 1 Pa-s vs. 0.2 Pa-s, respectively) capsules of an unacceptably high diameter of about 1 mm were formed if the capsule-shearing fluid was a coaxial air stream [105]. A decrease in capsule diameter to approximately 750 μm was achieved by replacing the coaxial airstream with a stationary nonsolvent hexadecane overlayer, from which the needle tip was removed repeatedly in an up-and-down motion (a submerged jet). The viscous hexadecane provided a higher shear force to separate each capsule droplet from the tip of the needle assembly [100].

The HEMA–MMA capsule droplets, consisting of the cellular core surrounded by the polymer solution, were produced by pumping the polymer solution (9 or 10% (w/v) in PEG-200) at a rate of 40–56 μl/min and cell suspension, comprised of various cell types (0.5–5 $\times$ 10^6/ml) in their complete tissue culture medium (i.e., containing added serum and antibiotics) at a rate of 21 μl/min, to the tip of a coaxial needle assembly. The polymer solution flowed through the outer needle and the cell suspension through the inner needle. For later encapsulations the cell suspension was augmented with the viscosity and density enhancer 20% (w/v) Ficoll-400 [106]. Each capsule droplet was sheared from the needle assembly as it was withdrawn from the hexadecane overlayer and then passed through the hexadecane overlayer into the PBS precipitation bath, containing 100 ppm of the Pluronic surfactant L101. The latter was added to facilitate the passage of the droplet through the hexadecane–PBS interface. In this precipitation bath, the polymer solvent was extracted, leaving behind a polymer wall surrounding the cellular core. After completion of the coextrusion, the capsules were kept suspended in the precipitation bath for 30 min, with a magnetic stirrer, and subsequently, for an additional 30 min in fresh PBS without added L101. Capsules were then transferred to petri dishes for incubation in complete tissue culture medium at 37°C in an atmosphere of 5% CO_2/95% air.

A consequence of the biocompatibility of HEMA–MMA is its failure to support the attachment, spreading, and growth of anchorage-dependent cells. The nonadherent HEMA–MMA intracapsule environment was modified by coencapsulating cell attachment and growth substrates, such as Matrigel or Cytodex beads, or cell immobilization matrices, such as agarose or chitosan. Matrigel is a commercially available (Collaborative Research, Bedford MD) preparation of mouse EHS tumor-derived, reconstituted basement membrane proteins, including laminin, collagen type IV, and heparan sulfate proteoglycan [107]. Inclusion of Matrigel in the capsule core was achieved by cooling the Matrigel–cell suspension to about 4°C as it was delivered to the coaxial needle assembly. Matrigel forms a gel at 37°C (gelling temperature ~ 22–35°C). Cytodex 2 beads are cross-linked dextran derivatized for N,N,N-trimethyl-2-hydroxyaminopropyl groups to provide a positive surface charge that promotes cellular attachment. Hydrated Cytodex beads, with preattached fibroblasts, were kept in suspension by using a microstirring bar in the cell suspension syringe. Agarose is a polysaccharide copolymer of 3-linked β-D-galactopyranose and 4-linked 3,6-anhydro-α-L-galactopyranose residues. Agarose forms a stable gel on cooling from physiological 37°C to room temperature. Hence, to facilitate its gelation within the capsule core, the precipitation bath was maintained at 4°–15°C, requiring that the hexadecane (melting point 19°C) overlayer be replaced with dodecane (melting point −5°C). Chitosan is a basic, amine polysaccharide [$\beta(1{\rightarrow}4)$ 2-amino-2-deoxy-D-glucan]; the deactylated form of chitin. Chitosan, dissolved in acidified PBS (pH < 6.5), was mixed with an equal volume of cell suspension for the capsule core solution. The chitosan immobilization matrix was formed by precipitation after an intracapsular increase in pH consequent to capsule droplet immersion in the PBS + L101 precipitation bath (pH 7.2).

A more recent development has been the submerged nozzle–liquid jet extrusion process, depicted schematically in Fig. 2, to produce HEMA–MMA microcapsules of even smaller diameter (300–600 μm). In this process, the polymer solution and cell suspension are delivered to the tip of a coaxial needle assembly. However, the needle assembly remains stationary, while the hexadecane (or dodecane) flows uniformly, coaxially to the needle assembly to shear off each capsule droplet [108]. It is this inherently higher capsule-shearing force that produces smaller capsules. Cells are suspended in medium containing either 15% (w/v) Ficoll-400, 25% (v/v) Matrigel, or 1.1 g/ml agarose (at flow rates of 5, 8.5, or 5.5 μl/min, respectively); the polymer solution was at 11.7, 18.7, or 12.5 μl/min, respectively. The hydrophobic shearing liquid was typically hexadecane; however, for Matrigel and agarose capsules, because of the need for cooling to control gelation, dodecane was used. The flow rate of hexadecane/dodecane affected both the diameter of the resultant capsules as well as their production rate; the typically used 130 ml/min resulted in capsule diameters of close to 400 μm at a rate of 96 capsules per minute. Capsule droplets were co-extruded into the PBS + L101 precipitation bath during a time period lasting up to 1 h, followed by two fresh PBS (no L101) washing steps of 20 min each.

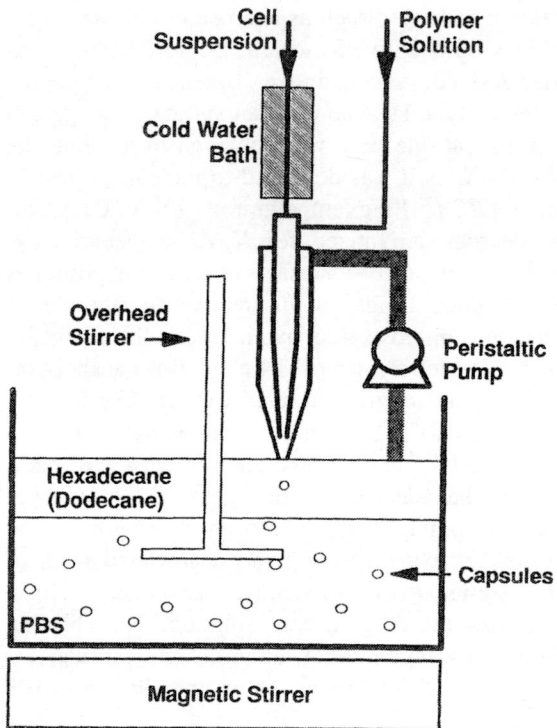

Fig. 2 Schematic drawing of the encapsulation apparatus. Cells and polymer solution are pumped by syringe pumps through a needle assembly. A peristaltic pump recirculates the hexadecane or dodecane. The capsules are kept in suspension by a magnetic overhead stirrer. (From Ref. 108.)

C. Hydroxyethyl Methacrylate–Methyl Methacrylate Microcapsules

1. *Membrane Structure and Formation*

The HEMA–MMA polymer capsules, prepared by either submerged jet process, resulting in large- or small-diameter microcapsules, were morphologically similar. The polymer walls were of an asymmetric morphology similar to ultrafiltration membranes, but in a spherical geometry, instead of the typical hollow fiber or flat-sheet geometry of the latter. The capsule walls were about 150- or 50-μm thick for large-diameter (Fig. 3A) or small-diameter (see Fig. 3B) capsules, respectively. The wall consisted of a thin outer skin, a macroporous sublayer, a thick, seemingly dense, layer, and an inner skin (see Fig. 3C). The dimensions of these regions for typical large- and small-diameter

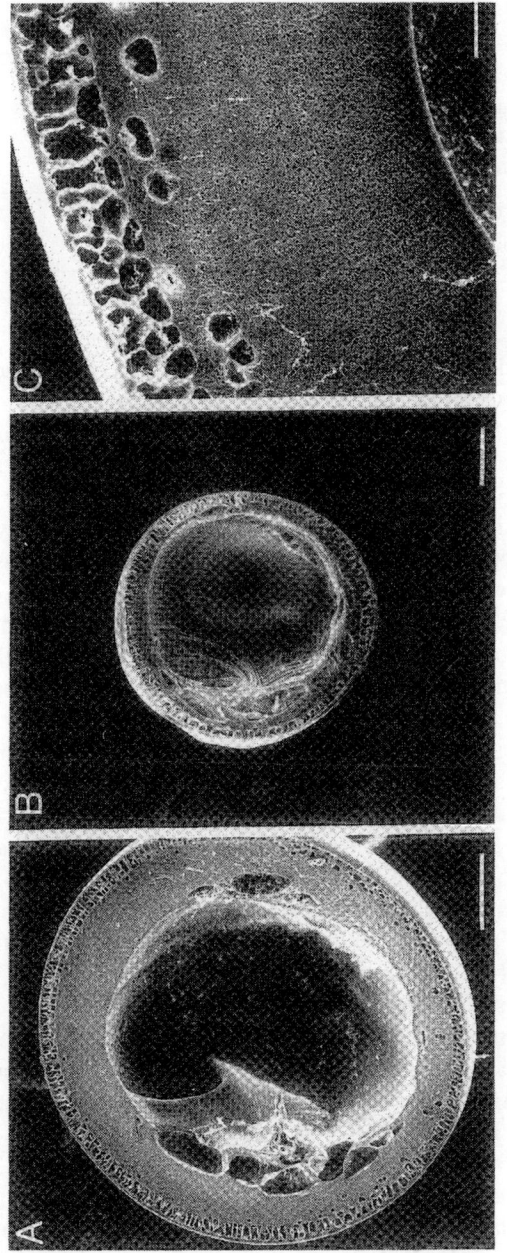

Fig. 3 Scanning electron micrographs of HEMA–MMA microcapsules: (A) large-diameter capsule (o.d., 660 μm); (B) small-diameter capsule (o.d., 450 μm); (C) wall structure of a large capsule. Magnification bars: (A) 100 μm, (B) 100 μm, (C) 25 μm.

Table 1 Dimensions of Polymer Capsule Regions

Capsule size (μm)	Wall thickness (μm)	Thickness of outer skin (μm)	Thickness of macroporous region (μm)	Thickness of dense region (μm)
750	150	5–8	30	110
300	50	2.5–6	20	30

capsules are presented in Table 1. The capsules were frequently eccentric, although this was dependent on the viscosity and density enhancer Ficoll-400, which was added to the capsule core solution [106]. For example, capsules prepared with a core consisting of medium containing 5% (w/v) Ficoll-400 showed a high level of eccentricity, with corresponding polymer wall thicknesses of 65 $\pm$ 30 μm to 340 $\pm$ 120 μm (n = 32 capsules). By increasing the amount of Ficoll-400 from 5 to 20% (w/v), a significant decrease in eccentricity was observed, with corresponding polymer wall thicknesses of 70 $\pm$ 25 to 160 $\pm$ 50 μm (n = 50 capsules). The inclusion of this additive in the capsule core presumably resulted in a reduction in the mixing of the core and polymer solutions, and the denser capsule core was more completely surrounded by polymer.

Large-diameter capsules, which were near 900–1000 μm in diameter immediately after preparation, shrank to about 750 μm in diameter over a 1-week period; a similar shrinkage ($\sim$ 15%) was observed in small-diameter capsules [109]. Corresponding to this decrease in capsule diameter were changes in the morphology of the polymer wall. One day after preparation, the thick inner layer of capsule polymer wall was microporous. This layer increased in density, over the 1-week period, as residual solvent PEG-200 was extracted from the polymer wall and as rearrangement of the microstructure of this region occurred to minimize surface area. Similar, but less noticeable effects, because of the smaller scale, may also occur in the skin layers. These morphological changes in capsule wall structure are presumed to have significant effects on the microcapsule permeability. The permeability to large molecular weight complement proteins, namely, C1q (400 kDa), as detected by the lysis of encapsulated, sensitized red blood cells, was observed for only up to 3 days after preparation; after this time no lysis was observed [110]. Similarly, the ability to retain high molecular weight solutes is impaired in the first few days after encapsulation (e.g., for solute-release studies). These complement results have implications for the immunoprotection of the encapsulated cells after their transplantation; presumably, a similar in vitro incubation period is needed before implantation to ensure that the capsule permeability can protect the transplanted cells from host complement-mediated lysis.

Each region of the HEMA–MMA capsule wall has a particular function. The skin layers are expected to provide selectivity to molecule permeation. The inner skin as well as an outer skin may provide a permeselective barrier, even if mechanical or chemical change occurs at the outer skin after implantation. The macroporous region is effective for the permeation of nutrients, cellular waste products, and cell-derived biomolecules through the capsule polymer wall, while it and the "dense" layer provide mechanical support. Capsules with well-centered cores are expected to be mechanically more stable because weak spots can be avoided. These capsules would withstand any internal pressure exerted by proliferating cells or by a swollen immobilization matrix, and would protect cells more completely, after their implantation, from cellular infiltration.

The outer skin and inner porous sublayer are formed by different mechanisms: gelation and liquid–liquid phase separation, respectively. When the polymer solution contacts the nonsolvent in the precipitation bath, polymer solvent is quickly depleted from the surface of the capsule, resulting in a high surface polymer concentration, but initially, a low nonsolvent concentration. This leads to gelation, rather than precipitation, as the two-phase region (on the polymer–solvent–nonsolvent ternary phase diagram) is not crossed. Contact in the overlayer with the nonsolvent hexadecane, may also increase the polymer surface concentration, even before contact with the precipitation bath, enhancing gelation. With time, the polymer solution increases in nonsolvent concentration as the nonsolvent diffuses in, crossing the two-phase boundary to form the rest of the capsule wall. The outer skin limits the inward diffusion of the nonsolvent, affecting the formation of the polymer wall by liquid–liquid phase separation. For solvent and nonsolvent combinations, in which phase separation occurs rapidly, finger-like macropores are formed, under the outer skin. Precipitation at a slower rate, owing to resistance by the newly precipitated polymer wall to nonsolvent diffusion, results in the denser microporous layer under the macroporous region.

2. Permeability

To provide an environment that is favorable for the maintenance of cellular viability and function, the capsules must be permeable to small molecules (nutrients and metabolites: glucose, oxygen, and lactate) and intermediate or large molecules (growth factors [~ 13–30 kDa] or transferrin [80 kDa]). Yet following implantation, the capsules must be impermeable to components of the host immune system (IgG antibodies [150 kDa] or complement components, such as C1q [410 kDa]) such that the encapsulated cells are not attacked by the host immune system on a molecular level. Furthermore, for a functioning cell-based biomolecule delivery system, the polymer wall must be permeable to the therapeutic product (e.g., dopamine [153 Da], insulin [6000 Da, if a monomer], or growth hormone [23 kDa]).

To estimate the permeability of HEMA–MMA capsules with an aqueous core, an aliquot of 200–300 one-week-old capsules were incubated in PBS solutions containing one of several molecules for 3 days, and its time-dependent release monitored in the extracapsular PBS sink [100]. The test molecules were [^{3}H]glucose (180 Da), [^{3}H]inulin (5.2 kDa), albumin (69 kDa), or alcohol dehydrogenase (ADH; 150 kDa). There was a decrease in the mass transfer coefficient with increasing molecular weight, and a significant drop between 69 and 150 kDa (log–log plot). This suggested that the molecular weight cutoff of the membrane was on the order of 100 kDa.

The degree of macroporosity of the microcapsule polymer wall was altered by changing the normal polymer precipitation conditions. The PBS/L101 precipitation bath was replaced with 0.3 M aqueous glycerol or 15% (v/v) water added to the PEG-200 polymer solutions or both [100]. Isotonic aqueous glycerol increased the precipitation rate, as the absence of salts increased solvent–nonsolvent compatibility. The wall thicknesses were not changed much, but the capsules precipitated in aqueous glycerol were more macroporous, with a corresponding doubling of the mass transfer coefficient. On the other hand, capsules precipitated from polymer dissolved in PEG-200 and water (bringing the polymer solution closer to the point of precipitation), showed a difference in wall structure that did not translate into a difference in the mass transfer coefficient. This result implied that the primary resistance to diffusion was the skin layer. Capsule permeability also increased with capsule diameter for the small-diameter process; the permeability coefficients to the model protein horseradish peroxidase increased from 2.1×10^{-10} cm^2/s to 9.0×10^{-10} cm^2/s for capsule diameters of 450 and 660 μm, respectively [109]. This was related to differences in the rate of precipitation associated with the differences in diameter, wall thickness, and eccentricity of the different capsules.

More recent capsule studies have been directed at examining the permeability of capsules at the single-capsule level, because of the observation of significant fibrinogen release from a subset of presumably defective capsules containing HepG2 cells (see later [111]). The broad distribution in fibrinogen release was consistent with the distribution in the permeability of similar capsules (without cells) to the model protein, horseradish peroxidasc (HRP; 40 kDa; Fig. 4). Here, the enzyme was loaded by diffusion from an external incubation solution, and the release was followed using 3,3',5,5'-tetramethylbenzidine (TMB) as a chromogenic substrate, with one capsule per well of a 96-well plate: the TMB provided the sensitivity needed to detect the protein release at the single-capsule level. Therefore, it is probable that the heterogeneity of cap-

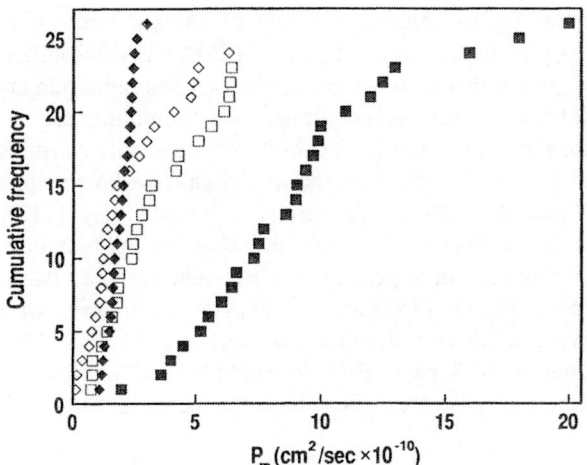

Fig. 4 Distributions in the estimated permeability coefficient of HEMA–MMA microcapsules with different outside diameters. Each point was calculated from the release curve of a single capsule. Outside diameters of microcapsules are 450 μm (open diamond), 500 μm (solid diamond), 580 μm (open square), and 660 μm (solid square). β was assumed to be the same for each capsule of the same batch. (From Ref. 109.)

sules to the cell-derived proteins is at least partly due to variability in the permeability of the capsules to proteins of similar molecular weight. The application of this technique, using a higher molecular weight enzyme ($\sim$100–300 kDa), would be useful for assessing of the importance of this effect for fibrinogen, in particular. The variation in HRP permeability was dependent on the care taken in setting up the encapsulation assembly. It was possible to significantly reduce this variability by careful processing. However, reengineering of the coextrusion nozzle will be needed to make the process more reproducible on a regular basis.

The capsules used in these single-capsule permeability studies differed slightly from those studied earlier [100] because, instead of containing an aqueous core, they contained a core of complete tissue culture medium and 20% (w/v) Ficoll-400 or 50% (v/v) Matrigel; in certain cases, there were suspended cells. Such changes in the capsule core solution change the rate of precipitation of the polymer; consequently, the permeation characteristics of the inner skin are presumably altered [109]. The consequences of this difference in precipitation rate are only beginning to be appreciated.

Recent studies have expanded our ability to control microcapsule permeability. The polymer solvent PEG-200 was replaced with triethylene glycol (TEG;

150 Da) with a lower diffusivity to enhance the rate of solvent removal at precipitation. Microcapsules prepared from a 9% (w/v) HEMA–MMA solution in TEG exhibited surface depressions that were greater in area and more numerous than those of the 10% (w/v) capsule preparations. This morphological observation correlated with an increase in the permeability of individual capsules from $0.39 \times 10^{-9} cm^2/s$ to 26×10^{-9} cm^2/s to the model enzyme horseradish peroxidase (HRP) for 10 and 9% capsule preparations, respectively [112]. These surface depressions do not appear to cross the skin, at least at high magnification in a SEM, so they do not appear to be pores. On the other hand, the individual capsule permeability also increased to 11.3×10^{-9} cm^2/s with the addition of a pore-forming additive poly(vinyl pyrrolidone) (PVP) to 10% HEMA–MMA in TEG solution (0.3 parts PVP to HEMA–MMA, w/w). It remains to be seen how these changes affect the molecular weight cutoff.

IV. BIOLOGICAL PROPERTIES: SPECIFIC EXAMPLES AND ISSUES

Microencapsulation processes using HEMA–MMA have been developed solely for living cells. Initially, our interest was in ensuring that enough cells survived the encapsulation process and, as appropriate, that they proliferated within the capsule and remained viable long term. The maintenance of differentiated functions by a variety of encapsulated cells was also studied, for this was the means by which the potentially therapeutically active biomolecule would be produced. Also important has been the effect of the intracapsular environment, as determined by the polymer wall on cell arrangement and their consequent viability and function.

A. Cell Survival and Proliferation

The success of encapsulation is initially dependent on the number of viable cells that are entrapped within the capsule on an absolute basis, or compared with the theoretical number of cells fed to each capsule—the encapsulation efficiency. It may be desirable in certain situations, such as when few cells are encapsulated, that cell proliferation occur within the microcapsule. Such proliferation also indicates a reasonable microenvironment within the capsule core for the cells. The rate of proliferation, however, may be lower than under normal tissue conditions, in part because the environment inside the capsule and the normal tissue culture dish are different. On the other hand, cells in an unproliferative state, may function in a more differentiated manner, with enhanced production of the desired product. For other cell types, such as pancreatic islets, which do not proliferate in culture, the initial encapsulation efficiency must be high to provide adequate cell mass within the capsule.

A variety of cell types, including the Chinese hamster ovary (CHO) fibroblast [106], rat pheochromocytoma, PC12 [113], human hepatoma, HepG2 [48,114], murine fibroblast L929 [115] cell lines, primary rat hepatocytes [116], and rat islet tissue [117], have been encapsulated within HEMA–MMA microcapsules. These studies have shown that the cells survive the encapsulation procedure, despite the exposure to shear forces and organic solvents or nonsolvents, and grow or function afterward for periods from 2 to at least 6 weeks.

Generally, only a fraction of the cells delivered to the needle assembly actually become enclosed by the polymer wall. For example, when HepG2 cells were encapsulated with a medium augmented with 20% Ficoll-400 (regular capsules), about 20–25% of theoretical number of cells actually were entrapped within the capsule core [114]. However, if the cell immobilization matrix, Matrigel, was used, the encapsulation efficiency for HepG2 cells increased to near 50%. The gelling properties of this cell attachment matrix may have protected the cells from the forces inherent in the encapsulation process (e.g., shear or PEG-200 contact), although fluid mechanic differences associated with two-component droplet formation were likely more important. The sensitivity of the cells to the encapsulation process also affects the encapsulation efficiency: CHO cells (in Ficoll) had a higher encapsulation efficiency ($\sim 65\%$ [106]) consistent with their more robust character. The encapsulation efficiency was also dependent on the quality of the encapsulation run, with better-centered capsules having higher efficiencies.

Cellular proliferation within microcapsules has been examined directly by counting the number of cells from an aliquot of about 25 capsules and indirectly by examining the metabolic activity of cells within a single capsule, using the MTT assay [118]. PC12 cells encapsulated, at a low cell density of 4×10^5/ml showed an initial decrease in MTT metabolic activity, followed by a quiescent period from days 7 to 21, with an increase from day 21 to day 28 [113]; there was no indication of a plateau. However, when these cells were encapsulated at a high cell density of 4×10^6/ml, there was an initial decrease (up to day 3) in the formazan absorbance, which then increased and reached a constant level after day 14 (at a level similar to that reached by the lower-density capsules; corresponding to about 3000–4000 cells per capsule). The HepG2 cells, encapsulated at 5×10^6/ml in complete tissue culture medium, without an attachment substrate (with only 20% [w/v] Ficoll-400) showed a slight, but insignificant, increase in cell number from day 1 to day 4 (860 $\pm$ 60, 1400 $\pm$ 330, respectively) and decreased thereafter (Fig. 5a). Correspondingly, the initially high formazan signal dropped by day 14 to a value that was similar to blank capsules without cells (see Fig. 5B). HepG2 cells, similarly encapsulated, but at a lower cell density of 1×10^6/ml, increased in number per capsule, over the initial 1-week period, to reach values similar to those in

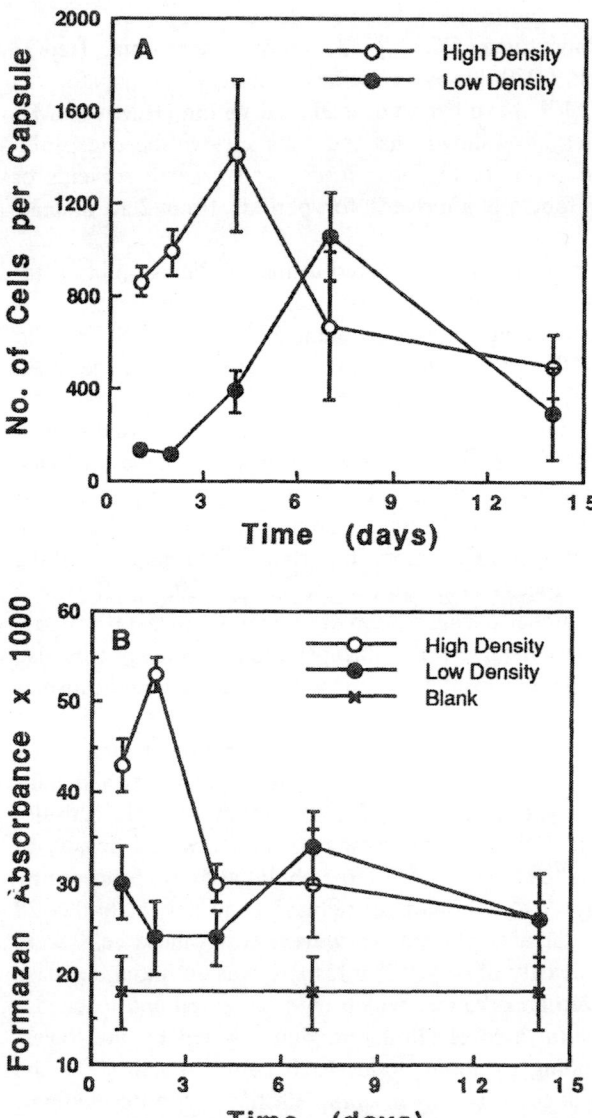

Fig. 5 Time-dependent changes in (A) the cell number and (B) metabolic activity of the HepG2 cells in regular capsules when encapsulated at low cell density (1×10^6/ml, closed symbols) or high cell density (5×10^6/ml, open symbols). The mean $\pm$ SD for three different batches of capsules is shown for both low- and high-density capsules. There was a significant ($p < 0.05$) decrease in the cell viability after cells numbers reached more than 1000 per capsule. (From Ref. 114.)

high-density capsules at this time (see Fig. 5A). These results suggest that the intracapsular environment (without a matrix) could maintain cellular proliferation and metabolic activity only up to a certain, limited extent (approximately 500 cells per capsule or 17×10^6/ml internal volume). Intracapsular space limitations may limit the number of cells per capsule, or the arrangement of cells within the capsule may limit diffusion of nutrients and metabolites (see Section III.C), such that more cells per capsule would have a lower metabolic activity per cell.

In contrast, when the cells were encapsulated along with an attachment substrate (Matrigel), cells proliferated over the 2-week period, reaching values of 4350 ± 860 cells per capsule by day 14 (Fig. 6A) [114]. However, whereas the number of cells per capsule increased, the formazan absorbance from these (large) capsules did not parallel this increase, but rather, decreased slightly after 2 weeks (see Fig. 6B). The relatively constant metabolic activity may indicate that the cells are correspondingly more differentiated while within Matrigel. HepG2 cells encapsulated at the high density within Matrigel in small-diameter capsules ($\sim 400 \ \mu m$) maintained a constant level of metabolic activity over 3 weeks, similar to the phenomenon described earlier for the large-diameter capsules, but not decreasing after 2 weeks as in the large-diameter capsules [108]. Cell numbers per small-diameter capsule showed that there was a increase in cell number for small-diameter capsules similar to that for large-diameter capsules (but fewer cells per capsule).

The picture that is emerging is that cells grow to fill the capsule space, but to the extent limited by the nutrient supply and intracapsule and intercellular diffusion gradients (see later discussion). They then may maintain a roughly constant cell number and cell activity level thereafter, provided the intracapsule environment is suitable (e.g., anchorage-dependent cells have the necessary attachment substrate). The rate of proliferation is likely less than that in normal tissue culture, as is the rate of metabolic activity on a per cell basis (as exemplified by MTT conversion). How long this steady cell number and activity is maintained is unknown, but MTT conversion has been observed for several months after encapsulation. Interestingly, cells are not found outside of the capsule, except on rare occasions, suggesting the capsule wall is strong enough to limit proliferation to the intracapsular space. Cells that do not receive adequate nutrients or that are apoptotic, will die within the capsule. These products of necrosis remain in close proximity to viable cells and may affect their viability. Eventually, proteases released at cell death will have sufficiently degraded the cellular debris such that it could be removed from the capsule core by diffusive transport. However, the cell cycle and cell death caused by an inadequate nutrient supply are dynamic processes, implying that the capsule core would presumably never be free of cellular debris. The implications of this are not yet understood.

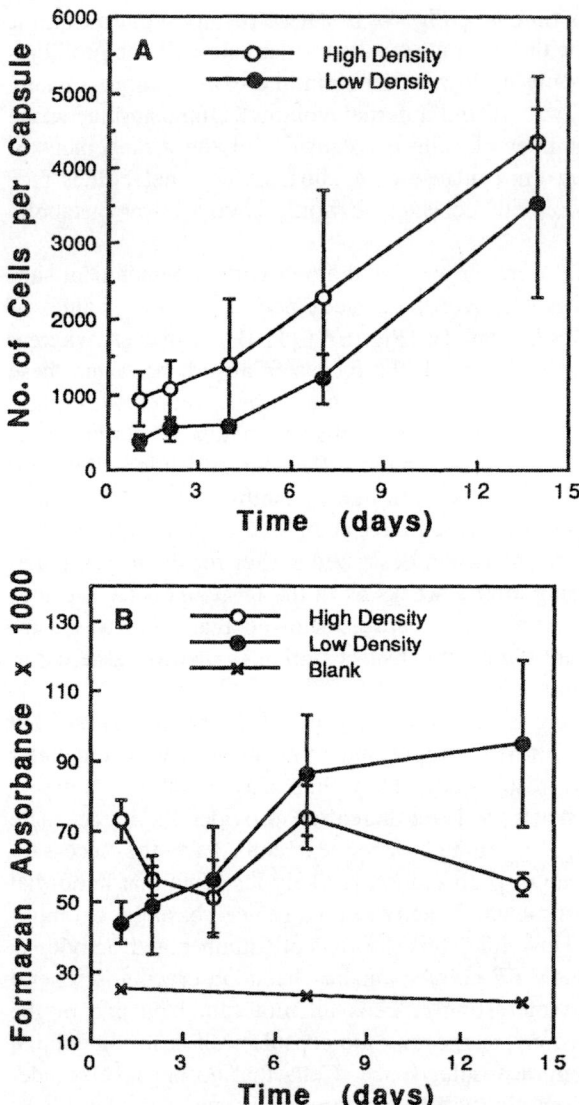

Fig. 6 Time-dependent changes in the mean ± SD of (A) cell number and (B) metabolic activity of the HepG2 cells in Matrigel capsules when encapsulated at a low cell density (1×10^6/ml, closed symbols) or a high cell density (5×10^6/ml, open symbols.) There was a steady increase in cell numbers in both low- and high-density capsules. Cell numbers in excess of 4000 per capsule were obtained. An active metabolic state was detectable in Matrigel capsules. (From Ref. 114.)

B. Biomolecule Release

For the ultimate application of encapsulated cells as a bioartifical organ or physiologically controlled biomolecule-delivery system, the cells must not only survive the encapsulation process and remain viable, but they must also express their differentiated functions. The ability of the HEMA–MMA microcapsule to support the functional state of the cells has been assessed by the quantification of encapsulated cell-derived biomolecule release into the extracapsular milieu. Protein release is described in more detail elsewhere [119].

Encapsulated PC12 cells were depolarized (i.e., incubated in a high potassium and tyrosine-release medium) after encapsulation (Fig. 7), and there was a statistically significant increase in amount of dopamine released from capsules from 7 to 28 days (logarithmic increase), consistent with an increase in cell number or MTT signal. Usually, there was no detectable release of dopamine in a basal RPMI medium, suggesting the presence of the intact physiological regulatory machinery.

As a protein-release example, human hepatoma (HepG2) cells, were used as a model for hepatocytes [114]. Four plasma proteins that span the molecular weight range of interest (α_1-acid glycoprotein [AG; 42 kDa], α_1-antitrypsin [AT; 52 kDa], haptoglobin [Hap; 98 kDa], and fibrinogen [Fbg; 340 kDa]) were released into the surrounding medium by HepG2 cells from an aliquot of about 100 Matrigel capsules. Protein release curves over a 2-week period

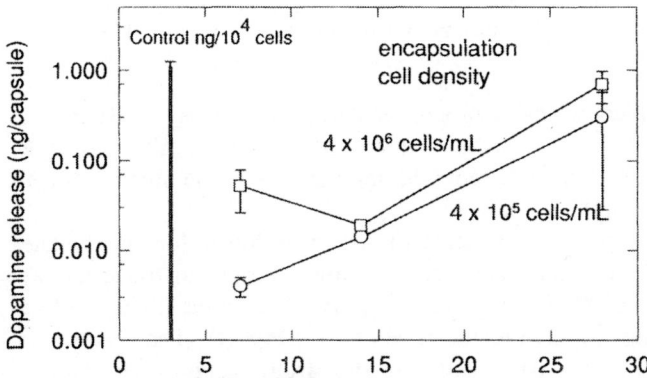

Fig. 7 Dopamine release within 1 h into K^+/Ca^+ release medium from encapsulated PC12 cells (ng/capsule) at two initial encapsulation cell densities—4×10^5/ml (open circles) and 4×10^6/ml (open squares)—in comparison with that from control PC12 cells (bar, ng/10^4 cells) in conventional tissue culture. Note different units for the capsules and the unencapsulated cells, but both use the same numerical scale; Mean ± SD, $n = 3$. (From Ref. 113.)

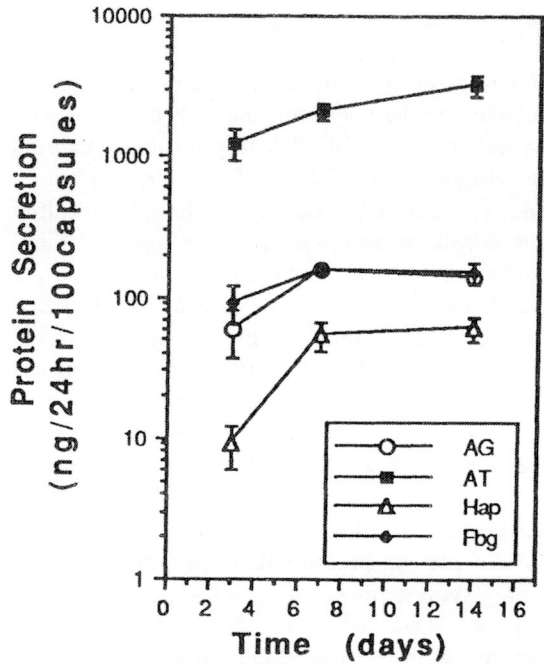

Fig. 8 Mean ± SD release rate of AG, AT, Hap, and Fbg by encapsulated HepG2 cells on days 3, 7, and 14 (ng/24 h/100 capsules; $n=3$). There was a significant increase in release rates for all proteins between days 3 and 7. Only AT secretion was increased significantly between day 7 and 14 in this set of experiments. (From Ref. 114.)

(Fig. 8) showed significantly more protein at day 7 than at day 3. There was no further increase in the amount secreted at day 14, except for AT. The higher amount of protein secreted is consistent with the increase in the number of cells per capsule (see foregoing).

The amount of Fbg released, relative to AG, was lower for encapsulated cells than for control, unencapsulated cells within Matrigel in tissue culture, consistent with a sieving effect of the polymer membrane on the higher molecular weight protein, Fbg, relative to the smaller AG (Fig. 9). The corresponding release rates for AT and Hap (relative to AG) from encapsulated and control cells were similar suggesting that, first, the cells from these two different environments may be compared in this manner and, second, that the result for Fbg is not likely due to an altered cell phenotype. The HepG2 cells within small-diameter capsules (~ 400 vs. ~ 800 μm) also showed significantly lower Fbg/AG secretion rate ratio when compared with unencapsulated cells [108].

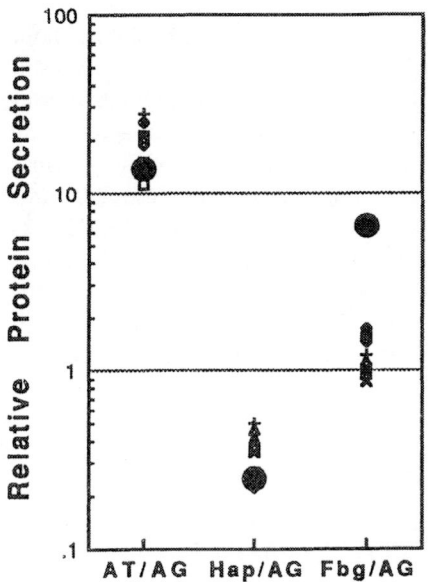

Fig. 9 Release of AT, Hap, and Fbg relative to AG (AT/AG, Hap/AG, and Fbg/AG, respectively) by the control (i.e., unencapsulated) and encapsulated HepG2 cells (large capsules). The relative release rates were obtained by dividing specific release rates by the AG release rate. Solid circles correspond to the average relative release rate of three independent control cultures (in duplicate), whereas the other symbols correspond to individual relative release rates from three batches of capsules on days 3, 7, and 14. Note that only the Fbg/AG was significantly lower for encapsulated when compared with control cells. (From Ref. 114.)

The dynamics of insulin release by microencapsulated pancreatic islets on glucose stimulation, although demonstrating the ability to encapsulate viable tissue fragments consisting of nonproliferating cells, also illustrated the effect of the polymer wall on stimulus-elicited protein release. The glucose enhancement of insulin secretion, expressed as a stimulation index (the ratio of insulin secretion for high to low glucose), was similar for both encapsulated and unencapsulated control islets as they responded to a higher external glucose concentration by increasing their insulin release rates. However, the rates of insulin release were significantly lower (3.5–5 times) for encapsulated islets, compared with the control unencapsulated islets. This reduction was attributed to the polymer capsule wall acting as a barrier to insulin diffusion, because on freeing encapsulated islets at the end of a 3-h static incubation in the low-glucose medium, there was a concomitant liberation of large quantities of insulin. Although encapsulated islets released insulin at rates of 35 ± 4 μU per islet per hour, the freed islets released insulin at rates of 62 ± 18 μU per islet per hour ($n = 2$); similar to that for control islets (75 μU per islet per hour). As a result of the diffusion resistance associated with the capsule wall, the high intracapsular insulin concentration is also expected to down-regulate insulin secretion by the encapsulated islets, exacerbating the diffusion barrier effect of the polymer wall. Because the stagnant water layer surrounding the islets in these 800-μm–diameter capsules was a significant part of the diffusion resistance, it is expected that the smaller capsules prepared using the submerged nozzle–liquid jet extrusion process, would be preferred. These are currently being investigated, and the initial experience has been promising. Following this trend toward diminishing the stagnant water layer, as well as the membrane thickness, even further, a conformal-coating process that results in islets coated with a very thin layer of HEMA–MMA is currently under development. It is expected that these ultrasmall "capsules" will have virtually no insulin diffusion limitations.

C. Intracapsule Environment

Diffusion limitations associated with the capsule (because of its wall and related to its size) affect not only the transport of products, such as insulin, but also the transport of metabolites (nutrients and waste products). The corresponding changes in the intracapsular environment (e.g., oxygen, nutrient, or metabolite concentrations) and the presence of products from dying or dead cells further influence cell behavior. These are expressed both through their morphology or three-dimensional arrangement and protein secretion. The three-dimensional arrangement further influences the transport limitations, as the intracapsular or intercellular space may be rate-limiting. The changes in cell arrangement and morphology are illustrated by anchorage-dependent cells, such as CHO cells [106] and HepG2 cells [48], and by anchorage-independent PC12 [113] cells placed into the nonadherent environment of the HEMA–MMA capsule. These grow as spherical or elliptical aggregates, of varying sizes, within intracapsular polymer pockets. This was consistent with several reports on unencapsulated cell (e.g., hepatocyte) culture under nonadherent conditions [120]. In the absence of a substrate for the cells to attach and grow as in a normal tissue culture situation, cells attach to each other or to extracellular matrix components, such as laminin, fibronectin, and collagen, to arrange into a spheroidal shape [121]. When HepG2 cells (without coencapsulated Matrigel) are arranged within large aggregates ($\sim 300 \times 100$ μm) in capsules after about 3–7 days in vitro, the cells at the center of these aggregates became necrotic, whereas cells in an outer shell, two to five cells in thickness (~ 30 μm) remained viable (Fig. 10). Similar outer shells of viable cells surrounding ne-

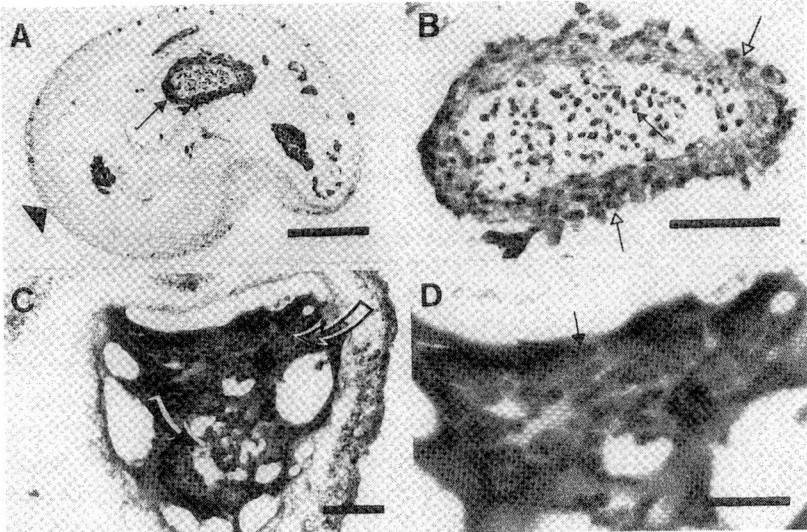

Fig. 10 Light micrographs of HepG2 cells in HEMA–MMA microcapsules (A,B) at day 7 without, or (C,D) at day 14 with Matrigel. (A) Large arrowhead indicates capsule outline and small arrow indicates cellular aggregate with necrotic core; (B) aggregate is shown in more detail. (C) Viable cells are shown with open arrow and necrotic region is indicated with closed arrow; (D) shows viable region in more detail. Magnification bars: (A) 150μm, (B) 50 μm, (C) 75 μm, (D) 50 μm. (From Ref. 133.)

crotic cores were observed within aggregates of encapsulated PC12 cells after about 1 week in vitro. This morphology was maintained throughout the 6-week observation period. Transmission electron microscopy (TEM) observations of PC12 cell spheroids within HEMA–MMA capsules showed the presence of dopamine granules in the healthy cells of the outer shell.

One consequence of the intracapsule microenvironment and cell morphology is the sensitivity of the cells to exogenous influences, such as cytokines. Murine connective tissue L929 cells, when encapsulated, were unaffected by tumor necrosis factor (TNF-α) in the external medium at concentrations up to 100 ng/ml over a 3-day culture period [115]. These cells are routinely used in monolayer culture for a cytotoxicity assay for TNF-α [122]. Tumor necrosis factor-α (Mw ~ 51 kDa as a trimer), presumably permeated the capsule, but the lack of a response by the normally TNF-sensitive L929 cells probably relates to the three-dimensional arrangement of the cells. In the nonadherent environment of the HEMA–MMA capsule, the cells possess a different morphology from the elongated fibroblast structure of the TNF-sensitive monolayer [123]. Furthermore, the interaction of cellular receptors with extracellular matrix components is key to their response to cytokines [124]. The TNF resistance of "tight" L929 colonies has been attributed to their inability to incorporate secreted fibronectin in their extracellular matrix, unlike TNF-sensitive "loose" colonies [125]. It is also possible that environmental stresses of the capsule interior (e.g., nonadherence) affect L929 cells to the extent that they produce heat-shock protein, hsp70 [126], which confers protection against the cytotoxic effects of TNF [127]. In vitro tests of the effects of cytokines on encapsulated cells are especially useful because they may predict the response in vivo on possible contact with these and other cytokines during an inflammatory reaction.

The environment within the HEMA–MMA capsule core may be altered by the coencapsulation of extracellular components, such as a cell attachment substrate, Matrigel, or the coencapsulation of an immobilization matrix, agarose. Whereas the former provides a substrate for cell attachment (which HEMA–MMA does not), the latter may simply improve the distribution of cells within the capsule and minimize intercellular diffusion limitations. Matrigel supports hepatocyte and other cell differentiation to a degree beyond that reported for individual extracellular matrix components [128–130]. Agarose has been used to enclose islets for their use in a bioartificial pancreas [83] and has enhanced the growth and phenotypical expression of immobilized chondrocytes [131].

Both anchorage-dependent HepG2 cells [48,114] or anchorage-independent PC12 cells [132] encapsulated within a Matrigel core, 1 day after encapsulation, were viable and uniformly distributed throughout the capsule as individual cells or small cellular aggregates. Although HepG2 or PC12 cellular aggregates, of sizes similar to those formed without Matrigel, were observed after 7 days in vitro, with Matrigel they did not develop necrotic cores. The benefits of Matrigel were realized through the more uniform distribution of individual cells and aggregates in which the cells attached not only to each other, but also to extracellular matrix components, such that cell–cell association was limited. In turn, this affected the local cell density and, hence, consumption of nutrients, production of wastes, and cellular metabolic state.

Within agarose-containing capsules, individual and small aggregates of viable PC12 cells were distributed even more uniformly than in Matrigel-containing capsules. For PC12 cells within agarose there was a significantly higher formazan absorbance on day 1 than within Matrigel. A higher level of metabolic activity was sustained by encapsulated PC12 cells immobilized within agarose, as compared with cells within Matrigel; the former also increasing over the 3-week culture period. It appeared, therefore, that the gelling conditions of agarose facilitated the entrapment of more cells within capsules that then proliferated. From the perspective of cell attachment, agarose is relatively inert. It remains to be seen what happens with anchorage-dependent cells or whether addition of extracellular matrix components, such as laminin, can enhance the beneficial effects of agarose, even for anchorage-dependent cells.

Because Matrigel is derived from a mouse sarcoma, its stability and biocompatibility in vivo may be questionable, as contact with inflammatory cell-derived proteolytic enzymes, such as collagenase or elastase could release Matrigel fragments the antigenicity of which (low molecular weight and xenogenicity) might exacerbate any inflammatory reaction. These concerns will likely preclude the use of Matrigel in vivo, despite its positive cellular effects in vitro. The issue of xenogenicity may be circumvented by using host species-specific extracellular matrix components, such as collagen that is derived from human umbilical cords, which is now commercially available. On the other hand, agarose, as a cell immobilization matrix in the capsule core, might elicit less of an immune and inflammatory response (if complement activation is not a problem).

V. FUTURE DIRECTIONS

We have reached the stage at which microparticles containing immobilized or encapsulated cells can be produced with natural (e.g., alginate) or synthetic polymers (e.g., HEMA–MMA) using water or organic solvents. Cell survival during the process and preserving cell viability or function afterward, at least in vitro, is not an issue. There is much to do to optimize the processes that are used, to improve their reproducibility, to be able to control permeability or molecular weight cutoff, or simply to reduce the capsule diameter. These are primarily technical problems that will become resolved as the processes are scaled-up and as further understanding of the processes are achieved.

There also is much to do in understanding how the capsule wall affects the encapsulated cell behavior. The complexities of the interrelations among wall permeability, extracellular matrix, cell density, and so forth, are beginning to be recognized, but we are still a long way from knowing how to design an encapsulated cell system. We have learned not to expect cell behavior that is comparable with what is seen in conventional tissue culture—the three-dimensional arrangement of cells inside a capsule creates a unique microenvironment and may even be more physiological than that on a polystyrene dish. We are at the stage where any cell can be encapsulated and, provided the conditions are reasonable, its differentiated function can be retained in vitro.

On the other hand, we are only beginning to understand what happens in vivo. After implantation, the tissue reaction, the implant site, the presence of cytokines, the release of antigenic proteins or cell debris, the presence of adsorbed protein, the capsule chemistry and durability, all intervene to affect the encapsulated cell behavior, its ability to secrete the desired product, and the duration of viability. Although it will take considerable effort to sort out these interrelations, the therapeutic benefit of cell transplantation warrants the effort.

ACKNOWLEDGMENTS

We acknowledge the financial support of the Medical Research Council of Canada, the Natural Sciences and Engineering Research Council, the Canadian Parkinson's Foundation, and the Canadian Liver Foundation, the latter in providing a fellowship to JEB. We also acknowledge the assistance of Mr. Vlad Horvath, and the advice of Professor U. DeBoni.

REFERENCES

1. D. W. Scharp, P. E. Lacy, J. V. Santiago, C. S. McCullough, L. G. Weide, L. Falqui, P. Marchetti, R. L. Gingerich, A. S. Jaffe, P. E. Cryer, C. B. Anderson, and M. W. Flye, Insulin independence after islet transplantation into type I diabetic patient, *Diabetes 39*:515 (1990).

2. W. J. Freed, Neural transplantation: Prospects for chemical use, *Cell Transplant. 2*:13 (1993).

3. S. R. Winn, J. P. Hammang, D. F. Emerich, A. Lee, R. D. Palmiter, and E. E. Baetge, Polymer encapsulated cells genetically modified to secrete human nerve growth factor promote the survival of axotomised septal cholinergic neurons, *Proc. Natl. Acad. Sci. USA 91*:2324 (1994).

4. K. J. Lafferty, Circumventing rejection of islet grafts: an overview, *Transplantation of the Endocrine Pancreas in Diabetes Mellitus* (R. van Schilfgarde and M. A. Hardy, eds.), Elsevier, New York, 1988, p. 279.

5. C. K. Colton and E. S. Avgoustiniatos, Bioengineering in development of the hybrid artificial pancreas, *Trans. ASME 113*:152 (1991).

6. A. Sun, W. Parisus, H. Macmorine, M. V. Sefton, and R. Stone, An artificial endocrine pancreas containing cultured islets of Langerhans, *Artif. Organs 4*:276 (1980).

7. H. Yang, H. Iwata, H. Shimizu, T. Takagi, T. Tsuji, and F. Ito, Comparative studies of in vitro and in vivo function of three different shaped bioartificial pancreases made of agarose hydrogel, *Biomaterials 15*:113 (1994).

8. J. J. Altman, D. Houlbert, A. Chollier, A. Leduc, P. McMillan, and P. M. Galletti, Encapsulated human islet transplants in diabetic rats, *Trans. Am Soc. Artif. Intern. Organs 30*:382 (1984).

9. R. P. Lanza, D. H. Butler, K. M. Borland, J. E. Staruk, D. L. Faustman, B. A. Solomon, T. E. Muller, R. G. Rupp, T. Maki, A. P. Monaco, and W. L. Chick, Xenotransplantation of canine, bovine and porcine islets in diabetic rats without immunosuppression, *Proc. Natl. Acad. Sci. USA 88*:11100 (1991).

10. P. E. Lacy, O. D. Hegre, A. Gerasimidi-Vazeou, F. T. Gentile, and K. E. Di-

onne, Maintenance of normoglycemia in diabetic mice by subcutaneous xenografts of encapsulated islets, *Science 254*:1782 (1991).

11. H. O. Jauregui and K. L. Gann, Mammalian hepatocytes as a foundation for treatment in human liver failure. *J. Cell. Biochem. 45*:359 (1991).

12. P. Aebischer, S. R. Winn, P. A. Tresco, C. B. Jaeger, and L. A. Greene, Transplantation of polymer encapsulated neurotransmitter secreting cells: Effect of the encapsulation technique, *J. Biomech. Eng. 113*:178 (1991).

13. P. Aebischer, P. A. Tresco, S. R. Winn, L. A. Greene, and C. B. Jaeger, Long-term cross species brain transplantation of a polymer-encapsulated dopamine secreting cell line, *Exp. Neurol. 111*:269 (1991).

14. P. Aebischer, L. Wahlberg, P. A. Tresco, and S. R. Winn, Macroencapsulation of dopamine secreting cells by coextrusion with an organic polymer solution, *Biomaterials 12*:50 (1991).

15. D. F. Emerich, S. R. Winn, L. Christenson, M. A. Palmatier, F. T. Gentile, and P. R. Sanberg, A novel approach to neural transplantation in Parkinson's disease: Use of polymer-encapsulated cell therapy, *Neurosci. Biobehav. Rev. 16*:437 (1992).

16. P. Aebischer, G. Panol, and P. M. Galletti, An intraperitoneal receptacle for macroencapsulated endocrine tissue, *Trans. Am. Soc. Artif. Intern. Organs 32*:130 (1986).

17. G. L. Warnock and R. V. Rajotte, Critical mass of purified islets that induce normoglycemia after implantation in dogs, *Diabetes 37*:467 (1988).

18. T. M. S. Chang, *Artificial Cells*, C. C. Thomas, Springfield IL, 1972.

19. F. Lim and A. M. Sun, Microencapsulated islets as a bioartificial pancreas, *Science 210*:908 (1980).

20. G. M. O'Shea, M. F. A. Goosen, and A. M. Sun, Prolonged survival of transplanted islets of Langerhans encapsulated in a biocompatible membrane, *Biochimica of Biophysica. Acta. 804*:133 (1984).

21. G. M. O'Shea and A. M. Sun, Encapsulation of rat islets of Langerhans prolongs xenograft survival in diabetic mice, *Diabetes 35*:943 (1986).

22. Z.-P. Lum, I. T. Tai, M. Krestow, J. Norton, I. Vacek, and A. M. Sun, Prolonged reversal of diabetic state in NOD mice by xenografts of microencapsulated rat islets, *Diabetes 40*:1511 (1991).

23. A. M. Sun, I. Vacek, Y.-L. Sun, X. Ma, and D. Zhou in vitro and in vivo evaluation of microencapsulated porcine islets, *Am. Soc. Artific. Intern. Organs J. 38*:125 (1992).

24. P. Soon-Shiong, E. Feldman, R. Nelson, R. Heintz, Q. Yao, Z. Yao, T. Zheng, N. Merideth, G. Skjak-Braek, T. Espevik, O. Smidsrod, and P. Sandford, Long-term reversal of diabetes by the injection of immunoprotected islets, *Proc. Natl. Acad. Sci. USA 90*:5843 (1993).

25. R. Mazaheri, P. Atkison, C. Stiller, J. Dupre, J. Vose, and G. O'Shea, Transplantation of encapsulated allogeneic islets into diabetic BB/W rats: Effects of immunosuppression, *Transplantation 51*:750 (1991).

26. C. J. Weber, S. Zabinski, T. Koschitzky, L. Wicker, R. Rajotte, V. D'Agati, L. Peterson, J. Norton, and K. Reemtsma, The role of CD4$^+$ helper T cells in the

destruction of microencapsulated islet xenografts in NOD mice, *Transplantation* *49*:396 (1990).

27. W. M. Fritschy, J. H. Strubbe, G. H. J. Wolters, and R. van Scilfgaarde, Glucose tolerance and insulin response to intravenous glucose infusion and test meal in rats with microencapsulated islet allografts, *Diabetologia 34*:542 (1991).

28. D. R. Cole, M. Waterfall, M. McIntyre, and J. D. Baird, Microencapsulated islet grafts in the BB/E rat: A possible role for cytokines in graft failure, *Diabetologia 35*:231 (1992).

29. P. Soon-Shiong, E. Feldman, R. Nelson, J. Komtebedde, O. Smidsrod, G. Skjak-Braek, T. Espevik, R. Heintz, and M. Lee, Successful reversal of spontaneous diabetes in dogs by intraperitoneal microencapsulated islets, *Transplantation 54*:769 (1992).

30. H. Wong and T. M. S. Chang, The microencapsulation of cells within alginate poly-L-lysine microcapsules prepared with the single step drop technique: Histologically identified membrane imperfections and the associated graft rejection, *Biomater. Artif. Cells Immobilization Biotechnol. 19*:675 (1991).

31. H. Gin, B. Dupuy, D. Bonnemaison-Bourignon, L. Bordenave, R. Bareille, M. J. Latapie, C. Baquey, J. H. Bezian, and D. Ducassou, Biocompatibility of polyacrylamide microcapsules implanted in peritoneal cavity or spleen of the rat: Effect on various inflammatory reactions in vitro, *Biomater. Artif. Cells Artif. Organs 18*:25 (1990).

32. K. M. Miller and J. M. Anderson, Human monocyte/macrophage activation and interleukin 1 generation by biomedical polymers, *J. Biomed. Mater. Res. 22*:713 (1988).

33. J. Brauker, L. A. Martinson, R. S. Hill, S. K. Young, V. E. Carr-Brendel, and R. C. Johnson, Neovascularization of immuno-isolation membrane; the effect of membrane architecture and encapsulated tissue (abstr 123), *Cell Transplant. 1*:163 (1992).

34. J. Folkman and M. Klagsbrun, Angiogenic factors, *Science 23*:442 (1987).

35. M. F. A. Goosen, G. M. O'Shea, H. M. Gharapetian, S. Chou, and A. M. Sun, Optimization of microencapsulation parameters: Semipermeable microcapsules as a bioartificial pancreas, *Biotechnol. Bioeng. 27*:146 (1985).

36. J.-P. Halle, S. Bourassa, F. A. Leblond, S. Chevalier, M. Beaudry, A. Chapdelaine, S. Cousineau, J. Saintonge, and J.-F. Yale, Protection of islets of Langerhans from antibodies by microencapsulation with alginate–poly-L-lysine membranes, *Transplantation 55*:350 (1993).

37. C. Weber, V. D'Agati, L. Ward, M. Costanzo, R. Rajotte, and K. Reemtsma, Humoral reaction to microencapsulated rat, canine and porcine islet xenografts in spontaneous diabetic NOD mice, *Transplant. Proc. 25*:462 (1993).

38. S. Darquy and G. Reach, Immunoisolation of pancreatic B cells by microencapsulation, *Diabetologia 28*:776 (1985).

39. K. A. Heald, T. R. Jay, and R. Downing, Assessment of the reproducibility of alginate encapsulation of pancreatic islets using the MTT colorimetric assay, *Cell Transplant. 3*:333 (1994).

40. G. S. Eisenbarth, Type 1 diabetes mellitus: A chronic autoimmune disease, *N. Engl. J. Med. 314*:1360 (1986).

41. R. K. Sibley, D. E. Sutherland, F. Goetz, and A. F. Michael, Recurrent diabetes mellitus in the pancreas iso and allograft: A light and electron microscopic and immunohistochemical analysis of four cases, *Lab. Invest. 53*:132 (1985).

42. A. Naji, W. K. Silvers, H. Kimura, A. O. Anderson, and C. F. Barker, Influence of islets and bone marrow transplantation on the diabetes and immunodeficiency of BB rats, *Metabolism 32*:62 (1983).

43. T. Mandrup-Poulsen, K. Bendtzen, J. Nerup, C. A. Dinarello, M. Svenson, and J. H. Nielson, Affinity purified human interleukin 1 is cytotoxic to isolated islets of Langerhans, *Diabetologia 29*:63 (1986).

44. T. Zekorn, U. Siebers, R. G. Bretzel, M. Renardy, H. Planck, P. Zschoke, and K. Federlin, Protection of islets of Langerhans from interleukin-1 toxicity by artificial membranes, *Transplantation 50*:391–394 (1990).

45. D. M. Bissell, J. M. Caron, L. E. Babiss, and J. M. Friedman, Transcriptional regulation of the albumin gene in cultured rat hepatocytes; role of basement-membrane matrix, *Mol. Biol. Med. 7*:187 (1990).

46. E. G. Schuetz, D. Li, C. J. Omecinski, U. Muller-Eberhard, H. K. Kleinman, B. Elswick, and P. S. Guzelian, Regulation of gene expression in adult rat hepatocytes cultured on a basement membrane matrix, *J. Cell. Physiol. 134*:309 (1988).

47. C. M. Alexander and Z. Werb, Proteinases and extracellular matrix remodeling, *Curr. Opin. Cell Biol. 1*:974 (1989).

48. J. E. Babensee, U. De Boni, and M. V. Sefton, Morphological assessment of hepatoma cells (HepG2) microencapsulated in a HEMA–MMA copolymer with and without Matrigel, *J. Biomed. Mater. Res. 26*:1401 (1992).

49. R. M. Sutherland, Cell and environment interactions in tumor microregions: The multicell spheroid model, *Science 240*:177 (1988).

50. R. M. Sutherland, Importance of critical metabolites and cellular interactions in the biology of microregions of tumors, *Cancer 58*:1668 (1986).

51. M. Erlanson, E. Daniel-Szolgay, and J. Carlsson, Relations between the penetration, binding and average concentration of cytostatic drugs in human tumour spheroids, *Cancer Chemother. Pharmacol. 29*:343 (1992).

52. J. Landry, D. Bernier, C. Ouellet, R. Goyette, and N. Marceau, Spheroidal aggregate culture of rat liver cells: Histotypic reorganization, biomatrix deposition, and maintenance of functional activities, *J. Cell Biol. 101*:914 (1985).

53. B. Van Der Schueren, C. Denef, and J.-J. Cassiman, Ultrastructural and functional characteristics of rat pituitary cell aggregates, *Endocrinology 110*:513 (1982).

54. A. M. Sun, G. M. O'Shea, and M. F. A. Goosen, Injectable microencapsulated islet cells as a bioartificial pancreas, *Appl. Biochem. Biotechnol. 10*:87 (1984).

55. A. S. Sawhney and J. A. Hubbell, Poly(ethylene oxide)-graft-poly(L-lysine) copolymers to enhance the biocompatibility of poly(L-lysine) alginate microcapsule membranes, *Biomaterials 13*:863 (1992).

56. G. M. R. Vandenbossche, M. E. Bracke, C. A. Cuvelier, H. E. Bortier, M. M. Mareel, and J.-P. Remon, Host reaction against empty alginate–polylysine microcapsules. Influence if preparation procedure, *J. Pharm. Pharmcol. 45*:115 (1993).

57. H. A. Clayton, N. J. M. London, P. S. Colloby, R. F. L. James, and P. R. F. Bell, The effect of capsule composition on the viability and biocompatibility of sodium alginate/poly-L-lysine encapsulated islets, *Transplant. Proc. 24*:959 (1992).

58. A. M. Sun, Z. Cai, Z. Shi, F. Ma, G. M. O'Shea, and H. Gharapetian, Microencapsulated hepatocytes as a bioartificial liver, *Trans. Am. Soc. Artif. Intern. Organs 32*:39 (1986).

59. V. Dixit, R. Darvasi, M. Arthur, M. Brezina, K. Lewin, and G. Gitnick, Restoration of liver function in Gunn rats without immunosuppression using transplanted microencapsulated hepatocytes, *Hepatology 12*:1342 (1990).

60. V. Dixit, M. Arthur, R. Reinhardt, and G. Gitnick, Improved function of microencapsulated hepatocytes in a hybrid bioartificial liver support system, *Artif. Organs 15*:272 (1991).

61. M. F. A. Goosen, G. M. O'Shea, H. M. Gharapetian, S. Chou, and A. M. Sun, Optimization of microencapsulation parameters: Semipermeable microcapsules as a bioartificial pancreas, *Biotechnol. Bioeng. 27*:146 (1985).

62. M. F. A. Goosen, G. A. King, C. A. McKnight, and N. Marcotte, Animal cell culture engineering using alginate polycation microcapsules of controlled membrane molecular weight cut-off, *J. Membr. Sci. 41*:323 (1989).

63. R. Calafiore, G. Basta, A. Falorni Jr., G. Gambelunghe, and P. Brunetti, An improved method for immunoisolation of pancreatic islet grafts within microcapsules (abstr), *Diabetologia 34*(suppl. 2):A169 (1991).

64. G. A. King, A. J. Daugulis, P. Faulkner, and M. F. A. Goosen, Alginate–polylysine microcapsules of controlled membrane molecular weight cut off for mammalian cell culture engineering, *Biotechnol. Prog. 3*:231 (1987).

65. M. Hommel, A. M. Sun, and M. F. A. Goosen, Droplet generation. Canadian Patent 458605 (1984).

66. T. Zekorn, U. Siebers, A. Horcher, R. Schnettler, G. Klock, R. G. Bretzel, U. Zimmermann, and K. Federlin, Barium–alginate beads for immunoisolated transplantation of islets of Langerhans, *Transplant. Proc. 24*:937 (1992).

67. P. Soon-Shiong, M. Otterlie, G. Skjak-Braek, O. Smidsrod, R. Heintz, R. P. Lanza, and T. Espevik, An immunologic basis for the fibrotic reaction to implanted microcapsules, *Transplant. Proc. 23*:758 (1991).

68. K. Tatarkiewicz, New membrane for cell encapsulation, *Artif. Organs 12*:446 (1988).

69. A. S. Sawhney, C. P. Pathak, and J. A. Hubbell, Interfacial photopolymerization of poly(ethylene glycol)-based hydrogels upon alginate–poly(L-lysine) microcapsules for enhanced biocompatibility, *Biomaterials 14*:1008 (1993).

70. J.-P. Halle, F. A. Leblond, J.-F. Pariseau, P. Jutras, M.-J., Brabant, and Y. Lepage, Studies on small (<300 µm) microcapsules: II. Parameters governing the production of alginate beads by high voltage electrostatic pulses, *Cell Transplant. 3*:5 (1994).

71. L. Levesque, P. L. Brubaker, and A. M. Sun, Maintenance of long-term secretory function by microencapsulated islets of Langerhans, *Endocrinology 130*:644 (1992).

72. O. Smidsrod and G. Skjak-Braek, Alginate as immobilization matrix for cells, *Trends in Biotechnology 8*:71 (1990).
73. J. Schrezezenmeir, B. J. Hering, L. Gero, J. Wiegand-Dressler, M. Solhdju, F. Velten, J. Kirchgessner, C. Laue, J. Beyer, R. Bretzel, and K. Federlin, Long-term function of porcine islets and single cells embedded in barium–alginate matrix, *Horm. Metab. Res. 25*:204 (1993).
74. K. Tatarkiewicz, E. Sitarek, P. Fiedor, M. Sadat, M. Morzycka-Michalik, and T. Orlowski, Successful rat-to-mouse xenotransplantation of Langerhans islets micro-encapsulated within a protamine–heparin membrane, *Transplant. Proc. 26*:807 (1994).
75. W. R. Gombotz, W. Guanghui, and A. S. Hoffman, Immobilization of poly(ethylene oxide) on poly(ethylene terephthalate) using a plasma polymerization process, *J. Appl. Polym. Sci. 37*:91 (1989).
76. N. P. Desai and J. A. Hubbell, Solution technique to incorporate polyethylene oxide and other water-soluble polymers into surfaces of polymeric biomaterials, *Biomaterials 12*:144 (1991).
77. H. Gin, B. Dupuy, C. Baquey, D. Ducassou, and J. Aubertin, Agarose encapsulation of islets of Langerhans: Reduced toxicity in vitro, *J. Microencapsul. 4*:239 (1987).
78. B. Dupuy, H. Gin, C. Baquey, and D. Ducassou, In situ polymerization of a microencapsulating medium round living cells, *Diabetologia 31*:54 (1988).
79. H. Gin, B. Dupuy, A. Baquey, and D. Ducasssou, Lack of responsiveness to glucose of microencapsulated islets of Langerhans after three weeks' implantation in the rat—influence of the complement, *J. Microencapsul. 7*:341 (1990).
80. H. Gin, B. Dupuy, D. Bonnemaison-Bourignon, L. Bordenave, R. Bareille, M. J. Latapie, and C. Baquey, Biocompatibility of polyacrylamide microcapsules implanted in peritoneal cavity or spleen of the rat. Effect on various inflammatory reactions in vitro, *Biomater. Artif. Cells Artif. Organs 18*:25 (1990).
81. B. Dupuy, C. Cadic, H. Gin, C. Baquey, B. Dufy, and D. Ducassou, Microencapsulation of isolated pituitary cells by polyacrylamide microlatex coagulation on agarose beads, *Biomaterials 12*:493 (1991).
82. H. Iwata, H. Amemiya, T. Matsuda, H. Takano, R. Hayashi, and T. Akutsu, Evaluation of microencapsulated islets in agarose gel as bioartificial pancreas by studies of hormone secretion in culture and by xenotransplanation, *Diabetes 38* (Suppl):224 (1989).
83. H. Iwata, T. Takagi, H. Amemiya, H. Shimizu, K. Yamashita, K. Koyayashi, and T. Akutsu, Agarose for a bioartificial pancreas, *J. Biomed. Mater. Res. 26*:967 (1992).
84. H. Iwata, K. Kobayashi, T. Takagi, T. Oka, H. Yang, H. Amemiya, T. Tsuji, and F. Ito, Feasibility of agarose microbeads with xenogeneic islets as a bioartificial pancreas, *J. Biomed. Mater. Res. 28*:1003 (1994).
85. T. Takagi, H. Iwata, H. Tashiro, T. Tsuji, and F. Ito, Development of a novel microbead applicable to xenogeneic islet transplantation, *J. Controlled Release 31*:283 (1994).

86. T. Takagi, H. Iwata, K. Kobayashi, T. Oka, T. Tsuji, and F. Ito, Development of a microcapsule applicable to islet xenotransplantation, *Transplant. Proc.* 26:801 (1994).

87. H. Gharapetian, N. A. Davies, and A. M. Sun, Polyacrylate membranes for encapsulation of viable cells, *Biotechnol. Bioeng.* 28:1595 (1986).

88. S. Wen, Y. Xiaonan, and W. T. K. Stevenson, Microcapsule through polymer complexation: I. By complex coacervation of polymers containing a high charge density, *Biomaterials* 12:374 (1991).

89. S. Wen, Y. Xiaonan, W. T. K. Stevenson, and H. Alexander, Microcapsule through polymer complexation: II. By complex coacervation of polymers containing a low charge density, *Biomaterials* 12:479 (1991).

90. M. C. Bano, S. Cohen, K. B. Visscher, H. R. Allcock, and R. S. Langer, A novel synthetic method for hybridoma cell encapsulation, *Biotechnology* 9:468 (1991).

91. S. Gupta, S. K. Kim, R. P. Vemuru, E. Aragona, P. R. Yerneni, R. D. Burk, and C. K. Rha, Hepatocyte transplantation: An alternative system for evaluating cell survival and immunoisolation, *Int. J. Artif. Organs* 16:155 (1993).

92. T. Yoshioka, R. Hirano, T. Shioya, and M. Kako, Encapsulation of mammalian cell with chitosan–CMC capsule, *Biotechnol. Bioeng.* 35:66 (1990).

93. H. W. T. Matthew, S. Basu, W. D. Peterson, S. O. Sallry, and M. D. Klein, Performance of plasma-perfused, microencapsulated hepatocytes: Prospects for extracorpeal liver support, *J. Pediatr. Surg.* 28:1423 (1993).

94. S. Gupta, S. K. Kim, R. P. Vemuru, E. Aragona, P. R. Yerneni, R. D. Burk, and C. K. Rha, Hepatocyte transplantation: An alternative system for evaluating cell survival and immunoisolation, *Int. J. Artif. Organs* 16:155 (1993).

95. G. M. Cruise, A. S. Sawhney, C. P. Pahak, K. M. Luther, and J. A. Hubbell, Poly(ethylene glycol) based conformal coatings as permselective membranes for transplantation, Transactions 19th Annual Meeting Society for Biomaterials, Birmingham AL, 1993, p. 205.

96. T. Zekron, U. Siebers, A. Horcher, R. Schnettler, U. Zimmermann, R. G. Bretzel, and K. Federlin, Alginate coating of islets of Langerhans: In vitro studies on a new method for microencapsulation for immuno-isolated transplantation, *Acta Diabetol.* 29:41 (1992).

97. J. Honiger, S. Darquy, G. Reach, E. Muscat, M. Thomas, and C. Collier, Preliminary report on cell encapsulation in a hydrogel made of a biocompatible material, AN69, for the development of a bioartificial pancreas, *Int. J. Artif. Organs* 17:46 (1994).

98. W. T. K. Stevenson, R. A. Evangelista, R. L. Broughton, and M. V. Sefton, Preparation and characterization of thermoplastic polymers from hydroxyalkyl methacrylates, *J. Appl. Polym. Sci.* 34:65 (1987).

99. W. T. K. Stevenson and M. V. Sefton, The equilibrium water content of some thermoplastic hydroxyalkyl methacrylates, *J. Appl. Polym. Sci.* 34:65 (1987).

100. C. A. Crooks, J. A. Douglas, R. L. Broughton, and M. V. Sefton, Microencapsulation of mammalian cells in a HEMA–MMA copolymer: Effects on capsule morphology and permeability, *J. Biomed. Mater. Res.* 24:1241 (1990).

101. A. H. Boag and M. V. Sefton, Microencapsulation of human fibroblasts in a water insoluble polyacrylate, *Biotechnol. Bioeng. 30*:954 (1987).
102. M. E. Sugamori and M. V. Sefton, Microencapsulation of pancreatic islets in a water insoluble polyacrylate (EUDRAGIT RL), *Trans. Am. Soc. Artif. Intern. Organs 35*:791 (1989).
103. R. L. Broughton and M. V. Sefton, Effect of capsule permeability on growth of CHO cells in Eudragit RL microcapsules: Use of FITC dextran as a marker of capsule quality, *Biomaterials 10*:462 (1989).
104. C. L. Mallabone, C. A. Crooks, and M. V. Sefton, Microencapsulation of human diploid fibroblasts in cationic polyacrylates, *Biomaterials 10*:380 (1989).
105. R. M. Dawson, R. L. Broughton, W. T. K. Stevenson, and M. V. Sefton, Microencapsulation of CHO cells in a hydroxyethyl methacrylate–methyl methacrylate copolymer, *Biomaterials 8*:360 (1987).
106. H. Uludag and M. V. Sefton, Metabolic activity and proliferation of CHO cells in hydroxyethyl–methacrylate (HEMA–MMA) microcapsules, *Cell Transplant. 2*:175 (1993).
107. H. K. Kleinman, M. L. McGarvey, J. L. Hassell, V. L. Star, F. B. Cannon, G. W. Laurie, and G. R. Martin, Basement membrane complexes with biological activity, *Biochemistry 25*:312 (1986).
108. H. Uludag, V. Horvath, J. P. Black, and M. V. Sefton, Viability and protein secretion from human hepatoma (HepG2) cells encapsulated in 400 μm polyacrylate microcapsules by submerged nozzle–liquid jet extrusion, *Biotechnol. Bioeng. 44*:1199 (1994).
109. J. R. Hwang and M. V. Sefton, Effect of microcapsule diameter on the permeability of individual HEMA-MMA microcapsules, *J. Controlled Release*: in press.
110. M. V. Sefton and W. T. K. Stevenson, Microencapsulation of live animal cells using polyacrylates, *Adv. Polym. Sci. 107*:145 (1993).
111. H. Uludag, J. R. Hwang, and M. V. Sefton, Microencapsulated human hepatoma (HepG2) cells: Capsule-to-capsule variations in protein secretion and permeability, *J. Controlled Release 33*:273 (1995).
112. J. R. Hwang, and M. V. Sefton, The effects of polymer concentration and a pore forming agent (PVP) on HEMA–MMA microcapsule permeability, *J. Membr. Sci. 108*:257–268 (1995).
113. T. Roberts, U. De Boni, and M. V. Sefton, Dopamine secretion by PC12 cells microencapsulated in a hydroxyethyl methacrylate–methyl methacrylate copolymer, *Biomaterials 17*:267 (1996).
114. H. Uludag and M. V. Sefton, Microencapsulated human hepatoma cells: In vitro growth and protein release, *J. Biomed. Mater. Res. 27*:1213 (1993).
115. D. Y.-P. Ung, The effect of cytokines on microencapsulated mammalian model cells, B.Sc. dissertation, University of Toronto, 1993.
116. G. D. M. Wells, M. M. Fisher, and M. V. Sefton, Microencapsulation of viable hepatocytes in HEMA–MMA microcapsules: A preliminary study, *Biomaterials 14*:615 (1993).

117. M. V. Sefton and L. Kharlip, Insulin release from rat pancreatic islets microencapsulated in a HEMA–MMA polyacrylate, *Pancreatic Islet Transplantation*, Vol. 3, *Immunoisolation of Pancreatic Islets* (R. P. Lanza, and W. L. Chick, eds.), R. G. Landes, Austin, TX, 1994, p. 107.

118. H. Uludag and M. V. Sefton, A colorimetric assay for cellular activity in microcapsules, *Biomaterials 11*:708 (1990).

119. J. E. Babensee and M. V. Sefton, Protein delivery by microencapsulated cells, *Controlled Drug Delivery: The Next Generation* (K. Park, ed.), ACS Books, Washington DC, in press.

120. N. Koide, K. Sakaguchi, Y. Koide, K. Asano, M. Kawagughi, H. Matsushima, T. Takenami, T. Shinji, M. Mori, and T. Tsuji, Formation of multicellular spheroids composed of adult rat hepatocytes in dishes with positively charged surfaces and under other nonadherent environments, *Exp. Cell Res. 186*:227 (1990).

121. J. Landry, D. Bernier, R. Ouellet, R. Goyette, and N. Marceau, Spheroidal aggregate culture of rat liver cells: Histotypic reorganization, biomatrix deposition, and maintenance of functional activities, *J. Cell Biol. 101*:914 (1985).

122. N. Matthews and M. L. Neale, Cytotoxicity assays for tumour necrosis factor and lymphotoxin, *Lymphokines and Interferons: A Practical Approach* (M. J. Clemens, A. G. Morris, and A. J. H. Gearing, eds.), IRL Press, New York, 1987, p. 221.

123. A. H. Ding, F. Porteu, E. Sanchez, and C. F. Nathan, Downregulation of tumor necrosis factor receptors on macrophages and endothelial cells by microtubule depolymerizing agents, *J. Exp. Med. 171*:712 (1990).

124. C. Nathan, and M. Sporn, Cytokines in context, *J. Cell Biol. 113*:981 (1991).

125. M. L. Neale, and N. Matthews, Colonial morphology of tumour cells and susceptibility to cytolysis by tumour necrosis factor. The role of cellular fibronectin deposition in the extracellular matrix, *Br. J. Cancer 6*:831 (1990).

126. R. Gopal-Srivastava, and J. Piatigorsky, The murine αB-crystallin/small heat shock protein enhancer: Identification of αBE-1, αBE-2, α-BE-3, and MRF control elements, *Mol. Cell. Biol. 13*:7144 (1993).

127. M. Jaattela, and D. Wissing, Heat-shock proteins protect cells from monocyte cytotixicity: Possible mechanism of self-protection, *J. Exp. Med. 177*:231 (1993).

128. D. M. Bissell, and M. O. Choun, The role of extracellular matrix in normal liver, *Scand. J. Gastroentrol. Suppl. 151*:1 (1988).

129. D. M. Bissell, D. M. Arenson, J. J. Maher, and F. J. Roll, Support of cultured hepatocytes by a laminin-rich gel. Evidence for a functionally significant subendothelial matrix in normal rat liver, *J. Clin. Invest. 79*:801 (1987).

130. E. G. Schuetz, D. Li, C. J. Omiecinski, U. Muller-Eberhard, H. K. Kleinman, B. Elswick, and P. S. Guzelian, Regulation of gene expression in adult rat hepatocytes cultured on basement membrane matrix, *J. Cell Physiol. 134*:309 (1988).

131. M. D. Buschmann, Y. A. Gluzband, A. J. Grodzinsky, J. H. Kimura, and E. B. Hunziker, Chondrocytes in agarose culture synthesize a mechanically functional extracellular matrix, *J. Orthopaed. Res. 10*:745 (1992).

132. A. De Castro, HEMA–MMA microencapsulated agarose, chitosan or Matrigel as a PC12 cell immobilization matrix, B.Sc. dissertation, University of Toronto, 1994.

133. M. V. Sefton, The good, the bad and the obvious, *Biomaterials, 14*:1127 (1993) [Clemson Award Lecture].

Index

521